POWDERS AND FIBERS

INTERFACIAL SCIENCE AND APPLICATIONS

SURFACTANT SCIENCE SERIES

FOUNDING EDITOR

MARTIN J. SCHICK
1918–1998

SERIES EDITOR

ARTHUR T. HUBBARD
Santa Barbara Science Project
Santa Barbara, California

ADVISORY BOARD

POWDERS AND FIBERS

INTERFACIAL SCIENCE AND APPLICATIONS

Michel Nardin
Centre National de la Recherche Scientifique
Mulhouse, France

Eugène Papirer
IGClab
Pulvershelm, France

CRC Press
Taylor & Francis Group
Boca Raton London New York

CRC Press is an imprint of the
Taylor & Francis Group, an informa business

CRC Press
Taylor & Francis Group
6000 Broken Sound Parkway NW, Suite 300
Boca Raton, FL 33487-2742

© 2007 by Taylor & Francis Group, LLC
CRC Press is an imprint of Taylor & Francis Group, an Informa business

First issued in paperback 2019

No claim to original U.S. Government works

ISBN 13: 978-0-367-45325-1 (pbk)
ISBN 13: 978-1-57444-513-8 (hbk)

Library of Congress Cataloging-in-Publication Data

Powders and fibers : interfacial science and applications / editors, Michel Nardin and Eugene Papirer.
    p. cm. -- (Surfactant science series ; 137)
    Includes bibliographical references and index.
    ISBN-13: 978-1-57444-513-8 (acid-free paper)
    ISBN-10: 1-57444-513-8 (acid-free paper)
    1. Powders--Surfaces. 2. Fibers--Surfaces. 3. Solid-liquid interfaces. 4. Surface chemistry. I. Nardin, Michel, Ph. D. II. Papirer, Eugène, 1935- III. Title. IV. Series.

QD509.S65P69 2006
620'.43--dc22
                                                    2006021794

Visit the Taylor & Francis Web site at
http://www.taylorandfrancis.com

and the CRC Press Web site at
http://www.crcpress.com

Preface

Powders and fibers are gaining increased interest for both fundamental and practical issues. A better understanding of their behaviors when brought in contact with gaseous, liquid, or solid (polymer) media requires knowledge of their surface characteristics (area, chemistry, interaction potential, etc.) and will allow enhancement of their performances. Yet, the number of surface characterization methods is rather limited and, moreover, only a small number of those solid surfaces have been scientifically investigated.

The object of this book is multiple:

- To present modern methods of surface characterization of powders and fibers
- To list a series of important applications of powders and fibers
- To discuss the formation and role of solid–gas, solid–liquid, and solid–polymer interfaces and interphases in relation to the performance of materials, including powders or fibers

Rather than try to make an exhaustive overview of all existing surface characterization methods, a limited number of methods have been selected either because they have already been shown to be effective or because they may be considered innovative or promising.

Indeed, the behaviors of powders and fibers in various environments are singular, depending on the nature of those associations. Clearly the peculiarities of the interface (the zone of molecular dimension in the vicinity of the solid surface) and interphase (the zone extending from the solid surface, influenced by the surface) largely determine the performance and life duration of materials submitted to stress or placed in a given environment. The mechanisms behind those behaviors are still under investigation using molecular, mechanical, computer simulation, and other approaches.

The book is divided into six sections preceded by an introduction.

In Chapter 1, J. Berg presents theoretical fundamentals—the concepts of interfacial energy, surface energy, and, in general, the impact of thermodynamics on the behavior of powders and fibers.

In Section I, which is devoted to the solid–gas and solid–rubber associations, those fundamentals will facilitate the understanding of a true surface characterization method: inverse gas chromatography. The chapter by E. Brendlé and E. Papirer gives an overview of the inverse gas chromatography (IGC) potential for a large panel of applications going from fillers to drugs, etc. The chapter by M.-J. Wang illustrates an interesting and actual application of IGC in the rubber industry where fillers (carbon black, silica) play a complex yet important role in the manufacturing of tires.

Section II concerns solid–solid interfaces. Additional methods for the surface characterization of powders and fibers are introduced. B. McCool and C. P. Tripp show how infrared spectroscopy gives access to molecular details that are of major importance for the formation of interfaces. The authors concentrate on actual issues, such as adsorption of biomolecules, biocompatibility, biochips, nanoparticles, etc. and, in general, the connection between the molecular chemistry of the interfacial region and macroscopic properties of the assemblies. The contribution of V. A. Tertykh relates the use of chemical methods. Not only does a chemical approach allow determining surface interactivity, but also it provides ways to modify it. Silica is the preferred surface in this instance since it plays a major role in chromatography and solid-phase extraction of ions and molecules. It is also a support of catalysts and furthermore an active filler for polymers. A. P. Legrand and H. Hommel examine the answers provided by magnetic resonance spectroscopies applied to silica, ceramics, carbon, and glass fibers composites, but also biological interfaces (bio-prostheses, bio-ceramics). The chapter by H. Van Damme introduces the present concepts concerning the behaviors of vapors, liquids and polymers in solid–solid contacts and solid–solid confinement conditions: cohesion of powders in wet conditions and polymer nanocomposites are examples of such confinement effects. Consequences of the aging of powders, vapor–liquid transition in confined media, polymer reinforcement, and liquid–glass transitions are also analyzed as well as the role of confined polymer layers in nanocomposites. M. Renner and M. A. Bueno present an overview of the characteristics of fibers used in the textile industry. After an introduction describing the various types of fibers, the authors analyze the fibers' morphologies and structures before demonstrating a careful investigation of the fibers' surface characteristics in relation to the wear and comfort properties of our modern fabrics.

Thereafter, in Section III, M. Nardin examines how to measure the wettability of powders and fibers, recalling the principles on which those measurements are based, the various experimental approaches, and the practical consequences of wettability in the case of composite materials. R. Al-Akoum, C. Vaulot, M. Owczarek, A. Vidal, and B. Haidar propose a direct method for the determination of the polymer–filler interaction potential: flow microcalorimetry that allows obtaining polymer adsorption isotherms and adsorption enthalpy. The effect of the form factor (particle, platelet, tube) is evidenced. The development of computer science results in a remarkable evolution of the electronic microscopy, giving not only a view of the fiber and particle surface, but allowing it also to appreciate interaction forces. B. Nysten provides a global description of the scanning probe microscopies applied to carbon and polymer fibers, and nanofibers (carbon and metallic nanofibers), but also to the study of interface properties in polymer composites and blends.

Section IV is devoted to interfaces and interphases in biological processes. P. Frayssinet and P. Laquerriere tackle a most timely issue: the biocompatibility of powdered materials. Clearly their surface properties are of utmost importance for cell adhesion, proliferation, transformation and death. The influence of size and surface area of the particles, wettability, surface crystallinity, and topography are examined by the authors.

Section V concerns the solid–liquid interface. F. Thomas, B. Prelot and J. Duval investigate the solid–electrolyte interfacial region, starting with the elucidation of the origin of the electrical surface charge formation. Different representations of the surface layer are proposed and the methods permitting measurement of the electrokinetic potential are described. The existence of soft interfaces as well as surface complexation mechanisms are discussed. E. Pefferkorn examines the complex problem of polymer dynamics at solid–liquid interfaces: formation and evolution (polymer layer re-conformation, relaxation) of the interface in aqueous aluminosilicate–polymer–water systems and cellulose–polymer–water systems. Whereas Pefferkorn favors aqueous media, H. Barthel, T. Gottschalk-Gaudig and M. Dreyer are interested in the interactions among silica, polymers and solvents. A series of techniques is applied for that purpose. Particle–particle (modifiable by controlled surface modification of silica), solvent–particle and silica–polymer interactions play an important role. The rheology data of the mixes are compared with the surface properties of silica and a semi-empirical computer model is proposed.

Section VI is an additional demonstration of the interest in computer simulation for the visualization of complex solid surfaces. V. A. Bakaev, C. G. Pantano, and W. A. Steele examine the surface heterogeneity of silicate glasses. Based on IGC measurements of the adsorption energy distribution of the various surface sites, the authors call on computer simulations by molecular dynamics of the glass surfaces and test their model with experimental results collected by IGC.

In conclusion, this book offers a broad overview of the interfacial science and applications of powders and fibers. Yet, that domain is too extensive to be treated exhaustively in this book. The reader might, nevertheless, find interest in the diversity of our approach where fundamental aspects stay close to applications. The newcomer will hopefully find a start for his own project and encouragement to enter into this exciting, timely and still progressing subject.

Editors

Michel Nardin is a research director at the Centre National de la Recherche Scientifique, Institut de Chimie des Surfaces et Interfaces (ICSI), Mulhouse, France, where he has worked since 1978. He is a chemical engineer from the National High School of Chemistry of Paris and received his PhD from the University of Mulhouse in 1984. His fields of expertise deal with the physical chemistry of surfaces and interfaces, fundamental and practical aspects of adhesion, thermodynamics and micromechanics of fiber–matrix interfaces in composite materials and, recently, interactions between surfaces and living matter. He is the author or co-author of more than 130 scientific articles and a few book chapters.

Eugène Papirer is a retired research director at the Centre National de la Recherche Scientifique, Institut de Chimie des Surfaces et Interfaces (ICSI), earlier Centre de Recherches sur la Physicochimie des Surfaces Solides, Mulhouse, France, where he has worked since 1966. He is the author of more than 250 scientific articles and numerous book chapters. He is also the editor of *Adsorption on Silica Surfaces* (Marcel Dekker, New York, 2000). He has organized several conferences, in particular, the first International Conference on SILICA, held in Mulhouse during 1998. Dr. Papirer received his PhD in 1964 from the University of Strasbourg, France.

Contributors

R. Al-Akoum
Institut de Chimie des Surfaces et
    Interfaces
Mulhouse, France

Victor A. Bakaev
Materials Research Institute
The Pennsylvania State University
University Park, Pennsylvania

Herbert Barthel
Wacker-Chemie GmbH
R&D and Process Development
Silicon Division
Werk Burghausen, Germany

John Berg
Department of Chemical
    Engineering
University of Washington
Washington, D.C.

Eric Brendlé
IGCLab SARL
Carreau Rodolphe
Pulversheim, France

Marie-Ange Bueno
Ecole Nationale Supérieure des
    Industries Textiles de Mulhouse
Mulhouse, France

Henri Van Damme
Laboratoire de Physico-Chimie
    Structurale et Moléculaire
Paris, France

Michael Dreyer
Wacker-Chemie GmbH
R&D and Process Development
Silicon Division
Werk Burghausen, Germany

Jérôme F.L. Duval
Laboratoire Environnement et
    Minéralurgie
Nancy, France

Patrick Frayssinet
Urodélia
St. Thomas, France

Torsten Gottschalk-Gaudig
Wacker-Chemie GmbH
R&D and Process Development
Silicon Division
Werk Burghausen, Germany

Bassel Haidar
Institut de Chimie des Surfaces et
    Interfaces
Mulhouse, France

Hubert Hommel
Laboratoire de Physico-Chimie
    Structurale et Macromoléculaire
Ecole Supérieure de Physique et de
    Chimie Industrielles de la Ville de
    Paris
Paris, France

Patrice Laquerriere
Interfaces Biomatériaux/Tissus
  Hôtes
Reims, France

André Pierre Legrand
Laboratoire Physique Quantique
Ecole Supérieure de Physique et de
  Chimie Industrielles de la Ville de
  Paris
Paris, France

Ben McCool
Department of Chemistry
University of Maine
Orono, Maine

Michel Nardin
Institut de Chimie des Surfaces et
  Interfaces
Mulhouse, France

Bernard Nysten
Laboratory of Polymer
  Chemistry and Physics
Research Centre on
  Micro- and Nanoscopic Materials
  and Electronic Devices
Université catholique de Louvain
Louvain-la-Neuve, Belgium

M. Owczarek
Technical University of Łódz
Institute of Polymers & Dye
  Technology
Łódz, Poland

Carlo G. Pantano
Materials Research Institute
The Pennsylvania State University
University Park, Pennsylvania

Eugène Papirer
IGCLab
Carreau Rodolphe
Pulversheim, France

Emile Pefferkorn
Institut Charles Sadron
Strasbourg, France

Bénédicte Prelot
Laboratoire des Agrégats
  Moléculaires et Matériaux
  Inorganiques
Université Montpellier
Montpellier, France

Marc Renner
Ecole Nationale Supérieure des
  Industries Textiles de Mulhouse
Mulhouse, France

William A. Steele
Materials Research Institute
The Pennsylvania State University
University Park, Pennsylvania

Valentin A. Tertykh
Institute of Surface Chemistry
National Academy of Sciences
  of Ukraine
Kyev, Ukraine

Fabien Thomas
Laboratoire Environnement et
  Minéralurgie
Nancy, France

Carl P. Tripp
Department of Chemistry
University of Maine
Orono, Maine

C. Vaulot
Institut de Chimie des Surfaces et
    Interfaces
Mulhouse, France

Meng-Jiao Wang
Cabot Corporation
Shanghai, China

Alain Vidal
Institut de Chimie des Surfaces et
    Interfaces
Mulhouse, France

Table of Contents

SECTION III   The Solid–Polymer Interface

SECTION IV   Interfaces and Interphases in Biological Processes

SECTION V The Solid–Liquid Interface

SECTION VI Computer Simulation of Solid Surfaces and Interfaces

1 The Thermodynamics of Interfacial Systems with Special Reference to Powders and Fibers

John Berg

CONTENTS

1.1 INTRODUCTION

Powders and fibers, or more generically, solids in a high degree of subdivision, exist in an enormous variety of chemistries and morphologies, and are used in a virtually limitless number of applications. The discrete entities or "kinetic units" of interest typically range in linear dimension from a few nanometers to a few micrometers (the colloid size range). Even in the nano-range, where powders and fibers take the form of quantum dots or nanowires, the objects are still usually sufficiently large to be considered macroscopic (i.e., large relative to the size of an ordinary molecule), and systems consisting of such particles are amenable to the descriptions afforded by macroscopic Gibbsian thermo-dynamics. It is thus the objective of this chapter to provide a brief summary of the thermodynamics of particulate systems. As with all of thermodynamics, it is useful in providing "rules" that govern the descriptions of such systems in terms of their properties.

Despite their diversity, one feature that all particulate systems have in common is their large area/volume ratio, guaranteeing that their properties and behavior are strongly influenced, and often dominated, by their surfaces (i.e., the outermost few molecular or atomic layers). It is thus logical that any survey of powders and fibers and their applications be couched in terms of interfacial science. The use of the term "interfacial," rather than "surface," is significant because it reminds us that all boundaries between macroscopic phases are "interfaces," whose properties depend on *both* phases. The "surface" of a silica micro-particle in contact with air, for example, is different from its "surface" in contact with water, or in contact with an apolar solvent, or in vacuo. In general, the "interface" dividing bulk phases is an inhomogeneous layer, usually no more than a few Ångströms in thickness. For purposes of macroscopic modeling, it is often convenient to define the boundary as one of *zero* thickness. Its properties cannot always be regarded as "autonomous" from those of the adjacent bulk material. Thus, the irreducible prototype of a thermodynamic system, in which material interfaces are considered, consists of a portion of an infinitesimally thin interfacial layer, together with macroscopic portions of the bulk phases on either side of it. The particulate systems of concern in this volume consist of solid particles in contact with a fluid phase, either gas or liquid. The fact that the system is, in part, a solid presents a number of challenges with respect to its thermodynamic description, as compared with the description required for fluid–fluid interface systems. In fluid systems, molecular mobility makes it reasonable to assume internal thermal, mechanical, and diffusional equilibrium, whereas in solids, nonequilibrium structures are frozen in place over the time scales of practical interest. Thus, fluid–solid interface systems are generally rough (as opposed to smooth), morphologically and energetically heterogeneous (as opposed to homogeneous) and possessed of unrelaxed internal stresses (as opposed to being free of all internal shear stresses when at rest). It is thus useful to first set forth the simpler thermodynamic description applicable to *fluid* interfacial (i.e., *capillary*) systems and to point out and discuss the ramifications of extending the description to fluid–solid systems. A system containing an interface of any type will be referred to as an *interfacial system*. Some of the major differences between fluid–fluid and fluid–solid interfacial systems are summarized in Table 1.1.

Only a brief overview of the thermodynamics of interfacial systems is given here. More thorough treatments have been given elsewhere (e.g., [1,2]).

1.2 A MODEL FOR INTERFACIAL SYSTEMS

1.2.1 Capillary Systems: Work, Heat Effects, and the Basic Laws

The prototype system studied in ordinary chemical thermodynamics is the *simple compressible system*, usually pictured as a single-phase compressible

---

TABLE 1.1
Contrast Between Attributes of Fluid–Fluid Interface and Fluid–Solid Interface Systems

Fluid–Fluid Interfaces	Fluid–Solid Interfaces
Internal equilibrium	Often internal disequilibrium
Smooth	Rough
Energetically and chemically uniform	Energetically and chemically heterogeneous
Subject to physical adsorption only	Often subject to chemisorption
Area extension by new surface formation	Area extension by formation and stretching
Work modes of volume compression and area extension only	Additional work modes for elastic deformation

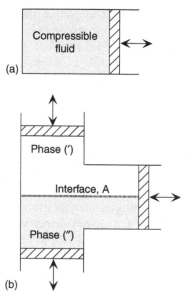

FIGURE 1.1 Thermodynamic systems. (a) *Simple compressible system,* consisting of a single-phase compressible fluid of volume *V*, subject only to the compression (i.e., *p–V*) work mode. (b) *Simple capillary system,* consisting of two bulk phase portions of volume V' and V'', respectively, (modeled as simple compressible systems) divided by an interfacial layer of area *A* (modeled as a zero-thickness membrane in uniform, isotropic tension, σ). In addition to the *p–V* work modes of bulk phase portions, the system is subject to the work of interfacial area extension.

material confined to a piston–cylinder enclosure as shown in Figure 1.1a. In the absence of external fields, such systems are subject only to the single work mode of compression:

$$\delta W = -p\, dV, \tag{1.1}$$

where *p* is the system pressure and *V* is its volume. *Simple capillary systems,* on the other hand, consist of an interfacial layer, modeled mechanically as a zero-thickness membrane in tension σ, plus portions of its two adjacent bulk fluid phases, which are themselves regarded as simple-compressible. As pictured in Figure 1.1b, in the absence of external fields, such systems are subject to the compression work modes of their bulk phase portions, plus the work associated with interfacial area extension. Work is associated with area extension because fluid interfaces always exist in a state of mechanical tension, manifest as measurable "surface tension" or "interfacial tension" in simple capillary systems. Thus,

$$\delta W = -p' \, dV' - p'' \, dV'' + \sigma \, dA, \tag{1.2}$$

where (') and ('') refer to the bulk phase portions of the system, A is the interfacial area, and σ is the interfacial tension.

The equilibrium pressure difference, $p'' - p'$, is related to the interfacial tension and the local curvature of the interface, κ, in accord with

$$p'' - p' = \sigma\kappa, \tag{1.3}$$

known as the Young–Laplace equation. For a spherical surface of radius r, the curvature is $\kappa = 2/r$, while for a surface of arbitrary shape, $\kappa = 2/r_m$, where r_m is the mean radius of curvature of the interface, or the radius of the "osculating sphere." The sign of κ is such as to require a higher pressure on the concave side of the interface. Unless the radius of curvature is less than a few hundred nanometers, one may assume $p'' \approx p' \approx p$, so that

$$\delta W \approx -p \, dV + \sigma \, dA. \tag{1.4}$$

The origin of the tension in the interfacial layer is the increased lateral intermolecular attraction that exists for the intermediate degrees of lateral molecular separation and/or composition that exists in the zone of inhomogeneity. In bulk fluids at rest, the state of stress is given simply by the scalar pressure, whereas in the inhomogenous interfacial layer, a tensor is required, and the components of the tensor tangential to the interface are less than those normal to the interface. The integration of this tangential pressure deficit with respect to distance across the thickness of the interfacial zone produces an energy per unit area (or force/length) which is assigned to the zero-thickness "membrane" located somewhere within the layer. Molecular mobility assures that the tension of the membrane is uniform and isotropic. For systems of moderate curvature, as suggested by Equation 1.3, the precise location of the membrane within the interfacial layer is not required to guarantee mechanical equivalence between the model and the actual system. If the interface is highly curved, the additional requirement that the moments of the tangential forces in the real system and the model be the same locates a specific "surface of tension."

Using Equation 1.4, together with the First Law of Thermodynamics, the heat effect for a quasi-static process in a closed capillary system is given by:

$$\delta Q = dU - \delta W = \left(\frac{\partial U}{\partial T}\right)_{V,A} dT + \left(\frac{\partial U}{\partial V}\right)_{T,A} dV + \left(\frac{\partial U}{\partial A}\right)_{T,V} dA$$

$$+ p dV - \sigma dA$$

$$= C_{V,A} dT + \left[\left(\frac{\partial U}{\partial V}\right)_{T,A} + p\right] dV + \left[\left(\frac{\partial U}{\partial A}\right)_{T,V} - \sigma\right] dA, \tag{1.5}$$

in which the internal energy U has been taken as a function of temperature, T, volume, V, and area, A. $C_{V,A}$ is the heat capacity of the system at constant volume and area. One may proceed from this point to the expression for entropy change, dS, for a quasi-static process, viz. $\delta Q/T$, and thence to the Helmholtz free energy function $F = U - TS$. Using standard reductions to eliminate the abstract derivatives in Equation 1.5, one may express the heat effect in terms of measurable quantities as:

$$\delta Q = C_{V,A} dT + T\left(\frac{\partial p}{\partial T}\right)_{V,A} dV - T\left(\frac{\partial \sigma}{\partial T}\right)_{V,A} dA, \tag{1.6}$$

from which the entropy effect is given as:

$$dS = \frac{\delta Q}{T} = \frac{C_{V,A}}{T} dT + \left(\frac{\partial p}{\partial T}\right)_{V,A} dV - \left(\frac{\partial \sigma}{\partial T}\right)_{V,A} dA. \tag{1.7}$$

Equation 1.4 and Equation 1.6, together with volumetric, calorimetric, and boundary tension measurements, suggest that the work and heat effects associated with even rather large area changes are quite small, so that for practical purposes, the requirement of constant A in the above expressions may be dropped. Furthermore, for a given (constant-composition) closed system, the boundary tension is a function of temperature only. Using the above simplifications, the thermodynamics of quasi-static processes in closed simple capillary systems may be summarized as follows:

$$
\left.\begin{aligned}
\delta W &= -p\,dV + \sigma\,dA \\[1.5em]
\delta Q &= C_V dT + T\left(\frac{\partial p}{\partial T}\right)_V dV - T\left(\frac{d\sigma}{dT}\right) dA \\[1em]
dU &= C_V dT + \left[T\left(\frac{\partial p}{\partial T}\right)_{V,A} - p\right] dV + \left[\sigma - T\left(\frac{d\sigma}{dT}\right)\right] dA \\[1em]
dS &= \frac{C_V}{T} dT + \left(\frac{\partial p}{\partial T}\right)_V dV - \left(\frac{d\sigma}{dT}\right) dA \\[0.5em]
dF &= -S\,dT - p\,dV + \sigma\,dA.
\end{aligned}\right\} \quad (1.8)
$$

Expressions of enthalpy, $H = U + pV$, and Gibbs free energy, $G = H - TS$, may also be defined, and are especially useful when the base set of independent variables is changed from (T,V,A) to (T,p,A). In terms of the latter variables, Equation 1.8 takes the form:

$$
\left.\begin{aligned}
dW &= -p\left(\frac{\partial V}{\partial T}\right)_p dT - p\left(\frac{\partial V}{\partial p}\right)_T dp + \sigma\,dA \\[1em]
\delta Q &= C_p dT - T\left(\frac{\partial V}{\partial T}\right)_p dp - T\left(\frac{d\sigma}{dT}\right) dA \\[1em]
dH &= C_p dT + \left[V - T\left(\frac{\partial V}{\partial T}\right)_p\right] dp + \left[\sigma - T\left(\frac{d\sigma}{dT}\right)\right] dA \\[1em]
dS &= \frac{C_p}{T} dT - \left(\frac{\partial V}{\partial T}\right)_p dp - \left(\frac{d\sigma}{dT}\right) dA \\[0.5em]
dG &= -S\,dT + V\,dp + \sigma\,dA
\end{aligned}\right\} \quad (1.9)
$$

where $C_p = (\partial H/\partial T)_p$. It is to be noted that, except for inclusion of the final term in each expression, dealing with area extension in Equation 1.8 or Equation 1.9, they are the same as those for simple compressible systems.[*]

---

[*] It is also possible to use the set of variables (T, p, σ), in which case alternate forms for the enthalpy and the Gibbs free energy are defined, viz., $H = U + pV - \sigma A$ and $G = U + pV - \sigma A - TS$, respectively.

The last of Equation 1.8 and Equation 1.9 provide thermodynamic definitions of the boundary tension, viz.:

$$\sigma = \left(\frac{\partial F}{\partial A}\right)_{T,V} \text{ or } \sigma = \left(\frac{\partial G}{\partial A}\right)_{T,p}. \tag{1.10}$$

It is tacitly assumed in writing all of the above equations that the system is in internal thermal, mechanical, and diffusional equilibrium. Under such conditions, but not otherwise, the boundary tension defined by Equation 1.10 is identical to the mechanical tension measured in the laboratory.

The above thermodynamic description suggests that the reversible heat and work effects associated with isothermal, reversible area extension in capillary systems are proportional to σ and $-T(d\sigma/dT)$, respectively. As noted above, actual heat or work effects associated with dispersion processes, even with capillary systems, are usually several orders of magnitude larger than the corresponding reversible values, although they are sometimes found to be proportional to boundary tension or its temperature derivative. It also shows that, since the boundary tension is positive, fluid interfaces spontaneously contract to a minimum value consistent with external constraints on the system.

1.2.2 EXTENSION TO FLUID–SOLID INTERFACIAL SYSTEMS

One must next consider how fluid–solid interfacial systems differ from the fluid–fluid (capillary) systems described above. Provided appropriate precautions are taken, such systems can often qualitatively, and even quantitatively, be described in terms of the thermodynamics of capillary systems [3]. The tension σ at a solid–fluid interface represents, analogous to a capillary system, the integral effect of a reduced lateral stress component across the zone of inhomogeneity between the fluid and solid phases. The difference is that the strain (and hence stress) in the bulk solid phase is a tensor quantity and varies from point to point. It is, in general, not possible to know precisely what the spatial stress distribution in the solid is. Strictly speaking, the boundary tension at a fluid–solid interface can be defined only when the total free energy density tensor in the solid phase reduces to a uniform value. Fortunately, this restriction may not be critical because the elastic (stress) energy density is usually only a small (often negligible) component of the total energy density [1, p. 290].

For the evaluation of the work effect for fluid–solid interfacial systems, it is important to recognize that interfacial area creation in such systems may occur not only through "formation," as is the case in capillary systems, but also by way of stretching, a process which changes the structure and properties of the interfacial layer. In capillary systems, such changes are instantly relaxed out, but in solid–fluid systems may be frozen in indefinitely, or at least for significant periods of time. The isothermal quasi-static mechanical work required to dilate the interfacial area is given in general by [4, pp. 166–168]:

$$\left(\frac{\delta W}{\mathrm{d}A}\right)_{\text{int. dilation}} = \sigma_{\mathrm{m}} = \frac{\mathrm{d}F}{\mathrm{d}A} = \frac{\mathrm{d}(f^{\sigma}A)}{\mathrm{d}A} = f^{\sigma} + A\frac{\mathrm{d}f^{\sigma}}{\mathrm{d}A} = \sigma + A\frac{\mathrm{d}\sigma}{\mathrm{d}A}, \quad (1.11)$$

where σ_{m} is the mechanical boundary tension, and f^{σ} is the surface free energy per unit area, equal to the thermostatic "tension," σ, defined below in Equation 1.10. The last term in Equation 1.11 must be included because the structure of the surface is being changed. Finally, work may be associated with both bulk and surface shear strain in the solid. The work terms associated with these processes would be [5]:

$$\text{Work of bulk shear strain} = V(s^{\beta}:\delta\varepsilon^{\beta}), \text{ and} \qquad (1.12)$$

$$\text{Work of surface shear strain} = A(s^{\sigma}:\delta\varepsilon^{\sigma}), \qquad (1.13)$$

where s^{β} and s^{σ} are the bulk and surface stress tensors, respectively, and $\delta\varepsilon^{\beta}$ and $\delta\varepsilon^{\sigma}$ are the corresponding bulk and surface displacement tensors. In processes involving surface area extension in fluid–solid interfacial systems, the work of surface shear strain is usually negligible, but the work associated with the accompanying bulk shear strain is often the dominant component of the total work effect. It is generally not possible to calculate this effect quantitatively, but it is one of the reasons that the energies associated with size reduction (crushing and grinding) of solids are usually orders of magnitude larger than those suggested by Equation 1.4.

1.2.3 Compound Interfacial Systems

What is described above is a *simple* capillary or interfacial system, in that only a single type of fluid interface is involved. It is exemplified by a liquid against its own equilibrium vapor, or against a second liquid phase. The descriptions used may readily be extended to systems with more than a single type of interface (i.e., *compound* capillary or interfacial systems). A compound capillary system might be exemplified by an oil drop floating at a water–air interface, as pictured in Figure 1.2a, whereas an example of a compound fluid–solid interfacial system might be a liquid drop in air resting on a solid surface, as shown in Figure 1.2b. The compound capillary system shown is made up of three simple capillary systems, involving the water–air, oil–air, and oil–water

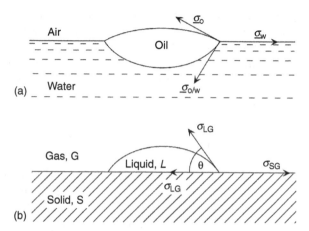

FIGURE 1.2 Examples of compound interfacial systems. (a) Oil drop at an air–water interface. Compound system consists of the water–air, oil–air, and oil–water simple capillary systems in mutual equilibrium. The magnitudes of the three boundary tensions determines the shape of the drop. (b) Sessile liquid drop at a solid–gas interface. Compound system consists of liquid–gas simple capillary system, and solid–gas and solid–liquid simple interface systems in mutual equilibrium. The magnitudes of the three boundary tensions determines the contact angle, θ.

interfaces, respectively, in mutual equilibrium. In such systems in general, the possibility of three-phase junctions, or "interlines," arises. Mechanical equilibrium is then manifest at any point along the three-phase interline in the example capillary system by the requirement that the boundary tension forces, $\overline{\sigma}_w$, $\overline{\sigma}_o$, and $\overline{\sigma}_{o/w}$ on a differential segment of the interline at that point sum (vectorially) to zero, i.e,

$$\overline{\sigma}_w + \overline{\sigma}_o + \overline{\sigma}_{o/w} = 0, \tag{1.14}$$

known as Neumann's triangle. Application of the mechanical equilibrium requirement fixes the relative size, shape, and location of the fluid interfaces in such systems. It is not always possible to satisfy Equation 1.14, as for example, when $\sigma_w(\sigma_o + \sigma_{o/w})$. In such a case, the oil drop would be drawn out into an ultra-thin film (possibly a monolayer) in a process termed spreading.

A compound interfacial system of the type shown in Figure 1.2b consists of a simple capillary system (containing the liquid–gas interface) plus two simple fluid–solid interfacial systems (containing the gas–solid and liquid–solid interfaces, respectively). If it is possible to regard the fluid–solid interfacial systems as being in full internal equilibrium (a questionable assumption in many cases), Neumann's triangle reduces to the set of scalar equations:

$$\sigma_{SG} - \sigma_{SL} = \sigma_{LG} \cos \theta, \text{ and} \qquad (1.15)$$

$$\sigma_{LG} \sin \theta = E_S. \qquad (1.16)$$

Equation 1.15, known as Young's equation, results here from a horizontal force balance on an element of the interline and involves the "tensions" σ_{SG} and σ_{SL} associated with the solid–fluid interfaces. The angle θ is known as the "contact angle" of the liquid L (in the gas G) against the solid S. Equation 1.16 shows that the vertical component of the interline force must be balanced by an elastic force E_S in the solid phase.

1.2.4 Configurational Entropy Effects of Subdivision

The process of area extension may involve only a change in the shape of the system, as occurs when a globular mass of liquid is extended into a thin film, or, on the other hand, may involve the subdivision of the system into an increased number of freely moving "kinetic units." In the latter case, an additional configurational entropy term must be added to the expression for dS in Equation 1.8 or Equation 1.9. This is sometimes important in describing processes of dispersion or condensation. Consider the total configurational entropy change in a system of total volume V, which occurs when a volume V'' ($\ll V$) of a phase ($''$) is subdivided into N particles, each of volume v_p in a medium of volume $V' \approx V$ under conditions of constant temperature and total volume. The process is one in which we mix n' *moles* of dispersion medium with n'' *moles* of particles to form a completely random (i.e., ideal or regular) "solution." The configurational entropy change is given by:

$$\Delta S^{\text{conf}} = R\left[n' \ln \frac{n' + n''}{n'} + n'' \ln \frac{n' + n''}{n''}\right], \qquad (1.17)$$

where R is the universal gas constant. We assume that the dispersion medium 1 is made up of material of moderate molecular weight so that:

$$n'' = \frac{N}{N_A} = \frac{V''}{N_A v_p} \approx \frac{\phi'' V}{N_A v_p} \ll n' = \frac{V'}{v'^0} \approx \frac{V}{v'^0}, \qquad (1.18)$$

where N_A is Avogadro's number, ϕ'' is the particle volume fraction, v_p is the particle volume, and v'^0 is the molar volume of the dispersion medium. Under the conditions of Equation 1.17:

$$\Delta S^{\text{conf}} \approx n'' R \left[1 + \ln \left(\frac{n'}{n''} \right) \right] = \frac{\phi'' V k}{v_p} \left[1 + \ln \left(\frac{v_p N_A}{v'^0 \phi''} \right) \right], \qquad (1.19)$$

where k is Boltzmann's constant. The total entropy change for the process is:

$$\Delta S = -\left(\frac{d\sigma}{dT} \right) \Delta A + \frac{\phi'' V k}{v_p} \left[1 + \ln \left(\frac{v_p N_A}{v'^0 \phi''} \right) \right]. \qquad (1.20)$$

The free energy change for the above process is given by

$$\Delta F = \Delta U - T \Delta S = \sigma \Delta A - \frac{\phi'' V k T}{v_p} \left[1 + \ln \left(\frac{v_p N_A}{v'^0 \phi''} \right) \right]. \qquad (1.21)$$

If the particles are spheres,

$$v_p = \frac{4}{3} \pi r_p^3, \text{ and } \Delta A = 4\pi r_p^2 N_2 = \frac{3\phi_2 V}{r_p}, \qquad (1.22)$$

so that

$$\Delta F = \frac{3\phi'' V}{r_p} \sigma - \frac{3\phi'' V k T}{4\pi r_p^3} \left[1 + \ln \left(\frac{4\pi r_p^3 N_A}{3 v'^0 \phi''} \right) \right]. \qquad (1.23)$$

Equation 1.23 shows that if, for a given volume fraction of material which is dispersed, the boundary tension is low enough, the effect of the configurational entropy gain can more than compensate for the free energy required to form the new surface area. Under such conditions, the total free energy change is negative, and the dispersion process would be expected to occur spontaneously. Insertion of realistic numbers into Equation 1.23 suggests that an ultra-low critical boundary tension is generally required for spontaneous emulsification. Taking $r_p = 50$ nm, $v'^0 = 18$ cm³/mole, $T = 298$ K, and $\phi'' = 0.001$, yields:

$$\sigma \leq \frac{kT}{4\pi r_p^2} \left[1 + \ln \left(\frac{4\pi r_p^3 N_A}{3 v'^0 \phi''} \right) \right] = 0.00322 \text{ mJ/m}^2. \qquad (1.24)$$

1.3 MULTICOMPONENT INTERFACIAL SYSTEMS

1.3.1 THE GIBBS DIVIDING SURFACE AND ADSORPTION

In multicomponent capillary systems, internal diffusional equilibrium generally produces different component concentrations in the two bulk phase portions, and a composition in the interfacial zone different from that of either of the bulk phases. The relative difference in concentration of a solute between the interfacial zone and an adjacent bulk phase is termed *adsorption*, and it may be either positive or negative, depending on whether the interfacial zone concentration is richer or leaner in the solute. The model as put forth so far, assumes that while the system is made to undergo temperature, volume and/or area changes, internal diffusional equilibrium is maintained, but it does not describe that internal equilibrium. To describe the equilibrium distribution of the various components of the system between the bulk phase portions and the interface, Gibbs extended the mechanical model by requiring that the interfacial layer be replaced by a "dividing surface," oriented normal to the density gradient in the interfacial zone, up to which *all* of the intensive variables describing the bulk phase portions of the system are taken to be uniform at their bulk-phase values. This is pictured schematically in Figure 1.3 for the concentrations of the solvent 1 and solute 2 in a binary simple capillary system. The bulk phase concentrations of these species are C_1', C_1'', C_1', and C_1''. Depending on the dividing surface location, the model requires an addition (or subtraction) of a particular number of moles of each species to the surface, n_1^σ and n_2^σ, respectively, to exhibit mass equivalence with the real system. Being dependent upon the (generally) arbitrary location of the dividing surface, n_1^σ and n_2^σ, called "surface excesses," are examples of "properties of the dividing surface," as distinct from properties of the capillary system. On a per unit area basis, $\Gamma_1 = n_1^\sigma/A$ and $\Gamma_2 = n_2^\sigma/A$, termed the *adsorptions* of components 1 and 2, respectively, Combining material balances for the two components between the model and the real system leads to an expression for the *relative* adsorption of the solute to that of the solvent, $\Gamma_{2,1}$, viz.,

$$\Gamma_{2,1} \equiv \Gamma_2 - \Gamma_1 \left(\frac{C_2' - C_2''}{C_1' - C_1''} \right)$$

$$= \frac{1}{A} \left[(n_2 - VC_2') - (n_1 - VC_1') \left(\frac{C_2' - C_2''}{C_1' - C_1''} \right) \right]. \tag{1.25}$$

It is invariant with respect to dividing surface location and therefore a property of the system. n_1 and n_2 are the total moles of the components in the

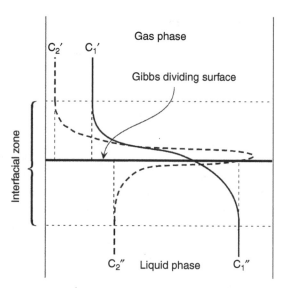

FIGURE 1.3 Gibbs Dividing Surface model of the component distribution in a binary gas–liquid simple capillary system in the vicinity of the interfacial zone. Actual concentration profiles for the solute 1 and the solute 2 show continuous variation across the interfacial zone from the liquid-phase values, C_1' and C_2'', respectively, to the gas-phase values, C_1' and C_2'', respectively. In the model, the component concentrations are constant at their respective bulk-phase values up to the location of the zero-thickness "dividing surface."

system, and V is the total volume of the system. It is clear how relative adsorptions may be defined for additional solute components if they are present.

When one of the phases, say phase ($'$), of the capillary system is a gas, the component concentrations in that phase are often negligible compared with their values in the liquid, so that

$$\Gamma_{2,1} \approx \Gamma_2 - \Gamma_1 \left(\frac{C_2''}{C_1'} \right), \tag{1.26}$$

which makes it clear that $\Gamma_{2,1}$ is positive when the component ratio ascribed to the interface, Γ_2/Γ_1, is higher than that existing in the bulk liquid, C_2''/C_1'.

The relative adsorption of a solute is difficult to measure in capillary systems. Although all of the individual terms on the right hand side of Equation 1.25 are easy to determine, their difference is generally several orders of magnitude smaller than the terms themselves. Despite this difficulty, the relative adsorptions of the solutes in a multicomponent capillary system must be added formally to the list of measurable descriptors of such systems. Thus, for a simple capillary system of c components, one has c independent variables (σ plus the $c-1$ solute relative adsorptions, $\Gamma_{i=2.c,1}$), in addition to those needed to describe a two-phase system in which the presence of the interface is ignored. The Gibbs phase rule remains unaltered, however, because one also has c additional equilibria, viz., the adsorption equilibria for all of the components. For example, for a binary simple capillary system, the number of independent intensive variables, F, required to fix the intensive state of the system is: $\mathcal{F} = c(=2) - \Phi(=2) + 2 = 2$. Such systems are usually described by equilibrium relationships at constant T, leaving only a single free variable. The most commonly used relationships are:

1. The surface (or interfacial) tension equation : $\quad \sigma = \sigma(c_2)$

2. The adsorption isotherm : $\qquad\qquad\qquad \Gamma_{2,1} = \Gamma_{2,1}(C_2)$ (1.27)

3. The surface equation of state : $\qquad\qquad \sigma = \sigma(\Gamma_{2,1})$

    These equations may apply to either vapor–liquid or liquid–liquid binary systems, with C_2 referring to the concentration in one of the bulk phases. (The concentration in the opposite bulk phase will assume its appropriate equilibrium value.)

    For capillary systems, the surface tension equation is easy to determine experimentally, whereas the adsorption isotherm and the surface equation of state are difficult to obtain, owing to the difficulty of measuring the relative adsorption. In contrast, as will be described in more detail below, the adsorption isotherm is often readily accessible for solid–fluid interfacial systems. The surface equation of state is readily measured only for the case in which the "solute" is effectively an insoluble monolayer spread at the interface [6].

    Gibbs derived a universal relationship for multicomponent capillary systems, which must exist between the variables σ, $C_{2,...,c}$, and $\Gamma_{i=2...c,1}$, analogous to the Gibbs-Duhem equation relating the chemical potentials of components in a given solution. The result is that Equation 1.27 is not independent. Gibbs' relationship, known as the *Gibbs adsorption equation*, for a binary simple capillary system at constant temperature takes the form [1, pp. 85–95]:

$$d\sigma = -\Gamma_{2,1}\, d\mu_2 = -RT\Gamma_{2,1}\, d\ln C_2, \qquad (1.28)$$

where μ_2 is the chemical potential of the solute, and in which in the last equation, the bulk liquid portion of the system is assumed to be an ideal solution. (This restriction is easily removed, if necessary, using an appropriate description of the bulk phase nonideality.) The Gibbs adsorption equation may be used to inter-convert the relationships in Equation 1.27. For example, the adsorption isotherm may be derived from the surface tension equation by differentiation:

$$\Gamma_{2,1}(C_2) = -\frac{1}{RT}\left(\frac{\partial \sigma}{\partial \ln C_2}\right)_T, \tag{1.29}$$

and the surface equation of state is obtained by elimination of C_2 between the surface tension equation and the adsorption isotherm. On the other hand, the surface tension equation may be derived from the adsorption isotherm by integration:

$$\sigma = \sigma_0^1 - \pi = \sigma_0^1 - \int_0^{C_2} \Gamma_{2,1}(C_2^*)\mathrm{d}\ln C_2^*, \tag{1.30}$$

where σ_0^1 is the boundary tension for the pure solvent 1 (against its equilibrium vapor) and π is known as the "surface pressure" of the adsorbate. When it is the adsorption isotherm which is known, the surface tension equation can be obtained only to within the additive constant, σ_0^1, which must be determined independently.

1.3.2 IMMISCIBLE INTERFACIAL SYSTEMS

One of the most important simplifications of the model of the simple interfacial system is that of the *immiscible* simple interfacial system. This occurs when the phases forming the interfacial system are completely immiscible, as pictured in Figure 1.4. The model applies in particular to most of the fluid–solid interfacial systems encountered in practice, and amounts to requiring that the solid be nonvolatile, or insoluble, in the adjoining fluid and that none of the components of the fluid phase dissolve into the solid. Capillary system examples would include an aqueous solution in contact with an oil or with mercury, in which all components of the solution are assumed to be completely insoluble, or a solution of completely nonvolatile components in contact with a gas that does not dissolve in the solution. For immiscible interfacial systems, the Gibbs dividing surface may be located without ambiguity so as to separate all of the atoms or molecules of the immiscible phases from one another. The relationships of Equation 1.27 may be extended essentially without change from a binary

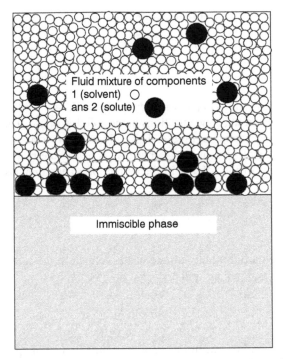

FIGURE 1.4 Immiscible interfacial system. The bulk phase portions of the interfacial system, are completely immiscible. The Gibbs Dividing Surface is made coincident with the actual boundary between the immiscible phases. In the system shown, the upper phase is a binary solution, whose components compete for space at phase boundary.

system to a ternary immiscible interfacial system if the third component constitutes of one of the two bulk phase portions of the interfacial system. As before, the relative adsorption of the solute is given by Equation 1.26, from which it is evident that the solute and solvent *compete* for space on the adsorbent surface, as pictured schematically in Figure 1.4. Under conditions in which the solution is dilute ($C_2 \to 0$) or for adsorption of a component from a nonadsorbing gas ($\Gamma_1 \to 0$), one may identify the relative adsorption of the solute with its actual adsorption, i.e., $\Gamma_{2,1} \approx \Gamma_2$. The tension σ_0^1 in immiscible interfacial systems refers to the tension of the solute-free interface between the substrate phase and the immiscible phase of the pure solvent component 1.

The relative adsorption of the solute in an immiscible interfacial system may be determined in principle from laboratory measurements using the appropriate form of the right-hand side of Equation 1.24:

$$\Gamma_{2,1} = \frac{1}{A}\left[(n_2 - VC_2) - (n_1 - VC_1)\left(\frac{C_2}{C_1}\right)\right], \tag{1.31}$$

where the concentrations in the immiscible phase, $C_1'' = C_2'' = 0$, and the $(')$ has been removed from the concentrations in the solution phase. The second term in the brackets is usually negligible. This follows from the assumption that the number of moles of solvent 1 adsorbed, n_1^σ, is negligible relative to the total number of moles of the solvent in the system, n_1. Then one has

$$C_1 = \frac{n_1 - n_1^\sigma}{V - \bar{v}_1 n_1^\sigma} \approx \frac{n_1}{V}, \tag{1.32}$$

where $\bar{v}_1$ is the partial molal volume of the solvent in the solution. The total number of moles of solute in the system may be expressed as:

$$n_2 = VC_2^0, \tag{1.33}$$

where C_2^0 is the initial concentration of the solute in the bulk phase (i.e., prior to any adsorption). Substitution of Equation 1.32 and Equation 1.33 into Equation 1.31 gives

$$\Gamma_{2,1} = \frac{V}{A}\left[C_2^0 - C_2\right], \tag{1.34}$$

where C_2 refers to the solute concentration in the bulk after adsorption equilibrium has been established. Even for immiscible capillary systems, measurement of the solute relative adsorption remains difficult due to the relatively small proportion of the solute, which is adsorbed relative to that which remains in the bulk phase. For solid adsorbents with large interfacial areas, however, a significant amount of solute may be in the adsorbed state, and the stoichiometric method of Equation 1.34 affords a convenient means for measurement of the relative adsorption at a liquid–solid interface. Adsorption at the gas–solid interface, for high surface area solids, may be measured directly by determining the weight gain of the solid specimen. A variety of other techniques is also available for the measurement of adsorption at the fluid–solid interface. These include gas and liquid chromatography, diffuse reflectance IR absorption, NMR, radio-tracer techniques, etc., all applicable to powders and fibers.

1.3.3 Adsorption as a Process: Heats of Adsorption

The formalism developed thus far for describing multicomponent interfacial systems assumes internal thermal and mechanical equilibrium, and in particular equilibrium spatial distribution of the components (i.e., adsorption). One may easily extend the description to systems in *partial equilibrium*, by which is meant internal thermal and mechanical equilibrium, but not necessarily equilibrium with respect to adsorption or chemical reaction. The present discussion will be limited to immiscible interfacial systems in which a "solute" B may adsorb either from a gas mixture with a nonadsorbing diluent gas D, or from solution in a solvent D. The two cases are different in that when adsorption occurs from a gas phase, the initial state may generally be regarded as one in which the adsorbent surface is considered empty (or in a modified state due to residual chemisorption of the diluent species), whereas in the latter case, the initial state is one in which the interface is completely occupied by the solvent liquid D. Thus, for adsorption of B from a gas mixture, the process may be represented by the statement:

$$B^g = B^\sigma, \tag{1.35}$$

but when B adsorbs from its solution in D, the statement is:

$$B^{soln} + N_B D^\sigma = B^\sigma + N_B D^{soln}, \tag{1.36}$$

where N_B is the number of moles of solvent D occupying the same area as one mole of B at the interface. (It is assumed for simplification here that this number does not change with the extent of adsorption). Adsorption from the gas phase is thus usually a simple addition process, while adsorption from solution is an exchange process.

With the chosen initial state, the extent of adsorption of the component B from the gas phase may be described by the variable ξ_B, which represents the instantaneous number of moles of B adsorbed, i.e.,

$$\xi_B = n_B^\sigma(t) = n_B^0 - n_B(t) = \Gamma_B(t)A, \tag{1.37}$$

where $n_B^\sigma(t)$, $n_B(t)$ and $\Gamma_B(t)$ are written as functions of time to remind us that they are not necessarily equilibrium values. Since in the case of adsorption from solution it can generally be assumed that the relative change in the total moles of the diluent liquid in the solution is negligible, Equation 1.36 remains valid for this case as well.

The adsorption process is generally accompanied with the evolution of heat. Thus, the expressions for δQ and dU in Equation 1.8 or δQ and dH in

Equation 1.9 must contain the additional terms

$$\left(\frac{\partial U}{\partial \xi_B}\right)_{T,V} d\xi_B = \left(\frac{\partial u_A}{\partial \Gamma_B}\right)_{T,V} d\Gamma_B \text{ or } \left(\frac{\partial H}{\partial \xi_B}\right)_{T,p} d\xi_B$$

$$= \left(\frac{\partial h_A}{\partial \Gamma_B}\right)_{T,p} d\Gamma_B, \text{ respectively,}$$

where u_A and h_A are system internal energy and enthalpy taken on a per-unit-area basis. These may be identified with the changes in the respective surface quantities u_A^σ and h_A^σ. The negative of the coefficients in the above expressions are "heats of adsorption," (positive numbers, since the adsorption process is exothermic). With reference to calorimetric experiments, the coefficient obtained under conditions of constant temperature and volume is:

$$-\frac{1}{A}\left[\frac{\delta Q}{\delta \Gamma_B}\right]_{T,V,\Gamma_B} = -\left(\frac{\partial u_A^\sigma}{\partial \Gamma_B}\right)_{T,V,\Gamma_B} = \Delta u_A, \qquad (1.38)$$

the "heat of adsorption at constant volume." For calorimetric measurements at constant pressure, we have

$$-\frac{1}{A}\left[\frac{\delta Q}{\delta \Gamma_B}\right]_{T,p,\Gamma_B} = -\left(\frac{\partial h_A^\sigma}{\partial \Gamma_B}\right)_{T,p,\Gamma_B} = \Delta h_A, \qquad (1.39)$$

the "heat of adsorption at constant pressure." These differ significantly only when adsorption occurs from a gas phase (by the amount RT, if the gas phase is ideal). For adsorption from solution, the difference is generally negligible.

One additional distinction needs to be made. Since adsorption is, in general, a cooperative process, the magnitude of the heats of adsorption depends on the extent of the process. This may be the result of lateral (i.e., adsorbate–adsorbate) interactions, or in the case of solid–fluid systems, the result also of adsorbent surface heterogeneity. The calorimetric quantities defined in Equation 1.38 and Equation 1.39 are thus termed *differential* heats of adsorption. They correspond to given specific values of the adsorption, Γ_B. It is more common (and easier) to measure the heat evolved when the extent of adsorption varies from 0 to some finite value of Γ_B. The resulting amounts of heat evolved are termed the *integral* heats of adsorption, viz.:

$$-\frac{1}{A} \int_{0}^{\Gamma_B} \left[\frac{\delta Q}{\delta \Gamma_B^*} \right]_{T,V,\Gamma_B} d\Gamma_B^* = \Delta U_A, \text{ or} \qquad (1.40)$$

$$-\frac{1}{A} \int_{0}^{\Gamma_B} \left[\frac{\delta Q}{\delta \Gamma_B^*} \right]_{T,p,\Gamma_B} d\Gamma_B^* = \Delta H_A, \qquad (1.41)$$

the integral heats of adsorption at constant volume and constant pressure, respectively.

One may alternatively obtain the differential heats of adsorption indirectly (i.e., noncalorimetrically), from knowledge of the equilibrium adsorption isotherms over a range of temperatures as follows. Adsorption equilibrium of component B at low concentration may be described in terms of an equilibrium constant

$$K_B = \frac{\Gamma_B}{C_B}, \qquad (1.42)$$

or more generally, the local slope of the adsorption isotherm, which is a function of temperature and the concentration of B, as suggested in Figure 1.5.

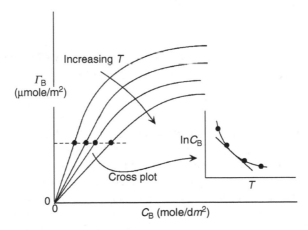

FIGURE 1.5 Determination of the *isosteric* heat of adsorption from a family of adsorption isotherms at different temperatures. A cross plot is prepared from values of the bulk adsorbate concentration, C_B, corresponding to a given adsorption, Γ_B at varying temperatures. The isosteric heat of adsorption is computed from the cross plot in accord with Equation 1.44.

The temperature dependence of the adsorption equilibrium at a constant extent of adsorption, Γ_B, is given by the van't Hoff equation:

$$\left(\frac{\partial \ln K_B}{\partial T}\right)_{\Gamma_B} = -\frac{\Delta h_A}{RT^2}. \tag{1.43}$$

Substitution of Equation 1.42 into Equation 1.43 leads to:

$$\Delta h_A = RT^2 \left(\frac{\partial \ln C_B}{\partial T}\right)_{\Gamma_B}. \tag{1.44}$$

Although the heat of adsorption obtained in this way is identical to the differential heat of adsorption, it is referred to as the *isosteric* heat of adsorption as a reminder that it refers to conditions of constant extent of adsorption. Knowledge of the differential or isosteric heat of adsorption as a function of the coverage (i.e., of Γ_B) provides useful insight into the site energy distribution of solid adsorbents.

A calorimetric quantity closely related to the integral heat of adsorption is the "heat of immersion," h_{imm}, the heat evolved per unit area when a solid initially in vacuum, its own equilibrium vapor, air or another nonadsorbing gas is immersed in a pure liquid. The quantity is generally small, but for solids of moderate to high specific area, is well within the measurability afforded by commercially available calorimeters.

1.4 THE THERMODYNAMIC CONSEQUENCES OF INTER-FACIAL CURVATURE: THE KELVIN EFFECT

One of the most important consequences of the Young–Laplace pressure jump that must exist across curved fluid interfaces, is the dependence of the bulk thermodynamic properties of small liquid drops (or gas bubbles in liquid) on their interfacial curvature. When considering the vapor pressure of a liquid droplet, for example, it must be recognized that the effective pressure exerted on the liquid (on the concave side of the curved interface) exceeds the external pressure (the vapor pressure) by an amount given by Equation 1.3. The curvature of a spherical droplet is given by $\kappa = 2/r$, where r is the droplet radius, so that for micro- or nanodroplets is very high. The change in chemical potential of a pure liquid μ^L, or the associated fugacity f^L, induced by a change in pressure at constant temperature is given by

$$d\mu^L = RT d \ln f^L = v^L dp, \tag{1.45}$$

where v^L is the molar volume of the liquid. If the saturated vapor is regarded as an ideal gas, the fugacity of the pure liquid may be replaced by the vapor pressure, p^s, so that

$$\mathrm{d}\ln p^s = \frac{v^L}{RT}\mathrm{d}p. \qquad (1.46)$$

Assuming the liquid to be effectively incompressible, Equation 1.46 may be integrated from the pressure p_1 equal to the vapor pressure of the liquid with a flat surface, p_∞^s (the handbook vapor pressure) to the pressure p_2 inside a liquid droplet of radius r. The latter will be the appropriate vapor pressure of the droplet of radius r, p_r^s, plus the Young–Laplace pressure jump, $2\sigma/r$, i.e., $p_2 = p_r^s + 2_\sigma/r$. For small values of r, the situation of interest, $p_r^s \ll 2\sigma/r$, so $p_2 \approx 2\sigma/r$. The integration of Equation 1.46 then gives

$$\ln\frac{p_r^s}{p_\infty^s} = \int_{p_\infty^s}^{\frac{2\sigma}{r}}\frac{v^L}{RT}\mathrm{d}p, \quad \text{or} \quad p_r^s = p_\infty^s, \exp\left(\frac{2\sigma v^L}{rRT}\right), \qquad (1.47)$$

first derived by Lord Kelvin and known as the Kelvin equation. For water droplets (at 25°C) of radius equal to 1 μm, the vapor pressure is enhanced over its handbook value by approximately 0.1%, but for a radius of 10 nm, the enhancement is about 12%. The Kelvin effect underlies the spontaneous size disproportionation process that occurs in polydisperse aerosols (and the formation of rain droplets), as well as the process of phase change through the nucleation and growth of nuclei of the new phase.

In addition to the result that small droplets of a liquid have a higher vapor pressure than large droplets, the Kelvin effect also leads to the result that small droplets vaporize (at a given pressure) at a lower temperature than larger droplets and have a lower heat of vaporization than larger droplets.

Curvature effects also apply to the condensation of vapors into the pores of a finely-porous solid. In this situation, if the condensate liquid wets the pore walls of the solid, the liquid in the pores will be on the *convex* side of the meniscus, and the appropriate vapor pressure is given by Equation 1.47 with the sign of the exponential argument reversed, and r referring to the pore radius. Thus, condensation occurs at partial pressures below the handbook vapor pressure (capillary condensation). The other Kelvin effects associated with small droplets are also reversed (i.e., the boiling point) and heat of vaporization of a liquids condensed in the fine pores of a solid will be higher than those for liquids with flat surfaces.

It is possible, with some precautions, to extend the above descriptions of curvature effects in capillary systems to fluid–solid interfacial systems. This is important in describing the thermodynamic properties of fine powders and fibers. Fine powders and fibers will have higher sublimation pressures and higher solubilities in solvents than the corresponding bulk solids. The latter result takes the form:

$$C_r^{sat} = C_\infty^{sat} \exp\left(\frac{2\sigma_{SL} v_S}{rRT}\right), \tag{1.48}$$

where σ_{SL} is the "tension" of the solid–liquid interface, and r is the effective radius of the solid particle. For solid particles with a finite molecular solubility in the dispersing liquid, any initial polydispersity in particle size will thus lead to a gradual size disproportionation in which the larger particles grow at the expense of the smaller ones, a process known as Ostwald ripening. Solid particles with rough or irregular surfaces under such circumstances will evolve into perfect crystals (for crystalline materials) or spheres (for amorphous materials) over time. Also, in analogy with liquid droplets, small crystals will exhibit lower melting points and lower heats of fusion than the corresponding bulk solids.

1.5 THE CONCEPT OF SURFACE OR INTERFACIAL ENERGY

1.5.1 The Definition of Interfacial Energy

It is useful to attempt to define an appropriate surface or interfacial energy to be associated with a given interfacial system for purposes of comparing different systems to one another and for formulating expressions for the driving forces for various processes involving interfacial systems. Thus, the definition and experimental determination of the interfacial energy of various interfacial systems is one of the major goals of the thermodynamic analysis of such systems. It is evident that more than one "interfacial energy" may be defined. First, it is generally assumed that the terminology refers to interfacial *free* energy, as opposed to interfacial internal energy or enthalpy. One measure of interfacial energy, particularly relevant to capillary systems, is afforded by Equation 1.10, defining the free energy change associated with unit interfacial area extension under conditions of constant temperature and total volume as the boundary tension, σ. For capillary systems, the boundary tension is readily measured, but the "boundary tensions" of solid–fluid interfaces found in the more general interfacial systems are not so easily accessible by direct measurement. It is important also to recognize that σ is a surface free energy derivative, rather than the surface energy itself. A surface free energy (per unit area) may be defined with reference to the Gibbs dividing surface as a surface excess quantity, viz.

$$\left(\frac{F^\sigma}{A}\right) \equiv f^\sigma = \frac{1}{A}[F - F' - F'']. \tag{1.49}$$

The above expression can be given for an arbitrary location of the dividing surface in terms of the component adsorptions and chemical potentials, viz. [1, p. 288]:

$$f^\sigma = \sigma + \sum_i \Gamma_i \mu_i. \tag{1.50}$$

The variable f^σ is clearly a property of the dividing surface, but if it is computed for the dividing surface location such that the adsorption of the solvent 1 is zero, one obtains the *relative* surface free energy:

$$f_1^v = \sigma_0^1 + \sum_{i=2} \Gamma_{i,1} \mu_i, \tag{1.51}$$

which is independent of dividing surface location, but of course dependent on the standard state values chosen for evaluating the component chemical potentials. It is clear that the relative surface free energy is equal to the boundary tension for a pure-component system or for a system with a solvent 1, in which solute adsorption is zero, designated as σ_0^1, but not otherwise. If the component standard states are chosen such as to make Equation 1.51 identical to Equation 1.30, we may identify f_1^σ with σ, so that for capillary systems, the interfacial free energy is identifiable with the measurable boundary tension. Determination of the interfacial free energy for fluid–solid interfacial systems is more challenging, as discussed below.

1.5.2 Combining Rules for Interfacial Tensions in Terms of Surface Tensions

Considerable effort over the years has been given to the attempt to develop procedures for the prediction of the interfacial tension between two liquids, σ_{AB}, in terms of their respective surface tension values, σ_A and σ_B. Most are applicable only to immiscible capillary systems and are based on a consideration of the intermolecular forces acting across the interface. One of the oldest and simplest is Antanow's rule, which states that

$$\sigma_{AB} = |\sigma_A - \sigma_B|. \tag{1.52}$$

Based on the idea that surface tension arises from unbalanced, inward-directed intermolecular forces, it reckons that the interfacial tension should be the *net* of unbalanced intermolecular forces directed toward the liquid of higher surface tension. Despite its simplicity, Antanow's rule is often quite successful. It even gives reasonable results for miscible capillary systems if the surface tensions of the pure liquids are substituted with the surface tensions of the mutually saturated liquids. Girifalco and Good [7] reasoned that interfacial tension should be the sum of the two surface tensions, minus the molecular attraction across the interface. If the molecular interactions across the interface were predominantly those of London (dispersion) forces, their energetics would be given by the geometric mixing rule. Thus they proposed the combining rule:

$$\sigma_{AB} = \sigma_A + \sigma_B - 2\Phi\sqrt{\sigma_A\sigma_B}, \tag{1.53}$$

where Φ was an adjustable parameter to account for the possible importance of nondispersive forces acting across the interface. For "apolar" liquids, $\Phi = 1$. Fowkes suggested a similar formula using the following reasoning [8]. If surface tensions are the direct consequence the intermolecular forces acting in the liquid, they should be separable into the sum of terms representing different contributions to the intermolecular forces, e.g.,

$$\sigma = \sigma^d + \sigma^p + \sigma^i + \sigma^h + \sigma^m + \cdots, \tag{1.54}$$

where the superscripts d, p, i, h, and m refer to dispersion, polar (permanent dipole), induced dipole, hydrogen bonding, and metallic bonding contributions. Fowkes argued that only dispersion forces were capable of acting across most liquid–liquid interfaces of interest, so the combining rule for interfacial tension took the form:

$$\sigma_{AB} = \sigma_A + \sigma_B - 2\sqrt{\sigma_A^d\sigma_B^d}. \tag{1.55}$$

Measured interfacial tensions were then used to develop a database for the σ^d values for a variety of liquids [9]. The Girifalco–Good and Fowkes combining rules have shown some success in predicting interfacial tensions, but also many failures, primarily when nondispersive forces are at play. It has been shown that the effect of permanent and induced dipoles may generally be neglected in condensed phase media [10], but that forces due to hydrogen bonding (or more generally, acid–base interactions) may not be neglected for systems capable of such interactions. For nonmetallic, nonmolten salt liquids, Equation 1.54 may generally be written as

$$\sigma = \sigma^d + \sigma^{ab}, \qquad (1.56)$$

where σ^{ab} refers to acid–base self-association (since many liquids are capable of acting as both acids and bases). If the liquids forming the interface possess complementary acid–base characteristics, one would expect such an interaction across the interface, and Fowkes' combining rule would need to be modified as:

$$\sigma_{AB} = \sigma_A + \sigma_B - 2\sqrt{\sigma_A^d \sigma_B^d} - I^{ab}, \qquad (1.57)$$

where I^{ab} refers to acid–base interactions across the interface. The decomposition of I^{ab} into terms associated with the pure liquids A and B has been the subject of much controversy, and it can probably be stated that no fully successful formulation has yet been proposed.

Combining rules based on the geometric mean mixing rule must be applied with caution. Lyklema points out that the mixing rule applies to energy quantities (such as internal energy or enthalpy), whereas surface and interfacial tensions are *free* energies [11]. Thus, equations such as those of Girifalco and Good, or Fowkes, ignore the entropy effect associated with bringing together or disjoining the phases.

The use of the Girifalco–Good and Fowkes combining rules have also been applied to liquid–solid interfacial systems. As an example, the Fowkes Equation (Equation 1.56) might be used to attempt to relate the solid–liquid interfacial energy to the gas–solid surface energy and the liquid surface tension:

$$\sigma_{SL} = \sigma_{SG} + \sigma_{LG} - 2\sqrt{\sigma_{SG}^d \sigma_{SL}^d}. \qquad (1.58)$$

The required condition of immiscibility is often better satisfied for such systems than for most liquid–liquid systems, but the inherent energetic heterogeneity and roughness of solid surfaces, and other manifestations of the lack of internal equilibrium in such systems, calls into serious question their applicability.

1.5.3 Determination of Interfacial Free Energies for Fluid–Solid Interfacial Systems

While surface or interfacial free energies in capillary systems are readily measured as the surface or interfacial tension, this is not the case for Fluid–Solid interfacial systems. It is thus useful to consider the various, often indirect, means that might be used to provide values or estimates for this property as the driving forces for processes such as wetting, spreading, and adsorption are conveniently formulated in terms of these values. For powders and fibers, wetting and spreading phenomena take the form of wicking, engulfment, and dispersion processes.

In examining the various approaches to determining fluid–solid interfacial energies, it is first assumed that the system is immiscible, and it is useful to distinguish three types of fluid–solid interfacial systems:

1. *Pristine* surface systems refer to surfaces formed and kept in vacuo. The surface energy for pristine systems is designated as:

$$(f^{\sigma})_{\text{prist}} = \sigma_0. \tag{1.59}$$

2. *Clean* surface systems are those in which the surface is formed and kept in a pure "solvent medium," which may be a pure gas or a pure liquid, designated as component 1. One may distinguish two types of "clean" surfaces, viz. "pure" and "modified" surfaces. In the case of pure clean systems, component is capable of only physical interaction (i.e., physical adsorption) with the solid. In the case of modified clean surface systems, some of component 1 is capable of chemisorbing to the solid surface, changing its surface chemistry and hence energy. Most high energy surfaces sustain such chemisorption by many species. For example, most metal surfaces chemisorb oxygen to form an oxide layer, and most mineral oxide surfaces chemisorb water to form a layer hydroxyl groups. The modified clean surface is then capable of sustaining subsequent physical adsorption of component 1. The surface energy for a pure clean system in the presence of fluid 1 is designated as σ_0^1, and is related to the pristine surface energy by

$$(f^{\sigma})_{\text{clean}} = \sigma_0^1 = \sigma_0 - \pi_{\text{eq}}^1, \tag{1.60}$$

where π_{eq}^1 is the "equilibrium spreading pressure" of pure liquid 1 physisorbed on the solid surface. It may be obtained, consistent with

Equation 1.30, by integrating the gas-phase adsorption isotherm for the physical adsorption of component 1 from zero to its saturation concentration (or partial pressure), i.e.,

$$\pi_{eq}^1 = \int_0^{C_1^{sat}} \Gamma_1\left(C_1^*\right) d \ln C_1^*, \quad \left(or \int_0^{p_1^{sat}} \Gamma_1\left(p_1^*\right) d \ln p_1^* \right). \qquad (1.61)$$

If the clean surface is modified by chemisorption either from the adjoining fluid phase 1 or from some previous contact with a chemisorbing component, one may write:

$$(f^\sigma)_{mod.\ clean} = \sigma_0^1 = \sigma_0^{mod} - \pi_{eq}^1 = (\sigma_0 - \pi_{chem}) - \pi_{eq}^1, \qquad (1.62)$$

where σ_0^{mod} is the energy of the surface modified by chemisorption, and π_{chem} is the surface energy reduction caused by the chemisorption. This reduction is often a significant fraction of the original pristine surface energy.

3. Finally, *practical* surface systems refer to the case where the solid is or has been in contact with "practical" (i.e., dirty) multicomponent fluid environments such as gas mixtures or solutions containing physically adsorbable components. The surface free energy for practical surface systems in which the fluid-phase portion consists of a diluent gas or solvent component 1 and one or more adsorbable components is given by:

$$(f^\sigma)_{pract} \equiv \sigma_{pract} = \sigma_0^1 - \pi = \sigma_0^1 - \sum_{i=2} \int_0^{C_i} \Gamma_{i,1} d \ln C_i^*, \qquad (1.63)$$

where π represents the surface pressure resulting from the combined adsorption of all of the solute species.

Pristine surface energies can be measured if fresh solid surface can be created in vacuo without surface stretching. This has been accomplished in two ways, as pictured schematically in Figure 1.6. The first, shown in Figure 1.6a, is that of cleaving a brittle solid and measuring the total reversible work required to open up a unit area of the crack. Subtracting from this the elastic strain energy yields the energy involved in creating new surface.

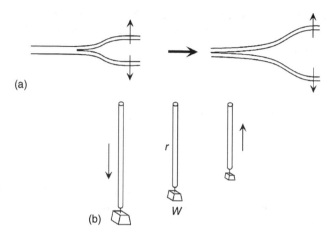

FIGURE 1.6 Methods for measuring the surface free energy of solids. (a) The fracture method for brittle solids. The total work required to open a unit area of the crack is measured, from which the elastic strain energy is subtracted, in accord with Equation 1.64. (b) The zero-creep method for ductile solids. Wires or fibers of radius r of the solid are stressed axially by a range of weights at a temperature close to, but below, the melting or softening point. Under high stress, elongation occurs; under low stress, contraction occurs due to the surface energy. At a particular intermediate critical stress, W_{cr}, no deformation occurs as the applied stress and the contractile stress of the surface are in balance. The surface energy is computed from W_{cr} in accord with Equation 1.65.

The pristine surface energy is given by:

$$\sigma_0 = \frac{1}{2}\left[\frac{\text{total work}}{\text{crack area}} - \frac{\text{strain energy}}{\text{crack area}}\right]. \tag{1.64}$$

This method has been applied to ionic and covalent crystalline materials, particularly at cryogenic temperatures, at which any ductility in the solid specimen is frozen out. It has also been applied to so-called van der Waals solids under these conditions. These are solids formed of simple molecules held together only by van der Waals forces. For ductile materials, like metals and polymers, a second technique, pictured in Figure 1.6b, and known as the "zero-creep" method has been developed. A series of wires of fixed radius r are hung with a range of different weights and brought to a temperature just below the melting point. Those with weights too large will become distended, while those with the smallest weights will contract upward due to the effect of surface tension forces. For a particular critical weight, W_{cr}, there will be zero creep, and this is used to determine the surface tension in

TABLE 1.2
Comparison of Ranges of Surface Energy Values for "Pristine Surfaces" (Formed and Kept *in Vacuo*) and "Clean" Surfaces (Formed and Kept in Air) for Various Types of Materials. The Reduced Values for the High-Energy Surfaces in Air are Generally Due to the Chemisorption of Oxygen

Solid Type	Surface Energy Range (mJ/m^2)	
	In Vacuo (σ_0)	In Air (σ_0^1)
van der Waals	20–60	20–60
Polymers	20–60	20–60
Ionic crystals	100–1000	60–300
Metals	500–2000	60–300
Covalent crystals	3000–9000	300–600

accord with [4, p. 167]:

$$\sigma_0 = \frac{W_{cr}}{\pi r}. \tag{1.65}$$

Clean surface energies may also be obtained by the cleavage and zero-creep methods when the processes are carried out in either a pure gas or liquid. Some qualitative ranges of results of measurements of this type are summarized in Table 1.2. It is reassuring that the numbers for the pristine surfaces are in reasonable good accord with theoretical calculations for these surface energies [4, p. 164]. Such calculations are easier to carry out than the experiments. The large differences in surface energy between those formed in vacuo and those formed in air are due largely to surface modification by chemisorption, particularly of oxygen, onto the surface and only negligibly by subsequent physical adsorption.

Another method for the determination of a clean surface energy σ_{SL} of a sparingly soluble solid S in a liquid L is afford by the Kelvin effect. If the solubility in the form of particles of known radius r is compared with its solubility in macroscopic form, the Kelvin Equation (Equation 1.48) yields:

$$\sigma_{SL} = \frac{rRT}{2v_S} \ln \frac{(C_S)_r^{sat}}{(C_S)_\infty^{sat}}. \tag{1.66}$$

Calorimetric measurement of the net heat, absorbed when particles initially in gas G are dissolved into liquid L, yield the difference between the actual heat of solution and the heat evolved due to the destruction of the SG interface. If a mass m of solid of specific area Σ is dissolved, the measured amount of heat absorbed is given by

$$Q_{\text{net}} = m\hat{\lambda}_{\text{soln}} - m\Sigma u_{\text{SG}} = m\hat{\lambda}_{\text{soln}} - m\Sigma\left(\sigma_{\text{SG}} - T\frac{\mathrm{d}\sigma_{\text{SG}}}{\mathrm{d}T}\right), \qquad (1.67)$$

where $\hat{\lambda}_{\text{soln}}$ is the heat of solution per unit mass. Calorimetric measurement of the heat evolved when an immiscible solid is immersed in a liquid yields information on the difference between the energy of the SG interface destroyed and the SL interface created:

$$\frac{Q_{\text{imm}}}{m\Sigma} = u_{\text{imm}} = u_{\text{SL}} - u_{\text{SG}} = (\sigma_{\text{SL}} - \sigma_{\text{SG}}) - T\frac{\mathrm{d}(\sigma_{\text{SL}} - \sigma_{\text{SG}})}{\mathrm{d}T} \qquad (1.68)$$

Equation 1.67 and Equation 1.68 are statements of the Gibbs–Helmholtz law relating energy quantities to the corresponding free energies and their temperature derivatives.

Other common methods for determining surface or interfacial energies are based on wetting, adhesion, or wicking measurements. This generally involves the direct or indirect determination of the contact angle θ, defined by Equation 1.15, solely in terms of the *mechanical*, rather than thermodynamic, properties of the interfaces involved. Discussion of these methods is thus deferred to the next section, following the discussion of the use of thermodynamic interfacial energy quantities to determine the equilibrium configuration of solid–fluid–fluid interfacial systems and to formulate the driving forces for various solid–liquid contacting processes.

1.6 FORMATION AND DESTRUCTION OF FLUID–SOLID INTERFACIAL SYSTEMS: WETTING, SPREADING, WICKING, AND ADHESION

1.6.1 THERMODYNAMIC SIGNIFICANCE OF THE CONTACT ANGLE

An interfacial system consisting of phase A and phase B (which for the moment may be considered immiscible, pure-component phases) may be formed by bringing together phase A and phase B, which are initially in vacuo, or in contact with some third immiscible fluid phase. Consider first the situation in which phase A is a liquid L, phase B is a solid S, and the inert fluid in which they are immersed initially is a gas G. Both the liquid and solid phases are assumed nonvolatile. Also, the system formed by bringing the liquid and solid phases together is assumed to conform to the *clean* surface system description given above. The thermodynamic driving force for the occurrence of such a process, or its reverse, may be formulated in terms of the surface and interfacial free energies described above. To simplify the presentation, the following notational identifications will be made:

$$\text{Surface tension of liquid L against gas G} = \sigma_{oL}^{G} \equiv \sigma_{LG}$$

$$\text{Surface free energy of gas–solid interface} = f_{SG}^{\sigma} = \sigma_{oS}^{G} \equiv \sigma_{SG}$$

$$\text{Surface free energy of liquid–solid interface} = f_{SL}^{\sigma} = \sigma_{oS}^{L} \equiv \sigma_{SL}$$

(This is consistent with the notation used in Equation 1.15 and Equation 1.16.)

Consider the system consisting of a fixed volume of liquid in contact with a solid–gas interface, as shown in Figure 1.7, subject to the process whereby the drop is permitted to spread or contract from its initial position. The free energy of the composite system is given by:

$$F = F_o + \sigma_{LG}A_{LG} + \sigma_{SL}A_{SL} + \sigma_{SG}A_{SG}$$

$$= F_o + \sigma_{LG}A_{LG} - (\sigma_{SG} - \sigma_{SL})A_{SL}. \qquad (1.69)$$

Minimization of the free energy requires that for a change dr in radius of the droplet base:

$$dF = \sigma_{LG}dA_{LG} - (\sigma_{SG} - \sigma_{SL})dA_{SL} = 0, \qquad (1.70)$$

subject to the restriction that the drop volume remain constant (i.e., $dV_{drop} = 0$). Solution of Equation 1.70 yields Young's equation, Equation 1.15, and appears

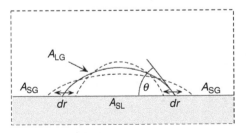

FIGURE 1.7 Thermodynamic derivation of Young's equation. The condition of the local minimum in the system free energy, in terms of the angle found as the base area (radius, r) of a circular drop of liquid at a solid–gas interface is varied under conditions of constant total system volume and constant total solid–fluid interfacial area, $A_{SG} + A_{SL}$. The result is Young's equation, Equation 1.70, giving the contact angle as a function of the interfacial free energies.

to give the contact angle θ the status of a thermodynamic property, viz.

$$\cos\theta = \frac{\sigma_{SG} - \sigma_{SL}}{\sigma_{LG}}, \qquad (1.71)$$

clearly, Equation 1.71 is valid only when the interfacial energy ratio on the right-hand side lies between -1 and $+1$. Under these conditions, and provided that the solid–fluid interfaces involved are smooth and energetically homogeneous (conditions generally *not* met), and the stated conditions of immiscibility exist, a measured finite value of the contact angle together with the liquid surface tension would provide a value for the solid interface energy difference: $\sigma_{SG} - \sigma_{SL}$. The usual failure to meet the conditions of smoothness and energetic uniformity (as well as other factors) leads to measured contact angle hysteresis, i.e., a difference between the angle observed when the liquid is advanced over the solid surface (the "advancing contact angle," θ_A) and that observed when the liquid is retracted (the "receding contact angle," θ_R) [12]. It is evident that the use of Young's equation with measured contact angles to infer solid interface free energies must be exercised with considerable caution.

1.6.2 DRIVING FORCES FOR SOLID–LIQUID CONTACTING AND SEPARATION PROCESSES

The joining of the liquid and solid surfaces to form and interface between them, or the converse disjoining process, may be envisioned to occur in three different ways, as pictured in Figure 1.8. In the first process, shown in Figure 1.8a, unit areas of the liquid–gas and solid–gas surfaces are brought together to form a unit area of solid–liquid interface. The driving force for the joining process is the negative of its free energy change, viz.

$$-\Delta F = -\Delta f^\sigma = -(\sigma_{SL} - \sigma_{SG} - \sigma_{LG}) \equiv (\sigma_{SG} + \sigma_{LG} - \sigma_{SL}). \qquad (1.72)$$

The driving force for the joining, or *adhesion,* process is seen to be the reversible work of the *dis*joining process, and is referred to as the "work of adhesion," W_A:

$$W_A = \sigma_{SG} + \sigma_{LG} - \sigma_{SL}. \qquad (1.73)$$

A relationship known as the Dupré equation. Figure 1.8b shows the *wetting* process, whereby the liquid moves over the solid surface, destroying SG interfacial area, but creating an equivalent SL interfacial area. The driving force for this process is again the reversible work of the reverse process, viz.

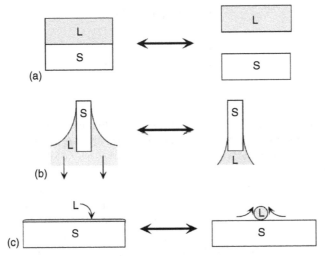

FIGURE 1.8 Liquid–solid contacting and disjoining processes. (a) Adhesion-disjoining. Disjoining process involves creation of liquid and gas surfaces and destruction of liquid–solid interface. (b) Wetting-de-wetting. De-wetting process involves creation of solid surface and destruction of solid–liquid interface. (c) Spreading de-wetting. De-wetting process involves creation of solid surface and destruction of liquid surface and solid–liquid interface.

$$W_{\mathrm{W}} = \sigma_{\mathrm{SG}} - \sigma_{\mathrm{SL}}, \tag{1.74}$$

which might be named systematically the "work of wetting," but is more commonly called the "wetting tension" or the "adhesion tension." Finally, Figure 1.8c shows the *spreading* process, whereby a consolidated liquid mass spreads out over a solid surface creating both LG and SL interfaces at the expense of SG interface. The driving force is again the reversible work of the reverse process

$$W_{\mathrm{S}} = \sigma_{\mathrm{SG}} - \sigma_{\mathrm{SL}} - \sigma_{\mathrm{LG}}, \tag{1.75}$$

the "work of spreading," or more commonly, the "spreading coefficient." In writing Equation 1.71 through Equation 1.75, it is important to recall that one has assumed the solid interfaces to be smooth and energetically uniform.

In the above driving force expressions, the solid interfacial energies occur only as their difference $\sigma_{\mathrm{SG}} - \sigma_{\mathrm{SL}}$, so that under the conditions outlined following Equation 1.71, these driving forces may all be put in terms of the measured contact angle and liquid surface tension, upon substitution of Young's equation viz.

$$W_A = \sigma_{LG}(1 + \cos\theta), \tag{1.76}$$

known as the Young-Dupré equation,

$$W_W = \sigma_{LG}\cos\theta, \text{ and} \tag{1.77}$$

$$W_S = \sigma_{LG}(\cos\theta - 1). \tag{1.78}$$

Equation 1.78 provides only the information that the contact angle must be zero for spontaneous spreading to occur, but does provide the magnitude of the spreading coefficient.

If the requirement that the liquid be nonvolatile is relaxed, it must be recognized that its vapors will adsorb at the SG surface. Under equilibrium conditions, adsorption will reduce the free energy of that surface by an amount equal to the equilibrium spreading pressure, π_{eq}, as given by Equation 1.61. Under these circumstances, π_{eq} must be added to the right hand side of Equation 1.71 through Equation 1.78. The importance of the equilibrium spreading pressure is frequently ignored, particularly for "low-energy surfaces," for which it is argued, the values should be negligibly small. There is evidence, however, that this assumption may not be justified [13].

Positive values for the driving forces, defined by Equation 1.72 through Equation 1.74, predict the spontaneity of the indicated process. A positive work of adhesion suggests that the liquid will spontaneously adhere to the solid, and that the larger the value, the stronger the adhesion. The work of adhesion will be zero or negative only when the contact angle is 180°. This is essentially never the case for liquids in gases put in contact with solid surfaces (i.e., there is always some adhesion). But liquid droplets immersed in another liquid often will not stick to a solid surface immersed in the liquid. One of the aspects of detergent action is the adsorption of the detergent (with consequent interfacial energy reduction, π_{det}) to the solid–water interface, such that the work of adhesion of a dirt particle D to the solid surface becomes negative, i.e.,

$$W_A = (\sigma_{SL} - \pi_{det}) + \sigma_{LD} - \sigma_{SD} < 0. \tag{1.79}$$

The dirt particle is then said to "roll up" and detach itself from the surface, as illustrated in Figure 1.9.

The work of adhesion may be measured directly under appropriate circumstances using the method of contact mechanics. The adhesive contact between two deformable (soft) solids or between a soft and a hard solid may be disjoined by the application of a tensile stress τ to the composite specimen, as pictured in Figure 1.10. The figure shows a soft sphere of material S_1 in

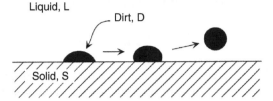

FIGURE 1.9 Roll-up of dirt particle from solid–liquid interface. Adsorption of surfactant from solution to the solid–liquid interface reduces solid–liquid interfacial energy to the point that the work of adhesion of the dirt particle to the interfaces becomes negative, as indicated in Equation 1.79.

contact with a flat surface of material S_2. The work of adhesion between S_1 and S_2 in a gas G, viz.

$$W_{A(S_1/G/S_2)} = \sigma_{S_1 G} + \sigma_{S_2 G} - \sigma_{S_1 S_2} \tag{1.80}$$

is given in terms of the tensile stress τ and the radius R of the contact circle by the theory of Johnson, Kendall, and Roberts (JKR theory) [14]:

$$W_A = \sqrt{\frac{2\tau}{3\pi R}}. \tag{1.81}$$

If the materials are the same, (i.e., $S_1 = S_2 = S$), the work of adhesion becomes the work of cohesion of S in the gas phase G:

$$W_A = (W_{coh})_G = 2\sigma_{SG}. \tag{1.82}$$

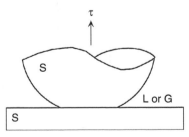

FIGURE 1.10 Contact adhesion between soft solids in liquid or gas. Detachment occurs upon application of tensile force τ, yielding the work of adhesion (or cohesion, if the solids are the same) in accord with JKR analysis, Equation 1.81.

Similarly, if the specimen is disjoined in a liquid phase L, one obtains:

$$W_A = (W_{coh})_L = 2\sigma_{SL}. \tag{1.83}$$

When both measurements are made, successful corroboration with the measured contact angle has been found using Young's equation, i.e., $(\sigma_{SG} - \sigma_{SL}) \approx \sigma_{LG} \cos \theta$ [15].

1.6.3 WICKING AND ENGULFMENT

The work of wetting, or the wetting tension, as given by Equation 1.74 or Equation 1.77, is the driving force for the wicking of a liquid through a dry bed of powder or mat (woven or nonwoven) of fibers, as pictured in Figure 1.11. The linear rate of wicking through a porous medium, v_{wick}, in the absence of gravitational effects, is given by the Washburn equation:

$$v_{wick} \equiv \frac{dx}{dt} = \frac{r_e}{4\mu x} \sigma_{LG} \cos \theta, \tag{1.84}$$

where r_e is the "wicking equivalent radius" of the open pore structure, μ is the viscosity of the liquid and x is the wicking distance from the liquid pool. Integration of Equation 1.84 gives the wicking distance x as a linear function of $t^{1/2}$. Wicking measurements constitute one of the commonest methods for determining the contact angle of a liquid against the surface of fine powders or fibers. For this purpose, the wicking equivalent radius is determined in a

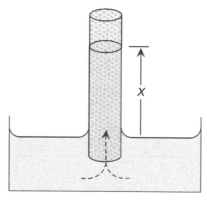

FIGURE 1.11 Wicking of liquid into a packed bed of particles or fibers. In the absence of gravitational effects, the wicking distance x varies as $t^{1/2}$.

separate measurement by using a volatile low surface tension liquid which is known to wet out the surface of the solid, i.e., $\theta = 0°$.

If the contact angle in a wicking event is finite, the Washburn equation takes the form:

$$v_{wick} = \frac{r_e}{4\mu x} W_W = \frac{r_e}{4\mu x}(\sigma_{SG} - \sigma_{SL}), \qquad (1.85)$$

which suggests that enhancement of wicking through use of adsorbing surfactant from the liquid phase will occur only insofar as σ_{LG} is reduced, and only to the point that the contact angle vanishes, i.e., $(\sigma_{SG} - \sigma_{SL}) = \sigma_{LG}$, or $\sigma_{SL} = (\sigma_{SG} - \sigma_{LG})$. Any further reduction in σ_{SL} will not enhance wicking, and in fact may reduce it by virtue of decreasing the liquid surface tension, σ_{LG}. Equation 1.84 and Equation 1.85 may be extended to describe the immiscible displacement of one liquid L_1 from a porous matrix by a second liquid L_2. For the simplest case, in which no external pressure drop is imposed, the equations take the form:

$$v_{wick} = \frac{r_e}{4[\mu_2 x_2 + \mu_1(X - x_2)]}\sigma_{L_1 L_2}\cos\theta_{12}, \qquad (1.86)$$

where x_2 is the wicking length in liquid 2 (the displacing liquid), X is total wicking length from liquid reservoir 2 to liquid reservoir 1, $\sigma_{L_1 L_2}$ is the interfacial tension between the liquids, and θ_{12} is the contact angle drawn in liquid 2. If the contact angle is finite, the Washburn equation becomes:

$$v_{wick} = \frac{r_e}{4[\mu_2 x_2 + \mu_1(X - x_2)]} W_W$$

$$= \frac{r_e}{4[\mu_2 x_2 + \mu_1(X - x_2)]}(\sigma_{SL_1} - \sigma_{SL_2}). \qquad (1.87)$$

The interaction of free particles (powders or fibers) with a fluid interface is governed by the work of spreading, or spreading coefficient. A dry solid particle in a gas–liquid system will be spontaneously imbibed by the liquid only if the spreading coefficient of the liquid over the particle surface is positive, as shown in Figure 1.12. If the spreading coefficient is negative, the particle will be trapped at the LG interface. For such cases, the smaller the contact angle, the greater the proportion of the particle volume that will reside in the liquid phase. A positive value of the spreading coefficient is thus the criterion for spontaneous dispersibility of a dry powder in a liquid. A particle in a system consisting of liquids L_1 and L_2 will find itself within L_1 if

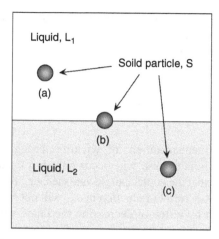

FIGURE 1.12 The interaction of solid particulates with fluid interfaces. (a) Particles are preferentially wet out by liquid L_1, cf. Equation 1.88. (b) Particles are preferentially wet out by liquid L_2, cf. Equation 1.89. (c) Particles are partially wet by both L_1 and L_2, cf. Equation 1.90.

the spreading coefficient is:

$$W_{S(L_1/S/L_2)} = \sigma_{SL_1} - \sigma_{L_1L_2} - \sigma_{SL_2} > 0, \tag{1.88}$$

and within L_2 if the spreading coefficient is:

$$W_{S(L_2/S/L_1)} = \sigma_{SL_2} - \sigma_{L_1L_2} - \sigma_{SL_1} > 0. \tag{1.89}$$

The particles will reside in the interface if both spreading coefficients are negative, which occurs when

$$\sigma_{L_1L_2} > \left| \sigma_{SL_1} - \sigma_{SL_2} \right|. \tag{1.90}$$

Electrostatic effects, not considered here, may also influence the attachment of particles to fluid interfaces [16].

1.7 THE EFFECT OF SOLID SURFACE ROUGHNESS AND ENERGETIC HETEROGENEITY ON WETTING AND ADSORPTION MEASUREMENTS

Most solid surfaces in practice are rough, as opposed to smooth, on at least the micro- or the nano-scale. Freshly cleaved Moscovite mica is one of the rare exceptions, and highly polished silicon wafers may be made to approach atomic smoothness. Thermodynamically, the roughness of real surfaces, in and of itself, affects only their area and requires no alteration of the format used for surface free energy. It does not affect adsorption until the conditions leading to capillary condensation are reached. With respect to wetting, however, minimization of the free energy of a fluid-(rough) solid interfacial system leads to a modified form of Young's equation, viz.

$$\cos \theta_{\text{obs}} = r \cos \theta_{\text{smooth}}, \tag{1.91}$$

known as Wenzel's equation. θ_{obs} is the observed macroscopic contact angle made by the liquid against the rough surface, while θ_{smooth} is the contact angle that would have made against a smooth surface of the same material. r is the "rugosity" of the surface (i.e, its actual area divided by its nominally flat surface area).

Most solid surfaces, in practice, are energetically heterogeneous, with crystalline solids possessing mixed populations of various surface defects, semi-crystalline solid surfaces having crystalline and amorphous patches, and amorphous materials exhibiting surface chemical and orientational hetero-geneity. Finally, the presence and distribution of chemisorbed contamination leads to energetic nonuniformity. Both wetting and adsorption measurements are affected by such heterogeneity. If it is assumed that the surface is made up of uniform patches of different surface energy, minimization of the free energy of a fluid-(heterogeneous) solid interfacial system leads to a second modified form of Young's equation, viz.

$$\cos \theta_{\text{obs}} = \sum_i \varphi_i \cos \theta_i, \tag{1.92}$$

known as the Cassie–Baxter equation. ϕ_i is the area fraction of the patches of type i, and θ_i is the contact angle made by the liquid against patches of type i.

As an illustration of the description of adsorption at a heterogeneous solid surface, consider the adsorption of a single gas 2, at a given temperature T, onto a patchwise heterogeneous surface, with the site energy characteristic of a given patch type i as ε_i. The adsorption of gas 2 on the patches of type i is given by the local adsorption isotherm (assumed to be the same functional format for all patch types):

$$\Gamma_i(p_2) = \Gamma_\infty \Theta_i(p_2, \varepsilon_i), \tag{1.93}$$

where Γ_∞ corresponds to full monolayer coverage, and Θ_i is the fractional monolayer coverage of patches of type i. The total adsorption is the sum of the adsorption on all of the different patch types, and as the latter approaches infinity, the summation may be replaced by an integral with respect to the site energy:

$$\Gamma_{\text{total}}(p_2) = \Gamma_\infty \int_{\varepsilon_{\min}}^{\varepsilon_{\max}} \Theta(p_2,\varepsilon)\chi(\varepsilon)\mathrm{d}\varepsilon, \tag{1.94}$$

where $\chi(\varepsilon)$ is the continuous site energy distribution function, normalized to unity:

$$\int_{\varepsilon_{\min}}^{\varepsilon_{\max}} \chi(\varepsilon)\mathrm{d}\varepsilon = 1. \tag{1.95}$$

The deconvolution of the total isotherm to obtain the site energy distribution is neither trivial nor unique [17]. One of the least complicated situations arises when the local adsorption isotherm is that of Langmuir, viz.

$$\Theta(p_2,\varepsilon) = \frac{Kp_2}{1 + Kp_2}, \text{ with} \tag{1.96}$$

$$K = K^0(T)\exp\left(\frac{\varepsilon}{kT}\right), \tag{1.97}$$

where K^0 is a characteristic constant at a given temperature, and k is Boltzmann's constant. Under these conditions, for a variety of different site energy distribution functions, the total adsorption isotherm conforms to the format of the empirical Freundlich equation:

$$\Gamma_{\text{total}} = \Gamma_\infty K^{\text{F}} p_2^\nu, \tag{1.98}$$

where K^{F} is a constant analogous to K in Equation 1.97, and ν is a constant with magnitude generally between 0 and 1. The format of Equation 1.98 is often observed for gas adsorption onto heterogeneous surfaces in the intermediate adsorbate concentration range. Additional information (or assumptions) would be needed, however, to convert the empirical parameters into a site energy distribution function. Measurements of the differential (or isosteric) heats of adsorption as functions of both temperature and coverage are useful in this task, but the details of such efforts are discussed elsewhere [17].

1.8 SUMMARY

Thermodynamics provides the framework for the description of the properties and behavior of interfacial systems, including powders and fibers. It applies most appropriately to fluid–fluid interfacial systems (i.e., capillary systems), in which internal physical equilibrium may be assumed to obtain. Despite their general lack of complete internal physical equilibrium, fluid–solid interfacial systems may be described by the same framework, provided certain precautions are observed. The model of the immiscible interfacial system, in which it is assumed that the fluid and solid phase portions of the system are completely immiscible, is especially useful in describing these types of systems. The thermodynamic properties of interfacial systems are found to depend on their degree of subdivision and the curvature of their interfaces. Processes characteristic of interfacial systems include area extension or creation, change in degree of subdivision, adsorption, wetting, spreading, adhesion and particle engulfment, and dispersion. Interfacial free energies provide the building blocks for the driving forces for many of the processes of interest in interfacial systems. While the interfacial free energy of a capillary system may be identified with its readily measurable boundary tension, the experimental determination of the interfacial free energy of a fluid–solid interfacial system generally more difficult and often indirect, requiring careful interpretation of the quantities which are measured.

REFERENCES

1. Defay, R., Prigogine, I., Bellemans, A., and Everett, D. H., *Surface Tension and Adsorption*, Longmans, London, U.K., p. 432, 1966.
2. Lyklema, J., *Fundamentals of Interface and Colloid Science, vol. 1*, Academic Press, London, U.K., pp. 2.1–2.103, 1991.
3. Herring, C., In *Structure and Properties of Solid Surfaces*, Gomer, R. and Smith, C. S., Eds., University of Chicago Press, Chicago, pp. 5–82, 1953.
4. Tabor, D., *Gases, Liquids and Solids and Other States of Matter*, 3rd ed., University of Chicago Press, Cambridge, U.K., p. 418, 1996.
5. Benson, G. C. and Yun, K. S., In *Solid–Gas Interface, vol. 2*, Flood, E. A., Ed., Marcel Dekker, New York, pp. 203–269, 1967.
6. Gaines, G. L. Jr., *Insoluble Monolayers at Liquid–Gas Interfaces*, Interscience Publishers, New York, p. 386, 1966.
7. Girifalco, L. A. and Good, R. J., *J. Phys. Chem.*, 61, 904–909, 1957.
8. Fowkes, F. M., *J. Phys. Chem.*, 66, 382, 1962.
9. Fowkes, F. M., Riddle, F. L. Jr., Pastore, W. E., and Weber, A. A., *Colloids. Surf.*, 43, 367–387, 1990.
10. Good, R. H. and Chaudhury, M. K., In *Fundamentals of Adhesion*, Lee, L. H., Ed., Plenum Press, New York, pp. 137–151, 1991.

11. Lyklema, J., *Colloids. Surf. A*, 156, 413–421, 1999.
12. Johnson, R. E. Jr. and Dettre, R. H., In *Surface and Colloid Science, vol. 2*, E. Matejevic, Ed., Wiley-Interscience, New York, pp. 85–153, 1969.
13. Jacob, P. N. and Berg, J. C., *J. Adhes.*, 54, 115–131, 1995.
14. Johnson, K. L., Kendall, K., and Roberts, A. D., *Proc. R. Soc. A*, 324, 301–313, 1971.
15. Chaudhury, M. K., *Mat. Sci. Eng. R*, 16, 97–159, 1996.
16. Paunov, V. N., Binks, B. P., and Ashby, N. P., *Langmuir*, 18, 6946–6955, 2002.
17. Jaroniec, M. and Madey, R., Physical Adsorption on Heterogeneous Solids, *Studies in Physical and Theoretical Chemistry 59*, Elsevier, Amsterdam, p. 351, 1988.

Section I

The Solid-Gas Interface

2 Surface Properties Characterization by Inverse Gas Chromatography (IGC) Applications

Eric Brendlé and Eugène Papirer

CONTENTS

2.1 INTRODUCTION

Chromatography is an analytical method that largely contributed to significant advances in chemistry and biochemistry. This is due to its unique detection sensitivity, but also to its ability to separate, and often to identify, minor amounts of constituents present in a complex mix. Furthermore, when compared to most sophisticated modern equipments, chromatography appears to be a nonexpensive method that is present in almost all analytical or preparative laboratories. In fact, chromatography is a general term covering a large family of possible methods. One may distinguish, in particular, gas (GC) and liquid chromatography (LC) methods.

The physical principles of GC are multiple. Molecules may be separated from a mix when they are passed over an "active" support (column filling) simply because they have different affinities or because their differences in shape or bulkiness will cause them to adsorb or to be excluded on supports having given surface textures.

Indeed, chromatography is practised since a long time, but it is still in development. Major efforts are presently spend to develop adequate chromatography supports, even more sensitive detectors, and computer-controlled apparatus, but also to reach a better understanding of the physics behind it.

In a classical GC experiment, a mixture of unknown solutes is injected in a GC column containing a chosen GC support. This mix is pushed through the column with a so-called carrier gas that has no (or limited) affinity for the GC support itself. Solutes will be retained, as they progress inside the column more or less strongly on the surface of the support. In other terms, they will progress more or less rapidly when crossing the column. At the column outlet, they will appear separately and will be detected. In an inverse gas chromatography (IGC) experiment, the main attention is focused on the GC support (the powder or fiber of interest placed inside the GC column). Here, chosen and well-defined solutes (molecular probes) are injected and their individual behavior inside the column is analyzed, providing information on the properties of the column filling itself. This is the inverse situation of what is done for analytical chromatography. When injecting small amounts of solutes, at the limit of detection of the most sensitive detectors, we are performing IGC at infinite dilution conditions (IGC-ID). The adsorbed molecules, at near zero solid surface coverage, may be considered as isolated from each other, avoiding thus interpretation complications originating from lateral interactions of neighboring adsorbed molecules. The relevant quantity to be determined is the net retention time (t_N): the time the solute spends on the solid's surface.

When measurable amounts of solutes are injected, we are performing IGC at finite concentration conditions (IGC-FC), allowing determining adsorption isotherms. Various methods are proposed in the literature for the measurement of adsorption and desorption isotherms. A favored method consists of taking advantage of the deformation of the chromatographic peak to evaluate principally desorption isotherms, from which a series of thermodynamic data may be obtained. Also, the specific surface area of the solid inside the column, and its adsorption potential, may be evaluated. The two IGC procedures deliver complementary information, as we shall see later on.

Before entering in the restricted subject of this chapter, it might be interesting to shortly recall the benefit of applying IGC to polymers. Over 250 IGC papers appeared in the last three years and a large number concern polymers. Indeed, the possibility to operate at infinite dilution conditions allowing the verification of the theories of thermodynamics of polymer solutions attracted the attention of Guillet[1] and his group already some 50 years ago. Here a polymer film is deposited on an inactive chromatography support and its affinity for various injected solutes is determined. Numerous other applications of IGC were developed to evaluate different properties of the polymer film (glass transition temperature, crystallinity, diffusion coefficients).

At the same time period, Kiselev and his group[2] obtained, by GC, in a very convenient way, a series of information on the surface characteristics of solid particles starting from adsorption isotherms. Indeed, the IGC measurements are rather rapid, the measuring temperature may be easily changed, and the choice of solutes is large. Hence, numerous adsorption isotherms allowing to determine adsorption enthalpies and entropies of solutes, polar surface properties of solids and "global" surface energy were recorded and interpreted.

But, it is only around 1980 that Gray[3] and his collaborators introduced modern surface energy concepts for the evaluation of the IGC data. From then on, numerous authors applied IGC to powders and fibers, a field of investigation that remains largely open today. Its principles, advantages, and limitations are described in several books and/or review papers.[4–10] Moreover, additional books analyze the types of physicochemical measurements that can be made by IGC.[11,12]

2.2 IGC AT INFINITE DILUTION CONDITIONS (IGC-ID)

2.2.1 FUNDAMENTALS

As indicated above, minor amounts of selected solutes are injected in the GC column containing the powder or the fiber. From a practical point of view, when the powders are too fine and susceptible to plug the column under the pushing of the carrier gas, it is indicated to first compact them. One may simply compress them in an IR dye and, thereafter, gently hand crush and sieve them so as to approach common GC filling sizes (100–400 μm). In fact, for IGC purposes, one does not seek high (separation) GC performance. Helium is often used as the carrier gas. Gaussian-shaped GC peaks are currently recorded, especially with n-alkanes, when injecting minor amounts of solutes. From the peak summit, one easily determines the total retention time (t_R), knowing the flow rate (D_c) of the carrier gas and taking into account the pressure drop inside the column. This time is then corrected for evaluating the net retention time (t_N): the time the solute spends on the solid's surface. Practically, one uses, as a reference, a detectable quasi-nonadsorbing solute (usually methane). The fundamental quantity (t_N), from which all IGC-ID calculations are made, is determined in the way illustrated on Figure 2.1.

Intuitively already, one understands that a long retention time (or retention volume: the volume of carrier gas necessary to convey the injected probe through the column) is indicative of a high "affinity" of the solute for the solid surface. This is translated, in thermodynamic terms, by the standard variation of the free energy of adsorption (ΔG_A) or desorption (ΔG_D), under equilibrium conditions. Fundamental equations, Equation 2.1 and Equation 2.2, were

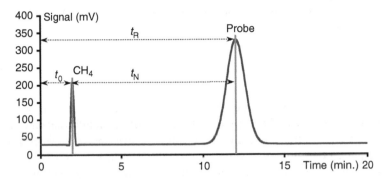

FIGURE 2.1 Typical IGC chromatogram, at infinite dilution conditions.

established (see Conder[13]):

$$\Delta G_A = -RT \operatorname{Ln}\left(\frac{K_S p_{S,g}}{\pi_S}\right) \tag{2.1}$$

or

$$\Delta G_D = -\Delta G_A = RT \operatorname{Ln}\left(\frac{V_N P_0}{Sm\pi_0}\right) \tag{2.2}$$

In these expressions, K_S is the surface partition coefficient, $p_{S,g}$ the adsorbate vapor pressure, π_S is the bi-dimensional spreading pressure of the adsorbed film in an arbitrarily defined adsorption standard state, S is the specific surface area, and m the weight of the column filling. The previous equations may be rewritten as:

$$\Delta G_A = -RT \operatorname{Ln}(K_S) + C \tag{2.3}$$

$$\Delta G_D = RT \operatorname{Ln}(V_N) + C' \tag{2.4}$$

where C and C' are constants, depending on the arbitrary choice of the reference state of the adsorbed solute. IGC readily provides ΔG_D and, from its temperature dependence, the standard entropy and enthalpy of desorption. However, care should be taken to approach the theoretical conditions allowing to make meaningful physico-chemical measurements.[13–15]

When progressing inside the GC column, the solute molecules undergo numerous adsorption–desorption cycles, being more or less strongly adsorbed depending on the intensity of the interactions exchanged with the solid surface. The surface residence time (t_N) depicts the intensity of the interactions the solute and the surface sites are able to exchange (adsorption energy (E_A)):

$$t_N = ke^{-E_A/RT} \tag{2.5}$$

In the case of an energetically heterogeneous surface (surface with different populations of adsorption sites), t_N will essentially depend on the most active adsorption sites. Almost all solid surfaces are heterogeneous from an energetic point of view. Even pure crystalline surfaces will possess planes of different orientations, allowing different adsorption possibilities. The origins of heterogeneities are numerous. Figure 2.2 naively illustrates different adsorption sites on an actual crystalline surface.

Various crystallographic planes, structure defects (dislocations, steps, etc.), sites occupied by pollutants (water, organics) are sketched on that scheme. It is evident that, not only the strength of a site will be of importance, but also its location that might or not facilitate the approach of an interacting molecule. Moreover, step-like defects will offer to a flexible molecule (n-alkane, for instance) the possibility to become attracted by sites located on both sides of the walls that constitute the step. In other terms, one may expect enhanced interaction (i.e., a larger t_N) in that special case. Conversely, this surface geometry will limit the access of bulky or stiffer molecules. This should not be ignored in the interpretation of the IGC results.

Before proceeding, it might be of interest to recall some surface energy concepts. Thermodynamics define the surface energy as the variation of free energy that accompanies the creation of a unit surface area under equilibrium

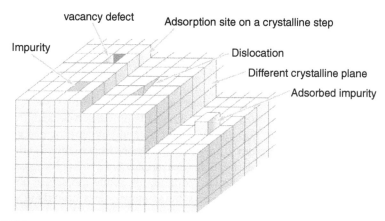

FIGURE 2.2 Different adsorption sites on an actual crystalline surface.

conditions. This applies fairly well to easily deformable surfaces, like those of liquids. But this is not evident for solids. Nevertheless, there are, in principle, some possibilities to evaluate surface energies by cleaving procedures. For instance, one may measure the cleavage energy of lamellar materials such as graphite, mica, or talc. The measurement should be made so as to minimize other energy losses due, for example, to noncontrolled deformation of the crystal itself. Furthermore, at the best, the surface energy of the basal cleavage plane, and not a global value, is obtained.

When performing the cleavage of, say, a mica crystal, one has to overcome the forces that assured its cohesion. Those forces are of different origins. Presently, one distinguishes nonspecific and specific interaction forces.[16] Nonspecific interactions do not depend on the nature of the partners. They are universal. Those forces are generated by the deformation of electron clouds creating instantaneous dipoles that appear even in nonpolar molecules, such as methane, for instance. Nonspecific (London, dispersive, or universal) interactions capacities define the dispersive component of the surface energy. The other component of surface energy is called "specific component." It comprises all types of interactions (polar, ionic, magnetic, metallic) except London interactions. Hence:

$$\gamma_S = \gamma_S^d + \gamma_S^{sp} \qquad (2.6)$$

and the work of adhesion between two partners is given by:

$$W_A = W_A^d + W_A^{sp} \qquad (2.7)$$

In the case of nonspecific interactions, and knowing the γ_S^d of a solid surface and the γ_L^d of a liquid, for example, the work of adhesion (W_A) between the two partners is simply given by:

$$W_A = 2\sqrt{\gamma_S^d \gamma_L^d} \qquad (2.8)$$

Hence, measuring W_A with different n-alkanes will allow to determine γ_S^d. The work of adhesion, W_A, may also be estimated by liquid contact angle measurements on flat surfaces. Obviously, this method becomes problematic with powders even though one may try to prepare "relatively" flat surfaces by compression; yet such a surface always presents defects (surface roughness) that need to be estimated before making corrections. IGC, on the other hand, allows to evaluate an "apparent dispersive component" of the surface energy for the reasons discussed below.

2.2.2 "Apparent" Dispersive Component of Surface Energy

According to the discussion relative to surface heterogeneity, it appears that the concept of surface energy is hardly applicable to solids. Let us take the example of a highly organized material, such as graphite. Graphite exhibits two very different adsorbing surfaces: the basal plane surface made of graphene structures and the lateral surface. The surface energy of the basal plane may be evaluated through the measurement of the cleaving energy of a large graphite crystal. It may also be calculated from the values of contact angles of liquid drops deposited on that smooth surface. Comparable results are thus obtained (around 80 mJ/m^2). Yet, those techniques do not apply for the measurement of the surface energy of the lateral surface, that is much more complex both from structural (existence of defects), but also, on actual samples, from chemical (existence of oxygenated surface groups) point of views.

Neither will IGC, at infinite dilution conditions, allow making such distinction. Moreover, depending on the graphite preparation method or processing, lateral surface defects, allowing a possible "insertion" of n-alkane chains chosen for the evaluation of γ_S^d, will appear on the periphery of the crystal; n-alkane molecules may then lodge between defect intermediate basal planes that form like "a molecular cradle" (i.e., a surface defect of molecular size in which alkanes may be lodged). Therefore, from a fundamental point of view, γ_S^d, determined essentially by those strong adsorption sites, is not at all be representative of a global component of the surface energy of graphite. It could at the best be considered as an "apparent" dispersive component value. Yet, for the sake of comparison, since almost all experimental values of the literature are expressed as γ_S^d values, we shall keep that denomination.

Dorris and Gray[3] pioneered the γ_S^d measurement by IGC when examining cellulose materials. They strongly defined the conditions of application of their method:

- The solid's surface should be flat, at the molecular level.
- The surface should be energetically homogeneous.
- The desorption energy should be equal (opposite sign) to the adsorption energy $\Delta G_A = -\Delta G_D$.
- ΔG_A and ΔH_A should be linearly correlated.

It is evident that those conditions are scarcely obeyed by powder and or fiber surfaces.

When injecting a homologous series of n-alkanes in the GC column containing cellulose and determining the net retention volume V_N, Dorris and

Gray observed, as did others before,[2] a linear relationship between $RT\,\mathrm{Ln}(V_N)$ (or $-\Delta G_A$) and the number of C atoms of the n-alkanes. As an example, Figure 2.3 shows such a plot obtained on silica samples.

From the slope of the straight line, one easily calculates an incremental value $\Delta G_A(CH_2)$, corresponding to the free adsorption energy of one CH_2 group. Moreover, $\Delta G_A(CH_2)$ does not depend on the arbitrary choice of a reference state of the adsorbed n-alkane molecules. Consequently, by applying Equation 2.8, it follows that:

$$W_A^{CH_2} = 2\sqrt{\gamma_S^d \gamma_{CH_2}} \qquad (2.9)$$

In this relation, γ_{CH_2} represents the dispersive component of a surface entirely made of CH_2 groups (i.e. poly(ethylene)), for which $\gamma_S^d = \gamma_S = 35.6 - 0.058 \times T$ (mJ/m^2).

$\Delta G_A(CH_2)$ is then related then to the work of adhesion by:

$$\Delta G_A(CH_2) = Na_{CH_2} W_A^{CH_2} \qquad (2.10)$$

where N is Avogadro's number and a_{CH_2} is the surface of an adsorbed CH_2 group (6 $\mathring{A}^2$, the cross sectional area of a CH_2 group in a parallel arrangement of n-alkane chains in the bulk liquid). This value is currently used for IGC purposes, but there is a debate in the literature: that value depends on the actual conformation the alkane adopts on the examined solid surface and its possible modification with temperature.[17]

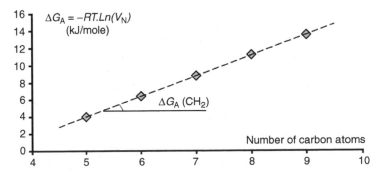

FIGURE 2.3 Variation of ΔG_A, with the number of carbon atoms of n-alkanes, and determination of $\Delta G_A(CH_2)$ (Aerosil 130 from Degussa, at 120°C).

Ignoring, this aspect, it follows that:

$$\gamma_S^d = \frac{1}{\gamma_{CH_2}} \left(\frac{\Delta G_A(CH_2)}{2Na_{CH_2}} \right)^2 \tag{2.11}$$

An alternative approach based on the same principle[18] consists to apply the general equation $\Delta G_D = RT\,Ln(V_N) + C' = 2aN\sqrt{\gamma_S^d\gamma_L^d} + C$ to a series of n-alkanes. γ_S^d is calculated from the plot of $RT\,Ln(V_N)$ versus the quantity "$a\sqrt{\gamma_L^d}$," where a is the surface area of the n-alkane and $\gamma_L^d = \gamma_L$ its surface energy. Both mentioned methods give concordant results. Concerning the latter approach, one might question the pertinence of the values "a" of the surface areas of n-alkanes.[19] Those values are either determined from the molecular weight and density of n-alkanes or they are experimentally evaluated, from liquid contact angle measurements on model surfaces, such as poly(tetrafluoroethane), that are not representative of say, mineral oxide or polar organic surfaces.

Table 2.1 displays a selection of γ_S^d values taken from the literature.

This list of γ_S^d values constitutes only a small selection of the large number of published data, leading to the following general remarks.

IGC measurements are, for a given sample, reproducible. The experimental errors, not mentioned in the Table 2.1, are small (a few percents only).

The γ_S^d values of a given solid, calcium carbonate for instance, greatly vary with the method of preparation of the sample or its natural origin. Several authors[41] recorded values close to 50 mJ/m^2, but values as high as 180 mJ/m^2 were measured under the same conditions on mined chalk.[42] This points to the necessity of reporting, whenever possible, the exact origin of the sample, its conditioning and storage conditions, and all other available characteristics (such as chemical purity) that may facilitate the interpretation of the experimental results. Also, when comparing γ_S^d of mineral oxides, some nonexpected high values do appear, especially with lamellar surfaces. Increased γ_S^d values are also recorded after a grinding process. Once more, since IGC is a most sensitive method, it is essential to specify the origin and history of the sample. In other terms, from an IGC-ID analysis point of view, each solid surface is almost unique; IGC is capable to detect minor differences amongst samples, even of a given family, or of samples stemming from the same production. IGC may thus be used as a remarkable instrument to monitor an industrial production process. For a save interpretation of the IGC results, it is recommended to combine that method with complementary surface analysis methods (the problem being that such methods are rather limited in number) to detect the origin of those differences (different surface chemistry, different impurities contents, different surface structures, etc.). Table 2.1 underlines,

TABLE 2.1
Typical γ_s^d Values Determined by IGC on Powders

Solids	S_{BET} (m²/g)	γ_s^d (Mj/m²)	Measurement (°C)	References
Calcium carbonate				
CaCO₃ (Socal Solvay)	9	52	96	20
CaCO₃ (Millicarb Omya)	3	48		20
Ca CO₃ (precipitated)	32	54	90	41
Ca CO₃ stearic acid treated	32	23	90	41
Chalk	2.9	180	100	42
Marble	2.1	55	100	42
Silica				
Aerosil (Degussa)	130	38	110	44
Aerosil (Degussa)	200	52	110	44
SiO₂ (Davison)				
Calcined at 875°C	288	39	202	21
SiO₂–Al₂O₃ (1% Al)	269	67	202	21
SiO₂–Al₂O₃ (10% Al)	240	53	202	21
Al₂O₃ (Calc. Bayerite)	192	72	202	21
Silica (Zeosil Rhodia)	165	26	180	22
Ion modified SiO₂ samples	152–239	31.6–64.5	373	23
Other mineral surfaces				
NiO	1.1	120	104	24
MgO (Merck)	7	95	25	25
TiO₂ (SICPA)		40	50	26
TiO₂ (JT Baker)		79	105	27
γ Al₂O₃ (Degussa)	100	92		28
Hematite	3–17	71–92	104	29
Goethite (Johnson Matthey)	13	125	104	30
Zircon ZrO₂	45	220	200	IGClab
ZnO	4.5	120	20	31
ZnO stearic acid treated	4.5	38	20	31
Hydroxyapatite	20.2	71.3	70	32
Hydroxyapatite		69.3	65	32
Hexacyanoferrate (Ni)	48	23.4	80	33
Hexacyanoferrate (Cu)	972	77.4	80	33
Chromia (Amorphous)	77	83.4	150	34
Lamellar minerals				
Muscovite (mica)	<1	38	132	35
Mica samples (grinded)	10–100	50–130	132	35

(continued)

Table 2.1 *(Continued)*

Solids	S_{BET} (m^2/g)	γ_s^d (Mj/m^2)	Measurement (°C)	References
H-magadiite (SiO$_2$)	48	178	60	36
H-kenyaite (SiO$_2$)	18	175	60	36
Talc samples	3–25	110–200		Nonpublished
Illite	30	177	80	37
Kaolinite	21	211	80	37
Carbon blacks				
Carbon black N110	130	400	200	38
Carbon black N220	110	450	200	38
Carbon black N330	78	240	200	38
Carbon black N550	41	195	200	38
Carbon black N773	30	95	200	38
Carbon black N339	89	304	180	22, 38
Carbon black N 234	122	365	180	22, 38
Carbon black (3.0% Si)	91	485	180	22
Carbon black (4.8% Si)	154	521	180	22
Carbon black (5.3% Si)	200	517	180	22
Organic powders				
Paracetamol		58.7	30	39
Carbamazepine		57.8	30	39
Tubing				
Passivated tubing		105	40	40
Electropolished tubing		49	40	40

also, some unexpected results observed on carbon black samples. Apparently, the γ_S^d values seem to be related to the extent of surface area values (and hence to the manufacturing parameters, such as temperature or residence time in the furnace). Furthermore, IGC highlights the consequences of surface treatments that largely influence the γ_S^d values; conversely, IGC may be used to monitor such treatments.

Sun and Berg[43] examined to what degree the surface energy of an heterogeneous solid, determined IGC at infinite dilution conditions, is averaged by carrying out IGC measurements on specimens of "known" energetic heterogeneity (in fact, on mixtures of a low surface energy solid such as poly(methylmethacrylate or PMMA) or poly(vinylchloride) and a high surface energy solid such as silica). They came to the conclusion that alkanes test both low and high adsorption energy surface sites and that the determined γ_S^d is dependent on the composition of the mix. In fact, silica by itself is already a heterogeneous surface and at the molecular size, it is not evident that the surface of the polymers may be considered as homogeneous. Furthermore, the authors suppose that the silica surface is smooth, but did not verify it. Also, silica and PMMA strongly interact. An alternative to this approach would be to modify the silica surface in a controlled way so as to change its surface heterogeneity (ratio of low and high adsorption energy sites). But this generates, again, a complex situation as shown elsewhere.[44]

Newell and Buckton[45] examined the view that IGC preferentially measures high-energy sites on a powder surface performing experiments on mixtures of amorphous (high-energy) and crystalline (lower energy) lactose particles, using nonpolar probes. The results were weighted averages of the surface energy for amorphous and crystalline material until the amorphous content exceeded 15% w/w of the sample, after which the surface energy become equivalent to that of the amorphous form. The amorphous content dominated when the surface area was 40% of the total area. Given that the amorphous particles were much smaller and adhered to the crystalline ones, it is reasonable to conclude that many (most) of the binding sites on the surface of the crystalline particles were masked by the amorphous particles by the time that the amorphous content dominated the surface energy measurements. IGC does not simply measure the high-energy sites in the packed column, but equally there is a complex process that results in measured data on mixtures not being a weighted mean of the surface energy of the two components.

Table 2.2 displays also some recent results obtained on fibers (a more exhaustive review of γ_S^d values of powders and fibers, obtained earlier, is offered by Belgacem and Gandini's review paper[9]).

γ_S^d values extend in a large range: 34 mJ/m^2 (at 50°C) for polyaramide fibers up to 120 mJ/m^2 (at 30°C) for ex PAN carbon fibers after O$_2$-plasma treatment. Here surface nano-structure and or surface nano-roughness may play a decisive role. Those parameters have not been systematically assessed yet.

2.2.3 Surface Nano-Roughness

To illustrate the possible origin of the significant variations of γ_S^d of solids, having the same chemical composition, let us take the example of an amorphous silica (Aerosil type prepared by pyro-hydrolysis of SiCl$_4$ in an H$_2$/O$_2$ flame) and of a crystalline silica (H-magadiite) obtained by an acid attack of pure synthetic magadiite. From a chemical point of view, both are of high purity. Yet, the one presents a γ_S^d value of 70 mJ/m^2 at a measuring temperature[36] of 60°C and the other exhibits a γ_S^d that amounts to 178 mJ/m^2, also at 60°C. Obviously, the major difference between the two silica samples is their amorphous, or crystalline, structure. But how will this influence the γ_S^d characterization by IGC? First of all, one should keep in

TABLE 2.2
Typical γ_s^d Values Determined by IGC on Fibers

Fibers	S_{BET} (m²/g)	γ_s^d (mJ/m²)	Measurement (°C)	References
Polymer fibers				
Kevlar 29 (DuPont de Nemours)		34	50	46
Twaron Akzo Nobel		81	79	47
Kevlar 29 (DuPont de Nemours)		39.5	20	48
Kevlar 29 (Acetone washed)		53.7	20	48
Kevlar 29 (O₂ Plasma treated)		57.8	20	48
Carbon fibers				
HTA Tenax		89	79	47
AS4 Hercules oxidized		50–49	40	49
AKZO Tenax 5000		104	50	50
Tenax Electochem. Ox		89–78	50	50
Ex PAN		103	30	51
Ex-PAN, O₂ Plasma treated		125	30	51
Glass fibers				
E-fiber		31.3	60	52
*0.3 wt 1% APS sized E fiber		31.9	60	52
Rockwool	0.3	44	65	53
Ashai Fiber Glass APS-sized fiber	0.1	40–33	25	54
Cellulose fibers				
Cotton Cellulose	1.1	49.9	40	3
**TMP		38.8	30	3
Kraft pulp		45.1	20	55
Norway spruce wood particles		42.3	20	56
Wood particles (extracted)		44.9	20	56

[a] Amino Propyl triethoxy Silone
[b] Thermomechanical pulp

mind that only a restricted number of probe *n*-alkane molecules are used to explore the surface of both silica samples. Secondly, *n*-alkane molecules are flexible molecules that may easily adapt their conformation to surface asperities for optimum interaction. Various types of defects are present on

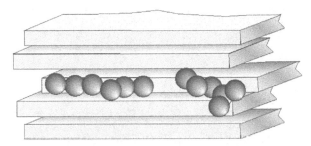

FIGURE 2.4 Schematic representation of the interaction of an n-alkane molecule with defects of a lamellar silica.

the periphery of the lamellar H-magadiite crystals as schematically sketched on Figure 2.4.

Here, we suppose that a peripheral part of an interstitial lamellar plane is absent and thus offers an original access to an n-alkane molecule. This molecule will experience a much stronger force field than the one existing on the planar H-magadiite surface. As a consequence, it will be retained much more strongly on this particular type of adsorption site leading to high values of γ_S^d. Obviously, this value does not correspond to a "true" value, representative of the dispersive component of the surface energy as discussed above. An indirect proof of this assertion consists to destroy, by heat treatment, the crystalline structure and to measure again γ_S^d on the resulting residue; γ_S^d is now similar to the one recorded on an amorphous silica sample.

It follows that a small number of structure defects will significantly influence the γ_S^d determination by IGC-ID. As pointed out before, defects of molecular dimensions, like those suggested on Figure 2.4 may not be seen by classical x-ray methods. However, the development of modern local field microscopy (Scanning Electron Tunneling (STM), Atomic Force Microscopy) will allow to check the correctness of IGC deductions. Figure 2.5 shows, as an example, STM pictures comparing two carbon black samples: initial N234 (Cabot) carbon black and a sample of similar surface area, but having on its surface SiO_2 inclusions of molecular size ("Duofiller" from Cabot). The latter sample is prepared by adding, during the incomplete combustion of hydrocarbon feedstuffs, some silicium derived products.

The surface of the carbon black N234 is rather rough and made of polyaromatic structures resembling scales. On the border of those scales, flexible alkane chains will adjust themselves so as to achieve optimum interaction. Computer simulation confirms that an adsorbed alkane molecule will move, when allowed, on a C surface until it encounters the most energetically active sites of adsorption. As a model surface, two superposed

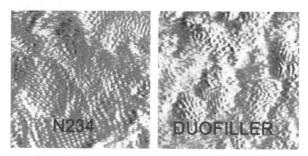

FIGURE 2.5 Scanning Tunneling Electron Microscopy (10×10 nm) of carbon blacks. Comparison of N234 (Cabot) and a carbon black of similar surface area, but containing silica inclusions. (Courtesy of Dr. M. J. Wang, Billerica, MA.)

graphite layers were taken. The superior layer does not entirely recover the layer beneath it, creating a step of atomic dimensions. When a pentane molecule is "deposited" on the uncovered bottom layer and is allowed to freely move, it invariantly finds its optimum interaction position when attached to the edge of the upper layer and centered above a ring of the lower layer. Bick[57] used the Cerius 2 simulation program, supposing that the surface atoms are fixed and that pentane is fully flexible.

The surface of N324 differs significantly from the "Duofiller" surface by the presence of SiO_2 inclusions. The extent and shape of the polyaromatic surface structures are quite different. γ_S^d values, determined on both samples at 180°C, are excessively high, respectively equal to 365 and 521 mJ/m^2.

The peculiar surface structures (nano-roughness) observed on carbon blacks should also lead to different interaction possibilities of linear or branched alkane probes.[20] In fact, a way to detect the existence of such a surface roughness, at the molecular level, is to compare the adsorption behaviors of *n*-alkanes and of their branched isomers having parent (London) interaction potentials, but different bulkiness. The following branched alkanes were selected to test this point:

1. Tetramethyl butane (TeMB)
2. 2,2,4-trimethyl pentane (224 TMP)
3. 2,3,4-trimethyl pentane (234 TMP)
4. 2,5-dimethyl hexane (25 DMH)

By definition, the morphology index (IM) is given by:

$$IM = 100 \times \frac{V_b V_n^{ref}}{V_n V_b^{ref}} \tag{2.12}$$

where V_n and V_b are the net retention volumes of n-octane and of the branched alkane respectively. V_n^{ref} and V_b^{ref} correspond to the retention volumes observed with the same alkanes, but on a flat reference solid surface, the reference surface being a pyrogenic silica that is flat at the molecular level, at least at the level of the probing n-alkane molecules, as demonstrated by different approaches;[58] Table 2.3 presents IM values determined on several powders.

Comparatively to pyrogenic silica, the precipitated silica sample (obtained by acidification of a sodium silicate solution) shows some nano-roughness. This is understandable since this oxide was prepared at rather low temperatures (drying temperature of the order of 200–300°C) and may be considered as being in an unstable (not finished) surface state. Several facts, in particular the large density of surface hydroxyl groups (i.e. of dangling noncondensed silanol groups similar to that of silicic acid surface chains[59]) are in favor of that hypothesis. For lamellar solids, as expected, the nano-roughness is clearly detected by the proposed method.

A more global approach relies to molecular probe topology concepts. Wiener[60] was the first proposing a descriptor of the topology of alkanes. He defined two parameters W and p. The Wiener index (W), is the total number of C–C links between all carbon atom pairs in the molecule, whereas p, the polarity number, represents the number of pairs of C atoms separated by three C–C links. This allowed him to correlate his index with the boiling temperature, the heat of isomerization, and the heat of vaporization of alkanes. Later on, he extended the application of his index for the evaluation of the molar volume, the surface tension, and saturating vapor pressure[61] of the same alkanes. Wiener's index allowed the determination of Quantitative Structure–Properties Relationships (QSPR) that are precious for the development of medicaments.[62] Numerous papers from the literature discuss molecular structure,[63] geometrical and physical meaning of indices[64,65] and ways to take advantages of them.[66,67] Others, as mentioned before, consider

TABLE 2.3
Morphology Index (IM) Determined on Silica and Talc Samples

Solids	TeMB	2,2,4 TMP	2,3,4 TMP	2,5 DMH
Pyrogenic silica	100	100	100	100
Precipitated silica	73	77	82	90
Talc	14	22	22	41
H-magadiite	0.6	1.0	1.1	2.1

the applications of selected indices[68] for the prediction of physical properties of hydrocarbon molecules. As an example, a single equation, Equation 2.13, relates the boiling temperature (T_b°) and the Wiener (W) index[69] demonstrating the potential of such approaches:

$$T_b^\circ = 77.93 \times W^{0.3089} - 3.35.10^{-5} \times W^3 - 164.24 \qquad (2.13)$$

The Randic connectivity index χ is the most widely used topological index in QSPR. It is given by:

$$\chi = \sum_{\text{all C–C links}} \sqrt{d(i)d(j)} \qquad (2.14)$$

where $d(i)$ and $d(j)$ represent the valences of carbons i and j entering in the link $(d(i) - d(j))$. For saturated hydrocarbons, the d values are 1, 2, 3, and 4 and there are only 10 possible types of $d(i) - d(j)$ bonds. Randic[70] established a correlation between his connectivity index and the Kovats index[71,72] currently used for analytical chromatography purposes. We used[73] the Wiener (W) and Randic indices (χ) for defining a new index (χ_T). The χ_T of a branched alkane represents the number (noninteger) of C atoms of a hypothetical linear alkane that would interact with a solid surface in the same way, as does the branched alkane. In a first step, the relationships between the numbers of C atoms of n-alkanes and the previous indices were established:

$$\chi_T(W) = 1.8789 \times W^{0.3271} \qquad (2.15)$$

$$\chi_T(\chi) = 2 \times \chi + 0.1716 \qquad (2.16)$$

The above method allowed us to calculate the topological indices of both linear and branched alkanes. To adapt Wiener's index to cyclic hydrocarbons, we need to introduce some rules for the calculation of W. First, we open the cyclic molecule and calculate the W value as for an ordinary n-alkane. The calculated index W is then weighed by the amount $n/(n-1)$ that is the ratio of C–C links in the cyclic alkane compared to the number of C–C of a noncyclic molecule having the same number of C.

For linear alkanes, the following relation holds:

$$t_N = \exp(an_C + b) \qquad (2.17)$$

Once the coefficients a and b are experimentally determined, we may calculate the parameter χ_T and use it for the prediction of the retention times

(t_N) of solutes on "flat" surfaces.

$$t_N = \exp(a\chi_T + b) \tag{2.18}$$

If the measured retention times on a given chromatographic support correspond to the calculated retention times, we conclude that the surface is planar at the molecular level. or that the surface accessibility is the same for either linear or branched alkanes. If not, we conclude that the surface is not flat at the molecular level. For the quantification of the nanorugosity concept, rather than calculating, as previously, a retention time difference for its evaluation, we shall evaluate the difference in the number of C atoms susceptible to interact with the solid surface. The relation (2.18) is transformed in:

$$\chi_{exp} = \frac{Ln(t_N) - b}{a} \tag{2.19}$$

By doing so, we compare the measured χ_{exp} to the χ_T determined using the Wiener's index corresponding to flat surfaces. Then, by definition, the morphology index (IM_{χ_T}) is given by:

$$IM\chi_T = 100 \times \frac{\chi_{exp} - \chi_T}{\chi_T} = 100 \times \frac{\Delta\chi_T}{\chi_T} \tag{2.20}$$

Figure 2.6 is an illustration of the method of evaluation of IM_{χ_T}.

The IM_{χ_T} values were measured on Aerosil 130 using a series of linear, branched, and cyclic alkanes. All alkanes' representative points of the plot, relating the variation of the alkanes free energy of adsorption with χ_T (measurements made at 110°C), fall on the same line; this demonstrates that their accessibility to the surface is not hindered. Supplementary measurements, made at 90, 104, and 120°C, lead to the same χ_T demonstrating the stability of the method. The IM_{χ_T} values measured on that silica sample are, as expected, very small ($-0,7\%$ for 2,2,4 TMP, for instance) whatever the injected alkane probe, and thus confirm that the silica surface is rather flat at the level of the probing molecules. This is not the case ($IM_{\chi_T} \approx -30\%$) for H-magadiite, a lamellar crystalline silica, as seen on Figure 2.7, the experimental points of branched alkanes lying well below the n-alkanes line.

Table 2.4 encloses IM_{χ_T} values determined on various solid surfaces. Graphite A is natural graphite from Madagascar, whereas graphite B originates from an industrial process (Lonza, Switzerland). The γ_S^d values of those samples differ significantly. One reason may be the fact that sample B contains traces of metallic oxides (possibly stemming from a ball milling

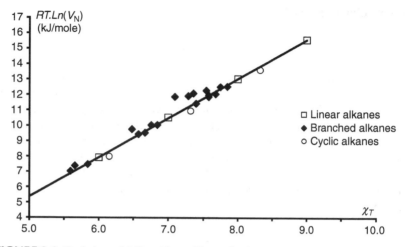

FIGURE 2.6 Variation of ΔG_A with χ_T (determined on a pyrogenic silica A130 from Degussa).

process). The carbon black samples were obtained from Columbian Carbon (U.S.A.). Both develop a surface area of 50 m^2/g, but come from different manufacturing batches. The difference between the samples is small. Yet, it seems that IGC is capable to detect minor differences that might have consequences in applications. In other terms, IGC may be used as an efficient method to control the industrial process.

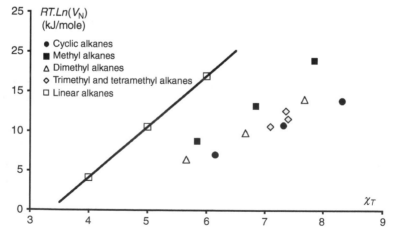

FIGURE 2.7 Comparison of the variation, with χ_T, of ΔG_A of alkanes on H-magadiite.

TABLE 2.4
Morphology Index (IMχT) of Various Powder Surfaces

Solid	γ_s^d (mJ/m^2)	IM$_{\chi_T}$ 2,2,4 TMP (%)	IM$_{\chi_T}$ cyclooc-tane (%)	References
Aerosil 130	44	−0.7	−1.8	73
H-magadiite	258	−30	−34	73
Graphite A	105	−6	12	143
Graphite B	279	−4.5	−13.5	143
Carbon back A	174	−4.0	−10	143
Carbon back B	204	−3.7	−10.8	143
Nickel oxide	170	−3.1	−10	24
Goethite	125	0.6	−7.3	73

Scanning tunneling electron microscopy indicates (Figure 2.8), as seen also by transmission electron microscopy, the reorganization of the carbon surface after a heat treatment at 2700°C. This will induce modifications of the surface characteristics, modifications that should also appear on the IGC results.

On the initial carbon black, a "scale" structure is evidenced whereas, upon heat treatment, extended graphite like zones do appear. However, the totality of the carbon black surface is not transformed. Defects still exist. The macroscopic γ_s^d value, determined on graphite basal plane, by liquid contact

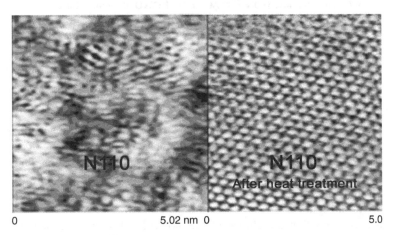

FIGURE 2.8 STM pictures of carbon black samples before and after heat treatment. (Courtesy of Prof. J. B. Donnet, UHA, Mulhouse, France.)

angle measurements, is close to 90 mJ/m^2; a value significantly lower than the one obtained by IGC on a graphitized carbon black sample. As an example, a value equal to 243 mJ/m^2 was recorded on the initial sample. After graphitization, a noticeable increase of γ_S^d is observed ($\gamma_S^d = 323$ mJ/m^2). This may be encountered for by a surface cleaning process leading to a limited number of new nano-scaled organized surface step structures, operating like highly energetic adsorption sites.

To sum up this part devoted to the determination of the dispersive component of surface energy, we may underline that:

- IGC-ID, using n-alkanes as probes, readily produces a large amount of data that are not straightforward interpretable in terms of dispersive component of surface energy (γ_S^d). The calculated value corresponds, at the best, to the dispersive forces activity of the most efficient adsorption sites.
- A given γ_S^d cannot be attributed to a solid surface without specifying its origin and history (conditioning for instance). Nevertheless, when measurements are made under controlled conditions, useful information is gained when comparing samples of a given family, as will be seen later.

IGC is, in some cases, a unique method in terms of simplicity of use, quantity, and quality of the generated results.

2.2.4 SPECIFIC INTERACTION CAPACITY OF POWDERS AND FIBERS

2.2.4.1 Experimental Determination

Graphical determination. When injecting polar probes in a column containing a polar support, both dispersive and specific interactions will take place. The total standard variation of adsorption free energy is then given by:
$\Delta G_A = \Delta G_A^D + \Delta G_A^{sp}$

The problem to be solved is to evaluate the part of each type of contributions. A possible solution, illustrated in Figure 2.9, consists in comparing the free energies of adsorption of n-alkanes with those of polar probes. By definition, ΔG_A^{sp} is given by the quantity corresponding to the departure, from the alkane line, of the representative point of the polar probe as shown on Figure 2.9.

It is obvious that the choice of the molecular descriptor (abscissa) may influence notably the value of the measured ΔG_A^{sp}. Mukhopadhyay and Schreiber,[8] Brendlé, and Papirer[74] paid a special attention to the choice of

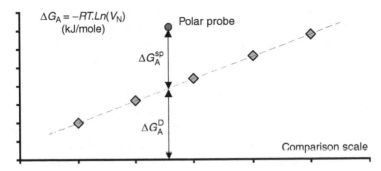

FIGURE 2.9 Principle of the determination of specific interaction energies.

the molecular descriptor, reviewing the different proposals made in the literature that are namely:

- $Log(P_0)$, the logarithm of the vapor pressure of the probes[75]
- $a\sqrt{\gamma_L^d}$, the product of the molecular area a by the square root of the dispersive component of the surface tension of the probe γ_L^d taken in the liquid state[18]
- T_b, the boiling point of the injected probe[76]
- ΔH_{vap}^d, the enthalpy of vaporization of the solute, corrected for the dispersive forces contribution to vaporization[77]
- $\alpha_0\sqrt{h\nu}$, the product of the polarisability α_0 of the probe and the square root of its electronic frequency $h\nu$, where h is the Planck constant[78,79]
- χ_T, a topological index[74] that takes into account both the geometry and the local electronic density associated with the presence in the polar probe structure of hetero-atoms, such as chlorine, oxygen or nitrogen atoms

$Log(P_0)$, was initially proposed by Saint-Flour and Papirer.[75] The choice of this parameter was made considering that the logarithm of the vapor pressure, which is closely related to the evaporation enthalpies (ΔH_{vap}), is representative of the interaction capacity of two identical molecules. This molecular descriptor offers several advantages, since P_0 values are either given in the literature or are computable, even at relatively high measurement temperatures, provided that these stay below the critical temperature of the probe. Yet, problems do appear with certain solids, possessing a high surface energy, since the representative points of some polar probes fall beneath the alkane-line. Chehimi and Pigois-Landureau[77] explained this observation in terms of the self-association of given polar probes. Moreover, they proposed to use the enthalpy of evaporation, ΔH_{vap}, instead of $a\sqrt{\gamma_L^d}$, correcting it by taking into account the contribution of the dispersive interactions to ΔH_{vap}. Of course, for alkanes, ΔH_{vap} and T_b are closely related physical parameters. Similar difficulties are encountered with the quantity $a\sqrt{\gamma_L^d}$ as seen in Figure 2.10.

The representative points of polar probes, that are more interactive with the polar surface, should always be located above the so-called "alkane-line." But this is not the case when using $Log(P_0)$ or $a\sqrt{\gamma_L^d}$ as variables as seen on Figure 2.10. The limitations are due to the fact that the value of the molecular area, "a," of the adsorbed probe varies with the nature of the solid,[19] the temperature, and the surface coverage of the solid by the probe molecules. Moreover, γ_L^d values are not always available from the literature at the temperature used for the IGC measurements. In fact, the previous molecular descriptors suffer from the same weakness; they all are based on parameters that are characteristic of interactions between identical molecules in the liquid phase, whereas ΔG_A is related to the interaction between an isolated molecule and a solid surface having generally a very different chemical nature. This observation instigated some authors to propose new molecular descriptors that are related to the molecular structure of the probe and not solely to the intermolecular interactions that take place between them in the liquid phase. The first attempt was made by Dong et al.[78] and refined by Donnet et al.,[79] who, starting from the equations describing the van der Waals interaction energy between two partners, proposed to adopt the product $\alpha_0\sqrt{h\nu}$ as the

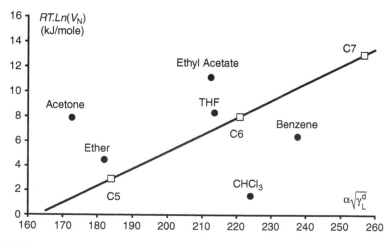

FIGURE 2.10 Specific interaction energy determination on natural graphite, at 53.3°C, using $a\sqrt{\gamma_L^d}$ as variable.

molecular descriptor. The main problem encountered by these authors was due to the fact that polarizability data, measured at low frequency, are rather scarce in the literature, whereas those computed from refractive indexes, according to the Debye relation, are rough approximations. An example of application of Donnet's method is illustrated, in the case of natural graphite, on Figure 2.11.

Later, Brendlé and Papirer[74] introduced topology indexes χ_T. The main advantage of that approach is that χ_T may be easily computed for each probe of interest, without any physical measurement requirement. Furthermore, both parameters ($\alpha_0\sqrt{h\nu}$ and χ_T) leading to the best representation of the IGC data, are closely related as seen on Figure 2.12.

Hence, it is recommended to use either the polarizability, or preferentially, the topology approach for the evaluation of specific interactions as shown by Dutschk et al.[52] who compared the various theoretical approaches for the determination of the specific interaction energy of (sized and unsized) glass fibers.

Before trying to exploit the ΔG_A^{sp} values determined in the way suggested above, it might be interesting to recall some earlier work on the evaluation of the "polarity" of chromatographic supports.

Kovats retention indices I_x. I_x values are commonly used in GC analytical work because their values are independent of parameters like flow rate, column size, dead volume, % stationary phase. I_x is only dependent on the analytical characteristics of the column filling and on temperature.[72] The difference between the Kovats index I_x of a polar probe injected in a column x,

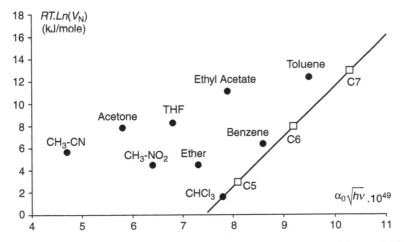

FIGURE 2.11 Specific interaction energy determination on natural graphite, at 53.3°C, using $\alpha_0\sqrt{h\nu}$ as a variable.

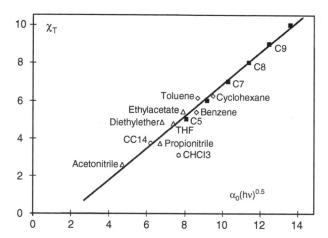

FIGURE 2.12 Relationship between the polarizability coefficient and χ_T of probes currently used for IGC measurements.

containing a polar filling and the Kovats index I_{ref} of the same probe on a nonpolar column, is a measure of the polar character of the column x. This method may be readily applied to fibers (carbon) and resins,[80] and active carbons.[81] In fact, the experimental results (specific interaction energy) are in satisfactory agreement with those given by the χ_T method (χ_T and I_x are closely related).

Acid–base index Ω. Numerous attempts were made to modify the surface properties of powders and fibers: preparation of more efficient GC phases; manufacturing of more effective fillers for increasing the performance (reinforcement) of say, tires; enhancement of the fiber-matrix adhesion; and consequently, the quality of composites materials, etc. An easy and rapid method to evidence qualitatively the consequences of the surface treatment is the acid–base index proposed by Schreiber,[82] measuring specific retention volumes V_g^0, on the one hand, with an acid (*n*-butanol) probe $(V_g^0)_a$, and on the other hand, with a base (butylamine) probe $(V_g^0)_b$. The following relation defines this index:

$$\Omega = 1 - \frac{(V_g^0)_b}{(V_g^0)_a} \qquad (2.21)$$

On an acidic surface, butylamine will be retained much more strongly than the acid probe (butanol); Ω takes large values. They applied this concept to follow the surface properties variations of glass fibers after treatment with functionalized silanes. The trends indicated by IGC followed those expectable from the functionalities of the grafted silanes.

π interactions. π interactions are special types of acid–base interactions. From Figure 2.13, we notice that the representative points of *n*-alkanes and *n*-alkenes form two parallel lines when plotting ΔG_A versus their number of C atoms.

The only difference between *n*-alkanes and *n*-alkenes having the same number of C, is the existence of a double bond rich in electrons (i.e., constituting a base). The distance between the two lines represents the π interaction capacity of the acidic silica surface in the mentioned example. Jagiello et al.[83] verified this method using model adsorbents such as anthracene, 9-phenantrol, anthrarobin, and anthracenecarboxylic acid. The latter exhibited a particular behavior being less interactive, a fact explained taking into account its tendency to self-association.[84] They applied this method with success, in particular, for the evaluation of the surface acidity of active carbons.

Acid–base scales. Whenever possible, acid–base interactions become prevalent[16] comparatively to dispersive interactions; hence, the importance of disposing of appropriate methods to evaluate those interactions. The methods presented so far allow essentially a description of the surface acid–base of the solid surface in contact with a given polar probe or a family of probes. The solutes used to assess the acid–base character, in the Lewis acidity or basicity definition, are acids like chloroform (possessing a C atom carrying a positive charge due the high electron attraction of the chlorine atoms), or

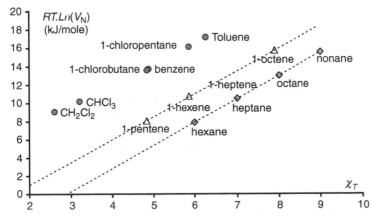

FIGURE 2.13 Specific interaction energy determination on silica (Aerosil 130) using χ_T as variable.

bases like ether (having an Oxygen atom with a high electron density), or amphoterics molecules (such as acetone). Amphoteric means that a molecule possesses a group of atoms showing, simultaneously, acid and base properties (the carbonyl group in the case of acetone). Up to now, there is no theoretical way to quantify those characteristics. Only empirical or semi-empirical acid–base scales are available:

- The hard–soft acid–base scale of Pearson[85]
- The E–C relation of Drago[86]
- The donor acceptor numbers of Gutmann[87]
- The solvatochromic parameters[88]
- The solubility parameters[89]
- The van Oss electron donor and acceptor parameters[90]
- The Abraham approach[91]

The hard–soft acid–base scale of Pearson [85] is based on molecular orbital theories where the absolute hardness of a molecule is defined as being equal to half the negative rate at which its electro-negativity changes with a change of its electron population, at constant potential. Pearson derived laws allowing to predict the orientation of organic reactions between electron donor and electron acceptor molecules or groups. So far, these concepts were not used for IGC purposes, owing to the rather unknown and complex nature of solid surfaces.

The four parameter semi-empirical relation of Drago[86] is based on an equation predicting acid–base reaction enthalpies in the gas phase or in poorly solvating solvents:

$$-\Delta H_{AB} = E_A E_B + C_A C_B \qquad (2.22)$$

The acid (A) and the base (B) are characterized by two values: an E value that describes the ability of A and B to participate in electrostatic bonding (hardness), and a C value (softness) that indicates their tendency to form covalent links. Considering iodine as a reference substance, for which $C_A = E_A$ and making a series of calorimetric measurements, mixing iodine with the substance of interest, allows by computer fitting of Equation 2.22, to calculate C and E values of some tens of molecules. The Drago approach has been used with success, starting from wettability measurements, for the evaluation of surface properties of oxides,[92] yet not through IGC. But, there is no peculiar reason for that. One major drawback of the Drago scale comes from the fact that a molecule is defined either as an acid or a base. In reality, most molecules exhibit amphoteric properties.

The Gutmann's electron acceptor (AN) and donor numbers (DN), that respectively stand for acid and base properties of a molecule or substance, are determined experimentally. AN is defined as the enthalpy of formation of a 1/1 molecular adduct of a given molecule with a reference Lewis acid ($SbCl_5$). DN is measured by the NMR shift of P when the given molecule is mixed with a solution of oxotriethylphosphine (Et_3PO) taken as a reference base; the shift being normalized by taking the value 0 for the solvent (1,2 dichloroethane) and 100 for $SbCl_5$. Knowing *AN* of an a "pure" acid and *DN* of a "pure" base, allows calculating the enthalpy of interaction using the following relation:

$$\Delta H_{AB} = ANDN/100 \qquad (2.23)$$

For the calculation of the enthalpy of interaction between two amphoteric molecules, having respectively $(AN)_1$ and $(AN)_2$ as acceptor numbers (expressed as %) and $(DN)_1$ and $(DN)_2$ (expressed as kcal/mol or kJ/mol) as donor numbers, we suggested to apply the equation:[36]

$$\Delta H_A^{sp} = [(AN)_1(DN)_2 + (AN)_2(DN)_1]/100 \qquad (2.24)$$

The method of Kamlet and Taft[88] is called the "solvatochromic method" since the measurements are usually performed by UV or visible spectrometry. Peak shifts are detected when the solid is mixed with solvatochromic probe molecules of known characteristics. This method is based on a general "linear solvation energy relationship":

$$XYZ = XYZ_0 + s(\pi^* + d\delta) + a\alpha + b\beta + h\delta_H \qquad (2.25)$$

XYZ is a given property of some solute in a series of solvents. A number of solvent properties are used as independent variables; π^* is the solvent dipolarity/polarizability ratio, δ is a polarizability correction term, α and β are the solvent H bond acidity and basicity, and δ_H is the Hildebrand solubility parameter. The coefficients in the previous equation, XYZ_0, s, d, a, b and h are calculated by a linear regression process. This approach has been extended for gas and condensed-phase processes[93] and used for the evaluation of IGC data (partition coefficient, retention factor) obtained on graphite[94], charcoals,[95] fullerene,[96] carbon fibers,[97] and bulk water adsorbed on a chromatographic stationary phase.[98] The contribution of "linear solvation energy relationship" approach for the determining of polar and nonpolar sites on carbonaceous adsorbents has been recently reviewed.[99]

Spange and Reuter[100] applied this method to a series of chemically modified silica samples and related the H bond, accepting property to the isoelectric points of silica samples, determined by zeta potential measurements. The silica hydrogen bond acidity and basicity, as well as its polarizability (London interactions), could also be evaluated by IGC.[101]

The Solubility Parameters Approach. The solubility parameter δ is given by the square root of the energy density.[89,102,103] For volatile materials, it is calculated from the following relation:

$$\delta = \sqrt{\frac{\Delta H_v - RT}{V^0}} \tag{2.26}$$

where ΔH_v is the enthalpy of vaporization, R the ideal gas constant, T the absolute temperature, and V^0 the molar volume. According to Hansen,[102] the total energy of vaporization ΔH_v may be considered as a sum of different contributions of dispersion forces (E_D), permanent dipole–dipole forces (E_P,) and hydrogen bonding forces (E_H):

$$E = E_D + E_P + E_H \tag{2.27}$$

Hence:

$$\delta_T^2 = \delta_D^2 + \delta_P^2 + \delta_H^2 \tag{2.28}$$

The Hansen approach has been applied to solids.[104] IGC allows readily determining the specific retention volume V_G (that is simply V_N divided by the mass of the solid inside the column). The enthalpy of adsorption ΔH_A is obtained from a plot of $\mathrm{Ln}(V_G)$ vs. $1/T$. ΔH_A is then related to the Hansen solubility parameter by Equation 2.29:

$$\Delta H_A = V^0(\delta_D^P \delta_D^S + \delta_P^P \delta_P^S + \delta_H^P \delta_H^S) \tag{2.29}$$

where the subscript P refers to the probe and the subscript S to the solid surface. By selecting probes of known Hansen's values, it becomes possible to solve the previous equation and have access to the dispersive, polar, and H bond formation capacity of the solid surface. Voelkel and Grzeskowiak[105] used solubility parameters for the characterization of surface modified silica gels whereas Tong et al.[106] applied it to pharmaceutical powders. The latter pointed out the necessity of considering entropy playing an important role.

Van Oss Approach. Van Oss et al.[90] suggested the following relation for the evaluation of acid–base free energy variation:

$$\Delta G_{12}^{AB} = -2[(\gamma_1^+ \gamma_2^-)^{1/2} + (\gamma_1^- \gamma_2^+)^{1/2}] \tag{2.30}$$

In this expression, γ^+ and γ^- are the electron acceptor and donor characteristics of the interacting partners 1 and 2. γ^+ and γ^- are evaluated from measurement of interfacial tension or liquid contact angles, taking water as a reference. Unfortunately, only a limited number of acid–base values is available. This method, designed for (macroscopic) liquid contact angle evaluation (wettability), is very useful to predict interactions in aqueous media, but is scarcely used so far for conventional IGC acid–base evaluations. Yet it shows promises.[17]

The Abraham Approach. Goss[107] made a critical review of the proposed acid–base or electron donor–acceptor (EDA) scales. In particular, he noticed that the Van Oss scale is not adequate, especially when adsorbed water molecules are present on the solid's surface, a situation that is current for actual surfaces placed in a humid atmosphere. This scale implies that the water molecules in the surface layer of a bulk water phase have the same capability to interact by EDA interactions with their horizontal neighbors as with other molecules across the interface. However, EDA interactions strongly depend on the orientation of molecules.

The method proposed by Goss is based on Abraham et al.[108] EDA scales having as variables the H-donor property and H-acceptor property of a compounds immersed in a liquid bulk phase whose molecules serve as interaction partners. Those relative scales were derived from fitting absorption data of the respective compound between various organic bulk phases and water or air.

Goss, using a whole set of experimental data, adapted the Abraham method with much success for the study of the air/surface adsorption equilibrium of organic compounds under ambient conditions, a major concern for the understanding of environmental problems. That method was also applied for the understanding of the interactions that occur in the adsorption process of a solutes and carbonaceous adsorbents, using IGC.[109]

Evaluation of acid–base interaction potential of powders and fibers. So far, the Gutmann's approach is very popular.[8] According to the Gutmann terminology,[87] the probes are categorized as acidic, basic, or amphoteric molecules by their electron-acceptor and their electron-donor numbers AN and DN, respectively (Table 2.5).

The specific enthalpies of interaction ΔH_A^{sp} are computed directly from specific free energies of adsorption, measured at different temperatures, according to the Clausius Clapeyron (Equation 2.31):

$$\Delta H_A^{sp} = -T^2 \frac{\partial(\Delta G_A^{sp}/T)}{\partial T} \qquad (2.31)$$

While the Gutmann's equation normally allows the calculation of the interaction enthalpy between a "pure" acid and a "pure" base, it was extended

TABLE 2.5
Properties of Current IGC Probes

Probes	a (Å^2)	γ_s^d mJ/m^2	χ_T	$\alpha_0\sqrt{hv}.1049$	DN (kcal/mol)	AN (%)	Character
n-Pentane	44.2	15.5 (25°)	5	8.1	0	0	Neutral
n-Hexane	51.5	17.9 (25°C)	6	9.2	0	0	Neutral
n-Heptane	57.0	20.3 (25°C)	7	10.3	0	0	Neutral
n-Octane	63.0	21.3 (25°C)	8	11.4	0	0	Neutral
n-Nonane	69.0	22.7 (25°C)	9	12.5	0	0	Neutral
n-Decane	75.0	23.4 (25°C)	10	13.2	0	0	Neutral
Cyclohexane	43.1	24.7 (25°C)	6.15	8.7	0	0	Neutral
CH$_2$Cl$_2$	30.2	27.2 (25°C)	2.60	5.6	0	20.4	Acid
CHCl$_3$	32.5	26.7(25°C)	3.21	7.8	0	23.1	Acid
CCl$_4$	31.9	26.4 (25°C)	3.94	8.4	0.0	8.6	acidic
Benzene	44.1	28.2 (25°C)	4.88	8.6	0.1	8.2	Acid
Methanol	19.1	22.1 (25°C)	1.88	3.5	19.1	41.5	+Acid
Ethanol	26.1	22.0 (25°C)	2.90	4.9	20.0	37.9	+Acid
n-Propanol	32.8	23.3 (25°C)	3.92	6.2	—	—	+Acid
Isopropanol	32.3	21.3 (25°C)	3.78	6.2	—	—	+Acid
Ethyl ether	39.4	16.5(25°C)	4.77	7.4	19.2	3.9	+Basic
THF	35.0	22.5 (25°C)	4.85	6.8	20.1	8.0	+Basic
1,4-Dioxane	37.1	33.5 (23°C)	5.82	—	14.8	10.3	+Basic
Pyridine	38.2	38.0 (23°C)	4.86	7.8	33.1	14.2	+Basic
Ethyl acetate	43.2	20.5 (25°C)	5.44	7.9	17.1	9.3	Amphoteric
Acetonitrile	28.1	19.4 (23°C)	2.60	4.4	14.1	19.3	Amphoteric
Acetone	32.0	20.5 (25°C)	3.61	5.8	17.0	12.5	Amphoteric
Water	11.9	21.1 (25°C)	—	—	18	54.8	Amphoteric

in order to allow the calculation of the specific interaction enthalpy (ΔH_A^{sp}) between an amphoteric (having both acid and base interactions capacities) probe (characterized by AN and DN) and an amphoteric solid surface (characterized by K_A and K_D) using Equation 2.32:

$$\Delta H_A^{sp} = ANK_D + DNK_A \qquad (2.32)$$

K_A and K_D are graphically determined as shown on Figure 2.14 for natural and heat treated graphite samples.[74]

The differences in surface acid–base properties are apparent. The K_A and K_D values, expressed in arbitrary units, are respectively equal, for the initial sample, to 1022 and 709, whereas after heating they drop to respectively, 408 and 735, and thus demonstrate a significant loss of polar surface acid and base like groups. Yet polar groups are still present, after heating and cooling; they are possibly reformed during air re-exposure of the treated graphite sample. It is also important to stress the fact that only the strength of acid–base groups, but not the amount of acid–base groups may be evaluated by this approach.

On an energetically homogeneous surface, the number of sites visited by the molecular probe depends on both the number (n_s) and the energies of interaction (e_s) of the surface sites. The measured retention time (t_r), may be considered as the product of local residence time (t_s), and the total number of visited sites (n_s), the two variables cannot be separately evaluated (Equation 2.33).

$$t_r = n_s t_s \qquad (2.33)$$

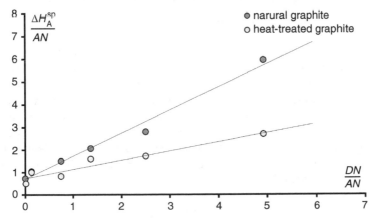

FIGURE 2.14 Graphical determination of K_A and K_D on natural and heat treated graphite samples.

with

$$t_s = t_0 . \exp\left(\frac{e_s}{RT}\right) \tag{2.34}$$

where t_0 is between 10^{-12} and 10^{-13} s.

The second reason that renders difficult a quantitative determination of acid–base characteristics of solid surfaces by using the reported methods, stems from the fact that the semi-empirical acid–base scales of solutes were established considering the interactions between small organic molecules in a solvent. In a solvent, the two interacting molecules are free to adopt the best relative orientation leading to the highest level of interaction. This certainly is not always the case on a solid surface on which defects, steps, pores, or crevices hinder the mobile probe molecule to adopt its optimal orientation, and thus decrease the energy of interaction. Keeping in mind that the ΔG_A^{sp} determination is based on a comparison of the adsorption behaviors of n-alkanes and polar probes, it becomes evident that if size exclusion effects intervene, only erroneous or apparent ΔG_A^{sp} will be obtained and, consequently K_A and K_B values will also be affected. For that reason, and also the fact that ΔG_A^{sp} may be obtained using various approaches, it is not of great interest to tabulate K_A and K_B values. Yet, having specified the previously evoked items, the determination and comparison of acid–base parameters is of major value for the understanding and control of the surface properties of powders and fibers.

2.3 NONCONVENTIONAL IGC-ID MEASUREMENTS

2.3.1 Specific Surface Area Determination

The determination of the surface area of coarse powders and fibers, having small specific surface area values, is difficult using classical adsorption methods. The method proposed by Jagiello[110] is, on the one hand, based on the linear relationship (alkane line) between the specific retention volume V_G (the net retention volume of carrier gas, under normal conditions, per unit of weight of solid inside the column) and the number of carbon atoms of the probing n-alkanes, and on the other hand, on the application of Henry's Law, since the adsorption occurs at infinite dilution conditions. The basic equation describes the "alkane line":

$$\log V_{G,n} = n \log V_{CH_2} + C \tag{2.35}$$

where $\log V_{G,n}$ is the specific retention volume of an n-alkane having n atoms of C. Furthermore, according to Henry's Law:

$$V_{G,n} = k_n S \tag{2.36}$$

k_n is Henry's Law constant and S the surface of the solid inside the column. Hence,

$$\log V_{G,n} = \log k_n + \log S \tag{2.37}$$

It follows that:

$$\log k_n + \log S = n \log V_{CH_2} + C \tag{2.38}$$

$$\log k_n = n \log V_{CH_2} + C - \log S = n \log V_{CH_2} + \log k_0 \tag{2.39}$$

$$\log V_{G,n} = n \log V_{CH_2} + \log k_0 + \log S \tag{2.40}$$

The constant C is equal to $\log V_{G,0}$ where $V_{G,0}$ would correspond of the specific retention volume of an hypothetical hydrocarbon having zero C atom and interacting with the surface with zero adsorption energy. The specific retention volume of such a gas is independent of the adsorption system and temperature as is the corresponding k_0. It follows that

$$\log V_{g,n} = n \log V_{CH_2} + \log A \tag{2.41}$$

by supposing that:

$$\log A = \log k_0 + \log S \tag{2.42}$$

$\log A$ is simply obtained together with $\log V_{CH_2}$ from the plot of $\log V_{G,n}$ versus n. This method has been verified on a series of mineral oxides of known surface areas and has been applied to the measurement of the surface area per length of tubing, developed by the surface oxides present on the inner surface walls of metallic tubing, an estimation that can be hardly made using any other method.

2.3.2 REVERSED-FLOW INVERSE GAS CHROMATOGRAPHY

In Reversed Flow Gas Chromatography (RF-GC),[111,112] the chromatographic process, being in a steady-state situation, is perturbed so that it deviates from equilibrium for a short time interval and, then after, it is allowed to return to its original equilibrium state. This is analogous to a relaxation technique. The perturbation is caused by a change of the direction of the carrier gas flow using a four ports valve as shown in Figure 2.15. This change permits to evaluate the gas phase concentration of a solute as a function of time.

The experimental arrangement necessitates a small modification of a commercial gas chromatograph, including a special cell placed inside the chromatographic oven. This T-shaped cell comprises diffusion and sampling columns and a gas injector port. The solids of interest are placed in the vicinity of that port. Small amounts of solutes are injected at the base of the solid bed. They successively adsorb and desorb from the solid surface and penetrate the diffusion column. Then, they migrate towards the sampling column and are carried to a highly sensitive GC detector. By making repeated carrier gas flow reversals (sampling of solutes leaving the diffusion column), of short duration (10–60 s), by means of the four-port valve, one obtains on the recorder a series of narrow, symmetrical peaks depending on the time t at which the flow rate was inversed. A typical chromatogram is illustrated on Figure 2.16.

The use of such an arrangement offers many advantages, in particular in the field of heterogeneous catalysis. Some of the measurable quantities are: chemical kinetics parameters in heterogeneous and homogeneous catalysis,[113,114] mass transfer and partition coefficients across gas–liquid and gas–solid boundaries,[115–117] thermodynamic and kinetic parameters for the interactions of gases and liquids with polymers,[118,119] overall, and differential experimental isotherms of adsorption.[120] The local adsorption rate constant, the desorption rate constant, the surface reaction rate constant,

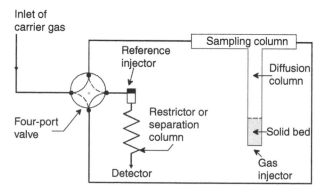

FIGURE 2.15 Experimental setup for reverse flow gas chromatography (RF-GC) measurements.

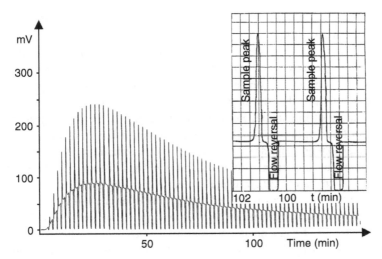

FIGURE 2.16 Typical RF-GC peaks (carrier gas: N_2; injected solute: propene; investigated solid: TiO_2).

the deposition velocity, the reaction probability, and the apparent gaseous reaction rate constant were measured, together with a simultaneous determination of the isotherm, for the adsorption of gases on solid surfaces.[121] The local adsorption energies, the local monolayer capacities, the local adsorption isotherm, and the adsorption energy distribution functions, for adsorption of gases on heterogeneous surfaces, may also be measured by time-resolved surface heterogeneity studies.[122] Recently, three types of adsorption sites of gases on heterogeneous surfaces were detected.[123] The energy of lateral molecular interactions and the surface diffusion coefficients for adsorbed species are also measurable using this time resolved procedure.[124,125] Molecular and dissociative adsorption, as well as the selective oxidation of, say, CO on Pt–Rh catalysts, may be evaluated.[126] Reverse flow gas chromatography is a most versatile method susceptible to find· applications in numerous dynamic systems and situations involving powders and fibers.

2.3.3 IGC-ID at Low Temperature

Operating IGC at low temperature presents some advantages, in particular, the possibility to use small-sized molecular probes, such as Ar, CH_4, N_2, CO, etc.[127] Those molecules should lead to a better "resolution" of the solid surface properties. So far, only exploratory studies were made some time ago.

As an example, Figure 2.17 illustrates the variation of log V_n with the reciprocal of temperature.

Measurements were performed on a silica sample on which C16 alkyl chains were grafted by esterification of the surface silanol groups. Three solutes were chosen: argon (a neutral probe), nitrogen (a polar probe), and methane. It is seen that the slopes of the lines (adsorption enthalpies) depend on the nature of the solute. Furthermore, several line breaks suggest diverse changes of adsorption processes or grafted chain behaviors. Now, with pure silica, such phenomena are not observed. A systematic study, comparing a series of silica samples, modified with alkyl groups differing by their number of C atoms (from 3 to 20), was performed. According to NMR examinations on poly(ethylene), it was proposed that the break evidenced by at the lowest temperature may be attributed to the freezing of the end methyl group's movements. The breaks positions, appearing in the intermediate temperature range, change with the length of the grafted chain and are associated with crankshaft movements freezing of the grafted chains. This example demonstrates at least that IGC allows evidencing the existence of events, in a given temperature range where interesting phenomena do appear, that deserve a more detailed investigation using independent physicochemical methods such

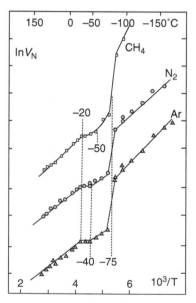

FIGURE 2.17 Variation of log V_n with $1/T$ for various solutes on silica grafted with C16 alkyl chains.

as NMR. Grafted or physically deposited polymer chains, on various substrates, were also investigated by IGC. IGC readily detects phase transition phenomena of glassy polymers related to chain dynamics.[128]

2.3.4 IGC-ID OF EMPTY GC COLUMNS AND METALLIC TUBING

The determination of the actual surface area value and of the interaction capacity of the oxides at the inner side of metallic tubing is of major importance for the estimation of the water adsorption capacity and of other contaminants. This information is important for high-tech applications, such as the transport of ultra-pure gaseous reagents in the electronic industry. Current gas adsorption methods are not sufficiently sensitive to detect those properties. IGC-ID renders that estimation possible, even on empty metallic tubes. Indeed injected *n*-alkanes and other probes are sufficiently long-retained to become separated and detectable at the outlet of the column. Thereafter, from the retention volumes, surface properties are evaluated in the way described above.[40] As an example, Figure 2.18 compares the chromatograms obtained on passivated (oxygen) stainless steel tubing and electropolished (chromium oxide) tubing used for high purity gas connections.

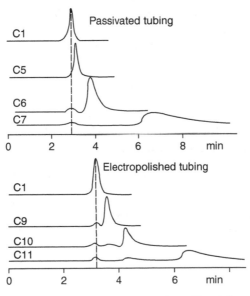

FIGURE 2.18 Separation of *n*-alkanes on "passivated" stainless steel tubing and "electropolished" tubing.

Both treatments were applied to obtain stable, nonporous, corrosion resistant, oxide layers. Whereas, the passivated layer exhibits a γ_S^d value (at 40°C) of 105 mJ/m^2, the γ_S^d of the electro-polished one amounts to 49 mJ/m^2 only. The specific interaction of some selected IGC probes (Isp) are collected in Table 2.6.

Important differences are noted between the tubes demonstrating the remarkable detection sensitivity of IGC. Furthermore, the linear inner surface area values of the tubes, determined with alkanes, are significantly higher than the geometrical surface areas of the tubes, expressed per unit of length (m^2/m). The surface developed by the inner oxide layers, (evaluated according to Jagiello's method) represents about 35 times the geometrical (S_g) surface area of the passivated tubing ($S_g = 0.22$ m^2/m) and 9 times the one ($S_g = 0.058$ m^2/m) of the electro-polished tubing. Once more, these results highlight the remarkable sensitivity of IGC and its interest for applications.

2.4 INVERSE GAS CHROMATOGRAPHY AT FINITE CONCEN-TRATION CONDITIONS (IGC-FC)

Kiselev and his coworkers[129] practiced intensively IGC-FC that offers a convenient way for the recording of adsorption and desorption isotherms of a multitude of solutes, in a large range of measurement temperatures. The interpretation of the isotherms leads to a wealth of information: specific surface area, enthalpies of adsorption in relation with surface coverage, and the evaluation of the surface energetic heterogeneity.

TABLE 2.6
Specific Interaction Energy (kJ/mol) Measured on Metallic Tubing (at 40°C)

Probes	Passivated	Electropolished
Chloroform	n.r.	n.r.
Benzene	1.8	n.r.
Di-ethyl ether	n.e.	n.r.
Methanol	n.e.	n.e.
Ethanol	n.e.	2.9
Propanol	n.e.	3.3
Butanol	n.e.	3.1

n.e., noneluted at 40°C; n.r., nonretained comparative to methane.

2.4.1 DETERMINATION OF THE ADSORPTION ISOTHERM

Several methods are proposed in the literature and reviewed in books, in particular by Conder.[13] The simplest and the most efficient method, from the point of view of the analysis duration, is "the elution characteristic point method" (ECP) that allows the acquisition of part of the desorption isotherm from a unique chromatographic peak. Figure 2.19 shows a typical example of a favorable situation where the peaks corresponding to increasing amounts of injected solute superimpose.

From the rear side of the GC peak, part of the adsorption isotherm is calculated. Each point corresponds to a given value of the solute vapor pressure (in relation with the height of that point) and the amount adsorbed is related to the shaded area shown on Figure 2.19.

The fundamental equation for the isotherm determination is:

$$\left(\frac{\delta N}{\delta P}\right)_{l,t_r} = \frac{1}{RT}\frac{JDt_r'}{m} = \frac{V_n}{m} \tag{2.43}$$

where N is the number of adsorbed molecules, P the pressure at the output of the column, l the column length, t_r' the net retention time corresponding to a characteristic point of the rear diffuse profile of the chromatogram, V_n the corresponding net retention volume, J the James-Martin[130] correction factor taking into account the compression of the gas inside the GC column, D the output flow rate, m the mass of adsorbent, R the gas constant for an ideal gas, and T the absolute temperature at which the measurement is made.

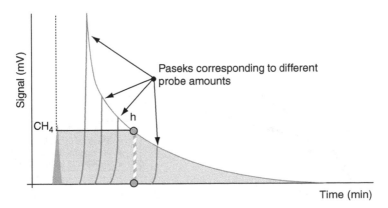

FIGURE 2.19 Typical superimposed GC peaks defining the rear profile of a chromatography peak.

The integration of the previous equation leads to:

$$N(P) = \frac{1}{m} \frac{S_{\text{ads}}}{S} n \qquad (2.44)$$

where S is the area of the GC peak (μVs), n the injected amount in moles

$$P = \frac{RT}{D_{\text{cor}}} \frac{n}{S} h \qquad (2.45)$$

where D_{cor} is the corrected flow rate ($\text{m}^3\,\text{s}^{-1}$) and h is the height of the characteristic point as shown on Figure 2.19. As an illustration, adsorption isotherms on a carbon black sample before and after heat treatments (graphitization), using two solutes, are shown on Figure 2.20.

The amount of adsorbed pyridine largely exceeds the adsorbed heptane amount, in the given relative pressure interval. Upon heat treatment, both amounts are lowered consecutively to a specific surface area diminution caused by the elimination of the surface nano-rugosity, as indicated by the morphology index variation.

Table 2.7 contains additional results obtained on other carbon materials: carbon blacks, graphite, and fullerene samples. The results obtained from the interpretation of the heptane adsorption isotherms concern the specific surface area, BET constant, and Henry's constant (related to the enthalpy of interaction near zero surface coverage).

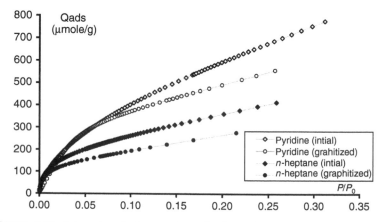

FIGURE 2.20 Adsorption isotherms determined by IGC. Comparison of n-heptane and pyridine adsorption isotherms on initial and graphitized carbon black samples.

TABLE 2.7
Specific Surface Area, BET, and Henry's Constants Determined on
Several Carbon Samples

Samples	S (m^2/g)	C_{BET}	C_{Henry}
Carbon black A	54	15	2.9
Carbon black B	49	26	4.2
Graphite A	1.7	10	0.06
Graphite B	5.2	27	0.48
Fullerene	1.1	9	0.02

The selected rubber reinforcing carbon black and graphite samples are those of Table 2.4. The fullerene C60 sample was purchased from Technocarbo (France).

The surface areas calculated from the IGC adsorption isotherms of heptane are in agreement with values calculated from N$_2$ adsorption isotherms, measured at liquid nitrogen temperature. The BET constant value is, at first approximation, in relation with the heat of formation of the monolayer, giving thus an overall indication of the adsorption energy. As expected from γ_s^D determinations (in the range of 170–200 mJ/m^2 at 180°C), the carbon black samples are by far the most interactive with alkane probes. Fullerene, on the opposite, is rather inactive ($\gamma_s^D = 57$ mJ/m^2 at 39°C) towards alkanes. The Henry's constant relates mostly to the most active sites that are first occupied during the adsorption process. Significant differences between the graphite samples are possibly connected with structure defects or mineral impurities introduced on the graphite surface during the grinding process.

A most important advantage of IGC is the possibility of obtaining directly the value of the first derivative of the adsorption isotherm as indicated by Equation 2.43. This will allow a quantitative evaluation of the energetic surface heterogeneity of powders and fibers.

2.4.2 ESTIMATION OF THE SOLID'S SURFACE HETEROGENEITY USING ADSORPTION ENERGY DISTRIBUTION FUNCTIONS

Actual solid surfaces comprise a variety of adsorption sites (i.e., sites that behave differently when they are contacted with a solute). The global adsorption energy is made of the contribution of all sites. Hence, a possible way for the characterization of the energetic status of the solid's surface consists to elaborate adsorption energy distribution graphs (graphs that display the number of sites, having discrete adsorption energy values, and the corresponding values of the adsorption energies).

The sites may be arranged in a random way or may be clustered in patches on the solid's surface.

All theoretical approaches described in the literature[131,132] for the evaluation of the distribution functions of the adsorption energies are based on a physical model that supposes that an energetically heterogeneous surface, with a continuous distribution of adsorption energies, may be described, in the simplest way, as a sum of homogeneous adsorption patches. Hence, the amount of adsorbed molecules (probes) is given by an integral equation:

$$N(P_m, T_m) = N_0 \int_{\varepsilon_{min}}^{\varepsilon_{max}} \theta(\varepsilon, P_m, T_m).\chi(\varepsilon).\delta\varepsilon \qquad (2.46)$$

where $N(P_m, T_m)$ is the number of molecules adsorbed at pressure P_m and at the temperature of measurement T_m, N_0 is the number of molecules needed for the formation of the monolayer, $\theta(\varepsilon, P_m, T_m)$ is the local isotherm, ε the adsorption energy on a given site, and $\chi(\varepsilon)$ is the adsorption energy distribution function (DF) of the sites seen by the probe. The range of adsorption energies is included between minimal (ε_{min}) and maximal (ε_{max}) adsorption energy values.

From a mathematical point of view, solving the former integral equation is not a trivial task because it has no general solution. The simplest way to solve this equation is to consider the condensation approximation θ_{CA} instead of the Langmuir isotherm as kernel of the integral equation. The condensation approximation supposes that the sites of adsorption of given energy are unoccupied below a characteristic pressure and entirely occupied above it.

The distribution functions using the condensation approximation, adsorption energy distribution functions (DFCA), are directly related to the first derivative of the isotherm which can be easily obtained from the desorption profile of the chromatographic peak, as stated before. This approximation is all the better when the temperature of measurement approaches the absolute zero. But, the usual chromatographic measurement temperatures are far away from those conditions. The actual distribution function can nevertheless be approached by using the extended approximation of Rudzinski-Jagiello[133] that assumes the knowledge of the even derivatives of the DFCA.

There is no problem for solving the adsorption integral equation. There is a general agreement that regularization/smoothing methods are appropriate for this kind of calculations. Analytical methods[133-138] are elegant, but less practical. With present computers, there is no problem with numerical solutions.[139–141]

The issue, which is still open, is the choice of the so-called local isotherm that represents actual adsorption on different adsorption sites. Other important aspects of adsorption are the effects of geometrical heterogeneity and adsorption in micropores. For instance, a relationship between adsorption energy distribution and micropore size distribution has been evidenced.[142]

The procedure developed by Balard[141] provides an efficient way for the separation of the signal from the experimental noise. This method solves the general DFCA equation using a Fourier Transform approach. Such methods are currently applied when the relevant weak signal is mixed with important background noise. This is often the case when spectroscopy methods (FTIR, NMR) are pushed toward their limit of detection. Without going into further details, two examples of adsorption energy distribution curves will be analyzed: the first concerns the carbon samples mentioned above and the second establishes a comparison of crystalline and amorphous silica samples. Figure 2.21 relates the number of active sites (having a discrete value of adsorption energy) to their adsorption energy. Significant differences amongst carbons show up.

It is of importance to define what is meant by the adsorption site concept. Obviously, this concept is not unique, but depends on the characteristics of the adsorbent/adsorbate couple: the chemical nature, surface structure and nanomorphology; and radius of attraction of a given site (dimension, topology, and flexibility) of the probe molecule. In other terms, this adsorption energy distribution function will not provide a global description of the solid surface heterogeneity, but rather a fingerprint typical of the chosen probe molecule.

As an example, clear differences do appear when comparing the adsorption energy distribution curves of heptane on graphite, carbon black samples and fullerene C60.[143] Yet, carbon blacks and graphite samples show some similitude; the maximum of adsorption energy of the first population of sites is at 18–19 kJ/mole and certainly corresponds to graphene layers. The same samples present also a population corresponding to adsorption sites in the 33–34 kJ/mole adsorption energy range. When comparing together the two graphite samples, a significant increase of the high adsorption energy population is noted after grinding, comparatively to the graphene population. It is argued that this population corresponds to structural defects located on the periphery of the graphene layers, the extent of the lateral surface being necessarily augmented by the comminution of the graphite crystals. Similar adsorption energy curves were observed with muscovite samples submitted also to grinding.[144] Finally, the intermediate population may be tentatively, but reasonably, attributed to the existence of polar surface groups, like the well known carboxylic or hydroxyl groups on carbon black; this number is

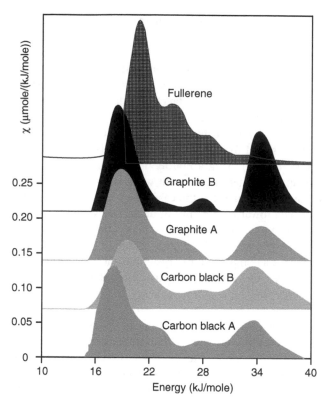

FIGURE 2.21 Comparison of heptane adsorption energy distribution curves determined on various carbon materials.

negligible on the native graphite, but increases significantly after grinding. Globally, these results are consistent with the observations done by STM and the model of carbon black and graphite surfaces proposed by Donnet and Custodéro.[145]

The curves of the fullerene sample, as expectable, are not connected with those of the other carbons. The first populations are possibly related to specific C (footballene) structures of C60. Yet, the chosen IGC probe (heptane) is unable to distinguish the discrete adsorption sites that may be identified on the individual fullerene crystal surface. By comparison with the other studied carbon samples, we may attribute the second population to oxygenated surface sites originating from oxidation of fullerene; the existence of oxygenated groups being demonstrated by ESCA.

The second example (Figure 2.22) compares three silica samples: two amorphous silica samples prepared according to different processes (in the gas

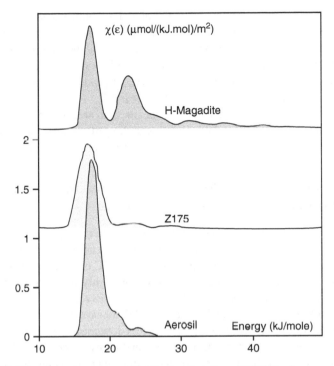

FIGURE 2.22 Adsorption energy distribution functions (DFCA) of octane measured on lamellar silica (H-magadiite), precipitated silica (Zeosil 175M from Rhodia, France) and on pyrogenic silica sample (Aerosil 130 from Degussa, Germany) at 40°C.

phase, Aerosil 130 from Degussa and in liquid phase by precipitation, Zeosil 175 M from Rhône Poulenc) and a lamellar crystalline sample (i.e., H-magadiite).

As expected, the crystalline sample exhibits a bimodal distribution of the adsorption energies of octane, attributed respectively to sites located on the basal (sites of lower adsorption energy) and lateral surfaces (sites of higher energy). The other silica samples show a first maximum close to the one of H-magadiite. By comparison, this family of surface sites corresponds to surface siloxane bridges, whereas the peaks at higher adsorption energy values may be attributed to surface polar groups such as hydroxyl groups. The precipitated silica sample presents a broader distribution, suggesting a lower degree of organization of that surface.

In short, IGC-FC appears as a complementary method of IGC-ID allowing to evaluate the number of surface sites. However, a true identification of the sites is only possible when associating IGC with complementary analytical methods. A comparison amongst samples of a given family, a study of samples issued from a same starting material, but modified in a controlled manner, will allow to make reasonable hypotheses. In all cases, the energy distribution curves may only be taken as "fingerprints" of solid surfaces. This may already be of importance when monitoring, with IGC, a production process or when trying to master a surface modification process.[146]

Obviously, much remains to be done in that area of research; on the fundamental side, to reach a better understanding of the adsorption phenomena and, on the experimental side, to be able to treat correctly the IGC data.

2.4.3 Evaluation of the Number of Strong Adsorption Sites by Combining IGC and Thermodesorption

This approach concerns solids exhibiting strong interacting surface sites, i.e., solids like carbon blacks possessing surface sites, leading to an "irreversible" adsorption process of the probes under ordinary IGC experimental conditions. The method, allowing the evaluation of those highly active sites, or of the irreversibly adsorbed probe amounts, consists of several steps.

First, the sites responsible for "irreversible" adsorption are "covered" with probe molecules up to saturation at a relatively low measurement temperature (at 40°C, for the present example). This is verified by successive injections of given amounts of probe, until reproducible chromatograms are observed.

Thereafter, by applying the usual isothermal IGC-FC method to those reproducible chromatograms, desorption isotherms are computed, from which the extent of surface area (reversible adsorption area) represented by the "reversibly adsorbing sites," as well as the adsorption energy distribution functions, are computed in the usual way.

This procedure is best described by taking the example of a study (nonpublished results from IGClab) of five carbon black samples obtained either by the furnace or the acetylene process. Table 2.8 gathers the main characteristics of those samples.

The carbon black samples exhibit surface areas between 70 and 150 m^2/g. Their γ_S^d values greatly vary going from 129 m^2/g (for a relatively low surface area furnace black) up to 520 mJ/m^2 for a furnace black having a surface area also equal to 150 m^2/g.

The "reversible" specific surface area values calculated from the isotherms are, in some cases, below the values obtained by the BET method using N$_2$, the difference being attributed to the part of surface area already occupied by the strongly irreversibly adsorbing sites. (This part may be evaluated by thermodesorption measurements in the way illustrated below.) Figure 2.23 presents FDCA determined with pyridine at 40°C.

TABLE 2.8
"Reversible" and "Irreversible" Adsorption Surfaces Determined from Pyridine Adsorption Measurements on Different Carbon Black Samples, Combining IGC and TPD

Samples Carbon Black	B1 Furnace	B2 Acetylene	B3 Furnace	B4 Furnace	B5 Furnace
γ_s^d (mJ/m^2)	520±21	200±8	480±20	129±5	144±6
N$_2$ BET area (m^2/g)	150–170	100	100	70	100
"Reversible area" (m^2/g)	140±5	93±3	80±2	66±2	100±3
"Irreversible area" (m^2/g)	17±1	1±1	9±1	1±1	1±1
Total area (m^2/g)	157±6	94±4	89±3	67±3	101±4

Significant differences show up. In general, all samples exhibit at least a bimodal distribution of adsorption sites, differing from one sample to the other by the ratio of adsorption sites displaying low and higher adsorption energies. These variations reflect possible differences in surface nano-structures (size and size distribution at the nano-scale), or in surface chemistry. But those sites belong to the family of "reversibly" adsorbing sites detectable by IGC. The

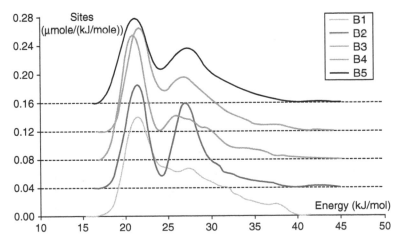

FIGURE 2.23 DFCA of pyridine (for reversible adsorption: classical exploitation of adsorption isotherm).

others "irreversibly adsorbing" sites may not be evaluated by IGC, but through thermodesorption measurements.

The thermal programmed desorption (TPD) protocol was developed referring to the work of Choudhary et al.,[147] using short columns and a high carrier gas flow. Under those conditions, the non-readsorption of the thermally desorbed molecules may be assumed, allowing significant simplifications of the thermogram analysis. The thermal programmed desorption of the irreversible adsorbed fraction was performed between 40 and 300°C, with a 3K/min heating rate. Figure 2.24 exhibits a typical thermogram (Variation of the GC signal, or amount of eluted pyridine, when applying a linear heating rate).

The "irreversibly" adsorbed probe fraction is progressively desorbed as the temperature is increased. The exploitation of the results stemming from the combination of IGC-FC and TPD is based on some simplifications. The first assumes the non-readsorption of the thermally desorbed molecules and a first order desorption kinetics. The second states that the solid's surface is made of domains characterized by discrete desorption energy values (E_D). Accordingly, desorption energies distribution functions may be calculated, providing additional information on the bonding strength capacity or bonding strength distribution of the irreversibly adsorbing sites. Knowing the total amount of thermodesorbed pyridine and the section of an adsorbed pyridine molecule, allows to compute the part of surface area of the carbon black occupied by those highly active sites. Major differences amongst the carbon black samples are now evidenced (Figure 2.25).

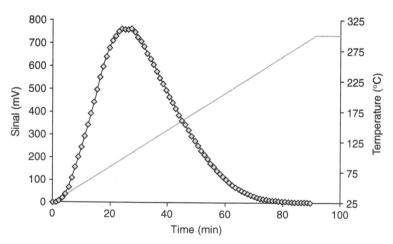

FIGURE 2.24 Example of a thermodesorption signal.

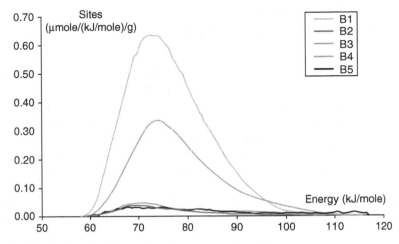

FIGURE 2.25 Typical desorption thermogram of pyridine.

Samples B1 and B3, possessing the highest γ_S^d values, have also the largest number of active sites, and hence the highest values of irreversible adsorbing area, respectively, 17 and 9 m^2/g. It is interesting to underline that addition of the "irreversible adsorbing" and "reversible adsorbing" surface areas, determined by TPD, results in a sum that is close to the BET urface area values determined by nitrogen adsorption. Those measured values are in line with "active surface area" values determined using a totally different approach.

Those types of studies underline the diversity of IGC approaches for the examination of adsorption processes on complex solid surfaces. Making a clever choice of test molecules will even allow a closer insight into the solid's surface characteristics. For instance, an adequate choice of hydrophilic and hydrophobic probe molecules allows to determine the hydrophobic and hydrophilic parts of say, pyrogenic silica samples.[148]

2.5 IGC APPLICATIONS

This part of the chapter is by no means an overview of the published work concerning the application of IGC for academic research and or industrial purposes. It is intended to give to a newcomer some indication of the remarkable potential of IGC to solve or assist solving both fundamental and practical aspects of the powder and fiber science.

2.5.1 ACADEMIC RESEARCH

2.5.1.1 Surface Modification of Powders

Silica Surface Modification with Alkali Ions. Amongst the numerous mineral oxide surfaces, silica has received much attention. Indeed, silica is the preferred support for chromatography, for catalysts and also silica has found a major application as filler for the reinforcement of rubbers. Numerous surface modification methods are described in literature: oxidation, grafting of organic phases to control the adsorption properties, or to enhance catalytic activity, modifications by metal ions. For instance, Milonjic[23,149] did a systematic study evaluating the variation of γ_s^d with the nature of the adsorbed metal ion. Significant variations are observed since the introduction of Li^+ (120 μmol/g), K^+ (270 μmol/g), Cs^+ (450 μmol/g), Na^+ (350 μmol/g), is accompanied by a γ_s^d decrease from 64.5 mJ/m^2 to 31.6 mJ/m^2 (at 100°C). This, of course, has consequences if the silica is used in a hydrocarbon medium. Moreover, the modified silica samples may lead to tailor-made adsorbents with a large range of selectivity.[149]

Trimethylsilylation (Decoration) of Silica. The "decoration" is based on a controlled chemical modification of a solid surface taking advantage of the reactivity and repartition of surface groups, or of the existence of chemical heterogeneities due, on crystalline materials, to the specific location of chemical groups on given crystalline planes or surfaces.[44]

The first case is well documented in the literature.[59] For instance, silica samples were systematically modified, by reacting the surface silanol groups with trimethylchlorosilane. Thereafter, the variation of the surface properties was evaluated by IGC using a homologous series oligomers of poly (dimethylsiloxane) or PDMS, having the general formula $(CH_3)_3Si-O-(Si(CH_3)_2O)_{n-2}-Si(CH_3)_3$ with $n = 2, 3, 4, 5,$ and 6.[150] Why did we choose those oligomers? Simply, because fumed silica is the preferred reinforcing agent of PDMS. Looking at the variation of the incremental free energy of adsorption of a monomer unit of PDMS (since the representative points of the siloxane probe molecules also align on a straight line when plotting the variation of standard free energy of adsorption versus the number of monomer units that constitute the probe), one notices unexpected results. The surface properties change in an abrupt manner when grafting only a limited number of trimethylsilyl groups. The surface energetic heterogeneity is increased. This is indicative of a percolation threshold. This is explained by considering the different adsorption possibilities on the silica surface after grafting. The threshold is reached when enough trimethyl groups are fixed so as to entirely hinder the free migration of the IGC probes on the still available or free surface of the modified silica. After having reached that threshold, adsorbed probe molecules will have to jump from one adsorption site to another that is still accessible. Molecular modeling and estimation (IGC-FC) of the different populations of adsorption sites support this mechanism. In practice, the threshold value determined by IGC is related to rheological behavior of silica/silicone oil mixtures, since the dynamic viscosity diminishes gradually until the threshold value is attained.

Surface Decoration of Talc by Impregnation with Polyethylene Glycol. Talc is a lamellar magnesium silicate showing an evident energetic surface heterogeneity; the lamellar surface is made of siloxane bridges of low interactivity whereas the peripheral surface of the talc lamellae carries both magnesium and silicium hydroxyl groups. Hence, when contacting talc with a polar polymer able to exchange strong hydrogen bonds, such as polyethylene glycol (M_W of 20.000), it is expected that the polymer will be preferentially hold on the lateral surface.[151] This may be verified using IGC. Figure 2.26 exhibits two curves: one corresponding to the variation of γ_s^d with the talc surface coverage (expressed as number of monomer units per Å^2 on the talc, with a specific surface area of 3 m²/g) and the other one illustrating the evolution of the morphology index (determined with cyclooctane).

Four regions may be identified. Region A corresponds to a low surface coverage, a rather high value of γ_s^d (measured at 100°C) and a large surface nano-roughness. This is attributable to the possibility offered to the flexible *n*-alkane probes to become "inserted" between slit-like defects on the lateral

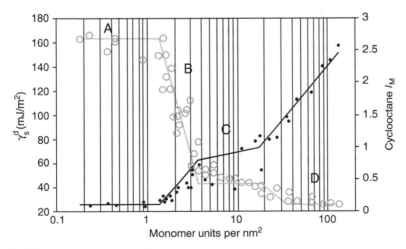

FIGURE 2.26 Variation of γ_s^d with the surface coverage and with the morphology index IM of talc samples progressively covered with PEG.

surface of talc. With increasing surface coverage (Region B), γ_s^d progressively decreases as the lateral surface defects become plugged by the polymer, as indicated also by the increasing IM value that approaches the value corresponding to an entirely accessible surface. Progressively, the basal surface of talc becomes covered with a monolayer of polymer, and finally an additional addition of polymer leads to a rather thick polymer layer, in which the IGC probes may dissolve. Apparent IM values augment consequently to this dissolution process and do no longer keep their physical meaning.[152]

Since the performances of composite materials, made by associating talc and polymers, are largely dependent on that shape factor, it is evident that this IGC procedure is of interest for monitoring applications.

Influence of Humidity on Surface Properties. The influence of adsorbed water on the surface energetic characteristics of powders and fibers has received much attention. Indeed, in practice, humidity is unavoidable in most cases and its presence significantly changes their surface properties and interaction potential. Studying adsorption of organic derivatives, at different relative humidities at 15°C, on quartz, calcium carbonate, and α alumina, Goss and Schwarzenbach[153] noticed an exponential decrease of the adsorption constants (K_S) of nonpolar solutes, between 4 and 97% relative humidity. At 100%, (value obtained by extrapolation), K_S corresponds to the one determined on a bulk water surface. With polar solutes, K_S values decrease also, up to 90% humidity. Thereafter, they augment again, a variation attributed to a change in the orientation of the water molecules on the oxides surface.

Goss,[154] still using IGC, made an interesting contribution for a closer understanding of the complex phenomena that govern the equilibrium of adsorption (quantitative model for the prediction of adsorption constants) and interaction (contribution to the environmental partitioning of organic compounds) in presence of humidity. The progress he made sheds light on environmental concerns and possibly will offer solutions to overcome them: "It provides a good explanation of the variability of more than thousand experimental adsorption constants and it facilitates a comparison of the surface properties of mineral and salt surfaces at various relative humidities with the surface properties of a bulk water surface". Certainly, his findings are applicable in other research domains.

The behavior of adsorbed water on solid surfaces, when increasing the temperature was also studied.

Upon moderate heating, the physically (or reversibly) adsorbed water is removed, but as the temperature is increased, part of the water coming from the oxide surface stems from a condensation mechanism of surface hydroxyl groups. When measuring the total amount of released water, it is not possible to differentiate between the reversibly fixed molecular water and water coming from the condensation of surface hydroxyl groups. Sun and Berg[27] examined high purity MgO, Al_2O_3, TiO_2, SnO_2, and SiO_2 samples, before and after moderate heat treatments (from 50 to 220°C), that produce different degrees of dehydration. Water contents of the treated samples were determined independently by the Karl Fischer titration method. As water coverage decreases, the γ_s^d increases. Oxide surfaces become energetically more heterogeneous. At the same time, the acid–base interaction potential ($-\Delta G_{AB}$) decreases. Compared to water, bare oxide surfaces have higher surface energy, but weaker acid–base interaction capability. Acid–base properties are particularly sensitive to changes in water coverage below a "critical value," below the water monolayer value. Possibly a combination of water and hydroxyl groups constitute, at the "critical value," a complete coverage of the oxide surface.

The thermal treatment of oxides has received much attention.[155] Dehydration of SiO_2,[36] γ alumina,[28] hematite,[29] and nickel oxide[24] were performed in a large temperature range (from about 100 to 600°C). The departure of water upon heating causes important changes of the surface properties, depending on the nature of the oxide.

For silica, a condensation of silanol groups is noted already after a moderate treatment (100°C). This condensation leads to the formation of strained three-membered siloxane rings (D3) whose surface concentration augments up to a treatment temperature of about 450°C. Thereafter, a surface reorganization occurs without measurable water loss; the D3 transform in the more stable four-membered siloxane ring (D4). The situation is quite different when heating either γ alumina or the hematite samples. Apparently, on those surfaces exists a fine layer of hydroxide, formed by an aging process during the storage of the samples at atmospheric conditions, that transforms into oxide upon heating; the γ_s^d values change in the same manner as they do when heat treating pure hydroxides. Moreover, when heating at even higher temperatures, surface reorganization or restructuration occurs at about 600°C. This restructuration is not accompanied by a water loss and results from the surface atom migration activated by heating.

These are just a few examples of application of IGC for the elucidation of the surface characteristics of powders submitted to various treatments.

2.5.1.2 Surface Modification of Fibers

Glass fibers. Glass fibers are currently surface treated to enhance their stability during mechanical processing, but also to increase their ability to reinforce composite materials; familiar processes are the applications of silane derivatives and of reactive polymers.

Already in 1983, Saint Flour and Papirer[156] examined by IGC glass fibers before and after surface sizing with γ-aminopropyltriethoxysilane (γ-APS). The influence of the treatment on the surface properties of the fibers could be clearly evidenced. The improvement of the mechanical performance of the composites made of glass fibers and phenolic resins was attributed to strong acid–base interactions. Since, numerous authors applied IGC for the study of fibers. For instance, Dutschk et al.[52] examined homemade E glass fibers treated either with γ-APS or with maleic anhydride grafted polypropylene (PPpm). When determining γ_s^d by IGC-ID, they did not observe significant variations when comparing the γ_s^d of the initial fiber (34.5 mJ/m^2 at 40°C) with the 0.06% silane modified sample ($\gamma_s^d = 36.0$ mJ/m^2) or with the silane (0.06%) + polymer (1.18%) modified fiber (35.3 mJ/m^2 at 40°C). The discrepancies between the published results possibly stem from the fact that the silane treatment itself depends on a series of parameters (nature of the solvent, silane concentration, pH, and age of the reagent solution). Those factors influence the nature, regularity, and thickness of the deposited silane surface layer. A noncontinuous layer still offers access to the fiber surface to the very few alkane probes that are injected when operating at infinite dilution conditions and this could explain why γ_s^d does not vary. Concomitant, the determination of the acid–base characteristics indicates clear differences. Now, the computed values (K_A and K_D) depend on the chosen acid–base evaluation method, as discussed earlier. Taking as a molecular descriptor, the topology index of the probes, an approach that appears to be more realistic than the other ones, it is seen that the initial fiber is rather acidic, whereas the silane treated one acquired base-like properties; a fact suggesting that the amine groups are still accessible to acidic probes. Also, the fiber treated by silane + polymer reagents, retains more longer the probes due to a partial dissolution of the polar probes in the rather thick surface layer, rendering the evaluation of acid–base parameters questionable.

In another study, polarity parameters of modified glass fibers, determined as above, were compared to results of solvatochromism.[157] The UV/Vis. absorption peak maximum of selected dye molecules is shifted when they are brought into contact with acid and/or base molecules Empirically derived equations relate the spectroscopic shift to the interaction of say, acid–base groups of a solid surface. The sizing of the fiber with γ-APS significantly decreases the surface acidity. Sizing the fiber with Ppm causes an increase of the surface acidity; this could not be seen by IGC for the reasons given above. Even though solvatochromism does not allow making a quantitative determination of acid–base properties, it provides additional information to the IGC results.

Cellulose fibers. Cellulose fibers are also interesting fillers for thermoplastics. The resulting composite material has the advantage of an easy elimination after its utilization, since it is entirely combustible. However, the high degree of dispersion required to achieve satisfactory performance of the composite is not always obtained, owing to the rather hydrophilic nature of the fibers and the hydrophobic nature of the organic matrix. Hence, surface modification methods need to be developed. Riedl et al.[158] modified cellulose fibers (bleached sulfite pulp) by esterification with different (C11 and C18) fatty acids at low and high esterification degrees. X-ray photon spectroscopy indicates that a gradient of esterification towards the center of the fiber occurs and that the surface is not entirely covered by the ester. Acid–base properties evaluated by IGC are intermediate between those of untreated fiber and polyethylene. Composites were prepared by compounding linear low-density polyethylene with unmodified and esterified fibers as reinforcing fillers. Composites made with fibers having the lowest values of γ_s^d (oleate modified fiber) gave the best mechanical properties, as expected. This study demonstrates that performing IGC, even with natural products (complex in structure, exhibiting amorphous and crystalline zones, hydrophilic, susceptible of swelling during treatment), leads to interesting information, even though all recorded data are not immediately exploitable or interpretable applying our present knowledge of IGC.

2.5.2 APPLIED RESEARCH

2.5.2.1 Airplane Construction and Space Industries

The ability of IGC to detect minor changes of the surface properties may be used to verify the stability and or reproducibility of powder and fiber surfaces. This is of importance, especially for products that may undergo an aging process during their storage: aging process caused by oxidation; surface carbonatation upon exposure to humid air; rearrangement of surface groups when possible, so as to minimize the surface energy; evolution, due to the fact that the manufactured material is not in an equilibrium state due to an abrupt quenching; etc.

IGC is routinely used as a surface characterization tool[47] for industrial carbon fibers, especially in the airplane construction and space industries. The characteristics are dependent on production parameters and quality of starting materials. Combining XPS with IGC often reveals the origin of manufacturing deviations. Even small variations of the acid–base characteristics may influence the adhesion of the fibers, and hence the performance and lifetime of composites comprising those fibers. Due to its capacity to readily produce the relevant information, IGC appears as a most suitable method to monitor the quality and reproducibility of industrial fibers.

2.5.2.2 Ceramics: Influence of Traces of Mineral Impurities on Surface Properties

The existence of mineral impurities has major influences on the surface characteristics of ceramics (and on their applications). This is illustrated in Table 2.9 on γ alumina samples (samples A, B, C), stemming from the current industrial process of the alumina used for the manufacturing of aluminium, having trace amounts of Si contaminants on their surface. It is demonstrated by minor changes of the composition of barium titanates (samples D, E, F, G).

The major mineral impurities of alumina samples are Si, beside traces of CaO, Na_2O, and MgO. The titanates contain also traces of Sr, Ca, and Na impurities. It is seen that γ_s^d significantly decreases with increasing yields of Si impurities. At the same time, the specific interaction capacity is augmented. In other terms, traces of mineral impurities will strongly affect the interaction potential, as evidenced by IGC-ID.[159] This offers also a rapid and convenient way to control the reproducibility of a process. But on the fundamental side, it points also to the absolute necessity of better defining the origin when trying, for instance, to compare samples having a global denomination.

2.5.2.3 Polymer Industry (Fillers)

Quality control. McMahon et al.[160] propose to apply temperature programmed-multiple probe IGC to compare carbon black samples that are difficult to characterize with other methods. Existing methods are time consuming, a requirement that is not acceptable for industrial processes where control of reproducibility is essential if strictly controlled surface

---

TABLE 2.9
Characteristics of Si Containing γ Alumina (A, B, C) Samples and Barium Titanates (D, E, F, G) Samples

Samples	S (m²/g)	SiO₂ ppm/Al₂O₃	γ_s^d (mJ/m²)	Isp CH₂Cl₂ (kJ/mol)	Isp Ethyl ether (kJ/mol)
A	10.6	45	100	4.2	2.1
B	9.3	630	65	5.1	3.1
C	0.7	1050	42	5.6	6.9
D	2	0.95	118	7.0	30
E	2.1	0.99	102	7.1	28
F	1.5	1.05	88	4.5	18
G	2.7		58	5.7	23

properties are a factor of performance. The proposed method is inexpensive and allows the examination of surface probe interactions even in the presence of background gas (i.e., in conditions closer to those in use in industry). Practically, a cocktail of selected probe molecules is injected to provide a chromatographic fingerprint of the surface. Two difficulties must be overcome: the selection of adequate probe molecules, and the possibility to control the interactivity of the IGC column filling.

The selected probes should not remain chemisorbed on the surface in the conditions of the IGC thermodesorption step. The probe selection should also cover a large range of functionalities (alkanes, nitriles, ketones, etc.). In the present case, 25 probes were chosen to examine carbon blacks. Kovats indices were determined that facilitate the choice of the probes for making IGC measurement in the right conditions.

The second problem concerns the adsorption capacity of the IGC support. Some carbon blacks are too interactive and a direct IGC comparison becomes problematic. A solution would be to mix them with an inert diluent. This would allow comparing solids having different surface areas, since what really counts, for parent supports, is the total surface area of the IGC supports present in the column. One possibility is offered by addition of surface deactivated silica. A useful method for comparing the experimental results is to plot the retention data against the boiling point of the probes.

The method does not yield quantitative thermodynamic adsorption parameters. Nevertheless, when having defined, in the laboratory, acceptable IGC conditions, the proposed method may then be safely transferred to the plant.

Modification and characterization of CaCO₃. Powders and fibers are currently used[161] to increase the performance and the time life of say, manufactured tires, filled thermoplastics, etc. The reinforcement of rubbers filled by carbon blacks and/or silica is described in the chapter of M.J. Wang. Other important industrial fillers are calcium carbonate, talc, clays, etc. Calcium carbonate is currently used as filler for thermoplastics,[162] and for papers and coatings for its optical and physical properties and relative low cost. As mentioned above, strongly adsorbed water strongly decreases γ_s^d. The presence of water and its liberation during the elaboration of filled polymer materials is highly deleterious, since this will create micro-holes (i.e.,defects that lower the mechanical performances of filled polymers). Surface treatment, which may be followed by IGC, is one way to understand and master the surface characteristics. CaCO₃ it is currently surface treated[163] so as to limit the water uptake, increase its ease of dispersion in a given matrix and to improve its adhesivity.[164] IGC appears as a most convenient method[165] to monitor the modification process and to assess the surface characteristics.

2.5.2.4 Composite Materials

A composite material is made of a resin and fibers (glass, carbon, polyaramid, and cellulose). IGC was applied to the study of glass fibers already in 1982.[53] IGC is particularly well-adapted to the study of surface modification that might even be performed inside the column or applied in the usual way during the processing of the fibers. Moreover, glass fibers are subjected to aging since they are, at the start, in nonequilibrium conditions. In the extreme surface layer ions will migrate whenever possible (upon heat treatment for instance), and some may react with the surrounding atmosphere (carbonation of Ca ions). All those processes may be followed by IGC.[166]

As indicated above, IGC has been extensively applied to evidence possible correlations between surface characteristics and wetability, adhesivity, interface formation and finally, mechanical performance of composite materials. Nardin et al.[49,167,168] for example, related the results of IGC surface characterization and interaction capacity of PAN fibers and of PEEE K, poly(ether–ether–ketone), with interfacial shear strengths of the fiber-resin interface, determined by fragmentation tests on single fiber model composites.

The surface chemistry and acid–base properties of PAN-based carbon fibers (Akzo Tenax 5000) were examined by Vickers et al.,[50] combining XPS, ToF-SIMS, and IGC. A comparison of the initial fibers and samples modified by electrochemical oxidation indicates that this treatment diminishes γ_s^d from 104.5 mJ/m^2 (a rather high value corresponding to a nonpolluted surface) to 78 mJ/m^2, typical of a surface rich in hydroxyl and carboxyl groups. XPS shows a steady increase in surface O content and SIMS detects the presence of mineral ions (Na, Cl, and Ca) stemming from the electrochemical oxidation. The increase of basicity and acidity (evaluated from the acid–base interaction energies taking chloroform and THF as reference acid and base probes) correlates well with the increase in heteroatom contents detected by XPS and SIMS, underlining once more the interest of combining IGC with adequate and complementary analytical tools.

The interfacial bond strengths in epoxy/resin composites depend on the kind of the sizing formulation and properties, and on the aging conditions.[169] A combination of analytical methods (Zeta potential, IGC, AFM, and wetting techniques) is able to reveal the characteristic parameters of the differently-aged glass fiber surfaces and enables to predict interfacial adhesion. The interface characterization, using pull-out and cyclic loading tests of epoxy/glass fiber composites, evidences the aging effects on the fiber surfaces and is in a good agreement with the surface characterization results, in particular those provided by IGC.

2.5.2.5 Carbon Industry

Active carbons. Active carbons are currently used as catalysts[170] for the purification of drinking water, for the atmosphere depollution, for the gold recovery, etc. Their adsorption power stems from their particular texture (micropores and mesopores) and surface characteristics. IGC has been used, since it offers a most convenient way to qualitatively detect surface peculiarities,[171] but also for plotting adsorption isotherms from which important information (adsorption enthalpy, entropy, and adsorption energy distributions) is extracted.[172] Bandosz et al. examined the adsorption of various solutes: acetaldehyde,[173] H_2S,[174] water, methanol, and diethyl ether,[175,176] trying, in particular, to estimate the contribution of surface chemistry in the adsorption process.[177] Whereas the major source of adsorption sites is located in micro- and meso-pores, surface functional (especially oxygenated groups) intervene. However, surface oxidation, by nitric acid for example,[84,178] may either hinder or perturb the occupancy of adsorbed molecules on their energetically most favorable position. This depends on the chemical nature of the molecules and on the active carbon interaction capacity determined by hydrophobic graphene groups and oxygenated hydrophilic groups allowing acid–base or polar type of interactions.[170] In other terms, when manufacturing an active carbon for special adsorption processes, one should consider both porosity and surface chemistry.

Carbon blacks. Carbon blacks are produced on a large scale and are mainly employed to enhance or reinforce the performance of rubber vulcanizates.[179]

Amongst the important characteristics that determine the reinforcing potential of carbon black, figures obviously the interaction capacity of the filler, a characteristic most difficult to quantify even with modern sophisticated methods. IGC, as already underlined, appears as a unique method to describe such properties.

Moreover IGC may be standardized so as to become an appropriate tool for the control of a carbon black manufacturing process.

2.5.2.6 Printing Industry

Printed-paper is, as many other materials, submitted to an aging process that alters its physical and optical qualities. Furthermore, this aging process has consequences on the recycling of paper products and may complicate the de-inking of printed papers. Offset inks contain vegetable oils (Soja bean, linseed oil) and alkyl resins that adhere strongly to the paper, for instance, after atmospheric oxidation, especially in summer time. Model experiences of oxidation may be conducted on GC supports impregnated with those natural compounds using a carrier gas containing oxygen and controlled humidity. Oxidation rates and activation energies may be readily obtained by IGC. The determination of Kovats indices provides information on the variation of surface polarity upon aging. Also, solute–oil interaction measurements, determined from heats of mixing, will lead to a better choice of possible solubilizers of the oil residue after oxidation.[180]

The printability of two commercial oil proof papers was also examined[181] by IGC to establish a correlation between their surface properties and their printability. The measurements showed substantial differences between the dispersive components of the surface energy of the two samples. Thus, paper A was found to have a γ_s^d three times higher than paper B. The polar characteristics of the two papers showed the same tendency. The data obtained from contact angle measurements were in good agreement with those determined from IGC data. Laboratory tests of printability, as well as those carried out using an industrial flexography press, showed that paper A displayed, as expected, better printability. The observed difference could be attributed to the presence of fluorine resins in both thickness and at the paper surface.

Also, model xerographic toner particles (pure polyester, polyester particles filled with magnetic iron, the same with a shell of C particles) were studied combining atomic force microscopy, that allows to measure the force an isolated particle feels when it is brought into contact with a solid surface, and ICG that provides further information on interaction potential at the molecular scale. Useful information is recorded, but a quantitative evaluation of the adhesion of toner particles to a surface remains difficult. The main reasons are: a broad distribution of adhesion forces and an undefined contact area due to particle size distribution, etc. Nevertheless, the combination of IGC with the possibility to directly evidence nano-morphologies on solid particle surfaces is much promising.[182,183]

2.5.2.7 Wood Industry

A better knowledge of lignocelluloses fiber surfaces will facilitate the development of wood–paper and wood–polymer association. Acid–base properties of wood were evaluated using contact angle measurements and IGC. The comparison of nonextracted and solvent extracted (Soxhlet extracted with ethanol, acetone, petroleum ether, and deionized water) wood samples shows that the acid–base properties depend largely from the wood extractives.[184] Those extractive constituents are generally hydrophobic in nature.[56] This, of course, will have consequences, since the existence of such a weak surface layer will handicap the adhesion properties of wood. Gardner et al.[185] examined wood particles using liquid contact angle (column wicking) measurements and IGC. Unfortunately, the electron donor–acceptor properties given by the two approaches cannot be directly compared. Possibly, the solvent–wood extractible products complicate the analysis of the wicking experiments. However, γ_s^d determined by both methods are in satisfactory agreement. The surface properties' modifications of wood pulp fibers, during wet–dry–rewet recycling, were also investigated using the techniques mentioned above.[186] Contact angle measurements show that the overall effect of such treatment is an increase of the nonpolar component of surface energy, an increase that is not really confirmed by IGC. The decrease of the polar component is accompanied by a decrease in the number of surface hydroxyl determined by acetylation, whereas IGC does not distinguish clearly between virgin and recycled fibers. This is surprising at first sight. In fact, the contact angle measurements technique is a macroscopic method that concerns the totality of the fiber surface, whereas IGC is a molecular approach restricted, as far as the nonpolar component is concerned, to the most active sites. Their number has changed according to the acetylation results, but not necessarily their quality. Furthermore, water plays a known role by promoting the orientation of polar groups, whereas dehydration in the chromatography column leads to the opposite effect.

2.5.2.8 Environment

Effect of pollutants on cultural heritage. RF-GC was applied to understand and to control the damage caused by air pollutants (SO_2, NO_2, etc.) on marbles and monuments. The kinetics of the action of SO_2, NO_2, and $(CH_3)_2S$ on pure $CaCO_3$ and marbles, as well as on surfaces of cultural and artistic interest, were established. The determined physicochemical data are, on the one hand, the rate constants of adsorption and desorption phenomena and chemical reactions of the gaseous pollutants with the solid surface, together with the coefficient of slow diffusion of the pollutants into the pores of the solid. In addition, deposition velocities and reaction probabilities for the action of air pollutants on the same surfaces were calculated. On the other hand, adsorption energies, local monolayer capacities, local isotherms, and probability density functions for the adsorption energies during the deposition of the pollutants on the heterogeneous surface of the cultural heritage objects were determined.[187] Synergistic effects of two gaseous substances or the effects of airborne particles deposited on the objects under investigation were also assessed from the above physicochemical parameters. Using all those parameters allows clarifying the action of air pollutants on cultural heritage objects. This places the actions to be taken by conservators and museum curators on a scientific basis, and not on empiricism.

Soil pollution. Steinberg et al.[188] studied the sorptive properties of soils, minerals, and organic matter, in order to reach a better understanding of vapor movement (contaminant migration) in soil, since, for instance, gaseous flux from facilities such as underground storage tanks and pipelines may provoke irreversible damage to the groundwater. For this purpose, they applied both classical static adsorption techniques, but also IGC to determine the adsorption isotherms of a series of volatile organic compounds (VOCs) on quartz,[188] calcareous soils,[189] and desert soil.[190,191] The extent of VOC adsorption depends on the soil conditions and on the characteristics of the contaminants. Soil organic contaminants dominate the sorption process (of VOC) in water-saturated soils. In dry soils, the sorption is primarily a function of the available mineral surface and water vapor. The effect of cationic surfactant treatment of the soil was also examined, and it was shown that such treatment attenuates the vapor phase migration of volatile hydrocarbons in low organic carbon desert sand. These results were confirmed by Garcia-Herruzo et al.[192] They established relationships between the retention properties of aromatic hydrocarbon and relative humidity. IGC appears as a most suitable method for performing this type of investigation, delivering readily adsorption isotherms, allowing to easily change the measuring conditions (temperature of preconditioning, temperature of measurement, humidity control, and choice of adsorbates).

2.5.2.9 Pharmaceutical Applications

The determination of the surface energy of pharmaceutical powders by IGC has gained much attention[193] in recent years. A few examples will illustrate the unique value of IGC in that domain.

Aerosolisation behavior of micronized and supercritically-processed powders. Supercritical fluid technologies, in particular solution-enhanced dispersion by supercritical fluid technology (SEDS), allow the preparation of micro-fine drug powders with the desired physical and surface properties for pulmonary delivery. Compared to the traditional sequential batch crystallization and fluid energy milling, SEDS provides highly pure, crystalline, and solvent free, uniform-sized particles. Chow et al. examined, by IGC, Salmeterol xinafoate (SX) particles obtained by SEDS.[194] SX may be produced in two polymorphs that are enantiotropically related. SX1 is, at ambient temperature and pressure, the thermodynamic stable form and SX2 is the meta stable form. SX1 has the larger γ_s^d, whereas the specific component is higher for SX2. The latter has been attributed to the existence of polar surface groups (–OH, –COOH and –NH) that are more exposed on SX2. IGC measurements were performed at 30, 40, 50, and 60°C, allowing the calculation of the standard free energy variation, the enthalpy and entropy of adsorption, in order to verify if entropy needs to be taken into account for the evaluation of the results. Indeed, the entropy plays an important role in the free energy change, particularly for the meta-stable form SX2. In that example, IGC detects subtle aspects, allowing to clearly distinguish between the two polymorphs of SX. Possibly, the application of molecular computer simulation would have been an interesting complement to this study. The evaluation of Hansen's solubility parameters points to the predominant role of H bond formation between polar probes and the SX surface.

Chow et al.[195] further investigated, by IGC at finite concentration conditions, four salmeterol xinafoate samples. From the octane (nonpolar IGC probe), chloroform (acidic probe), and tetrahydrofuran (base like probe) isotherms, they evaluated (through the calculation of the Henry's constants) the interaction energy of the most active sites. The results are in agreement with earlier data obtained by IGC at infinite dilution conditions.

A direct application of the surface energetic evaluation concerns the aerosol formation of powders (in lactose formulation) in dry powder inhalers. A typical formulation consists of micronized drug mixed with a carrier of larger particle size. Upon inhalation of the powder blend, the drug and the carrier ideally disaggregate through mechanical and shear forces. Smaller particles of drug can change direction in the inhaled stream and deposit deep in the lung, whereas larger carrier particles impact in the throat and are swallowed. This is a complex problem involving particle size distributions, surface energetics, adhesion, and aerodynamics (shape factor) of particles.[196] The SEDS-produced particles perform much better than do particles obtained in the classical way. The main factor for superior performance is the dispersibility of the powder at low airflow, which is mostly defined by the low bulk density, small sphericity, and reduced surface energy of SX particles.

Predicting the quality of powders for inhalation from surface energy and area. Cline and Dalby[197] correlated the surface energy, determined by IGC, of active and carrier components in an aerosol powder to in vitro performance of a passive dry powder inhaler. IGC was used to assess the surface energy of active (albuterol and ipratropium bromide) and carrier (lactose monohydrate, trehalose dehydrate, and mannitol) components of a dry powder inhaler formulation. Blends (1% w/w) of drug and carrier were prepared and evaluated for dry powder inhaler performance by cascade impaction. In vitro performance of the powder blends was strongly correlated to surface energy interaction between active and carrier components. Plotting fine particle fraction vs. surface energy interaction yielded an R2 value of 0.9283. Increasing surface energy interaction between drug and carrier resulted in greater fine particle fraction of drug. This relationship is potentially useful for rapid formulation design and achieving efficient dry powder inhalers.

Drug processing control. IGC was applied for the monitoring of a milling process of pharmaceutical powders (DL-propanolol hydrochloride) in conjunction with molecular modeling[198,199] that describes the functional groups of DL-propanolol hydrochloride and allows also to predict dominant crystal faces with the lowest attachment energy (cleavage plane) where fracture will possibly occur during milling: a π electron-rich naphthalene moiety of dl-propanolol hydrochloride is present on that face, supporting the indications of IGC. IGC and computer simulation were also associated to explain the differences in surface energetics of two optical forms of mannitol. The modeling agreed with the trends indicated by the IGC values.[200] Furthermore, the influence of moisture on the surface properties of carbamazepine and paracetamol was established by IGC and the preferential sites for water adsorption identified by molecular modeling. It appears that if the water molecules compete with polar probes for the same adsorption sites, the interaction of those probes is diminished. It may be concluded that the computer modeling and IGC association is a most efficient way to progress in the understanding of the adsorption mechanisms on solid surfaces. However, this supposes that the solid surface itself needs to be modeled, a task that may be performed on pure and well-organized surfaces, which is usually not the case for actual powder and fiber surface commonly polluted by organic and/or mineral substances, rich in structural defects, etc.

An easier task is to take IGC for just identifying differences, say between samples originating from a given manufacturing process. In other terms, IGC may be an accessible and time-saving procedure for process control. Batch variations could be detected during the manufacturing of α-lactose, for instance.[201] Subtle changes in surface properties, that may induce important secondary processing properties, such as powder flow, were evidenced upon micronization of salbutamol sulphate.[202]

IGC was also used to study a different inhalation formulation comprising an active agent (albuterol and ipratropium bromide) and a carrier (lactose monohydrate, manitol and trehalose dihydrate) component.[203] It is shown that there exists a linear correlation between the fine particle fraction of drug (blended with carrier) and surface interaction energy, expressed by unit weight. This finding will facilitate the selection of drugs and carrier agents.

2.5.2.10 Cosmetics

IGC may be used to determine the surface properties of hair as shown by Tielmann.[204] The surface of hair is quite complex, both from chemical and morphological points of view. Furthermore, hairs may be submitted to various treatments, such as bleaching with hydrogen peroxide, "perming" with thioglycolate, etc. The variations of surface characteristics were assessed, by plotting hexane and water adsorption isotherms, from which adsorption energy distribution functions were calculated. Differences become apparent. But, those functions are complicated and not well resolved, rendering the interpretation difficult. Moreover, the authors did not identify the influence of noise (experimental and computational) and its importance on their results. But, as expected, IGC is certainly capable to detect the influence of the drastic treatments applied to hair.

2.5.2.11 Food Industry

IGC, in food science, is currently used to measure the sorption of water in dry foods[205] or the sorption of aroma compounds on food packaging polymers.[206] Owing to the dynamic nature of the IGC process, it is also possible to study more complex problems, such as the kinetics and nature of interactions of IGC-chosen solutes with food and/or the starch surfaces. Combined with mass spectrometry (MS), IGC allows to determine the flavor release from a food matrix as a function of water uptake.[207] Indeed, the adsorption of water, monitored by IGC, is one cause of the flavor loss (determined by MS). Adsorbed water, from atmosphere, competes with flavor molecules already adsorbed (on buttery-flavored crackers, for instance) that are progressively eliminated. This permits to estimate the onset of the flavor release and the total flavor loss, as well as the nature and order of release. Ducruet et al.[208,209] examined the influence of the nature and treatment of starch on aroma retention. They investigated also the influence of humidity on the aroma-corn starch interactions. Specific retention volumes (Vg) of selected solutes (1-hexanol, 2-hexanol, octanal, ethyl hexanoate, and D-limonene) were measured. Differences are observed according to the nature (functionality) of the injected solute. H bond formation and dipole–dipole interactions govern the retention mechanism. Moreover, the particular structure of starch, made of amorphous and crystalline zones needs to be taken into account when adsorption occurs in presence of water.

Obviously, the knowledge of the aroma retention mechanisms is of major interest for the optimization of the conditions increasing the flavor retention in foods.

2.5.2.12 Petrochemical Industry

Even complex substances, such as the residues of oil distillation, are susceptible to be analyzed by IGC.

Puig et al.[210] determined the glass transition temperature and surface energy of various bituminous binders by IGC. Bitumens are most complex mixtures made of different molecular species, originating from oil refinery residues. Their constituents are classified according to their solubility in given solvents. If the bitumen is diluted with *n*-heptane or *n*-pentane, a precipitate known as "asphaltenes" is formed. The soluble part refers to maltenes. Bitumens may be compared to composite materials, where asphaltenes play the role of fillers. IGC readily allows evaluating the glass transition temperature of those complex mixtures when operating at subambient temperatures. Furthermore, IGC indicates that asphaltenes present higher surface energies and surface polarity in accordance with earlier work.[211]

The glass transition temperature variations of the selected samples correlate with the surface characteristics of bitumens that depend on the asphaltene contents.

2.6 CONCLUSION

The aim of this chapter was to shortly introduce the principles and practice, to underline the advantages, but also limitations, and to point out possible applications of Inverse Gas Chromatography. This may be useful for newcomers and may be encouraging for those who have in charge the difficult task to analyze, for fundamental or practical purposes, the surface of powders and fibers. Indeed, a satisfactory knowledge of the surface properties is a prerequisite for a better understanding of their behavior in presence of different environments. Presently, there are not many methods to achieve this goal, especially when the solid's surface area is small, forbidding the classical gas adsorption approaches. In fact, in some cases IGC appears as a unique method to readily detect minor changes resulting, for instance, from small parameter's variations during the elaboration process or say, from aging processes during storage. New applications of IGC are currently developed, one reason being that IGC is not a "finished" method, but is largely open to the researcher's imagination. Only a limited number of solid surfaces have so far been examined, but the domains of applications are multiple, going from material science to biology. Moreover, each solid surface has its own history that should be taken into account when evaluating the IGC results, since IGC belongs certainly to the most sensitive and true surface analysis methods. Often IGC detects unexpected and subtle "surface events" that may then induce a complementary research based on modern, more sophisticated spectrometric and or microscopic methods.

In conclusion, it is the authors' opinion, that IGC remains an interesting research and application domain both for academic, but also for practical concerns. It is always worth to try it since the required basic equipment is present in most laboratories. Experimental IGC results are readily obtained, yet their interpretation is far from being evident. Nevertheless, surface properties' variations are easily evidenced: major information for process control, or for monitoring a controlled surface modification procedure that is currently required for attaining the optimal surface properties for specific applications of powders and fibers.

ACKNOWLEDGMENTS

We are indebted to all authors who did send us the reprints of their recent work in the IGC field. Unfortunately, space was missing to describe in depth all the interesting information we received.

REFERENCES

1. Smidsrod, O. and Guillet, J. E., *Macromolecules*, 2, 272, 1969.
2. Kiselev, A. V., *Advances in Chromatography*, Marcel Dekker, New York, 1967.
3. Dorris, G. M. and Gray, D. G., *J. Colloid Interface Sci.*, 77, 353, 1980.
4. Derminot, J., *Physicochimie des Polymères et Surfaces par Chromatographie en Phase Gazeuse*, ANRT, Paris, 1981.
5. Lloyd, D. R., Ward, T. C., Schreiber, H. P., *Inverse Gas Chromatography*, A.C.S. Symp. Series, No. 391, A.C.S., Washington, D.C., 1989.
6. Voelkel, A., *Crit. Rev. Anal. Chem.*, 22, 411, 1991.
7. Papirer, E. and Balard, H., In *Adsorption on New and Modified Inorganic Sorbents*, Dabrowski, A. and Tertykh, V. A., Eds., Elsevier, Amsterdam, p. 479, 1995.
8. Mukhopadhyay, P. and Schreiber, H. P., *Colloids Surf.*, A, 100, 47, 1995.
9. Belgacem, M. N. and Gandini, A., In *Interfacial Phenomena in Chromatography*, Pefferkorn, E., Ed., Marcel Dekker, New York, p. 41, 1999.
10. Williams, D. R., First International on Conference on Inverse Gas Chromatography, (London IGC, 2001), *J. Chromatogr. A*, 969, 2002.
11. Snyder, L. R., *Principle of Adsorption Chromatography*, Marcel Dekker, New York, 1968.
12. Lamb, R. J. and Pecsok, R. L., *Physico-Chemical Applications of Gas Chromatography*, Wiley, New York, 1987.
13. Conder, J. R. and Young, C. L., *Physico-Chemical Measurement by Gas Chromatography*, Wiley, New York, 1979.
14. Conder, J. R., *J. High Resolut. Chromatogr.*, 5, 341, 1982.
15. Conder, J. R., *J. High Resolut. Chromatogr.*, 5, 397, 1982.
16. Fowkes, F. M., *Chemistry and Physics of Interfaces*, ACS Symposium on Interface, A.C.S., Washington, D.C., p. 1, 1997.
17. Goss, K. U., *J. Colloid Interface Sci.*, 190, 241, 1997.
18. Schultz, J., Lavielle, L., and Martin, C., *J. Adhes.*, 23, 45, 1987.
19. Hamieh, T. and Schultz, J., *J. Chromatogr. A*, 969, 17, 2002.
20. Balard, H. and Papirer, E., *Prog. Org. Coat.*, 22, 1, 1993.

21. Contescu, C., Jagiello, J., and Schwarz, J. A., *J. Catal.*, 131, 433, 1991.
22. Wang, M. J., Tu, H., Murphy, J. L., and Mahmud, K., *Rubber Chem. Technol.*, 72, 666, 1999.
23. Milonjic, S. K., *Colloids Surf., A*, 149, 461, 1999.
24. Papirer, E., Brendlé, E., Balard, H., and Dentzer, J., *J. Mater. Sci.*, 35, 3573, 2000.
25. Papirer, E. and Kuczinski, J., *Eur. Polym. J.*, 27, 653, 1991.
26. Belgacem, N. M., Blayo, A., and Gandini, A., *J. Colloid Interface Sci.*, 182, 431, 1996.
27. Sun, C. H. and Berg, J. C., *J. Chromatogr. A*, 969, 59, 2002.
28. Papirer, E., Ligner, G., Balard, H., Vidal, A., and Mauss, F., In *Chemically Modified Oxide Surfaces*, Leyden, D. E. and Collins, W. T., Eds., Gordon and Breach Science Publishers, New York, p. 15, 1990.
29. Brendlé, E., Dentzer, J., and Papirer, E., *J. Colloid Interface Sci.*, 199, 63, 1998.
30. Brendlé, E. and Papirer, E., *J. Chim. Phys.*, 95, 1020, 1998.
31. Zaborski, M., Slusarski, L., Donnet, J. B., and Papirer, E., *Kautsch. Gummi Kunstst.*, 47, 730, 1994.
32. Smiciklasl, I. D., Milonjic, S. K., and Zec, S., *J. Mater. Sci.*, 35, 2825, 2000.
33. Onjia, A. E., Milonjic, S. K., Todorovic, M., Loos-Neskovic, C., Fedoroff, M., and Jones, D. J., *J. Colloid Interface Sci.*, 251, 10, 2002.
34. Onjia, A. E., Milonjic, S. K., and Rajakovic, L., *J. Serb. Chem. Soc.*, 66, 259, 2001.
35. Papirer, E., Roland, P., Nardin, M., and Balard, H., *J. Colloid Interface Sci.*, 113, 62, 1986.
36. Ligner, G., Vidal, A., Balard, H., and Papirer, E., *J. Colloid Interface Sci.*, 133, 200, 1989.
37. Saada, A., Papirer, E., Balard, H., and Siffert, B., *J. Colloid Interface Sci.*, 175, 212, 1995.
38. Donnet, J. B. and Lanzinger, C. M., *Kautsch. Gummi Kunstst.*, 45, 263, 1992.
39. Sunkersett, M. R., Grimsey, I. M., Doughty, S. W., Osborn, J. C., York, P., and Rowe, R. C., *Eur. J. Pharm. Sci.*, 13, 219, 2001.
40. Papirer, E., Balard, H., Brendlé, E., and Lignières, J., *J. Adhes. Sci. Technol.*, 10, 1401, 1996.
41. Papirer, E., Schultz, J., and Turchi, C., *Eur. Polym. J.*, 20, 1155, 1984.
42. Keller, D. S. and Luner, P., *Colloid Surf., A*, 161, 401, 2000.
43. Sun, C. and Berg, J., *J. Colloid Interface Sci.*, 260, 443, 2003.
44. Papirer, E., Balard, H., and Vergelati, C., In *Adsorption on Silica Surfaces*, Papirer, E., Ed., Marcel Dekker, New York, p. 205, 2000.
45. Newell, H. E. and Buckton, G., *Pharm. Res.*, 21 (8), 1440, 2004.
46. Rebouillat, S., Escoubes, S., Gauthier, R., and Vigier, A., *J. Appl. Polym. Sci.*, 58, 1305, 1995.
47. Van Asten, A., Van Veenendaal, N., and Koster, S., *J. Chromatogr. A*, 888, 175, 2000.
48. Montes-Moran, M. A., Paredes, J. I., Martinez-Alonso, A., and Tascon, J. M. D., *Macromolecules*, 35, 5085, 2002.

49. Nardin, M., Asloun, E. M., and Schultz, J., *Surface Interface Anal.*, 17, 485, 1991.
50. Vickers, P. E., Watts, J. F., Perruchot, C., and Chehimi, M. M., *Carbon*, 38, 675, 2000.
51. Montes-Moran, M. A., Martinez-Alonso, A., Tascon, J. M. D., Paiva, M. C., and Bernardo, C. A., *Carbon*, 39, 1057, 2001.
52. Dutschk, V., Mader, E., and Rudoy, V., *J. Adhes. Sci. Technol.*, 15, 1373, 2001.
53. Saint Flour, C. and Papirer, E., *Ind. Eng. Chem. Prod. Res. Dev.*, 21, 337, 1982.
54. Tsutsumi, K. and Ohsuga, T., *Colloid Polym. Sci.*, 268, 38, 1990.
55. Gurnagul, N. and Gray, D. G., *Can. J. Chem.*, 65, 1987, 1935.
56. Walinder, M. E. P. and Gardner, D. J., *Wood Fiber Sci.*, 32, 478, 2000.
57. Bick, A., *Molecular Simulation*, Cambridge, UK, Personal Communication.
58. Tuel, A., Hommel, H., Legrand, A. P., Balard, H., Sidqi, M., and Papirer, E., *Colloids Surfaces*, 58, 17, 1991.
59. Vidal, A. and Papirer, E., In *The Surface Properties of Silica*, Legrand, A. P., Ed., Wiley, Chicester, UK, p. 285, 1998.
60. Wiener, H., *J. Am. Chem. Soc.*, 69, 2636, 1947.
61. Wiener, H., *J. Phys. Chem.*, 52, 425, 1948.
62. Basak, S. C., Gieschen, D. P., Magnuson, V. R., and Hariss, D. K., *IRCS Med. Sci.*, 10, 619, 1982.
63. Randic, M., *J. Chem. Educ.*, 69, 713, 1992.
64. Randic, M., *J. Math. Chem.*, 4, 157, 1990.
65. Randic, M., *J. Math. Chem.*, 7, 155, 1991.
66. Rouvray, D. H., *Discrete Appl. Math.*, 19, 317, 1988.
67. Rouvray, D. H., *J. Comput. Chem.*, 8, 470, 1987.
68. Motoc, I. and Balaban, A. T., *Rev. Roum. Chim.*, 26, 593, 1981.
69. Mihalic, Z. and Trinajstic, N., *J. Chem. Educ.*, 69, 701, 1992.
70. Randic, M., *J. Am. Chem. Soc.*, 97, 6609, 1975.
71. Kovats, E., *Helvetica Chim. Acta*, XLI, 1915, 1958.
72. Kovats, E., *Z. Anal. Chem.*, 181, 351, 1961.
73. Brendlé, E. and Papirer, E., *J. Colloid Interface Sci.*, 194, 207, 1997.
74. Brendlé, E. and Papirer, E., *J. Colloid Interface Sci.*, 194, 217, 1997.
75. Saint Flour, C. and Papirer, E., *J. Colloid Interface Sci.*, 91, 69, 1983.
76. Sawyer, D. T. and Brookman, D. J., *Anal. Chem.*, 40, 1847, 1968.
77. Chehimi, M. M. and Pigois-Landureau, E., *J. Mater. Chem.*, 4, 741, 1994.
78. Dong, S., Brendlé, M., and Donnet, J. B., *Chromatographia*, 28, 85, 1989.
79. Donnet, J. B., Park, S. J., and Balard, H., *Chromatographia*, 31, 434, 1991.
80. Guttierrez, M. C., Rubio, J., and Oteo, J. L., *J. Chromatogr. A*, 845, 53, 1999.
81. Grajek, H., Witkiewicz, Z., and Jankowska, H., *J. Chromatogr. A*, 782, 87, 1997.
82. Osmont, E., Schreiber, H. P., In *Inverse Gas Chromatography Characterization of Polymers and other Materials*, A.C.S. Symposium Series No. 391, p. 230, 1989.
83. Sidqi, M., Lignier, G., Jagiello, J., Balard, H., and Papirer, E., *Chromatographia*, 28, 588, 1989.

84. Jagiello, J., Bandosz, T. J., and Schwartz, J. A., *Chromatographia*, 33, 441, 1992.

85. Pearson, R. G., *Acc. Chem. Res.*, 26, 250, 1993.

86. Drago, R. S., Vogel, G. C., and Needham, T. E., *J. Am. Chem. Soc.*, 93, 6014, 1971.

87. Gutmann, V., *The Donor–Acceptor Approach to Molecular Interactions*, Plenum Press, New York, 1978.

88. Kamlet, M. J., Abboud, J. L. M., Abraham, M. H., and Taft, R. W., *J. Solution Chem.*, 13, 485, 1981.

89. Hildebrand, J. H. and Scott, R. L., *The Solubility of Nonelectrolytes*, Rheinhold, New York, 1936.

90. Van Oss, C. J., Good, R. J., and Chaudhury, M. K., *Langmuir*, 4, 884, 1988.

91. Abraham, M. H., Chadha, H. S., Whiting, G. S., and Mitchell, R. C., *J. Pharm. Sci.*, 83, 1085, 1994.

92. Joslin, S. T. and Fowkes, F. M., *Ind. Eng. Chem. Prod. Res. Dev.*, 24, 369, 1985.

93. Abraham, M. H., *Chem. Soc. Rev.*, 22, 73, 1993.

94. Grate, J. W., Abraham, M. H., Matt, M., and Shuely, W. J., *Langmuir*, 11, 2125, 1995.

95. Abraham, M. H. and Walsh, D. P., *J Chromatogr.*, 627, 294, 1992.

96. Abraham, M. H., Du, C. M., Grate, J. W., Mcgill, R. A., and Shuely, W. J., *J. Chem. Comm.*, 24, 1993, 1863.

97. Park, J. H., Lee, Y. K., and Donnet, J. B., *Chromatographia*, 33, 154, 1992.

98. Roth, C. M., Goss, K. U., and Schwartzenbach, P., *J. Colloid Interface Sci.*, 252, 21, 2002.

99. Burg, P., Abraham, M. H., and Cagniant, D., *Carbon*, 41, 867, 2003.

100. Spange, S., Reuter, A., and Vilsmeier, E., *Colloid Polym. Sci.*, 274, 59, 1996.

101. Spange, S., Vilsmeier, E., and Zimmermann, Y., *J. Phys. Chem. B*, 104, 6417, 2000.

102. Hansen, C. M., *Hansen Solubility Parameters—A User's Handbook*, CRC Press, Boca Raton, FL, 2000.

103. Barton, A. F. M., *Handbook of Solubility Parameters and Other Cohesion Parameters*, CRC Press, Boca Raton, FL, 1991.

104. Karger, B. L., Snyder, L. R., and Eon, C., *Anal. Chem.*, 50, 2126, 1978.

105. Voelkel, A. and Grzeskowiak, T., *Chromatographia*, 51, 608, 2000.

106. Tong, H. Y., Shekunov, B. Y., York, P., and Chow, A. H. L., *Pharm. Res.*, 19, 640, 2002.

107. Goss, K. U., *Crit. Rev. Environ. Sci. Technol.*, 34, 339, 2004.

108. Abraham, M. H., Chadha, H. S., Whiting, G. S., and Mitchell, R. C., *J. Pharm. Sci.*, 83, 1085, 1994.

109. Burg, P., Abraham, M. H., and Cagniant, D., *Carbon*, 41 (5), 867, 2003.

110. Jagiello, J. and Papirer, E., *J. Colloid Interface Sci.*, 142, 232, 1991.

111. Katsanos, N. A., *J. Chem. Soc., Faraday Trans. 1*, 78, 1051, 1982.

112. Katsanos, N. A. and Karaiskakis, G., *Adv. Chromatogr.*, 24, 125, 1984.

113. Gavril, D., Koliadima, A., and Karaiskakis, G., *Chromatographia*, 49, 285, 1999.

114. Gavril, D., Katsanos, N. A., and Karaiskakis, G., *J. Chromatogr. A*, 852, 507, 1999.
115. Katsanos, N. A., Agathonos, P., and Niotis, A., *J. Phys. Chem.*, 92, 1645, 1988.
116. Gavril, D. and Karaiskakis, G., *Instrum. Sci. Technol.*, 25, 217, 1997.
117. Rashid, K. A., Gavril, D., Katsanos, N. A., and Karaiskakis, G., *J. Chromatogr. A*, 934, 31, 2001.
118. Agathonos, P. and Karaiskakis, G., *J. Appl. Polym. Sci.*, 37, 2237, 1989.
119. Koliadima, A., Agathonos, P., and Karaiskakis, G., *J. Chromatogr.*, 550, 171, 1991.
120. Sotiropoulou, V., Vassilev, G. P., Katsanos, N. A., Metaxa, H., and Roubani-Kalantzopoulou, F., *J. Chem. Soc., Faraday Trans.*, 91, 485, 1995.
121. Abatzoglou, Ch., Iliopoulou, E., Katsanos, N. A., Roubani-Kalantzopoulou, F., and Kalantzopoulos, A., *J. Chromatogr. A*, 775, 211, 1997.
122. Katsanos, N. A., Arvanitopoulou, E., Roubani-Kalantzopoulou, F., and Kalantzopoulos, A., *J. Phys. Chem. B*, 103, 1152, 1999.
123. Roubani-Kalantzopoulou, F., Artemiadi, Th., Bassiotis, I., Katsanos, N. A., and Plagianakos, V., *Chromatographia*, 53, 315, 2001.
124. Katsanos, N. A., Roubani-Kalantzopoulou, F., Iliopoulou, E., Bassiotis, I., Siokos, V., Vrahatis, M. N., and Plagianakos, V. P., *Colloid Surf., A*, 201, 173, 2002.
125. Bakaoukas, N., Koliadima, A., Farmakis, L., Karaiskakis, G., and Katsanos, N. A., *Chromatographia*, 57, 783, 2003.
126. Gavril, D. J., *Liq. Chromatogr. Related Technol.*, 25, 2079, 2002.
127. Kessaissia, Z., Papirer, E., and Donnet, J. B., *J. Colloid Interface Sci.*, 79, 257, 1981.
128. Hamieh, T., Rezzaki, M., Grohens, Y., and Schultz, J., *J. Chim. Phys.*, 95 (9), 1990, 1964.
129. Kiselev, A. V., *Gas-Adsorption Chromatography*, Plenum Press, New York, 1969.
130. James, A. T. and Martin, A. J. P., *Biochem. J.*, 50, 627, 1952.
131. Jaroniec, M. and Madey, R., *Physical Adsorption on Heterogeneous Solids*, Elsevier, Amsterdam, 1988.
132. Rudzinski, W. and Everett, D. H., *Adsorption of Gases on Heterogeneous Surfaces*, Academic Press, London, 1992.
133. Rudzinski, W., Jagiello, J., and Grillet, Y., *J. Colloid Interface Sci.*, 87, 478, 1982.
134. Rudzinski, W. and Jagiello, J., *J. Low Temp. Phys.*, 48, 307, 1982.
135. Jagiello, J., Ligner, G., and Papirer, E., *J. Colloid Interface Sci.*, 137, 128, 1990.
136. Tijburg, I., Jagiello, J., Vidal, A., and Papirer, E., *Langmuir*, 7, 2243, 1991.
137. Papirer, E., Li, S., Balard, H., and Jagiello, J., *Carbon*, 29, 1135, 1991.
138. Jagiello, J. and Schwarz, J. A., *J. Colloid Interface Sci.*, 146, 415, 1991.
139. Jagiello, J., *Langmuir*, 10, 2778, 1994.
140. Jagiello, J., Bandosz, T. J., Putyera, K., and Schwarz, J. A., *Proc. Fifth International Conference on Fundamentals of Adsorption, Boston, MA*, 417, 1996.

141. Balard, H., *Langmuir*, 13, 1260, 1997.
142. Jagiello, J. and Schwarz, J. A., *Langmuir*, 9, 2513, 1993.
143. Papirer, E., Brendlé, E., Ozil, F., and Balard, H., *Carbon*, 37, 1265, 1999.
144. Balard, H., Aouadj, O., and Papirer, E., *Langmuir*, 13, 1251, 1997.
145. Donnet, J. B. and Custodéro, E., *Carbon*, 30, 813, 1992.
146. Balard, H. and Brendlé, E., *Kautsch. Gummi Kunstst.*, 55, 464, 2002.
147. Choudhary, V. R. and Mantri, K., *Microporous Macroporous Mater.*, 40, 127, 2000.
148. Unpublished data from IGClab, 2005.
149. Kopecni, M. M., Milonjic, S. K., and Laub, R. J., *Anal. Chem.*, 52, 1032, 1980.
150. Balard, H., Papirer, E., Khalfi, A., and Barthel, H., *Compos. Interfaces*, 6, 19, 1999.
151. Comard, M. P., Calvet, R., Dodds, J. A., and Balard, H., *J. Chromatogr. A*, 969, 93, 2002.
152. Comard, M. P., Calvet, R., Balard, H., and Dodds, J. A., *Colloids Surf. (Physicochemical Eng. Aspects)*, 232, 269, 2004.
153. Goss, K. U. and Schwarzenbach, R. P., *J. Colloid Interface Sci.*, 252, 31, 2002.
154. Goss, K. U., *Crit. Rev. Environ. Sci. Technol.*, 34, 339, 2004.
155. Davydov, V. Y., In *Adsorption on Silica Surfaces*, Surfactant Sci. Series No. 90, Papirer, E., Ed., Marcel Dekker, New York, p. 63, 2000.
156. Saint Flour, C. and Papirer, E., *J. Colloid Interface Sci.*, 91 (1), 69, 1983.
157. Dutschk, V., Prause, S., and Spange, S., *J. Adhes. Sci. Technol.*, 16, 1749, 2002.
158. Jandura, P., Riedl, B., and Kokta, B. V., *J. Chromatogr. A*, 969, 301, 2002.
159. Papirer, E., Perrin, J. M., Siffert, B., and Philipponneau, G., *J. Phys. III*, 1, 697, 1991.
160. McMahon, A. W., Kelly, D. G., and McLaughlin, P. J., *Analyst*, 127, 17, 2002.
161. Wypych, G., *Filler Hanbook*, Chem. Tech. Publication Inc., Toronto-Scarborough, 2000.
162. Shui, M., Reng, Y. L., Pu, B. Y., and Li, J. R., *J. Colloid Interface Sci.*, 273 (1), 205, 2004.
163. Price, G. J. and Ansari, D. M., *Polym. Int.*, 53 (4), 430, 2004.
164. Keller, S. and Luner, P., *Colloids Surf. A*, 161, 401, 2000.
165. Fekete, E., Moczo, J., and Pukansky, B., *J. Colloid Interface Sci.*, 269 (1), 143, 2004.
166. Papirer, E. and Brendlé, E., unpublished data, 1998.
167. Nardin, M., Asloun, E. M., and Schultz, J., *Polym. Adv. Technol.*, 2, 1091, 1991.
168. Nardin, M. and Schultz, J., *Compos. Interfaces*, 1, 177, 1993.
169. Plonka, R., Mader, E., Gao, S. L., Bellmann, C., Dutschk, V., and Zhandarov, S., *Compos. Appl. Sci. Manuf.*, 35 (10), 1207, 2004.
170. Domingo-Garcia, M., Lopez-Garzon, F. J., and Perez-Mendoza, M., *Langmuir*, 16, 7012, 2000.
171. Grajek, H., Swiatkowski, A., Witkiewicz, Z., Pakula, M., and Biniak, S., *Adsorpt. Sci. Technol.*, 19, 565, 2001.
172. Grajek, H., *J. Chromatogr. A*, 986, 89, 2003.
173. El-Sayed, Y. and Bandosz, T. J., *J. Colloid Interface Sci.*, 242, 44, 2001.

174. Bagreev, A. and Bandosz, T. J., *J. Phys. Chem. B*, 104, 8841, 2000.
175. Salame, I. I. and Bandosz, T. J., *Langmuir*, 17, 4967, 2001.
176. Salame, I. I. and Bandosz, T. J., *Mol. Phys.*, 100, 2041, 2002.
177. Bandosz, T. J., Jagiello, J., and Schwarz, J., *Langmuir*, 9, 2518, 1993.
178. Jagiello, J., Bandosz, T. J., and Schwarz, J. A., *J. Colloid Interface Sci.*, 151, 433, 1992.
179. Donnet, J. B. and Bansal, R., Eds., *Carbon Black Science and Technology,* 2nd ed., Marcel Dekker, New York, 1993.
180. Castro, C., Dorris, G. M., and Daneault, C. J., *J. Chromatogr. A*, 969, 313, 2002.
181. Valera, N., Chaussy, D., and Tourron, J. L., *Cellul. Chem. Technol.*, 38 (1–2), 95, 2004.
182. Segeren, L. H., PhD Thesis, Univ. Twente, The Netherlands, 2002.
183. Segeren, L. H., Wouters, M. E. L., Bos, M., Van den Berg, J. W. A., and Vancso, G. J., *J. Chromatogr. A*, 969, 215, 2002.
184. Walinder, M. E. P. and Gardner, D., *J. Adhes. Sci. Technol.*, 16, 1625, 2002.
185. Gardner, D. J., Tze, W. T., and Shi, S. Q., Eds., *Advances Lignocellulosics Characterization*, Tappi Press, Atlanta, p. 263, 1999.
186. Tze, W. T. and Gardner, D. J., *J. Adhes. Sci. Technol.*, 15, 223, 2001.
187. Katsanos, N. A. and Karaiskakis, G., *J. Chromatogr.*, 395, 423, 1987.
188. Kreamer, D. K., Oja, J. K., Steinberg, S. M., and Philips, II., *J. Environ. Eng.*, 120, 348, 1994.
189. Steinberg, S. M. and Kreamer, D. H., *J. Environ. Sci. Technol.*, 27, 883, 1993.
190. Steinberg, S. M., Schmeltzer, J., and Kreamer, H. D., *Chemosphere*, 33, 961, 1999.
191. Steinberg, S. M., Swallow, C., and Ma, W. K., *Chemosphere*, 38, 2143, 1999.
192. Garcia-Herruzo, F., Rodriguez-Maroto, J. M., Garcia-Delgado, R. A., Gomez-Lahoz, C., and Vereda-Alonso, C., *Chemosphere*, 41, 1167, 2000.
193. Grimsey, I. M., Feeley, J. C., and York, P., *J. Pharm. Sci.*, 91, 571, 2002.
194. Tong, H. H. Y., Shekunov, B. Yu., Kordikowski, A., and Chow, A. H. L., *Pharm. Res.*, 18, 852, 2001.
195. Tong, H. H. Y., Shekunov, B., York, P., and Chow, A. H. L., *J. Pharm. Sci.*, 94 (3), 695, 2005.
196. Henry, H. Y., Shekunov, B. Yu., Feely, J. C., Chow, A. H. L., Tong, H. H. Y., and York, P., *Aerosol Sci.*, 34, 553, 2003.
197. Cline, D. and Dalby, R., *Pharm. Res.*, 21 (9), 1718, 2004.
198. York, P., Ticehurst, M. D., Osborn, J. C., Roberts, R. J., and Intern, *J. Pharm.*, 174, 179, 1998.
199. Grimsey, I. M., Osborn, J. C., Doughty, S. W., York, P., and Rowe, R. C., *J. Chromatogr. A*, 969, 49, 2002.
200. Grimsey, I. M., Sunkersett, M., Osborn, J. C., York, P., and Rowe, R. C.*Intern, J. Pharm.*, 191, 43, 1999.
201. Ticehurst, M. D., York, P., Rowe, R. C., Dwivedi, S. K.*Intern, J. Pharm.*, 141, 93, 1996.

202. Feeley, J. C., York, P., Sumby, B. S., and Grimsey, J. C.*Int, J. Pharm.*, 172, 89, 1998.

203. Cline, D. and Dalby, R., *Pharm. Res.*, 19, 1274, 2002.

204. Tielmann, F., Pearce, D., and Kamath, Y., *IFSCC Mag.*, 5, 190, 2002.

205. Helen, H. J. and Gilbert, S. G., *J. Food Sci.*, 50, 454, 1985.

206. Acucejo, S., Pozo, M. J., and Gavara, J., *J. Appl. Polym. Sci.*, 70, 711, 1998.

207. Castellano, J. and Snow, N. H., *J. Agric. Food Chem.*, 49, 2496, 2001.

208. Boutboul, A., Giampaoli, P., Feigenbaum, A., and Ducruet, V., *Food Chem.*, 71, 387, 2000.

209. Boutboul, A., Lenfant, F., Giampaoli, P., Feigenbaum, A., and Ducruet, V., *J. Chromatogr. A*, 969, 9, 2002.

210. Puig, C. C., Meijer, H. E. H., Michels, M. A. J., Segeren, L. H. G. J., and Vancso, G. J., *Energy Fuels*, 18, 63, 2004.

211. Papirer, E., Kuczinski, J., and Siffert, B., *Chromatographia*, 23, 401, 1987.

3 Application of Inverse Gas Chromatography to the Study of Rubber Reinforcement

Meng-Jiao Wang

CONTENTS

3.1 INTRODUCTION

Fillers represent a major component in the manufacture of rubber products, with consumption second only to rubber itself. Among others, carbon black and silica are by far the most active rubber reinforcing agents, due to their unique ability to enhance the physical properties of rubbers.

Besides chemical composition, the fillers are characterized by thier morphology, and the physical nature of the surface, which is determined by their microstructure and surface functionality. The primary dispersable unit of fillers, carbon black and silica in particular, is referred to as an "aggregate," which is a discrete, rigid, and colloidal entity. It is the functional unit in well-dispersed systems. For most fillers, the aggregate is composed of spheres that are fused together. These spheres are generally termed primary "particles." Morphology is a set of properties related to the average magnitude and frequency distribution of the particle dimensions which, with surface porosity, determines the surface area and "structure." The term "structure," used widely in filler and rubber communities, describes the ensemble of aggregates that is a stochastic distribution of the number and arrangement of the primary particles that make up the aggregates.

For the same types of filler, the behavior of different grades in rubber is dominated mainly by their morphology, such as surface area and structure, and by their surface activity, which is related to physical chemistry of the surfaces. All these parameters play a role in rubber reinforcement through different mechanisms, such as occlusion of the polymer in the internal voids of the aggregates, interfacial interaction between rubber and fillers, and the agglomeration of filler aggregates in the polymer matrix.

The effect of filler structure on the rubber properties of filled rubber has been explained by the occlusion of rubber by filler aggregates. When structured fillers are dispersed in rubber, the polymer portion filling the internal voids of the aggregates, or the polymer portion located within the irregular contours of the aggregates, is unable to participate fully in the macrodeformation. The partial immobilization in the form of occluded rubber causes this portion of rubber to behave like the filler, rather than like the polymer matrix. Due to this phenomenon, the effective volume of the filler, with regard to the stress–strain behavior and viscoelastic properties of the filled rubber, is increased considerably.

Upon incorporation of fillers into a polymer, an interface between a rigid solid phase and a soft elastomer phase is created. For fillers whose surfaces exhibit very little porosity, the total interfacial area depends on the loading and specific surface area of the filler. Due to the interaction between rubber and filler, two phenomena are well documented: the formation of bound rubber, and a rubber shell on the carbon black surface. Both are related to the restriction of the segmental movement of polymer molecules.

The filler aggregates in the polymer matrix have a tendency to associate to agglomerates, especially at high loadings, leading to chain-like filler structures or clusters. These are generally termed secondary structure or, in some cases, a filler network, even though the latter is not comparable to the continuous polymer network structure. The formation of a filler network is dependent on the intensity of the interaggregate attractive potential, the distance between aggregates and the polymer-filler interaction.

Technically, structure and surface area have been well characterized and their effects on rubber properties have has been well documented [1,2].

Surface activity is also an important factor in performance. This factor can, in a chemical sense, be related to different chemical groups on the filler surface. In a physical sense, variations in surface energy determine the adsorptive capacity of the fillers and their energy of adsorption. However, compared with morphology, a satisfactory description of surface properties of fillers is still lacking, due to the fact that only a limited number of tools have been available to assess the filler surface in terms of ensemble properties. In this regard, the inverse gas chromatography (IGC) has been developed very rapidly over the last two decades, and a lot of work on its application to the studies of rubber reinforcement has been published. This review will start with a brief summary of what is known on surface energies of fillers, and the application of IGC to address the filler surface characteristics that are related to polymer-filler, filler–filler, and filler-ingredient interactions. Then, the role of the filler surface characteristics in rubber reinforcement will be discussed based on the IGC results.

3.2 SURFACE ENERGY OF SOLIDS AND INTERACTIONS BETWEEN MATERIALS

For nonelectrolyte solid systems, such as hydrocarbon elastomers and conventional fillers, the interaction between two materials is determined by their surface energies. It is generally established that the surface energy of fillers has a much greater effect on the mechanical properties of filled elastomers than their chemical compositions, particularly when general-purpose hydrocarbon rubbers are concerned. The surface energy of a material, γ, is defined as the energy necessary to create a unit new surface. This energy is comprised of different types of cohesive forces, such as dispersive, dipole–dipole, induced dipole–dipole, acid-base, and hydrogen bonding. Since all these cohesive forces can be involved in independent ways, the surface energy can be expressed as the sum of several components, each corresponding to a type of molecular interaction (dispersive, polar, acid-base, hydrogen bond, etc.). The dispersive component of the surface free energy, γ^d, is particularly important since the effect of the dispersive force is universal. If a solid substance can have only a dispersive interaction with its environment, the surface energy of the solid, γ_s, is identical with its dispersive component, γ_s^d. For most substances, the surface energy of a solid is the sum of γ_s^d and γ_s^{sp} that is the sum with the other components of surface energy of the solid, termed "specific component" or "polar" component, therefore,

$$\gamma_s = \gamma_s^d + \gamma_s^{sp} \tag{3.1}$$

It is known that the possible interaction between two materials, 1 and 2, is determined by their surface energies. When only dispersive forces are responsible for the interaction, according to Fowkes' model [3], the energy of adhesion between these two materials would correspond to the geometric mean value of their γ_s^d:

$$W_a^d = 2(\gamma_1^d \gamma_2^d)^{1/2} \tag{3.2}$$

where W_a^d is the dispersive component of the adhesion energy. Similarly, the polar component of the adhesion energy, W_a^p, can be described by the polar components of their surface free energies [4,5]:

$$W_a^p = 2(\gamma_1^p \gamma_2^p)^{1/2} \tag{3.3}$$

Hence, the total adhesion energy, W_a, can be given by:

$$W_a = W_a^d + W_a^p + W_a^{ab} + W_a^h + \cdots; \tag{3.4}$$

or

$$W_a = 2(\gamma_1^d \gamma_2^d)^{1/2} + 2(\gamma_1^p \gamma_2^p) + W_a^{ab} + W_a^h + \cdots\cdots, \tag{3.5}$$

where W_a^{ab} and W_a^h are the adhesion energy due to acid-base interaction and hydrogen bonding. It is therefore concluded that polymer-filler interaction, filler-ingredient interaction, and filler–filler interaction in a given polymer system are determined by the filler surface energy, particularly where physical interaction is concerned.

3.3 SURFACE ENERGIES OF FILLERS AND ADSORPTION ENERGIES OF CHEMICALS ON FILLER SURFACE MEASURED BY IGC

Several techniques have been used to estimate filler surface energy, e.g., different contact angle measurements [6–8], and calorimetry for measuring heats of immersion [9–11]. Surface energy can also be evaluated from the adsorption behavior of fillers, especially using the thermodynamic parameters of gas adsorption. IGC has obvious advantages over the conventional methods since it is easy to operate, less time consuming, highly accurate, and provides ample information regarding filler surface characteristics [12–14].

In IGC, the filler to be characterized is used as the stationary phase and the solute injected is called a probe. When a chemical is eluted at infinite dilution, the adsorption energy of the chemical on the filler surface, ΔG°, can be calculated from the net retention volume [15,16]. The dispersive component of the filler surface energy, γ_s^d, can be derived from the adsorption free energies of a series of normal alkanes [17] and the specific component of the surface energy and its nature can be estimated from the specific (or polar) component of adsorption free energies of polar chemicals, I^{sp} [15,16]. The calculation of I^{sp} is based on the adsorption energies of alkanes and polar chemicals as a function of their surface area, σ_m (Figure 3.1). If, however, the surface is energetically heterogeneous, the values of parameters obtained from

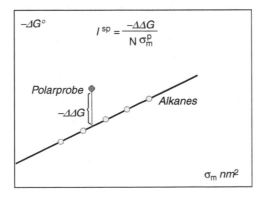

FIGURE 3.1 Schematic principle of estimating specific component of adsorption energy (N–Avogadro number; σ_m^p–area covered by a molecule of the polar chemical).

IGC measurement are average values over the whole surface of the fillers, but they are "energy-weighted" (i.e., the high-energy sites play a more important role in the determination of the adsorption parameters measured [18]). When the probe is eluted at finite concentration, the adsorption isotherms of the probes on the filler surface can be generated from the pressure dependence of the retention volume and, hence, a distribution of free energy of adsorption of the probe chemicals can be derived. In this paper, the polymer-filler and filler–filler interactions are characterized by the adsorption parameters of different chemicals on the filler surfaces and thus filler surface energies, measured at infinite dilution [18].

3.4 POLYMER-FILLER INTERACTION VS. SURFACE ENERGY OF FILLERS

According to Equation 3.5, the polymer-filler interaction is determined by the surface energies of the polymer and filler, their chemical natures and, in some cases, chemical linkages between filler surface and polymer chains. It has been demonstrated that for hydrocarbon rubbers and conventional fillers such as carbon black and silica, the polymer-filler interaction is essentially physical in nature [19–21]. So, for given hydrocarbon rubbers such as natural rubber (NR), polybutadiene (BR), and styrene–butadiene-copolymer (SBR) that are non- or low-polar polymers, the dispersive interaction is predominant. Therefore, the polymer-filler interaction characterized by the adhesion energy between the polymer and filler, W_a^{pf}, is determined by the dispersive component of filler surface energy, γ_s^d, by

$$W_a^{pf} = 2\sqrt{\gamma_p^d \gamma_s^d}, \tag{3.6}$$

where γ_p^d is the dispersive component of the polymer surface energy. The higher the γ_s^d of the filler, the stronger is the polymer-filler interaction.

The filler-polymer interaction can also be characterized by measuring adsorption energies of low molecular weight-analogs of polymers on the filler surface [16,22] with IGC. In this case, the contribution of the functional groups on polymer chains, such as the benzene ring in SBR or the -CN group in nitrile rubber (NBR), to the polymer-filler interaction can be estimated from the specific (or polar) components of the adsorption free energies, I^{sp}.

3.5 FILLER NETWORKING (AGGLOMERATION) VS. SURFACE ENERGY OF FILLERS

It is known that a certain amount of energy input is needed during mixing to break down the agglomerates and to disperse the aggregates in the polymer matrix. However, it has been demonstrated that when the filler is well dispersed in the polymer, the aggregates tend to reagglomerate during storage and vulcanization of the uncured compound, especially at high loading, leading to chain-like filler structures or clusters [23,24]. These are generally termed secondary structure or, in some cases, filler network. This imparts a significant effect on the properties of filled rubber, especially rheological properties of the uncured compounds and viscoelastic properties of the vulcanizates. Obviously, filler network formation in the polymer matrix is mainly determined by the attractive force between filler particles or aggregates and the interaction between polymer molecules, as well as the interaction between filler and polymer. In a given polymer system, the driving force for filler networking originates from polymer-filler interaction and filler–filler interaction which is determined by filler surface energy and chemical nature [25].

As shown in Figure 3.2, in the filled compounds the agglomeration process may be simulated as:

$$\boxed{\text{F}\,\text{P}} + \boxed{\text{F}\,\text{P}} \longrightarrow \boxed{\text{F}\,\text{F}} + \boxed{\text{P}\,\text{P}}$$

In this system, when two filler particles agglomerate, a pair of filler particles and a pair of polymer units will be formed. This process can be dealt with as a kinetic process in which the change in potential energy is the driving force for filler networking. According to Wang, the change in potential energy can be estimated from total change in adhesion energy, ΔW, in the agglomeration process which is given by [25]

$$\Delta W = 2[(\gamma_f^d)^{1/2} - (\gamma_p^d)^{1/2}]^2 + 2[(\gamma_f^p)^{1/2} - (\gamma_p^p)^{1/2}]^2 + 2[W_f^h + W_p^h$$

$$- 2W_{fp}^h] + 2[W_f^{ab} + W_p^{ab} - 2W_{fp}^{ab}], \tag{3.7}$$

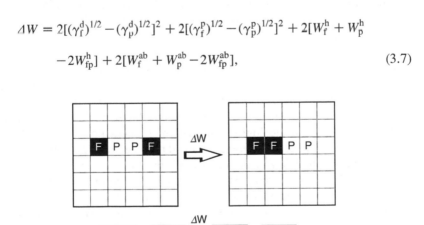

FIGURE 3.2 Change in energy associated with filler agglomeration process.

where γ_f^d and γ_f^p are the dispersive component and polar component of the surface energy of the filler, γ_p^d and γ_p^p are the dispersive component and polar component of the surface energy of the polymer, W_f^h, W_p^h, and W_{fp}^h the hydrogen bonding work, and W_f^{ab}, W_p^{ab}, and W_{fp}^{ab} the work from acid-base interactions between filler surface, polymer surface and between filler surface and polymer surface, respectively.

If $\gamma_f^d = \gamma_p^d$, $\gamma_f^p = \gamma_p^p$, $W_f^h = W_p^h = W_{fp}^h$, and $W_f^{ab} = W_p^{ab} = W_{fp}^{ab}$, then $\Delta W = 0$, the attractive potential between filler particles would disappear. This suggests that the driving force of filler networking is the difference in surface energies, both in the intensity and nature, between filler and polymer. In other words, from the thermodynamic point of view, the dispersed filler in a polymer matrix is stable only provided the energetic characteristics of the filler and polymer surfaces are identical, or if the adhesion energies between polymer and filler surface due to hydrogen bonding, acid-base interaction and/or other specific interactions are so high that they are able to offset the effect of the difference in surface energies between polymer and filler and cohesion energies of filler and polymer themselves. The greater the difference in the surface energies and the lower the specific interactions in terms of hydrogen bonding and acid-base interaction between filler and polymer, the higher is the tendency for filler networking in the polymer.

While the driving force for filler networking originates from the differences in surface characteristics between polymer and filler, the relative polarities of fillers also play an important role in filler networking in hydrocarbon rubber systems. Generally, while fillers have polar surfaces, the hydrocarbon rubbers are non- or very low-polar materials. The higher adsorption energy of a nonpolar chemical such as heptane, resulting from a higher γ_s^d of the filler surface, would indicate a greater ability of the filler to interact with hydrocarbon rubbers. However the higher adsorption energy of a highly polar probe, such as acetonitrile, resulting from the high γ_s^{sp}, would be representative of higher filler–filler interaction in hydrocarbon rubbers. It can be understood that when γ_s^d is constant, the increase of γ_s^{sp}, would promote incompatibility of the filler with hydrocarbon rubbers, thus enhancing aggregate agglomeration. On the contrary, at the same level of γ_s^{sp}, the higher γ_s^d would indicate higher filler-polymer interaction, hence a lesser degree of association of the aggregates would be expected. This is because, on the one hand, the higher γ_s^d will enhance polymer-filler compatibility. On the other hand, due to the strong adsorption of hydrocarbon rubber on the filler surface, the immobilized polymer chains on the filler surface may be treated as part of the filler, enhancing the similarity between filler surface and polymer surface. Based on this consideration, we have defined a specific interaction factor, S_f, as the adsorption energy of a given chemical, ΔG°, over that of an alkane (real or hypothetical), the surface area of which is identical with that of

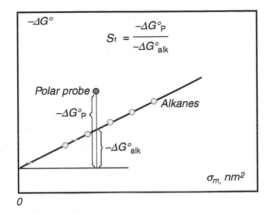

FIGURE 3.3 Definition of the S_f factor.

the given chemical, ΔG°_{alk} (Figure 3.3) [22]:

$$S_f = \frac{\Delta G^\circ}{\Delta G^\circ_{alk}} = f(\gamma_s^d, \gamma_s^{sp}) \tag{3.8}$$

In this definition, the reference state for calculation of ΔG° is the ΔG° of a hypothetical alkane with zero surface area, which is extrapolated from ΔG° of a series of n-alkanes. For a highly polar chemical, the higher the S_f of a filler, the more developed is the aggregate agglomeration in hydrocarbon rubbers.

3.6 POLYMER-FILLER AND FILLER–FILLER INTERACTIONS OF DIFFERENT FILLERS

3.6.1 CARBON BLACK

3.6.1.1 Surface Activity vs. Surface Area

Technically, the rubber grade carbon blacks are classified by their surface area and structure. With regard to surface activity in terms of polymer-filler interaction, the argument arises as which carbon black is more active: higher-surface-area (reinforcing) blacks or lower-surface-area (semi-reinforcing) blacks. Conventionally, polymer-filler interaction has been measured by bound rubber content, which is the portion of polymer in uncured compound being unable to be extracted by good solvent. It is generally caused by the adsorption of polymer chains on carbon black surface. While the total bound rubber content of the carbon black-filled rubber increases with increasing surface area at practical loadings, the *specific* bound rubber on unit surface area of carbon black decreases [21,26] as illustrated in Figure 3.4. Therefore, it had been concluded that the low-surface-area carbon blacks possess higher surface activity. This contradicts the results from surface energy measurements. The dispersive component of surface energy, γ_s^d, and the specific component of benzene, I^{sp}, measured by IGC at infinite dilution, are presented in Figure 3.5. Obviously, the polymer-filler interaction of low-surface-area carbon blacks is much lower than that of carbon blacks with high surface areas.

The contradiction can be explained based on the effect of multiple segment adsorption of polymer, and filler agglomeration on bound rubber formation [21]. It can be imagined that once one segment of a polymer chain is attached on the filler surface, the whole molecule becomes part of the bound rubber. In the case of multiple segment adsorption, two or more active sites of carbon black are occupied by the same molecule without increasing bound rubber, leading to a reduced effectiveness of the filler surface for bound rubber formation. Multiple segment adsorption can also take place between neighboring aggregates. In fact, it is the interaggregate multiple segment adsorption that, probably in combination with chain entanglement, keeps the carbon black and bound rubber together, forming a coherent mass. In addition, as mentioned before, carbon black aggregates may be associated with each other, forming agglomerates. Consequently, the bound rubber content per unit surface, calculated based on the assumption that no contacts among aggregates exist, would be underestimated. Moreover, multiple attachments and agglomeration are highly dependent on the distance between aggregates, i.e. the shorter the distance the higher would be the number of interaggregate multi-attachments and so would be the more developed agglomeration. On the other hand, the distance between filler aggregates in a compound is dependent on the degree of loading and carbon black surface area. At the same loading, the interaggregate distance is inversely proportional to the surface area [27]. As a result, compared with low-surface area carbon blacks, the high-surface area blacks should give a lower surface effectiveness for bound rubber formation, hence their specific bound rubber content will, therefore, be more underestimated. In fact, there are critical loadings below which a coherent mass of the swollen gel is unable to form during extraction for all carbon blacks. This loading decreases with increasing surface area of the black. When the comparison is made at the critical coherent loading where multiple attachments per unit surface can be assumed to be similar for all blacks, and to be free from effects of agglomeration, the specific bound rubber value increases with carbon black surface area (Figure 3.6) and is, to a first approximation,

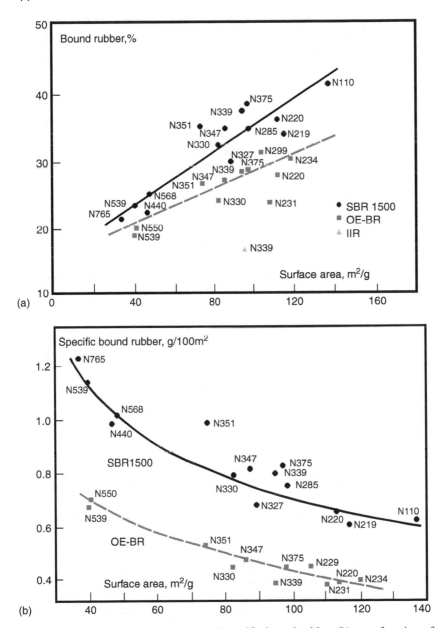

FIGURE 3.4 Total bound rubber (a) and specific bound rubber (b) as a function of surface area of carbon blacks.

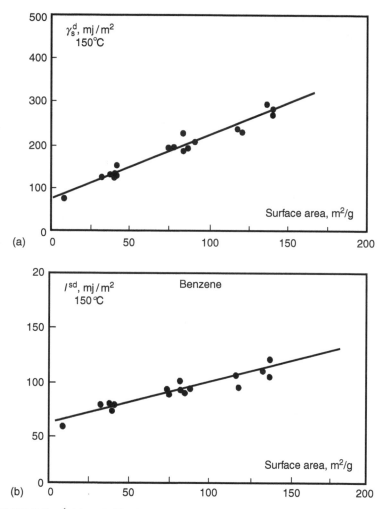

FIGURE 3.5 γ_s^d (a) and I^{sp} of benzene (b) as a function of surface area of carbon blacks.

directly proportional to the dispersive component of carbon black surface energy, γ_s^d, indicating that high-surface-area blacks are more active for rubber adsorption. This is in agreement with the direct observation of aggregate separation stress from the polymer matrix in carbon black-filled SBR vulcanizates by means of electron microscope [28]. The adhesion force between polymer-carbon black surface, at which a given percentage of carbon black separated from the vulcanizates increases in the order carbon black MT, GPF, FEF, and HAF, that is the same as surface area (Figure 3.7).

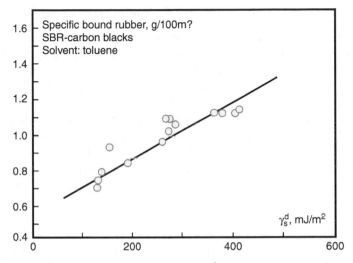

FIGURE 3.6 Specific bound rubber content *vs.* γ_s^d.

This suggests that while bound rubber is an important parameter in the mechanism of rubber reinforcement, it has little meaning per se as a criterion for the specific activity, except in cases when differences in filler morphology are absent or small. In this regard, IGC has shown a great advantage for estimation of filler surface activity.

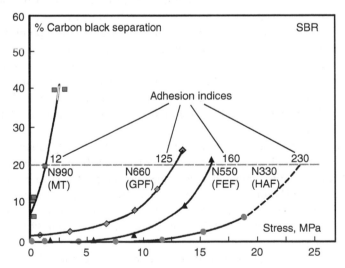

FIGURE 3.7 Carbon black separation as a function of stress in SBR.

3.6.1.2 Surface Activity vs. Oxygen-Containing Functional Groups

With IGC performed at finite concentration, the surface of carbon black has been identified to be energetically heterogeneous, and high energetic sites are found to be rich on the surface of high-surface area carbon black [18]. With regard to polymer-filler interaction, it is necessary to address ourselves to one specific question: what are the active centers on the carbon black surface? This question, much debated in the literature, can be answered by analyzing the results of a variety of carbon blacks by different approaches, but IGC provides lots of valuable information about the nature of the active centers.

The graphitization of carbon black at temperatures beyond 1500°C, leads to formation of a more regular graphitic crystal structure with only moderate alteration of gross morphology. Consequently, the γ_s^d and the adsorption energies of analogs of hydrocarbon rubbers measured by IGC are drastically reduced [22], and so are the bound rubber contents in hydrocarbon rubbers [29]. Therefore, it has been concluded that the edges and the defects of basal plans are mainly responsible for the high surface activity of carbon blacks. This conclusion was supported by the investigation of surface microstructure with scanning tunneling microscope (STM) [30]. From STM images it was found that, statistically, the dimension of the graphitic crystallites decreases with surface area of blacks, and upon graphitization the graphitic structure is so developed that there are only a small number of defects left on the surface.

If the active centers of carbon black surface are assigned to the defects and edges of surface crystallites, the statement may be challenged by the roles played by oxygen-containing groups that have been treated as active centers for rubber reinforcement [31]. This argument could also be supported by the facts that the oxygen groups on furnace carbon blacks generally increase with surface area and can be substantially eliminated during graphitization. However, the question about the function of oxygen groups on carbon black can be safely answered based on the surface energy measurements of oxidized and thermally treated carbon blacks with IGC, at least for the general purpose hydrocarbon rubbers.

For the carbon black N234, with surface area of 120 m^2/g, upon oxidation with nitric acid (acid dosage: 30 part of 65% acid per 100 part carbon black by weight), the volatile content that is a measure of oxidation increases from 2.4 to 5.2% and pH drops from 6.8 to 2.0. As a consequence of the introduction of more surface oxygen groups, the γ_s^d of the carbon black is reduced by 26%, and adsorption free energies $\Delta G°$ of heptane decreases from 38.5 kJ/mol to 32 kJ/mol at 180°C.

When the carbon black is heated at different temperatures beyond 200°C in an inert atmosphere (helium or nitrogen), the γ_s^d and $\Delta G°$ of alkanes increases considerably with temperature [22,32]. The γ_s^d of carbon black N234 as a function of temperature of heat treatment in an oven under nitrogen flow for four hours is shown in Figure 3.8. It can be seen that over the range of temperatures from 200 to 900°C, that is far below the graphitization temperature, the γ_s^d increases monotonously. It may be argued that this may be due to the removal of some substance adsorbed on the filler surface, which has blocked the active centers. This, however, does not appear to be a valid argument since, firstly, when the carbon black surface is heated below 200°C, where the oxygen-containing groups on black surface are stable, only minor change in surface energy of the blacks is observed. Secondly, the results of XPS and SIMS indicate that the concentrations of the oxygen groups, and all types of hydrogen on the black surface, decreases gradually with increase of temperatures. Moreover, Garten and Weiss [33] demonstrated that the number of acid groups of carbon blacks starts to decrease above 200°C, and other authors [34–37] showed that the oxygen groups decompose during thermal treatment (between 200 and 800°C) accompanied by the evolution of CO_2 and CO. Therefore, it is logical to assume that after decomposition of the oxygen groups during heat treatment, high energy sites would be left, and/or that some original high-energy spots that were screened by the oxygen groups would be revealed. This suggests that with regard to polymer-filler interaction, the oxygen-containing groups are much less active, especially in less- or nonpolar polymers. As a matter of fact, the results obtained from

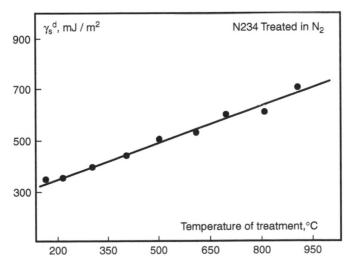

FIGURE 3.8 γ_s^d as a function of heat-treatment temperatures in nitrogen for carbon black N234.

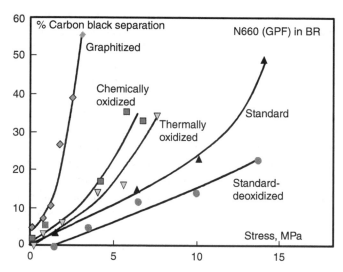

FIGURE 3.9 Carbon black separation as a function of stress for treated N660 blacks in BR.

IGC are in good agreement with the adhesion strength measured by electron microscopy [28]. The adhesion of carbon black GPF (N660) for BR is reduced by thermal and chemical oxidation, as illustrated in Figure 3.9. This effect is even more pronounced by graphitization of the black. When the surface of carbon black is deoxidized, the adhesion between polymer and carbon black is improved.

3.6.2 SILICA

3.6.2.1 Silica vs. Carbon Black

Silica is another important reinforcing agent for rubber with volume of consumption second only to carbon black. For hydrocarbon rubbers, the silica widely used is precipitated silica, which has been applied to passenger tire tread compounds to replace a significant volume of carbon black in the last decade. However, the compounding and processing techniques of the two fillers are quite different, as is their performance in tires. While their gross morphology, surface area, and structure, can be manipulated by changing the production parameters to match each other, in terms of rubber reinforcement, the difference between them is, beyond chemical composition, their surface characteristics. In Figure 3.10, the free energies of a series of normal alkanes and a variety of polar chemicals are presented as a function of the surface areas of their molecules, σ_m, for carbon black N234 and precipitated silica ZeoSil

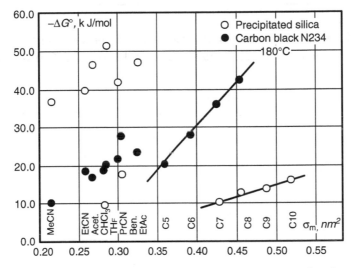

FIGURE 3.10 Adsorption energies of a variety of chemicals as a function of their surface areas for carbon black and silica.

1165. Both fillers represent the most popular fillers used in tread compounds of passenger tires. Their γ_s^d and I^{sp} for the polar chemicals are listed in Table 3.1. The specific interaction factor, S_f, of acetonitrile on a series of rubber blacks and precipitated silicas is also presented in Figure 3.11 as a function of their surface areas. As can be seen, regardless of the difference in their surface areas, the silica is characterized by higher free energies of adsorption, higher I^{sp}, and higher S_f for the polar chemicals, but the γ_s^d and the adsorption free energies of alkanes are substantially lower, relative to carbon black. However, when the comparison is made among rubber blacks, while the γ_s^d and I^{sp} increase with increasing surface area (see Figure 3.5), the S_f seems to be independent of morphology.

3.6.2.2 Silica vs. Modified Silica with Silane Coupling Agents

The reinforcing silicas, with respect to carbon black at comparable surface areas and structures, are characterized by a much lower polymer-filler interaction and higher tendency to form filler networking which, as will be discussed later, is not desirable for rubber reinforcement. However, when the silica surface is chemically modified with silanes, the specific component of surface energy is reduced drastically, whereas γ_s^d is only slightly decreased [38]. The effect of silica modification with *bis*(3-triethoxysilylpropyl)

TABLE 3.1

γ_s^d and Specific Components of Adsorption Energies of Some Polar Chemicals on Carbon Black and Silica (ZeoSil 1165) and TESPT-Modified Silica

	Carbon Black N234	Silica	Silica	Silica/TESPT
Test temperature, °C	180	180	130	130
γ_s^d, mJ/m^2	382	28	47	36
I^{sp}, mJ/m^2, Benzene	93	73	111	40
I^{sp}, mJ/m^2, CH$_3$CN	173	278		
I^{sp}, mJ/m^2, THF	74	271		
I^{sp}, mJ/m^2, Acetone	86	264		
I^{sp}, mJ/m^2, Ethyl estate	48	206		
I^{sp}, mJ/m^2, CHCl$_3$	72	39		

tetrasulfane (TESPT), a bi-functional silane, on its surface energies is also listed in Table 3.1, taking benzene as the polar chemical for estimation of surface polarity. With this modification, while the strong filler–filler interaction, or the tendency for filler networking, can be significantly depressed due to the reduction of surface polarity, the poor polymer-filler interaction of the modified silica, indicated by the low level of γ_s^d, can be compensated for by creating covalent bonds between the filler surface and elastomer matrix through sulfur linkages [39].

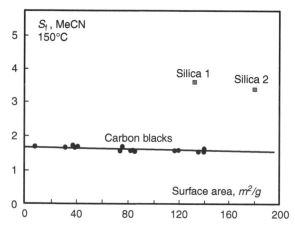

FIGURE 3.11 S_f of acetonitrile as a function of surface areas of carbon blacks and precipitated silicas.

3.7 EFFECT OF SURFACE ENERGIES OF FILLERS ON RUBBER REINFORCEMENT

In the next part of this chapter, the role of surface energy of the filler measured by IGC, hence polymer-filler interaction and filler–filler interaction in reinforcement of hydrocarbon rubber, will be demonstrated with different fillers having similar morphologies.

3.7.1 PROCESSABILITY OF FILLED RUBBER

The main properties of compounds that govern their downstream processability, such as extrusion and building of rubber products, tires in particular, are viscosity, die swell, and green strength.

3.7.1.1 Viscosity of Compounds

Fillers are known to change the rheological properties of compounds by several mechanisms:

- Hydrodynamic effect, related to a decrease in the volume fraction of the flow medium.
- Geometric effect, associated with the anisometry of the filler aggregates and occluded flow medium, which in turn increases the effective hydrodynamic volume of the filler [40].
- Adsorption of rubber molecules on the filler surface, forming an immobilized rubber or increasing polymer gel in the flow medium. It also causes a reduction of the molecular weight of the free rubber, as large molecules are adsorbed preferentially [41].
- Agglomeration of the aggregates, forming a filler network or interaggregate structure, which causes highly non-Newtonian and thixotropic behavior [42].

The latter two terms can affect the rheological properties mainly through polymer-filler interaction, and especially filler–filler interaction.

The Mooney viscosities measured at 100°C for a typical passenger tire tread compound, filled with the same loading of carbon black N234, silica, and TESPT-modified silica, are given in Table 3.2, along with the bound rubber contents determined by toluene extraction. Obviously, compared with the carbon black-filled compound, silica gives substantially higher viscosity. Although there are some differences in their morphologies, such a large difference in viscosity can be mainly, if not entirely, attributed to the highly developed and strong network of silica aggregates, as expected from its high

TABLE 3.2
Processability of Filled Compounds

	Carbon Black N234	Silica	Silica/TESPT
Bound rubber, %	42	27	58
Mooney viscosity, ML1+4 @ 100°C	82	150	59
Die swell, %	19	16	28
Green strength, MPa	0.41	0.62	0.34
Extrudate appearance, Garvey die @ 110°C	Excellent	Good	Very poor

Basic formulation: SSBR 75, BR 25, filler 80, oil 32.5, TESPT (only for silica) 6.4.

polarity. One of the consequences of filler networking is that, similar to occluded rubber, the rubber trapped in the network or agglomerates will be largely immobilized in the sense that this rubber portion is shielded from strains/stresses seen by the bulk rubber, as long as the agglomerates are not be broken down under the stress applied. As a result, the effective volume of the filler would be increased due to the filler agglomeration, hence the high viscosity of the compounds. This statement can also be strengthened by comparison of their bound rubber contents. Generally, high content of bound rubber would lead to high viscosity, as high resistance to the shearing strain would be created. This is because once a segment is anchored on the filler surface, the movement of the whole molecule in the flow field would be restricted, causing a reduction in the portion of free flowing polymer in the medium. Consequently, the effect of higher bound rubber content of black, due to its strong interaction with the polymer, would partially compensate for the difference caused by the less developed filler network in the black-filled rubber.

The conclusion about the effect of surface energies on the viscosity can be further verified by the silanization of silica. Upon TESPT-modification *in situ* during mixing, the viscosity of silica compound drops considerably to a level comparable to that of carbon black compounds. As the morphology of silica cannot be changed by the surface modification, and the bound rubber contents are significantly higher than the unmodified silica compound, which is probably due to the formation of chemical linkages between silica surface and polymer chains via coupling reaction during mixing, the decrease in filler–filler interaction due to the silane modification is apparently responsible for the lower compound viscosity.

3.7.1.2 Die Swell of Compounds

In practice, the elastic response of rubber compounds is reflected in their processing behavior in terms of die swell (extrusion shrinkage) and surface roughness of the extrudate. Die swell is defined as the ratio between the area of the cross-section of the extrudate and that of the die and is generally greater than unity. This phenomenon, which is associated with elastic recovery, is caused by the incomplete release of long-chain molecules orientated by shear in the die (or capillary) and occurs in the rubber phase alone [43]. It is, therefore, obvious that besides the test conditions such as temperature, extrusion rate, and geometrical features of the die, the primary factor influencing the die swell of unfilled elastomers is the entanglement of elastomer molecules, which, in turn, is determined by their molecular weight and molecular weight distribution. Die swell is improved by the addition of fillers, due to the reduction of the elastic component of the compound and the decrease in effective relaxation time [44]. It has been recognized that the primary filler parameter to govern the die swelling is its structure. With increasing structure, more rubber will be occluded within the aggregates, which is at least partially "dead," so that the effective polymer volume, hence die swell, will be reduced. Indeed, the die swell of filled compounds has been adopted as a measure of carbon black structure [45]. The data in Table 3.2 show the die swell and extrudate appearances for passenger tire tread compounds containing silica, TESPT-modified silica, and carbon black. The die swell is measured with a Plasti–Corder Extruder operating at extrusion output rate of 87 mL/min and die temperature of 70°C (die diameter of 4.75 mm). While the die swell is much lower for the unmodified silica-filled compounds, the highest value is found for modified silica. Apparently, the high die swell of modified silica compounds cannot be attributed to the silica structure. It may be due to its high bound rubber content, related to chemical reaction via the coupling agent. In this case, the filler particles behave like multiple crosslinks, increasing the elastic memory of the polymer matrix. It may also be related to filler agglomeration, as the agglomerates can be treated as highly structured fillers and the trapped rubber can be treated as occluded rubber, so long as they cannot be broken down under the sheer rate and temperature used for extrusion. Therefore, the lower die swell of the silica compound would also be expected from its highly developed filler network. Compared with TESPT-silica compounds, the more developed filler networking with less bound rubber content leads to a much lower die swell.

Accordingly, due to the lower filler networking and high bound rubber which, as will be discussed later, is most desirable for tread compounds in terms of tire performance, the surface modification of silica results in high compound elasticity that yields very poor appearance of the extrudate surface. Therefore, one of the deficiencies of silica tread compounds is its poorer processability related to extrusion, as the silane coupling agent has to be used to improve polymer-filler interaction and to depress filler–filler interaction for the most desirable tire performances.

3.7.1.3 Green Strength of Compounds

Green strength is important during the building of rubber products. For example, in the construction of radial-ply tires, the uncured compounds between the cords in the carcass may be subject to extensions of up to three times their original dimension during the building process [46]. Moreover, a high green strength is necessary to prevent the uncured tires from creep and distortion prior to molding. Compared to NR, the low green strength of certain types of synthetic rubber such as SBR and BR has been attributed to their lack of strain-induced crystallization. However, fillers also have an important effect on the stress–strain behavior of uncured compounds [47,48]. It was found that besides filler morphology, the filler surface energy is another important factor influencing green strength. The green strengths of filled compounds, based on a blend of SBR and BR, are shown in Table 3.2. While the green strength is higher for the silica-filled compound, its yield strain is much lower. Whereas the high green strength of carbon-black-filled rubber is related to higher polymer-filler interaction, the high green strength of the silica-filled compound originates from a strong filler network that is rigid, brittle, and characterized by very low yield strain. The lower green strength and yield strain of the compound filled with the TESPT-modified silica is, of course, associated with weaker polymer-filler and filler–filler interactions (i.e., the low specific and dispersive components of surface energy of this silica). This effect is partially compensated for by its high bound rubber content and possible precrosslinking between silica surface and polymer during mixing.

3.7.2 STRESS–STRAIN BEHAVIOR OF FILLED VULCANIZATES

Stress–strain curves of SSBR–BR vulcanizates with 80 phr fillers and 32.5 phr oil are shown in Figure 3.12. At relatively low strain, the stress of the N234-filled vulcanizate is lower than for the unmodified silica, but increases rapidly with increasing strain and reaches the level of the silica vulcanizate at $\lambda = 1.2$. This phenomenon was also expected from their surface characteristics. As reflected by γ_s^{sp}, the strong filler–filler interaction of the silica leads to a well-developed filler network, not only in the uncured compounds, but also in the vulcanizates. The rubber trapped in this secondary network is prevented from efficiently sharing the strain of the polymer matrix, which, as mentioned before, leads to an apparent increase in the filler volume fraction, and ultimately to a higher modulus. As the filler network is highly strain dependent, it is destroyed at high levels of strain, releasing the trapped rubber. This results in lower apparent filler volume loading and, hence,

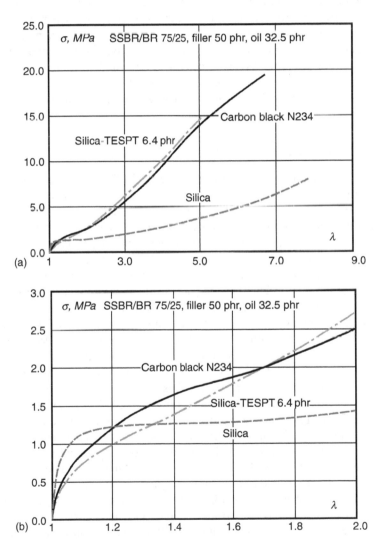

FIGURE 3.12 Stress–strain curves for SSBR/BR vulcanizates filled with carbon black N234, silica (ZeoSil 1165), and silica/TESPT.

lower modulus. At higher strain, the filler network will have disappeared, and filler-polymer interaction will be the dominant parameter. The much lower modulus of the silica-filled vulcanizate in comparison to carbon black at high elongation can be explained by weaker polymer-filler interaction leading to slippage between rubber molecules and the silica surface [49,50].

After silanization of the silica surface, the depressed filler network should result in lower stress at low strain, even compared with that of carbon black. This is clearly apparent from Figure 3.12b. At high strain, it is much higher than that of its unmodified counterpart, and close to carbon black. This phenomenon is certainly attributed to covalent polymer-filler bonds created by the coupling reaction. As mentioned earlier, the modulus of the vulcanizate at high strain is greatly influenced by polymer-filler interaction. This chemical interaction will prevent slippage of the rubber molecules on the filler surface.

3.7.3 DYNAMIC PROPERTIES OF FILLED VULCANIZATES

One of the consequences of incorporation of filler into a polymer is a considerable change in the dynamic properties of the rubber, both modulus and hysteresis. This phenomenon has been investigated in depth, especially in relation to rubber products. It has been recognized that, for a given polymer and cure system, the filler parameters influence dynamic properties in different ways. Among others, filler networking, both its architecture and strength, seems to be the main (though not the only) parameter to govern the dynamic behavior of the filled rubber. The impact of fillers on viscoelastic behavior of rubber has been reviewed in depth in several articles [25,51–53].

Figure 3.13 shows a plot of elastic modulus, G', measured at 0°C and 70°C and 10 Hz vs. the logarithm of the double strain amplitude (DSA) for the vulcanizates filled with different fillers at the same loading. As can be seen, over the range of DSA tested, the modulus decreases with strain amplitudes, showing a typical nonlinear behavior. This phenomenon was studied extensively by Payne, after whom the effect is often named [52,54]. The decrease in elastic modulus upon increasing strain amplitude was attributed by Payne to "the structure of the carbon black, and may be visualized as filler–filler linkages of physical nature which are broken down by straining" [55]. This structure was further clarified by Medalia as that "interaggregate association by physical forces, not the 'structure' or aggregate bulkiness," as generally termed in the rubber industry [53]. This suggests that the Payne effect is related to the filler network formed in the polymer matrix. As discussed before, the increased effective volume of fillers due to filler networking will result in increased modulus. The breakdown of the filler network by increasing strain amplitude would release the trapped rubber so that the effective filler

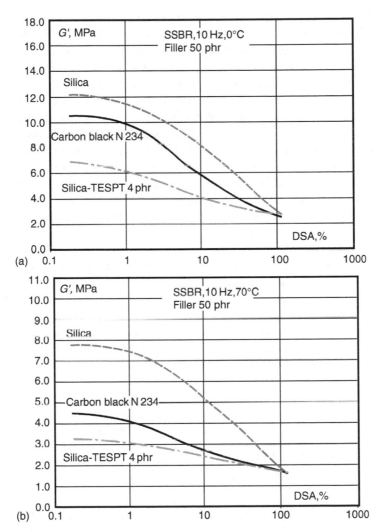

FIGURE 3.13 Strain dependence of elastic moduli at 0 (a) and 70°C (b) for solution SBR filled with carbon black N234, silica (HiSil 210), and silica with 4 phr TESPT.

volume fraction, and hence the modulus, would decrease. In fact, the Payne effect, measured by the difference in moduli, $\Delta G'$, at low and high strain amplitudes, has been used as a measure of filler networking. Bearing this in mind, the decreases in elastic moduli at low strain amplitudes, and the Payne effects in the order from silica-, carbon black- to silane-modified-silica-filled compounds, are expected from their surface characteristics.

The dynamic hysteresis of the rubber can be measured by loss tangent, or tan δ. By its definition, tan δ is a ratio of loss modulus G'' to G', which is representative of work converted into heat (or the work absorbed by the compound) to that recovered, for a given work input during dynamic strain. Besides the hydrodynamic effect, the influence of filler on G' and G'' involves different mechanisms and different strain dependencies, with both mechanisms influencing tan δ. While G' is mainly related to the filler network which subsists during dynamic strain, G'' is related to the breakdown and reformation of these structures [56]. Consequently, beyond the contribution of the polymer matrix, the factors predominantly determining tan δ would be the state of filler-related structures, or more precisely, the ratio between the portion capable of being broken down and reconstituted, and those remaining unchanged during dynamic strain. The change in tan δ, due to changing temperature, would reflect the ratio of these two processes.

The tan δ obtained from temperature sweeps at DSA of 5% and frequency of 10 Hz are shown in Figure 3.14. It can be seen that the silica compounds are quite distinctive in their temperature dependence of loss factors. Relative to the carbon black vulcanizates, the silica compounds show very low tan δ in the transition zone of the polymer and high hysteresis at high temperature. These features can be interpreted as resulting from different functions of the filler network in different temperature regimes, relative to the polymer which can be summarized as follows [57]:

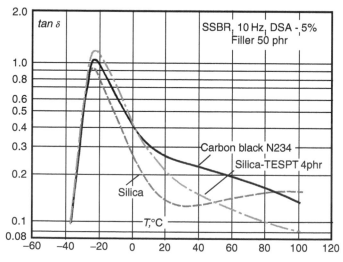

FIGURE 3.14 tan δ as a function of temperature for solution SBR filled with carbon black N234, silica (HiSil 210), and silica with 4 phr TESPT.

- To the extent that the filler network cannot be broken down under the applied strain, the formation of a filler network would substantially increase the effective volume fraction of the filler due to the rubber trapped in the agglomerates, leading to high elastic modulus and low hysteresis.
- The breakdown and reformation of the filler network would cause energy dissipation, hence higher hysteresis during cyclic strain would be expected.
- In the transition zone of the polymer at low temperature, where the main portion of the composite for energy dissipation is polymer matrix, and the filler network may not be easily broken down, the hysteresis may be significantly attenuated by the filler networking, due to the reduction of the effective volume of the polymer.
- Also in the transition zone, once the filler network can be broken down and reformed under a cyclic deformation, the hysteresis can be substantially augmented through release of polymer to participate in energy dissipation. The breakdown and reformation of the filler network would cause an additional energy dissipation, hence higher hysteresis during cyclic strain would be expected.

Accordingly, the low hysteresis of silica compounds in the transition zone of the polymer is certainly associated with the feature of more and stronger filler agglomerates. The large volume of trapped rubber is unable to participate in energy dissipation, as polymer, per se, gives very high hysteresis. In addition, the stronger filler network would also result in depressing another energy dissipation process involving the disruption and reformation of the filler network. Obviously, the increase of tan δ with temperature in the rubbery state beyond 20°C, where the hysteresis of the polymer is very low, represents the increase in the portion of filler network being involved in internal energy dissipation. This portion increases continuously, reaching a maximum in the temperature range of 85–95°C. This is indicative that the network is more developed and/or stronger than that of the carbon blacks. Consequently, there seems to be two peaks in the tan δ-temperature curve for the silica-filled rubber; one is related to polymer, the other corresponds to a filler network in which filler–filler interaction and polymer-filler interaction, hence mobility of polymer segments is also involved. Depending on the number and strength of the agglomerates, these two peaks may be distinctly separate as for silica, or merged together, leading to a broad peak, as is the case for carbon black.

Based on the same argument, the low filler–filler interaction, hence less and weaker filler network of silane modified silica, manifests itself primarily in a tan δ-temperature curve, featuring high hysteresis at low temperature, lower hysteresis at high temperature, and with a narrower peak in the polymer transition zone.

It should be pointed out that with regard to tire applications, it has been well established that repeated straining of the compound due to rotation and braking can be approximated as a process of constant energy input involving different temperatures and frequencies [58,59]. Rolling resistance, for example, is related to the movement of the whole tire, corresponding to deformation at a frequency of 10–100 Hz and a temperature of 50–80°C. In the case of wet skid, the stress is generated by resistance from the road surface and movement of the rubber at the surface, or near the surface of the tire tread. The frequency of this movement depends on the roughness of the road surface, but should be very high, probably around 10^4–10^7 Hz at room temperature [59,60]. It is, therefore, obvious that any change in dynamic hysteresis of the compounds at different frequencies and temperatures will alter the performance of the tire. Since certain tire properties, such as wet skid resistance, involve frequencies that are too high to be measured, these frequencies are reduced to a measurable level at lower temperatures by applying the Time-Temperature Equivalence Principle (or WLF Temperature-Frequency Conversion). The reduced temperature for different tire properties at low frequency has been used as the criterion for polymer and filler development for tire compounds. From the point of view of dynamic hysteresis, an ideal material, which is able to meet the requirement of a high-performance tire, should give a low tan δ value at a temperature 50–80°C in order to reduce rolling resistance and save energy. High hysteresis at lower temperature, for example from −20 to 0°C, is also needed in order to obtain high wet grip, even though the factors involved skid resistance are recognized to be more complex than a single compound property. It can be seen from Figure 3.14 that the tread compound of silane-modified silica would give lower rolling resistance and better wet skid resistance compared to carbon black compound. This, with an acceptable wear resistance due to increased polymer-filler interaction *via* chemical linkages, which will be discussed later, is the basis of successful application of silica in the "green tire."

The conclusion described above can be further demonstrated with a typical passenger tire tread compound consisting of a blend of solution SBR (SSBR) and BR in a ratio of 75/25, 75 phr filler, 25 phr oil with 6 phr TESPT for silica. As can be seen, the Payne effect in the silica compound is substantially lower than that in the carbon black compound (Figure 3.15). Accordingly, an improved balance of tan δ at different temperatures can be achieved with surface-modified silica (Figure 3.16).

3.7.4 ABRASION RESISTANCE OF FILLED VULCANIZATES

Abrasion, or wear, is one of the rubber properties that is strongly affected by fillers. It is generally recognized that the mechanism of rubber abrasion is

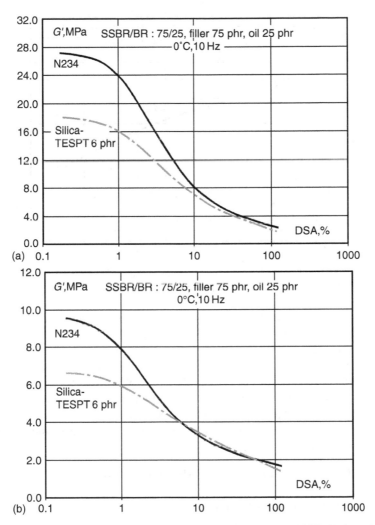

FIGURE 3.15 Strain dependence of elastic moduli at 0 (a) and 70°C (b) for solution SBR/BR filled with carbon black N234 and silica (ZeoSil 1165)/TESPT.

highly complex, involving not only mechanical failure of the material, such as fatigue, tearing, etc., but also mechano-chemical and thermo-chemical processes. Irrespective of the complex mechanism of abrasion, which is not yet fully understood, an enormous amount of work has been carried out with regard to the wear resistance of tires, and a wealth of knowledge concerning the effects of basic properties of carbon blacks on abrasion resistance has been accumulated. This has been the subject of several review papers [61,62]. Most

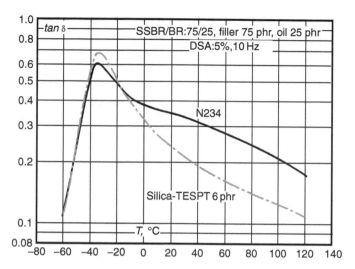

FIGURE 3.16 tan δ as a function of temperature for solution SBR/BR filled with carbon black N234 and silica (ZeoSil 1165)/TESPT.

of the studies have been focused on the filler morphology and filler dispersion in the polymers and very few studies have dealt with the effect of surface activity, even though the effect of the surface activity of carbon blacks on abrasion has long been recognized. Here again, the reason is a lack of effective tools to address the surface activity of the fillers. With IGC, the role of surface activity in abrasion resistance can be comprehensively investigated, some basic phenomena that have been observed in the practice of tire wear resistance can be better understood and, most importantly, the abrasion resistance of the filled compounds can be further optimized.

3.7.4.1 Effect of Carbon Black on Abrasion Resistance

3.7.4.1.1 Effect of Surface Area

Surface area (or particle size) is generally considered to be one of the most important parameters of carbon blacks contributing to abrasion reinforcement [63,64]. Rubber filled with high-surface-area carbon black usually shows a higher abrasion resistance. It was reported that above a certain surface area limit, abrasion resistance cannot be increased further [65,66]. In fact, poor abrasion resistance was observed for carbon blacks with very high surface areas. At a constant loading of 65 phr carbon black with 35 phr oil, the optimum surface area for an SBR/BR blend system lies between 130 and 150 m²/g (Figure 3.17). At lower loading, the increase in abrasion resistance with increasing surface area, may be attributed to several mechanisms related

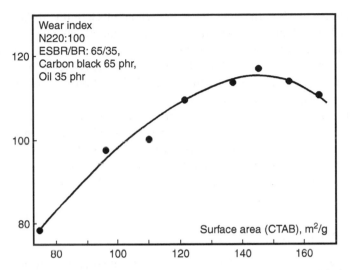

FIGURE 3.17 Wear resistance as a function of surface areas of carbon blacks.

to increasing effective interfacial area between filler and polymer. It may also be related to polymer-filler interaction, as the γ_s^d and I^{sp} of benzene increase with surface area (see Figure 3.5). The lower abrasion resistance of rubber filled with very fine carbon blacks may be attributed to poor carbon black dispersion caused by the reduced interaggregate distance that is inversely proportional to the surface area of carbon blacks [27]. This is confirmed by the increase in agglomerate size with increasing surface area. It is also confirmed by the dependence of abrasion on loading, which is similar to that on surface area [66]. For a given carbon black, the rate of abrasion passes through a maximum, as the black loading is increased to very high concentrations at which the dispersion of the filler in the polymer matrix becomes difficult, thus resulting in poor abrasion resistance. The poorer dispersion of the high-surface-area blacks may also be associated with the increased filler–filler interaction. When considering the adhesion energies between different materials, while the polymer-filler interaction determined by W_a^{pf} varies with the square root of the γ_s^d of the filler (see Equation 3.6), according to Equation 3.2 the filler–filler interaction characterized by cohesive energies between filler surface, W_c^{ff}, increases linearly with γ_s^d:

$$W_c^{ff} = 2\gamma_s^d \tag{3.9}$$

This suggests that with increasing surface area of carbon black, and hence γ_s^d, the filler–filler interaction increases more rapidly than polymer-filler interaction, making the fine-particle carbon black more difficult to be dispersed. In addition, rubber filled with very fine blacks may give rise to higher heat generation due to increased hysteresis, causing high surface temperatures, which accelerate the thermo-oxidative degradation of the polymer. All these effects promote crack initiation and fatigue crack growth, leading to an increased abrasion rate.

The statement about the effect surface energies on abrasion resistance for different surface area blacks seems to be farfetched at first glance, as it is hardly to be separated from the effect of morphology. However, an approach to the problem can be made by the investigation of simple heat-treatment of blacks on the abrasion resistance, from which the effect of change in morphology has been omitted.

3.7.4.1.2 Effect of Heat Treatment

When the carbon black was heated in an inert atmosphere (nitrogen in this case), the γ_s^d of the carbon black increases with increasing temperature of heat treatment up to 900°C, which is far below the graphitization temperature (about 1500°C) (Figure 3.18). When these treated blacks (N234) are compounded into emulsion SBR with 50 phr loading, the abrasion resistance is shown to improve with the temperature up, to a level around 600°C, thereafter a gradual deterioration of the property is observed, even though the γ_s^d steadily goes up. Since over the range of the temperatures used, the

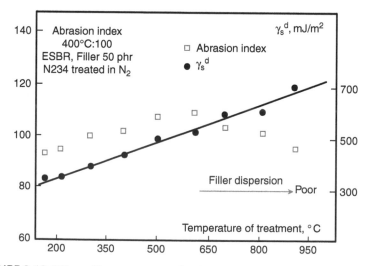

FIGURE 3.18 Effect of heat-treatment of carbon blacks on abrasion resistance.

morphologies of the carbon black cannot be changed, the variation of abrasion resistance of the filled compounds upon heat treatment can only be interpreted in terms of their surface characteristics. As discussed before with regard to the effect of surface area, the increase in abrasion resistance of the blacks treated at relatively low temperature is certainly attributed to the enhancement of polymer-filler interaction, as shown by γ_s^d. The negative impact of further temperature increase appears to be also related to the poorer dispersion of the filler, due to the stronger filler–filler interaction originated from the very high γ_s^d. In fact, for the carbon black treated at temperatures beyond 600°C, the undispersed area of black in the vulcanizates measured by dispersion analyzer increases with treatment temperatures. It seems that the positive impact of the improved polymer-filler interaction on the abrasion resistance is more than offset by the deterioration of filler dispersion. It may also be possible that when the γ_s^d or polymer-filler interaction is too high, the abrasion resistance may not be further improved, thus, the increased abrasion rate is mainly caused by poor dispersion.

3.7.4.1.3 Effect of Chemical Adsorption on Filler Surface

The impact of filler surface characteristics on abrasion resistance may be further verified using physical adsorption of chemicals. In a comprehensive study, Dannenberg, Papirer, and Donnet [67] modified the carbon black N339 with surfactant cetyltrimethyl ammonium bromide (CTAB), either by adding this chemical to the mix or by preadsorption onto the carbon black surface. The modification results in a reduction of bound rubber content and a significant drop in abrasion resistance, measured by the Akron Angle Abrader (see Table 3.3) even though the dispersion of the filler was actually improved. Although the surface activity can be estimated from bound rubber, the reduction of polymer-filler interaction by CTAB adsorption can be clearly demonstrated by IGC. Shown in Figure 3.19 are the adsorption energies of heptane, measured at 180°C for a series of modified carbon blacks as a function of the amount of CTAB adsorption. It can be seen that even a small amount of CTAB adsorbed on the surface, which is less than a monolayer, deteriorates the surface activity of carbon black significantly, reducing the abrasion resistance.

The correlation of surface characteristics, obtained using a series of protein-treated blacks with abrasion resistance, also leads to recognition of the usefulness of IGC in understanding rubber reinforcement. When the carbon black was treated in aqueous solutions of bovine albumin, a model compound used for protein in natural rubber, the reduced polymer-filler interaction, indicated by adsorption energies of heptane, results in significant loss in abrasion resistance (Figure 3.20).

TABLE 3.3
Effect of CTAB Adsorption on Abrasion Resistance

Carbon Black		Carbon Black N339		
CTAB	No	2.25 phr Added to Mix	0.9% Adsorbed	4.3% Adsorbed
Bound rubber, %	32.6	31.3	24.5	18.0
Abrasion index[a], %	100.0	67.2	63.8	52.9

Formulation: SBR 1500 100, carbon black 50, zinc oxide 3, stearic acid 1.5, antioxidant 1, oil 8, accelerator CBS 1.25, sulfur 1.25.

[a] Akron angle abrasion.

3.7.4.1.4 Effect of Carbon Black Mixing Procedure

Based on the results generated from IGC, the mixing practice of carbon black can be optimized to further improve the abrasion resistance of rubber products.

For chemicals having the same cross-sectional area, polar chemicals give higher adsorption energies than alkanes that can be taken as model compounds of hydrocarbon polymers and oils (Figure 3.10). This suggests that when

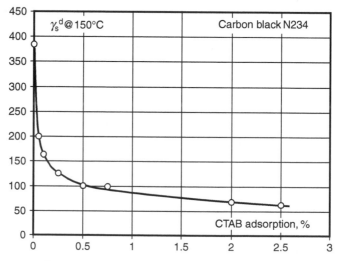

FIGURE 3.19 γ_s^d as a function of CTAB adsorption on carbon black N234.

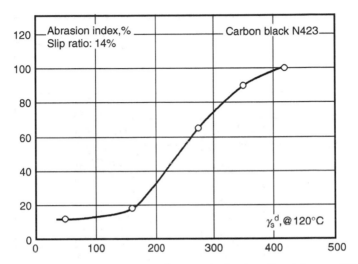

FIGURE 3.20 Abrasion resistance *vs.* γ_s^d of protein (bovine albumin) treated N234 for NR compounds.

carbon black comes in contact with polar ingredients, such as antioxidants and stearic acid, the amount of available surface area and the available number of filler active centers for polymer adsorption on filler surface can be substantially reduced. Once molecules of these ingredients are adsorbed on the filler surface they are very difficult to be replaced by polymer chains. For oil, which has better compatibility with polymer and low polarity, the competition for the higher energetic sites on the filler surface between polymer molecules and oil would make the mixing sequence equally important, especially for a carbon black which has higher interaction with hydrocarbon materials. Therefore, from the point of view of polymer-filler interaction, the mixing procedures of carbon black-filled compounds should be optimized in such a way that the oil and other ingredients are added after the filler is incorporated into the polymer. For example, for a typical passenger tire tread compound using a blend of SSBR/BR (75/25), 75 phr carbon black N234 and 25 phr oil, the bound rubber content was 40% when the oil was added with carbon black into the masticated polymers, followed by the addition of zinc oxide, stearic acid, and antioxidants. However, when the carbon black was added to the polymer first and oil was added after carbon black had been incorporated, the bound rubber content increased to 47%. Consequently, the abrasion resistance was significantly improved for the compound prepared with delayed addition of oil and other small ingredients as shown in Figure 3.21.

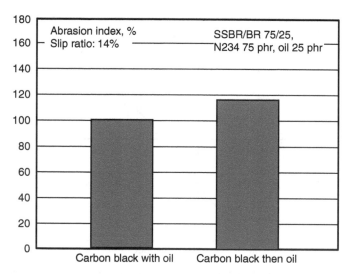

FIGURE 3.21 Effect of mixing sequences of ingredients on abrasion resistance of carbon black-filled vulcanizates.

3.7.4.2 Effect of Silica on Abrasion Resistance

3.7.4.2.1 Silica vs. Carbon Black

Since the beginning of the industrial-scale production of fine-particle silicas and silicates in 1948, silica manufacturers have been seeking to have their products used in all rubber products. Whereas silicas were rapidly able to replace up to 100% of carbon black in shoe sole materials, and also made their way into the mechanical goods sector, mostly as blends with carbon blacks, their use in tires in any quantities worth mentioning has long been limited to two types of compounds. One is the application to off-the-road tread compounds, containing 10–15 phr of silica blended with carbon black, in order to improve tear properties. Another is in textile and steel cord bonding compounds containing 15 phr of silica, again blended with carbon black, often N326, in combination with resorcinol/formaldehyde systems. The reasons why it could not be used as principal filler in tire tread compounds until recently can be understood again from its surface characteristics, determined by IGC (see Figure 3.10 and Table 3.1). Compared to carbon black, the silica has been shown to possess a strong tendency to agglomerate, forming a developed filler network. This results in poorer compound processability and unacceptable dynamic properties, as discussed before. The lack of strong polymer-filler interaction is an equally important factor contributing to the lower failure properties, such as ultimate strength and abrasion resistance, as shown in Figure 3.22. Without a coupling agent, the abrasion resistance of the silica in a typical passenger tire tread compound is reduced by about 70%, as compared with carbon black. These, in combination with poorer dynamic properties and poor cure characteristics, prevent the use of silica in tread compounds [68,69].

In 1992, a new tread compound for passenger tires was patented by Michelin [70] and the tire made with this compound was soon commercialized with the name "green tire." It was claimed that, besides the increased fuel efficiency which is generally related to the low rolling resistance, improved wet and snow traction, the tread life, and durability is comparable to those of the conventional tire. This suggests that the tread wear resistance is equal to that of carbon black compounds. From the material point of view, this compound is characterized as follows:

- The main component of the polymer blend is solution SBR.
- The compound contains silica as the majority component of the filler system. Only 6.4 phr carbon black was used as the carrier for the coupling agent.
- The bifunctional silane TESPT is used as a coupling agent between polymer and filler.

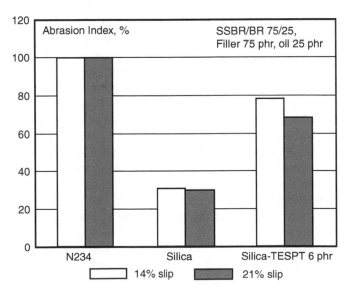

FIGURE 3.22 Abrasion resistance of vulcanizates filled with carbon black N234, silica (ZeoSil 1165) and silica/TESPT.

Among these, the application of sulfur containing silane-coupling agent is the key technology. By organo-silane-surface modification, filler networking can be effectively depressed via reduction of the specific component of the silica surface energy which, as discussed before, leads to significantly improved compound processability, and improved dynamic properties. The poorer polymer-filler interaction can also be significantly compensated for by chemical linkages between polymer and filler via sulfide group. As a result, the abrasion resistance of the compound is considerably promoted (Figure 3.22). Therefore, due to the surface modification, the compound properties can be improved in such a way that it gives low rolling resistance, better wet skid resistance (see Section 3.7.3), and acceptable wear resistance, which is the foundation of the silica-tread compound [71].

3.7.4.2.2 Silica in Emulsion SBR Compounds

As mentioned above, one of the features of the silica-tread compound described in the Michelin patent is the polymer system: the principal rubber is solution SBR. However, for typical passenger tread compounds, the dominant polymer is emulsion SBR (ESBR), even though SSBR had been commercialized for a few decades and its advantages in rolling resistance and skid resistance, though not significant, have been recognized [59]. Since the introduction of the silica-based tread compounds, although great effort has been made by tire manufacturers to use ESBR in silica compounds due to its better processability and low cost, the solution polymerized rubbers (including BR) are still the exclusive polymers for silica tires. Among other reasons, poor abrasion resistance is the key deficiency of silica in ESBR. While this inferior property is related to its weak polymer-filler interaction, unlike that of SSBR, it cannot be compensated for by a coupling reaction, as illustrated in Figure 3.23. Even with a high dosage of coupling agent, the abrasion resistance is far from competing with that of carbon black compounds.

Several factors may be responsible for the poor abrasion resistance of ESBR-silica compounds. Firstly it may be related to the efficiency of the coupling agent. Compared with carbon black, the silica surface has a stronger interaction with polar chemicals, characterized by the higher adsorption-free energies of polar probes such as acetonitrile, as measured by IGC (see Figure 3.10 and Table 3.1). In the SSBR-based passenger tire tread compound, the coupling agent can be driven to the filler surface by polar interaction. This facilitates the coupling reaction, and the poor polymer-filler interaction can be effectively offset by chemical bonds between polymer and filler via the coupling agent. However, in the case of ESBR-based tread compounds, the nonrubber impurities originating from the production process

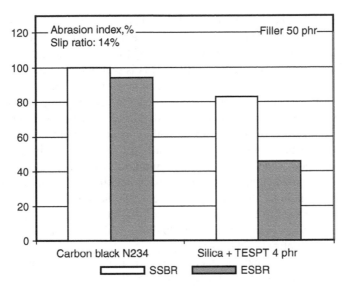

FIGURE 3.23 Abrasion resistance of carbon black N234 and silica (ZeoSil 1165)/TESPT in solution and emulsion SBRs.

of emulsion SBR interfere with the coupling reaction. It is known that in ESBR, there are about 5–8% nonrubbers that are mainly surfactants. As the silica surface is highly polar in nature, it readily accepts the polar groups of these substances, leaving their non- or less-polar groups to interact with the polymer. In other words, the surfactants in ESBR can be easily adsorbed on the filler surface, weakening polymer-filler interaction and filler–filler interaction. They also block the silanols on the silica surface, preventing chemical coupling reaction. In addition, the polarity of polysulfide groups in TESPT and/or their decomposition products is higher than that of the hydrocarbon rubber. In the compound they may preferably associate themselves with the nonrubber substances, reducing the reaction between coupling agent and polymer chains. Consequently, the poor polymer-silica interaction cannot be efficiently compensated for by the application of coupling agents, even at very high dosages. Such an argument can be further strengthened by the issues raised from development of silica compounds for truck tires.

3.7.4.2.3 Silica in NR Compounds

Although silica has been used with solution SBR in passenger tread compounds to improve rolling resistance and wet traction, it has been reported that the application of silica to truck tire compounds has not been successful [72]. The key deficiency of silica in truck-tire-tread compounds as a principal filler is also poor abrasion resistance that is, again, polymer-related. In the last few decades, the radialization of truck tires has led to a higher proportion of natural rubber usage [73]. This is first due to its excellent green strength and tackiness, which are very desirable properties for steel cord ply to eliminate the problem of open "tie in" for radial tire building. Secondly, the low-angle steel cord breaker system of radial truck tires makes the crown area stiffer, leading to a deficiency in its ability to envelop sharp objects. This makes the tire more prone to cutting and chipping, especially under over-loading and poor road conditions. This again drives higher usage of natural rubber for the tread component of the radial truck tire [74]. Consequently, the proportion of natural rubber used in the tire is significantly higher for the radial truck tire. For example, the usage range of NR in the tread compound for a standard heavy cross-ply truck tire is 40–90%, while for a similar size radial tire, 98–100% natural rubber is used [73].

Multiple factors may be accountable for the inferior properties of the truck-tire-tread compounds of silica, but probably it is mainly related to the surface characteristics of the silica.

Similar to ESBR, in which the surfactants can weaken the polymer-filler interaction, in the case of NR-based truck tire tread compounds, the nonrubbers originating from the natural rubber latex interfere with the polymer-filler interaction. In dry NR, there are more than 5% nonrubbers which are mainly proteins, fatty acids, phospholipids, other ester-like substances, and their degradation products. As they are easily adsorbed on the silica surface, these nonrubbers play the same role as the surfactants in ESBR in deteriorating polymer-filler interaction and reducing the effectiveness of coupling agents. For example, even with a ratio of TESPT to silica as high as 15/100 by weight, the enhancement of abrasion resistance is insufficient [75,76].

Also, due to the surface characteristics of silica, more curatives, which are generally polar substances, can be adsorbed on the filler surface. In addition, compared with carbon blacks, the surface acidity of the silica, which can be estimated by the adsorption energies measured by IGC for a base chemical such as THF and a acidic chemical such as chloroform [22,77], is much higher. These result in poor cure characteristics, leading to a lower cure rate and lower yield of crosslinks during vulcanization. Therefore, increased amounts of curatives are necessary in order to be able to obtain reasonable crosslink density and improved cure kinetics. For all-silica-filled NR compounds, the dosage of sulfenamide accelerators used has been as high as 2.8–3.6 phr with 1.7–2.0 phr sulfur, compared to 1–1.4 with 0.8–1.2 phr for carbon black-filled NR compounds [76,78]. This, along with the high level of TESPT, will lead to high levels of chain-modification of the polymer due to increased cyclic

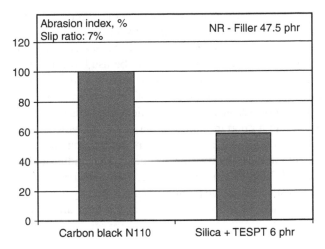

FIGURE 3.24 Abrasion resistance of NR vulcanizates filled with carbon black N110 and silica (ZeoSil 1165)/TESPT.

sulfides, pendant-S-accelerator groups, and pendant coupling agent moieties. This, in turn, will significantly reduce the flexibility of polymer chains, resulting in a deterioration of abrasion resistance. Moreover, due to the poorer processability of silica, especially during mixing, the filler incorporation time is increased. This results in a drastic drop in molecular weight of polymer caused by severe chain scission, as NR is very sensitive to mechanoxidative degradation. This is particularly true when heat treatment is performed for the precoupling reaction. Consequently, the abrasion resistance is further impaired. As can be seen in Figure 3.24, the abrasion resistance of the vulcanizates filled with 50 phr silica and 6 phr TESPT is about 40% lower than its carbon black counterpart, even though a high dosage of coupling agent was applied [79].

3.8 SUMMARY

It has been well known that filler properties play a vital role in terms of rubber reinforcement. The most important parameters of the fillers are particle size, structure, and surface activity. Of the three main filler parameters, particle size and structure are today sufficiently characterized, and their determination is carried out in routine tests for the purpose of quality control, both in the filler and rubber industries. The relationships between these two parameters and the properties of the filled rubber have also been studied in depth. On the other hand, although their importance was recognized long ago, a satisfactory description of the surface characteristics of the fillers is still lacking and present knowledge of the relationships between surface activity and rubber properties is far from complete. One of the reasons is that there are only a limited number of tools available which can be used to characterize the filler surface energetically, and that most of these methods are not accurate enough. However, inverse gas chromatography (IGC) has recently been demonstrated to be one of the most sensitive and convenient methods for studying the filler surface, providing important information about polymer-filler interaction, filler–filler interaction, and filler-ingredient interaction, which is important for filler development, quality improvement, and understanding of rubber reinforcement. Taking carbon black, silica, and modified fillers as examples, this chapter has shown how the parameters measured by IGC can be applied to address some issues in rubber reinforcement, which are related to surface activities of fillers.

ACKNOWLEDGMENTS

The author is grateful to Dr. Y. Kutsovsky for his support during writing this chapter. The help from Dr. M. Morris, Mr. R. Dickinson, and Mrs. H. Tu are also appreciated. Special thanks are extended to Cabot Corporation for permission to publish this paper.

REFERENCES

1. Kraus, G., Ed., *Reinforcement of Elastomers*, Interscience, New York, 1965.
2. Donnet, J.-B., Bansal, R. C., and Wang, M. J., Eds., In *Carbon Black, Science and Technology,* 2nd ed., Marcel Dekker, Inc., New York, 1993.
3. Fowkes, F. M., *J. Phys. Chem.*, 66, 382, 1962.
4. Owens, D. L. and Wendt, R. C., *J. Appl. Polym. Sci.*, 13, 1741, 1969.
5. Kaelble, D. H. and Uy, K. C., *J. Adhes.*, 2, 50, 1970.
6. Fowkes, F. M., In *Treatise on Adhesion and Adhesives*, Patrick, R. L., Ed., Vol. 1, Marcel Dekker, New York, 1967.
7. Bernett, M. K. and Zisman, W. A., *J. Colloid Interf. Sci.*, 29, 413, 1969.
8. Kessaissia, Z., Papirer, E., and Donnet, J.-B., *J. Colloid Interf. Sci.*, 82, 526, 1981.
9. Dubinin, M. M., *Progr. Surface Membrane Sci.*, 9, 1, 1979.
10. Dubun, B. V., Kiselev, V. F., and Aleksandrov, T. I., *DAN, SSSR*, 102, 1155, 1955.
11. Harkins, W.D. *The Physical Chemistry of Surface Films*, p. 255, 1954.
12. Kiselev, A. V. and Yashin, Y. I., *Gas-Adsorption Chromatography*, Plenum Press, New York, 1969.

13. Conder, J. R. and Young, C. L., *Physical Measurement by Gas Chromatography*, Interscience, New York, 1979.
14. Laub, R. J. and Pecsok, R. L., *Physicochemical Application of Gas Chromatography*, Interscience, New York, 1978.
15. Brendlè, E., Balard, H., Papirer, E., In *Powders and Fibers: Interfacial Science and Applications*, Ch. 1.2, Nardin, M. and Papirer, E., Eds., Marcel Dekker, 200X.
16. Wang, M. J., Wolff, S., and Donnet, J. B., *Rubber Chem. Technol.*, 64, 559, 1991.
17. Dorris, G. M. and Gray, D. G., *J. Colloid Interf. Sci.*, 77, 353, 1980.
18. Wang, M. J. and Wolff, S., *Rubber Chem. Technol.*, 65, 890, 1992.
19. Ban, L. L., Hess, W. M., and Papazian, L. A., *Rubber Chem. Technol.*, 47, 858, 1974.
20. Ban, L. L. and Hess, W. M., *Interactions Entre les Elastomeres et les Surfaces Solides Ayant Une Action Renforcante*, Collogues Internationaux du C.N.R.S., 1975.
21. Wolff, S., Wang, M. J., and Tan, E. H., *Rubber Chem. Technol.*, 66, 163, 1993.
22. Wang, M. J., Wolff, S., and Donnet, J. B., *Rubber Chem. Technol.*, 64, 714, 1991.
23. Bulgin, D., *Trans. Inst. Rubber Ind.*, 21, 188, 1945.
24. Böhm, G. G. A. and Nguyen, M. N., *J. Appl. Polymer Sci.*, 55, 1041, 1995.
25. Wang, M. J., *Rubber Chem. Technol.*, 71, 520, 1998.
26. Dannenberg, E. M., *Rubber Chem. Technol.*, 59, 512, 1986.
27. Wang, M. J., Wolff, S., and Tan, E. H., *Rubber Chem. Technol.*, 66, 178, 1993.
28. Hess, W. M., Lyon, F., and Burgess, K. A., *Kauts. Gummi, Kunstst.*, 20, 135, 1967.
29. Medalia, A. I. and Kraus, G., In *Science and Technology of Rubber*, Ch. 8, Mark, J. E., Erman, B., and Eirich, F. R., Eds., Academic Press, San Diego, P. 396, 1994.
30. Wang, M. J., Wolff, S., and Freund, B., *Rubber Chem. Technol.*, 67, 27, 1994.
31. Gessler, A. M., *Rubber Chem. Technol.*, 42, 850, 1969.
32. Wolff, S., Wang, M. J., and Tan, E. H., *Kauts. Gummi, Kunstst.*, 47, 780, 1994.
33. Garten, V. A. and Weiss, D. E., *Aust. J. Chem.*, 10, 309, 1957.
34. Donnet, J. B. and Voet, A., *Carbon Black, Physics, Chemistry, and Elastomer Reinforcement*, Marcel Dekker, New York, 1976.
35. Anderson, R. B. and Emmett, P. H., *J. Phys. Chem.*, 56, 753, 1952.
36. Coltharp, M. T. and Hackerman, N., *J. Phys. Chem.*, 72, 1171, 1968.
37. Rivin, D., *Rubber Chem. Technol.*, 44, 307, 1971.
38. Wang, M. J. and Wolff, S., *Rubber Chem. Technol.*, 65, 715, 1992.
39. Wolff, S., *Kautsch. Gummi Kunstst..*, 30, 516, 1977 (34, 280, 1981).
40. Medalia, A. I., *J. Colloid Interf. Sci.*, 32, 115, 1970.
41. Kraus, G. and Gruver, J. F., *Rubber Chem. Technol.*, 41, 1256, 1968.
42. Wang, M. J., Vidal, A., Papirer, E., and Donnet, J. B., *Colloid and Surfaces*, 40, 279, 1989.
43. Spencer, R. S. and Dillon, R. E., *J. Colloid Sci.*, 3, 163, 1948.

44. Minagawa, N. and White, J. L., *Appl. Polymer Sci.*, 20, 501, 1976.
45. Dannenberg, E. M. and Stokes, C. A., *Eng. Chem.*, 41, 821, 1949.
46. Buckler, E. J., Briggs, G., Dunn, J. R., Lasis, E., and Wei, Y. K., *Rubber Chem. Technol.*, 51, 872, 1978.
47. Cho, P. L. and Hamed, G. R., *Rubber Chem. Technol.*, 65, 475, 1992.
48. Hamed, G. R., *Rubber Chem. Technol.*, 54, 403, 1981.
49. Dannenberg, E. M., *Trans. Inst. Rubber Ind.*, 42, T26, 1966.
50. Dannenberg, E. M., *Rubber Chem. Technol.*, 48, 410, 1975.
51. Payne, A. R., In *Reinforcement of Elastomers*, Ch. 3, Kraus, G., Ed., Interscience, New York, 1965.
52. Payne, A. R. and Whittaker, R. E., *Rubber Chem. Technol.*, 44, 440, 1971.
53. Medalia, A. I., *Rubber Chem. Technol.*, 51, 437, 1978.
54. Payne, A. R., *J. Polym. Sci.*, 6, 57, 1962.
55. Payne, A. R., *Rubber Plast Age*, Aug., 963, 1961.
56. Kraus, G., *J. Appl. Polym. Sci., Appl. Polm. Symp.*, No. 39, 75 (1984).
57. Wang, M. J., *Rubber Chem. Technol.*, 72, 430, 1999.
58. Medalia, A. I., *Rubber Chem. Technol.*, 51, 437, 1978.
59. Saito, Y., *Kautsch. Gummi Kunstst.*, 39, 30, 1986.
60. Bulgin, D., Hubbard, D.G., and Walters, M.H., *Proc. Fourth Rubber Technol. Conf.*, London, p. 173, 1962.
61. Medalia, A. I., *Kautsch. Gummi Kunstst.*, 47, 364, 1994.
62. Wolff, S. and Wang, M. J., In *Carbon Black, Science and Technology*, Donnet, J. B., Bansal, R. C., and Wang, M. J., Eds. 2nd ed., Marcel Dekker, Inc., New York, 1993 (Chapter 9).
63. Studebaker, M. L., In *Reinforcement of Elastomers*, Kraus, G., Ed., Wiley Interscience, New York, 1965 (Chapter 12).
64. Dannenberg, E. M., *Rubber Age*, 98, 82, 1966.
65. Cotton, G. R. and Dannenberg, E. M., *Tire Sci. Technol., TSTCA*, 2, 211, 1974.
66. Shieh, C.H., Mace, M.L., Ouyang, G.B., Branan, J.M., and Funt, J.M., presented at a meeting of the Rubber Division, ACS, Toronto, Canada, May, pp. 21–24, 1991.
67. Dannenberg, E. M., Papirer, E., and Donnet, J. B., *Interactions Entre les Elastomeres et les Surfaces Solides Ayant Une Action Renforcante*, Collogues Internationaux du C.N.R.S., 1975.
68. Wolff, S., Wang, M. J., and Tan, E. H., *Kautsch. Gummi Kunstst.*, 47, 873, 1994.
69. Wolff, S., Görl, U., Wang, M. J., and Wolff, W., *Eur. Rubber J.*, 136, 16, 1994.
70. Rauline, R., EP 0 501 227 A1 (to Michelin & Cie, 1992).
71. Wolff, S., Görl, U., Wang, M.J., and Wolff, W., *Silica-Based Tread Compounds: Background and Performance*, Paper presented at the TIRETECH '. 93 Conference, Basel, Switzerland, pp. 28–29, 1993.
72. Freund, B., presented at a workshop of DIK Hannover, 1998; Eur. Rubber J., Sept, 34, 1998, Hess, M.G., presented at *Carbon Black World 99*, Padova, Italy, April, pp. 19–21, 1999.
73. Watson, P.J., Presented at TireTech'99.

74. Knill, R.B., Shepherd, D.J., Urbon, J.P., Endter, N.G., *Proc. Int. Rubb. Conf.* 1975, Volume V, RRIM, Kuala Lumpur, 1976.
75. Wolff, S., Paper presented at the meeting of Rubber Division, ACS, New York, April, pp 8–11, 1986.
76. Wolff, S. and Panenka, R., Paper presented at IRC'85, Kyoto, Japan, Oct., pp. 15–18, 1985.
77. Wang, M. J., Tu, H., Murphy, L. J., and Mahmud, K., *Rubber Chem. Technol.*, 72, 666, 1999.
78. Bomal, Y., Cochet, Ph., Dejean, B., Gelling, I., and Newell, R., *Kauts. Gummi Kunstst.*, 51, 259, 1998.
79. Wang, M. J., Zhang, P., and Mahmud, K., *Rubber Chem. Technol.*, 74, 124, 2001.

Section II

The Solid-Solid Interface

4 Probing the Molecular Details of the Surfaces of Powders and Fibers Using Infrared Spectroscopy

Ben McCool and Carl P. Tripp

CONTENTS

4.1 INTRODUCTION

Infrared spectroscopy (FTIR) is a versatile and widely used technique for obtaining molecular details of the interfacial region of powders and fibers. Its use has been primarily directed to the identification of surface functionalities on powders/fibers and to monitor the reaction of molecules with these materials. Most powders and fibers have relatively high surface area (greater than $10 \, m^2 \, g^{-1}$), and this has been a key enabler for detecting infrared (IR) bands due to surface functionalities. Part of the versatility of this technique stems from its ability to record spectra at the solid/gas interface, solid/liquid interface, (both aqueous and nonaqueous) and solid/supercritical fluid interface. This provides a wide latitude in adsorbates for analysis by FTIR ranging from small gaseous molecules to large polymers and biological material. As a result, there are thousands of articles and several books[1–6] and reviews[7–16] detailing its use for analyzing the surface sites and surface reactions on powders and fibers.

Given the number of in-depth reviews of this topic, the first question is, why another review? Our survey of the literature shows that the majority of the published work in this area employs this technique as one of many characterization tools. In this case, IR spectroscopy is mainly used to simply verify the chemical composition or extent of surface modification of the synthesized or chemically modified powders. This is certainly the case in highly active research areas such as sol–gel synthesis of templated oxide powders and atomic layer deposition (ALD) of metal oxide films on powders. When used as a characterization tool, there exists an abundant pool of reference material to aid in the interpretation of the spectra in terms of identification of surface sites and surface species. However, this represents a fraction of the potential information that can be gleaned from the use of IR spectroscopy, and in particular, that can be obtained using in situ FTIR measurements during powder synthesis or surface treatment. It is this aspect that we address in this review, as there is a need to provide a resource that describes the additional information that can be gleaned through the use of more advanced sampling protocols.

IR spectroscopy is notorious for its various sampling approaches, and virtually all have been applied to studies of the interfacial regions of powders and fibers. Standard methods such as transmission,[14] reflectance–absorbance (R–A),[17] attenuated total reflectance (ATR),[18–20] diffuse reflectance (DRIFT),[21,22] photoacoustic,[23–26] and emission[27,28] have been employed in one form or another. Powders and fibers come in various sizes and shapes, and this is one factor dictating the choice of sampling approach used to extract FTIR data from the interfacial region. Other important factors are the reaction conditions such as whether adsorption is conducted from gas phase, nonaqueous and aqueous solutions or supercritical fluids. Our bias in selecting the choice of sampling approach is to keep the spectroscopy as simple as possible. In other words, transmission works best for powders and fibers and use of other infrared methods should be employed only when transmission fails. This bias percolates throughout this manuscript.

Our approach is to first describe (Section 4.2) the criteria for selecting the sample approach from the point of view of the material properties, experimental conditions, and the type of information desired. We do not cover FTIR instrumentation, as this lies outside of the scope of our study.[29] In Section 4.2.1, we examine the general approaches used to extract molecular details for measurements conducted at the gas/solid, and in Section 4.2.2, extend this to the gas liquid interface. However, given the number of articles in which IR spectroscopy has been used to analyze powders and fibers at the solid/gas and solid liquid interface, a thorough review of all the work in this area would be a daunting task. Instead, in Section 4.3 and Section 4.4, we follow the general descriptions of the approaches given in Section 4.2 with a more detailed analysis of the application of IR-spectroscopic measurement to the synthesis and surface characterization of sol–gel oxide powders and their surface modification using ALD processes. Both of these topics represent areas that are current and have garnered much interest. It is noted, however, that the IR methods and information described for sol–gel oxide powders should provide a guide for their use with other types of powders and fibers.

4.2 SAMPLING CONSIDERATIONS

The large number of sampling methods in IR spectroscopy can be traced to two fundamental aspects of this technique. First, despite its popularity for probing surfaces and interfacial regions, IR spectroscopy is not a surface technique. This means that the infrared radiation is adsorbed by everything in its path. This includes the bulk of the powder, the interfacial region of interest, and the deposition medium (i.e., solvent and excess adsorbate). Thus, it is important to use sampling methods that are able to extract the spectral features of the interfacial region from those of the bulk modes and deposition mediums. The difficulty arises when the bands due to surface modes overlap with bands due to adsorption by the bulk or the deposition medium. For example, it would be difficult to monitor by IR spectroscopy the adsorption of a long alkyl chain surfactant on polyethylene fibers because the strong CH_2 modes of the alkyl chain appear at the same frequency as the polymer fiber. In this case, substitution of a deuterated surfactant is an example of a sampling approach that removes this overlap condition, as the bands due to CD_2 would shift in frequency.

In the case of oxide powders, the overlap of bands due to adsorbates with oxide occurs to a lesser degree. Oxide bulk modes lie in the low-frequency region and thus provide good transparency over a wide range enabling the detection of surface sites such as hydroxyl groups, as well as those of adsorbates. The lower limit of transparency varies for each oxide. For silica, the fundamental bulk modes render the region below 1300 cm^{-1} opaque whereas for metal oxides such as Al_2O_3, TiO_2, ZrO_2, MgO, ZnO, and ThO_2, the opacity appears near or below 1000 cm^{-1}.[7] Typical adsorbates are organic in nature (polymers, silanating agents, etc.), and the IR bands appear in the frequency range above the bulk modes.[30,31]

The second fundamental aspect of IR spectroscopy is that the light adsorbed at the interfacial region is not measured directly, but rather, inferred from the light transmitted or reflected from the material. Therefore, the collection of the transmitted light is very much dependant on the geometric and optical properties of the powders and fibers. This also means that it is very difficult to detect bands due to species adsorbed on low-surface-area materials. In essence, the attenuation of light in the sample beam by species adsorbed on low-surface-area powders and fibers is small relative to the reference beam. This results in a dynamic range limitation with the measurement of a small difference between two large signals of the sample and reference spectra.[29] There are techniques such as optical null FTIR, photoacoustic, and emission, which overcome the dynamic range limitation, but these have seen limited use in surface studies.[29,32]

The importance of sample morphology is clearly seen in IR adsorption studies on silica. The term 'silica' encompasses a variety of silicon oxides ranging from low-surface-area glass plates and the oxide layer on silicon to high-surface-area powders such as gels, fumed silica, and templated mesoporous structures. As described in the previous paragraph, IR studies of surface reactions on low-area silica are, at best, difficult. The hydroxyl bands are weak in intensity and have only recently been observed on oxidized silicon wafers by attenuated total reflection (ATR) techniques.[33,34] Glass is not a suitable material for infrared ATR due to low transmission, and thus, detection of the bands due to surface silanols has not been reported. It is for this reason that an initial justification for using high-surface-area silica was that it provided a model surface for understanding the chemistry occurring on glass and other low-surface-area siliceous materials.[10]

Much of the FTIR work with silica has been directed to studies of the chemistry of high-surface-area gels and fumed powders.[1,2,7,35] The high surface area of these silicas enables ease in detection of surface functionalities from bulk features. However, even in this case, different sampling approaches are needed because the differences in material morphology. For example, fumed silicas and silica gels have been the primary focus for FTIR studies because they both have high surface areas ($10–1000 \ m^2 \ g^{-1}$) but require different sampling methods because they differ in both particle size and porosity. Silica gels are usually studied by DRIFT[22] or photoacoustic spectroscopy[36] because the large particle size (greater than 1 μm) leads to unacceptable levels of scattering of the infrared beam for analysis by transmission methods. On the other hand, the smaller particle sizes of fumed silica (7–50 nm) render this material amenable to transmission studies.[7]

Absorption of the infrared beam by the solvent can also interfere with detection of bands due to adsorbed species. In ex situ measurements, this problem is minimal as the solvent can be removed by standard separation procedures. The dried powder or fiber is then usually dispersed with a suitable salt such as KBr and pressed into a disk for transmission or recorded as a powder using DRIFT. While our focus is on the use of FTIR for interfacial studies, this represents a subset of the general use of this technique with powders and fibers. As mentioned earlier, FTIR is used mainly as a characterization tool to verify the synthesis or surface treatment of the material, and in this case, the usual sampling approach is to use a pressed salt disk containing the powder or to suspend the silica gels in Nujol.

For example, FTIR has proven to be a valuable tool in understanding the structural evolution of sol gels during consolidation. Bertoluzza et al.[37] used FTIR to assign silica framework vibrations based on the known vibrational spectra of nonporous vitreous silica. They synthesized a silica xerogel from tetraethylorthosilicate (TEOS) using a single-step, acid-catalyzed process in a large excess of water. They then recorded the transmission IR spectra of the dried gel as a function of temperature to 800°C, presented in Figure 4.1. From

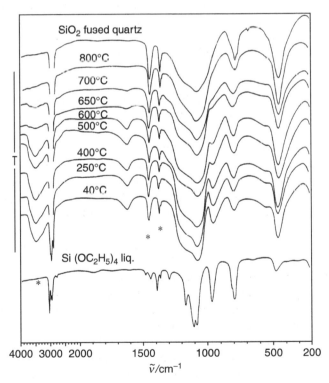

FIGURE 4.1 FTIR spectra of silica xerogels in Nujol emulsion heated at different temperatures (Nujol band marked by asterisks). (From Bertoluzza, A. et al., *Journal of Non-Crystalline Solids*, 48, 117, 1982. With permission.)

the spectra, they were able to monitor the formation of SiOH species and the removal of excess ethanol. At temperatures above 600°C, loss of structural water and SiOH species was observed, and at 800°C a spectrum similar to SiO_2 fused quartz was obtained.

4.2.1 Gas/Solid Interface

In situ FTIR studies on adsorption of gases on powders are most often performed using transmission or DRIFT, although other techniques such as ATR and photoacoustic have been reported.[7,19] In transmission, the powder is usually pressed into self-supported disks or spread onto an IR window or metal grid and mounted in an evacuable and temperature-adjustable cell.[38] In DRIFT, the powder is usually dispersed with a powdered salt and placed inside a temperature-controlled and evacuable chamber.[21,39] In both DRIFT and transmission studies, absorption of the infrared radiation by the excess gas phase adsorbate is avoided by simple evacuation. The temperature/evacuation control is used in performing the gas phase reaction and as a pretreatment protocol to control the number and type of surface sites on the oxide powders. The use of IR spectroscopy to identify surface sites as a function of evacuation temperature for a variety of oxide particles has been thoroughly reviewed by Morrow.[7] For illustrative purposes, we describe a typical use of transmission IR to identify surface sites and the physisorption/chemisorption of gaseous molecules on fumed silica powders.

A typical infrared spectrum of a self-supported disk of fumed silica (Aerosil 380) recorded in an evacuable transmission cell is shown in Figure 4.2a. Assignment of the various infrared bands of fumed silica is well established.[1,2,7] The region below $1300 \, \text{cm}^{-1}$ is opaque due to strong Si–O bulk modes. Clearly, by using a pressed silica disk it is not possible to detect bands below $1300 \, \text{cm}^{-1}$ arising from the interfacial region because all the IR radiation is absorbed by the bulk oxide. However, the important surface groups of silica are the hydroxyl groups, and these lie in the transparent region between 3750 and $3200 \, \text{cm}^{-1}$. The silica sample in air also contains a layer of adsorbed water, which gives rise to the broad band centered about $3450 \, \text{cm}^{-1}$ and a deformation mode at $1620 \, \text{cm}^{-1}$. The band at $1620 \, \text{cm}^{-1}$ is super-imposed on a region between 2000 and $1300 \, \text{cm}^{-1}$, which contain various Si–O combinations and overtone modes.

In the hydroxyl region, the sharp band at $3747 \, \text{cm}^{-1}$ is due to isolated (single and geminal) hydroxyl groups. Evacuation of the silica at room temperature removes the adsorbed water from the surface, exposing an underlying layer of hydrogen-bonded silanols represented by a broad band at $3550 \, \text{cm}^{-1}$ (Figure 4.2b). An additional band at $3650 \, \text{cm}^{-1}$ is due to inaccessible silanols located in interparticle points of contact. Evacuation at

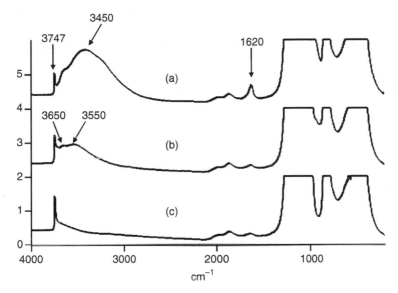

FIGURE 4.2 Pressed disk of fumed silica (a) room temperature, (b) evacuated at room temperature, and (c) evacuated at 450°C.

temperatures between 150 and 450°C leads to the reduction of the band at 3550 cm^{-1} and an increase in the isolated peak at 3747 cm^{-1}. At 450°C, there remain single/geminal hydroxyl groups on the surface represented by the sharp band at 3747 cm^{-1} (Figure 4.2c). The spectra in Figure 4.2 clearly shows that IR spectroscopy can readily distinguish between different surface hydroxyl groups on silica and can be used to monitor changes in the relative number and type of these groups with thermal treatment.

The spectra in Figure 4.2 show that the frequency position of the hydroxyl bands depends largely on the strength of the perturbation with neighboring groups. This is also the case when these surface silanols interact with gaseous adsorbate molecules. It is well known that the physisorption of molecules on the silica leads to a shift of the band at 3747 cm^{-1} to a lower frequency.[40] A stronger interaction with the adsorbate leads to a larger shift in the band. Table 4.1 is a compilation of the shift obtained for various molecules.[40–43]

Table 4.1 shows that weak Van der Wall's-type interactions observed between simple alkanes and silica lead to a shift in the band at 3747 cm^{-1} to near 3700 cm^{-1}, whereas the much stronger hydrogen bonding that occurs with amines shift this same band to below 3000 cm^{-1}. Some caution is needed in interpreting the relative strength of the interactions to the strength of adsorption of a particular molecule on the surface. For example, contact of a

TABLE 4.1
Shift in Silanol Peak Caused by Adsorption of Various Molecules on Fumed Silica

Compound	υ_{OH}	Reference	Compound	υ_{OH}	Reference
n-C_6H_{14}	3701	40	Acetone	3402	42
Cyclohexane	3699	40	Cyclepentanone	3372	42
CCl_4	3690	40	Cyclohexanone	3348	42
Chlorosilane	3690	40	1,4-dioxane	3327	40
CH_3NO_2	3683	40	Alkoxysilanes	3325	43
CH_3CN	3670	40	2-cyclohexene-1-one	3324	42
H_2CO	3493	42	TMP	3262	41
TCP	3488	43	$(C_2H_5)_2O$	3230	40
CH_3OH	3470	41	DMMP	3223	41
CH_3CHO	3447	42	Tetrahydrofuran	3205	40
MDCP	3425	41	Pyridine	2830	40
$CH_3CO_2C_2H_5$	3411	42	Triethylamine	2667	41

silica surface with a 1:1 ratio of dimethyl methyl phosponate (DMMP) and triethylamine leads to a surface covered only with DMMP.[41] Both molecules are not covalently bound to the surface, and from Table 4.1, one would anticipate that triethylamine would easily displace DMMP from the surface. However, DMMP possesses two P-OCH$_3$ groups, and upon contact with the surface, adsorbs through both groups with two surface silanols. The adsorption strength of DMMP adsorbed through two H-bonds per molecule is greater than the single H-bond between triethylamine and surface silanols.

In the case of chemisorbed molecules, there is a reaction with the surface silanols, and this results in a decrease in the intensity of the IR bands due to the hydroxyl groups. However, while changes in the hydroxyl spectral region that occur with chemisorption show that there is participation of hydroxyl groups in the reaction, it provides, at best, circumstantial evidence of surface species formed on the surface. An example is shown in the gas phase reaction of methylchlorosilanes at 400°C with a silica that was pretreated by evacuation at 450°C.[38] This pretreatment condition leads to isolated silanols on the surface, and the reaction proceeds monofunctionally,

$$Si_sOH + R_nSiCl_{4-n} \rightarrow Si_sOSiR_nCl_{3-n} + HCl, \qquad (4.1)$$

or difunctionally,

$$2Si_sOH + R_nSiCl_{4-n} \rightarrow (Si_sO)_2SiR_nCl_{2-n} + 2HCl \qquad (4.2)$$

and

$$Si_s(OH)_2 + R_nSiCl_{4-n} \rightarrow Si_sO_2SiR_nCl_{2-n} + 2HCl, \qquad (4.3)$$

with these groups, where Si_s refers to a surface silicon atom.

The reaction at 400°C results in the disappearance of the band at 3747 cm^{-1}, and this is accompanied by the appearance of methyl bands in the 3000–2900 cm^{-1} region. However, all three surface species shown in Equation 4.1 through Equation 4.3 could account for these spectral changes. The methylchlorosilanes all react in the same manner with the single/geminal hydroxyls, leading to the same predicted decrease in the band at 3747 cm^{-1}. Furthermore, the methyl bands do not help in identifying the surface species because the position of these bands is not structure sensitive. The key to the identification of the surface species are the infrared bands due to the Si–O–Si and Si–Cl modes, since the products described in Equation 4.1 through Equation 4.3 differ in the number and types of these bonds. These bands lie below 1300 cm^{-1} in the region that is opaque due to the strong Si–O bulk modes. However, access to this region is possible by using a simple modification of the sampling method.

The method first developed by Morrow et al.[44] uses a thin film of silica spread onto an IR window material. A typical spectrum of a thin film of fumed silica showing partial transparency over the entire region between 4000 and 200 cm^{-1} is plotted in Figure 4.3a.[45] The thickness of the silica probed by the infrared beam is reduced to a level where the strong Si–O bulk modes are not greater in intensity than one absorbance unit. This corresponds to a uniform amount of silica of about 0.25 mg cm^{-2}, and fabrication of a self-supported disk of fumed silica with this amount is not possible.

The spectrum obtained after a gas phase reaction of dichlorodimethylsilane at 300°C with a fumed silica pretreated by evacuation at 450°C is shown in Figure 4.3b.[45] In this representation, negative bands refer to bonds removed and positive bands to bonds formed on the surface. Above 1300 cm^{-1}, there is a decrease in the band at 3747 cm^{-1}, and this is accompanied by the appearance of the methyl bands at 2984 and 2918 cm^{-1}. In addition, several new bands are now visible below 1300 cm^{-1} and are assigned as follows: 1260 cm^{-1} (CH$_3$ bending), 1060 cm^{-1} (Si$_s$–O–Si stretching), negative 973 cm^{-1} (Si$_s$–OH stretching), 860 and 805 cm^{-1} (CH$_3$ rocking), and 485 cm^{-1} (Si–Cl stretching). With this low-frequency data it is possible to identify and differentiate between various surface species. More

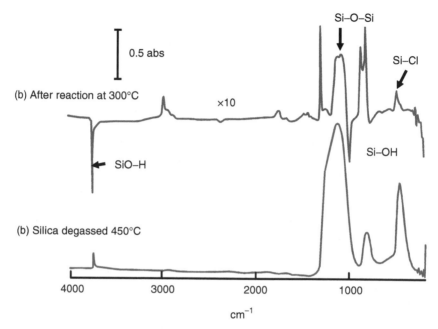

FIGURE 4.3 SiO$_2$ before and after gas phase reaction with dichlorodimethylsilane. (From Tripp, C. P., *Journal of Molecular Structure*, 408/409, 133, 1997. With permission.)

importantly, it is now possible to monitor the reaction of these surface species with other gaseous molecules. Thin oxide films for transmission IR studies have also been prepared on wire grid supports,[46] and extension of this thin film approach to other powders[47] and for in situ studies in nonaqueous,[48] aqueous[19] and supercritical fluids[49] have been demonstrated.

4.2.2 SOLID/NONAQUEOUS INTERFACE

As with gas adsorption studies on oxide powders, it is imperative to control the surface water content and the number and type of surface sites. The common way of accomplishing this is by thermal pretreatment of the oxide in a vacuum. In FTIR studies at the solid/nonaqueous interface, the oxide powder should also be pretreated in a vacuum, and the quality of the surface must be maintained before contact with the solvent. In addition, the IR cell usually requires narrow pathlengths because the solvents are strong IR absorbers. Furthermore, the adsorbate has to be mixed thoroughly with the silica, and there must be some means for discriminating between bound and unbound molecules. As a result, the number of IR adsorption studies on silica at the solid/nonaqueous liquid interface is few in comparison to measurements at the solid/gas interface.

The sampling approaches that have been used to address these constraints can be broadly divided into three categories. In the first approach, the oxide is pressed into self-supported disks similar to those used in solid/gas studies. Several methods and IR cells have been developed, and the most elegant and versatile are those based on an IR cell design reported by Rochester and colleagues,[16,50] as shown in Figure 4.4. In this cell, the oxide pellet is pretreated under a vacuum and then lowered into a narrow region containing IR transmitting windows. Without breaking the vacuum, the solvent and adsorbate are pumped across the face of the silica disk. Spectra can be acquired during this procedure, and the unbound adsorbate can be measured by recording a spectrum of solution after raising the silica disk from the

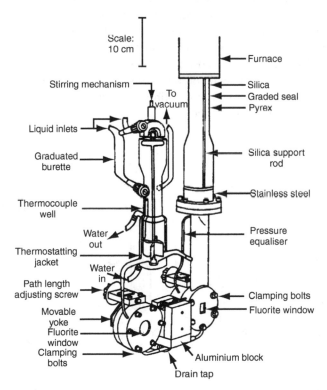

FIGURE 4.4 IR cell for in situ monitoring of oxide/nonaqueous interface. (From Buckland, A. D. et al., *Journal of the Chemical Society, Faraday Transactions 1*, 74, 2393, 1978. With permission.)

window region. Although pressed oxide disks are suitable for small-molecule adsorption, they are not useful for studying polymer adsorption. Polymers would have difficulty penetrating the pressed disk, and interparticle adsorption would be more prevalent because of the near proximity of oxide particles.[51–53]

For FTIR studies of polymers adsorbed on oxides, a second approach has been used.[54–60] A schematic for this type of experiment is illustrated in Figure 4.5. In a typical experiment, the oxide is stirred with the solvent to form a colloidal suspension. The polymer is mixed with the oxide suspension, and after attaining equilibria, the suspension is centrifuged and the polymer concentration in the supernatant is measured. Centrifugation and extraction are needed to distinguish between bound and unbound polymers. The adsorbed amount is calculated by subtracting the polymer concentration measured in the supernatant from the total amount of polymer added. The interaction of the polymer with the oxide is measured by recording a spectrum of the sedimented oxide. The sedimented oxide is transferred to a suitable IR liquid cell or is dried and recorded as a powder. There is a difficulty in controlling the surface quality using this method because the numerous manipulations expose the oxide surface to potential contamination, rehydroxylation, and rehydration. In most cases, the variance in the hydroxyl group distribution and density with pretreatment is neglected. Furthermore, all measurements are performed at equilibria so that polymer dynamic studies are not possible with this method.

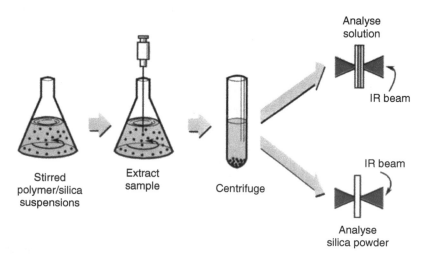

FIGURE 4.5 Schematic of multistep experimental procedure for measuring polymer adsorption on silica powders.

These limitations can be circumvented by again changing the sampling method. Figure 4.6 presents an infrared cell developed to enable recording of spectra of the suspension during mixing and thus enabling measurement of the dynamic adsorption of molecules and polymers on oxide powders.[61,62] The oxide pretreatment, solvent and adsorbate addition, and all spectroscopic measurements, are performed in situ in the evacuated cell, and therefore, strict control over surface quality is maintained throughout the experiment. The dynamic adsorbed amount can be deduced by recording spectra of the unstirred suspension. Spectra are recorded as the oxide particles settle from the beam area and the adsorbed amount is deduced from a measure of the decrease in intensity of bands due to the oxide and adsorbate. The oxide is never removed from the cell to measure the adsorbed amount.

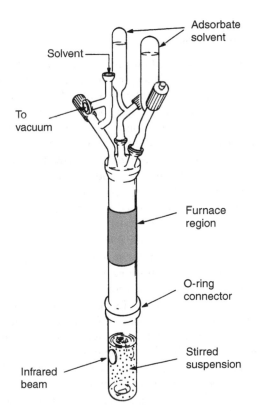

FIGURE 4.6 Cell for measuring polymer sorption on oxide in situ. (From Tripp, C. P. and Hair, M. L., *Langmuir*, 8, 1961, 1992. With permission.)

The third approach developed by Harris et al.[63] involves dispersing an oxide powder film on an ATR crystal and mounting the crystal in a flow-through ATR cell. The technique simply involves forming a suspension of suitable concentration of the metal oxide particles in a solvent. After thoroughly mixing and sonicating to achieve maximum dispersion, the crystal is dip-coated in the oxide suspension. Alternatively, a small amount of the suspension is deposited to the surface of the ATR crystal.[19] After drying, and depending on the surface energetics of the system, a stable oxide powder film is formed on the ATR crystal. The film thickness is chosen to match the sampling depth of the evanescent wave, and the finite penetration of the evanescent wave into the solution is such that spectral transparency is obtained over much of the infrared region. Furthermore, by working at dilute concentrations, the spectral contributions due to excess adsorbate in the solution are small in comparison to the spectral features due to adsorbed species.[64]

4.2.3 SOLID/WATER INTERFACE

FTIR-based studies on oxide particles in contact with water are particularly difficult because water is a strong absorber of infrared radiation. Transmission studies are not practical as the pathlength would have to be about 25 μm to gain transparency in the IR region. The common approach to overcome the strong absorption of infrared radiation by water is to use ATR spectroscopy. The simplest approach is to deposit a film of oxide similar to the manner used in nonaqueous studies. This technique has been used to prepare water-stable, high surface area, TiO_2 layers for studying the adsorption of small molecules,[19] surfactants,[65,66] polymers,[67] and biological membranes.[68] However, this method does not work with silica, as contact with water removes the silica from the ATR crystal. For studies on silica, the powder can be glued to the surface using a suitable polymer binder[19] or alternatively, the surface of the TiO_2 particles may be modified by depositing SiO_2 using ALD.[69] An alternate method to depositing powders is to grow a porous sol–gel oxide film directly on top of an ATR crystal.[70,71]

In the area of self-assembly, ATR has found use in measuring the orientation of molecules on surfaces. Information of the orientation of surfactant tails can be obtained by employing IR polarization techniques with the oxidized silicon wafers.[34] One potential drawback to powder-coated ATR crystals is that orientation measurements are not possible because the powder surface is randomly oriented with respect to the flat ATR crystal. However, Ninness et al.[20] demonstrated that the frequency location and width of methylene stretching modes are sensitive to the gauche/trans conformers ratio of the chains and thus provide structural information for the packing density of the surfactants. Moreover, the high surface area of the powders enables detection of the weaker, yet structure informative, headgroup modes of the surfactants.[65,66] It is found that the relative intensity of bands due to cetyltrimethylammonium bromide surfactant headgroups provides a unique glimpse of the structure of the surfactant layer and for elucidation of the various structures formed with mixed surfactant[66] and mixed surfactant/ polymer[67] systems.

The above brief survey demonstrates a wide variety of sampling approaches that have been developed to study the interfacial region of powders and more important, to extract information on dynamic adsorption processes that cannot be captured using ex situ measurements. During the last decade, powder based technology has experienced a tremendous growth and interest in the fabrication of sol–gel based powders with unique properties and architectures. As mentioned earlier, our survey of this field shows that IR spectroscopy is widely used as a characterization tool. It is our opinion that this field is ripe for the application of in situ IR sampling approaches that we outlined in this section. With this in mind, we now focus our survey on the topic of sol–gel based powders with the view of providing the general readership a glimpse of the potential additional information that could be obtained by applying in situ FTIR studies for the analysis of the sol–gel process and surface reactions.

4.3 SOL–GEL MATERIALS AND FTIR

The sol–gel process is a versatile solution process for making ceramic and glass materials. Due to the low temperature requirements and ease of processing, sol–gel techniques have become widely used for the fabrication of many types of ceramics and glasses. In general, the sol–gel process involves the transition of a system from a liquid sol into a solid gel phase. A sol is defined as a colloidal suspension of solid particles in a liquid. A gel is formed by condensing or growing a sol to the point where one polymerized molecule reaches macroscopic dimensions so that it extends throughout the solution.[72] By applying the sol–gel process, it is possible to fabricate ceramic or glass materials in a wide variety of forms: ultra-fine or spherical-shaped powders, thin-film coatings, ceramic fibers, microporous inorganic membranes, monolithic ceramics and glasses, or extremely porous aerogel materials.[72] An overview of the sol–gel process is presented in Figure 4.7.

The starting materials used in the preparation of the sol are usually inorganic metal salts or metal organic compounds such as metal alkoxides. In a typical sol–gel process, the precursor is subjected to a series of hydrolysis and condensation reactions to form a stable sol. Further processing of the sol

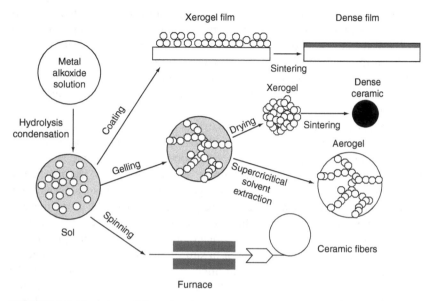

FIGURE 4.7 Overview of the sol–gel process.

enables one to make ceramic materials in different forms. Thin films can be produced on a piece of substrate by spin-coating or dip-coating. When the sol is cast into a mold, a wet gel will form. With further drying and heat-treatment, the gel is converted into dense ceramic or glass articles. If the liquid in a wet gel is removed under a supercritical condition, a highly porous and extremely low density material called aerogel is obtained. As the viscosity of a sol is adjusted into a proper viscosity range, ceramic fibers can be drawn from the sol. Ultra-fine and uniform ceramic powders are formed by precipitation, spray pyrolysis, or emulsion techniques.[72]

The application of FTIR provides a great deal of information about the chemical structure of aerogels and xerogels fabricated by the sol–gel process. For example, Almeida et al.[73] monitored the evolution of TEOS to silica gel and, finally, glass by monitoring the FTIR spectra of a silica xerogel as a function of temperature. The FTIR spectra were recorded ex situ by pressing the silica gels between two silicon wafers (see Figure 4.8). By using the technique of immobilizing the gels between two windows of temperature inert IR transmitting material, the researchers were able to record spectra more easily. This technique also eliminated the need to prepare a new sample at each temperature increment. In Figure 4.8, the authors use FTIR to study the room temperature consolidation of the gelling silica network with respect to time for a period from 30 min to 28 h. We note the disappearance to

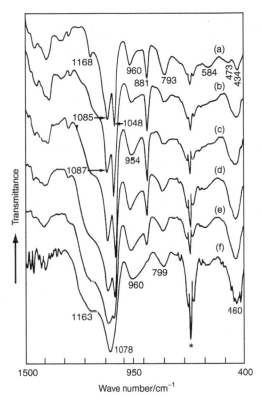

FIGURE 4.8 Infrared spectra taken during TEOS to silica gel conversion. Timescale is: (a) 30 min, (b) 2 h 10 min, (c) 5 h 30 min, (d) 7 h 10 min, (e) 8 h, and (f) 28 h. The bands marked by asterisks are due to difficulties with CO_2 subtraction. (From Matos, M. C. et al., *Journal of Non-Crystalline Solids*, 147, 232, 1992. With permission.)

TEOS-related bands 473 cm^{-1} (O–C–C deformation) and 1168 cm^{-1} (CH$_3$ rock) after the first half hour of gelling. After the first half hour, ethanol bands appear at 881 and 1048 cm^{-1} as a result of the hydrolysis and condensation of the TEOS. The ethanol bands persist for 8 h. By 28 h, all of the ethanol has evaporated from the gel. At 8 h, a band at 1078 cm^{-1} begins to appear. This band is attributed to the transverse optic component of the Si–O–Si asymmetric stretch. This band becomes the dominant spectral feature after 28 h of gellation. At 28 h, a weaker shoulder on the 1078 mode appears at 1163 cm^{-1}. This shoulder is attributed to the longitudinal optic component of the same Si–O–Si asymmetric stretch. This peak and shoulder is typically the strongest feature in the IR spectrum of silica xerogels and indicates a high degree of consolidation in the gelled network.

FTIR has also been shown to provide information as to the morphology of sol–gel-derived materials. In an earlier work, Almeida et al. used specular reflectance FTIR to study the effects of drying temperature on the structure of silica compared to vitreous silica.[74] Using Kramers–Kronig analysis of the near-normal infrared reflectivity spectra, they compared porous silica xerogels to porous vycor glass and vitreous silica. They found that the Si–O bulk modes at 1237 and 485 cm^{-1} of porous silica xerogel were red shifted relative to the vycor and vitreous silica (see Figure 4.9). This shifting in frequency and the appearance of a shoulder on the low-frequency side of the 1237 cm^{-1} peak were related to strained Si–O–Si linkages and a longer Si–O bond in the porous material.

Gallardo et al. have also used FTIR to analyze the structure of sol–gel films.[75,76] In their studies, they were able to relate a shoulder on the primary bulk Si–O–Si mode at 1080 cm^{-1} to the porosity of silica and methyl functionalized silica films. By using FTIR to qualitatively asses the porosity of silicas, they were able to relate the sintering process of methylated silica to differences in porosity when compared to un-methylated silica. Takada et al. later used a similar analysis of bulk mode peak shouldering to correlate temperature-dependent changes in the silica framework with increases in Young's elastic modulus and hardness of sol–gel silica films.[77] They were

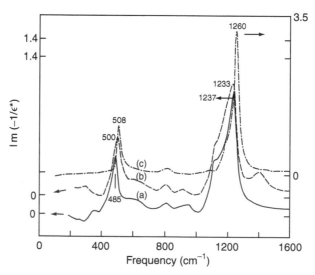

FIGURE 4.9 Energy loss function modes of (a) silica gel, (b) porous Vycor glass, and (c) vitreous silica. (From Almeida, R. M. et al., *Journal of Non-Crystalline Solids*, 121, 193, 1990. With permission.)

able to relate peak shifting of the bulk mode shoulder to the degree of crosslinking in the silica framework. These authors also found that the degree of crosslinking was in direct relation to the hardness and Young's modulus of the resulting materials.

Following the initial hydrolysis and condensation reactions involved in the gelling process, powders and fibers are calcined in order to consolidate the gel network into a temperature- and pressure-stable solid. Calcination refers to the temperature-curing process where organic components of the sol–gel precursors are removed from the gel network and the network is condensed. This is an extremely important step in the formation of sol–gel-derived materials. Infrared spectroscopy is a valuable tool for providing information about the removal of organics from these materials during the calcination process.

A typical example of the use of FTIR to aid in the understanding of the calicination process is found in the synthesis of barium-titanate sols prepared with inorganic precursors.[78] In this work, the researchers prepared barium titanate using 1,3 propanediol, barium acetate, and titanium isopropoxide modified by acetyl acetonate. The IR spectra of the dried gel were recorded at temperatures from 25 to 1000°C. At room temperature, the gel showed typical peaks due to acetate groups, and these peaks persisted to 400°C. Peaks due to TiO_2 began to appear at 400°C and persisted to 700°C. At 400°C carbonate peaks appeared and persisted to 900°C. It was determined by the authors that a calcination temperature of 1000°C was required to obtain pure barium-titanate powders.

The calcination process of nanofibers has also been studied using FTIR. Wang and Santiago-Aviles used specular reflectance FTIR to study the calcination process of lead zirconate titanate (PZT) nanofibers.[79] They synthesized the PZT fibers through a modified sol–gel electrospinning process. Starting sols were prepared using lead 2-ethylhexanoate, zirconium n-propoxide, and titanium isopropoxide. The sol was then passed through a syringe filter and metal tip held at a high voltage and deposited as fibers on a stationary collector screen. The calcination process was studied using FTIR by recording the reflectance spectrum of the nanofibers deposited on Pt and Si wafers after calcination in air for 2 h at 300, 300, 500, 600, 700, 800, and 850°C.

The transmission spectrum of the starting sol was recorded in transmission using a liquid cell (see Figure 4.10). Bands attributed to O–H stretching (band A, 3380 cm^{-1}), C–H stretching (band B, 2900 cm^{-1}), C=O stretching (band C, 1550 cm^{-1}), O–H bending (band D, 1400 cm^{-1}), and C–O stretching (band E, 1300 cm^{-1}) were observed and attributed to the organo-metallic precursors in the sol. After deposition, the O–H contributions disappear indicating the evaporation of alcohol in the gelled fibers. At 400°C all of

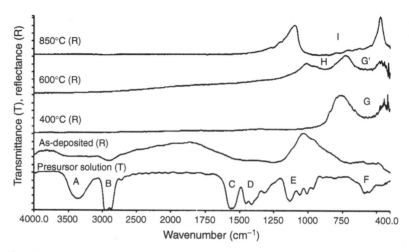

FIGURE 4.10 FTIR spectra of PZT precursor sol and deposited fibers at temperatures up to 850°C. (From Wang, Y. and Santiago-Aviles, J. J., *Nanotechnology*, 15, 32, 2004. With permission.)

the organic bands have disappeared from the spectrum, and a broad band attributed to a combination of Ti–O (619 cm^{-1}) and Zr–O (548 and 559 cm^{-1}) appears between 400 and 725 cm^{-1}, band G. At 600°C another broad band appears between 750 and 1050 cm^{-1} (band H), which was found to correspond to the appearance of PZT peaks in the x-ray diffraction spectrum. At 850°C the IR spectrum shows only one very broad peak between 500 and 1050 cm^{-1}, band I. This peak is attributed to the combined vibrations of the two co-existing crystalline phases (pyrochlore and perovskite) within the PZT nanofibers.[79]

As shown above, FTIR is a very common tool used to aid understanding of the calcination process; Table 4.2 presents a short list of materials where FTIR has been used to that end. The list is by no means exhaustive and only represents the most recent examples of FTIR's use in monitoring the calcination process of sol–gel materials.[78,80–97]

With the increase in our understanding developed from the continual use of ex situ FTIR measurements to characterize sol–gel powders, we anticipate new applications in terms of correlations to material properties. One such example is the recent development of a simplified technique using FTIR spectroscopy to estimate the surface area of silica powders (both porous and nonporous). This technique was extended to estimate the surface area of mesoporous silica membranes and supported films, which cannot be measured using conventional nitrogen-sorption techniques.

TABLE 4.2
Studies of Powders and Fibers Using FTIR as a Characterization Tool in the Calcination Process

Material	Morphology	Reference
Barium titanate	Powder	78
Ceria/silica composite	Powder	80
$BaCoO_3$-y perovskite	Powder	81
Ceria	Powder	82
$(Y,Gd)_2O_3$	Powder	83
YBCO superconductor	Powder	84
Co substituted aluminophospate	Powder	85
Pd and Pt doped Mg/Al mixed oxides	Powder	86
$LaMnO^{3+}g$	Powder	87
$MgAl_2O_4$	Powder	88
$LiNiVO_4$	Powder	89
Vanadium/niobium mixed oxides	Powder	90
MoO_3/MCM-41 composite	Powder	91
PZT	Fiber	79
Al_2O_3/SiO_2	Fiber	92
Mullite	Fiber	93
Polyaniline/V_2O_5	Fiber	94
γ-Al_2O_3	Fiber	95
Carbon	Fiber	96
As (Ge/Ga)-S	Fiber	97

Transmission or DRIFT spectra were recorded, and the surface area was estimated by comparing the integrated intensity of the band due to isolated silanol at 3747 cm^{-1} normalized to the intensity of a bulk mode at 1870 cm^{-1} for different silica samples. The assumption is that the isolated hydroxyl group density, or silanol number, is constant, the value obtained for the integrated intensity of the band at 3747 cm^{-1} for a given quantity of silica is then proportional to the surface area of this material. This is a reasonable assumption as it has been shown that the silanol number is about 4.9 OH nm^{-2}, irrespective of the type of silica.[98] It should be noted that there are silicas that fall outside of this range with values as low as 2 and as high as 9 OH nm^{-2} having been reported.[99] Nevertheless, it was found that an accuracy better than 7% in the surface area of several silica materials is obtained by using the FTIR approach for materials already characterized by BET.[100] Table 4.3 summarizes the 1870/3747 ratio as well as BET-measured and FTIR-calculated surface area for several silica samples.

In this section, we have provided a review of the use of IR spectroscopy in the synthesis and characterization of powders and fibers. We have also provided insight into how this powerful tool can be used to gain more information about the nature of powders and fibers than is typically found in the current literature. In Section 4.4 we turn to the surface modification of powders and fibers. While physical vapor deposition has been used to modify powders and fibers, it suffers from line-of-sight deposition. Chemical vapor deposition is another approach but has the disadvantage in that it is often difficult to control the thickness of the deposited compound. For these reasons, we will focus on ALD. As in Section 4.3, we will provide an introduction to the technology and a review of how FTIR fits into the current state-of-the-art research in the area. We will then discuss the limitations of the current use of FTIR and provide examples of how FTIR can be used to gain more information about the modification of powders and fibers by ALD.

4.4 ATOMIC LAYER DEPOSITION AND FTIR

Atomic layer deposition (ALD) is a vapor phase chemical reaction technique used to deposit thin films. ALD employs sequential saturative surface reactions that are separated by a purge to remove excess reactants. This sequential reaction technique allows for monolayer growth at the substrate surface. ALD has found great interest in the past 10 years as a film growth technique due to its high level of film thickness control and the conformality of the resulting films.[101] Figure 4.11a presents a simplified surface reaction

TABLE 4.3
Summary of FTIR Surface Area Estimations for Several High-Surface-Area Silicas

Material	Overtone/SiOH Area (± 0.03)	Estimated Surface Area (m^2 g^{-1} $\pm$ 4%)	Reported Surface Area (BET ± 1%)
Aerosil 380	0.791	—	358.1
CAB-O-SIL® L-90	3.300	85.8	89.2
CAB-O-SIL® LM-150	1.770	160.0	150.0
CAB-O-SIL® M5	1.370	206.7	200.0
CAB-O-SIL® HS-5	0.911	311.0	314.0
MCM-48	0.342	828.2	886.0
SiO$_2$ membrane	0.436	649.0	—
SiO$_2$ film on Ge	0.183	1640.0	—

Source: McCool, B.A. and Tripp, C.P., unpublished results, 2004.

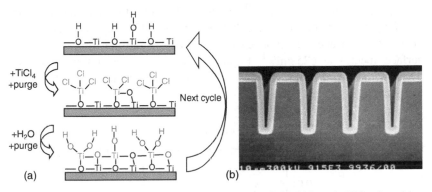

FIGURE 4.11 (a) Surface reactions of TiO_2 ALD and (b) 100-cycle TiO_2 deposition on Si substrate.

scheme for the ALD of TiO_2 using $TiCl_4$ and water. Figure 4.11b shows a scanning electron micrograph of 100 ALD cycles of TiO_2 grown on a high-aspect-ratio silicon substrate.

The ALD of silica can be accomplished by using sequential surface reactions of $SiCl_4$ and water. The overall reaction is represented by the equation:

$$SiCl_4 + 2H_2O \rightarrow SiO_2 + 4HCl. \tag{4.4}$$

This reaction can be broken down into two successive half reactions, where Si was deposited by $SiCl_4$ reaction with the surface, followed by the water half reaction where new silanol groups are formed at the surface:

$$(A) \ SiOH^* + SiCl_4 \rightarrow SiOSiCl_3^* + HCl \tag{4.5}$$

$$(B) \ SiCl^* + H_2O \rightarrow SiOH^* + HCl \tag{4.6}$$

where * indicates a surface species.

ALD of silica has been widely reported.[102–107] Many of these works used FTIR to monitor the cycle-to-cycle surface attachments at each reaction step. This is done by recording the IR spectrum of the $SiCl_4$ reaction step and noting the appearance of Si–Cl vibrations at around 625 and 473 cm^{-1}. The H_2O half reaction can be monitored by noting the appearance of isolated hydroxyl peaks located at 3747 cm^{-1}. The overall growth rate can be monitored by determining the growth rate of bulk Si–O–Si modes located at 1060–1080 cm^{-1}. Figure 4.12 illustrates the in situ monitoring of silica

ALD grown on a silicon wafer.[102] Figure 4.12a shows the high-frequency and low-frequency regions of the substrate during $SiCl_4$ exposure, noting the decrease in surface silanol species at 3747 cm^{-1} and the increase in Si–Cl surface species at 625 cm^{-1}. Figure 4.12b, shows essentially the reverse trend, with silanol peaks growing and Si–Cl peaks disappearing during the water cycle.

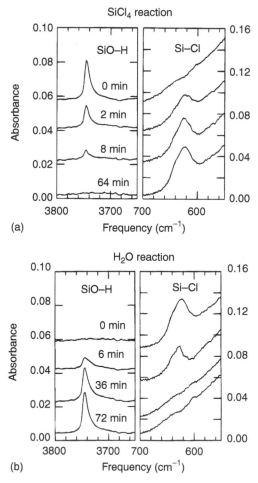

FIGURE 4.12 (a) High- and low-frequency FTIR spectra during the $SiCl_4$ reaction at 600 K and 10 Torr on silicon wafer from initial exposure to approximately 1 h and (b) FTIR spectra of the water cycle under the same conditions from initial exposure to 72 min. (From George, S. M. et al., *Journal of Physical Chemistry*, 100, 13121, 1996. With permission.)

In addition to SiO_2, ALD has been used for the deposition of a wide variety of films including but not limited to: Al_2O_3,[103,108–111] TiO_2,[104,112–114] V_2O_5,[115] GaAs,[116,117] and many other metal oxides and semiconductor materials. Ritala and Leskela have written two fine reviews on the subject.[101,118]

More recently, researchers have been focusing on ALD for modification of powders and porous materials. When performed on a small scale as a thin film of powder on an IR transparent window, FTIR analysis can provide a great deal of information about the specific surface reactions of ALD. ALD of TiO_2 on thin films of silica and kaolin mineral surfaces has been performed and monitored in situ using FTIR.[112,114] By monitoring the surface silanol species and Si–O–Ti bands, the researchers found that it takes a total of three ALD cycles to form a fully covered TiO_2 surface. Similarly, the ALD of TiO_2 and AlN on catalytic SiO_2 was performed by Haukka et al.[119,120] Due to the highly scattering nature of the SiO_2 powders used in these studies, the researchers were unable to monitor the ALD reactions in situ and used DRIFT after reaction cycles to verify the degree to which Si–OH species had reacted at various temperatures.

The most complete use of integrated FTIR with bulk powder modification has been performed by George's group. They have performed ALD of SiO_2 on boron nitride particles,[121,122] Al_2O_3 on polyethylene particles,[123] hafnium silicate on zirconia powders,[124] and ZnO on zirconia and barium titanate powders[125] in order to change the surface characteristics of the respective powders. In these studies, the FTIR analysis was done separate from the bulk powder modification. The FTIR chamber was exposed to the same conditions as the bulk powder, but the substrate powders were immobilized on a tungsten grid for transmission FTIR measurement. Similar to the strategies for FTIR data collection with ALD processes noted above, the FTIR spectrum of the substrate powders is measured, and a subtraction spectrum is taken in between each AB cycle. The bands associated with the reactive precursors are noted to determine completeness of the reactions. The bands associated with the deposited film are integrated at each complete cycle, and a bulk growth rate is estimated.

Figure 4.13 illustrates the ALD growth of alumina on low-density polyethylene (LDPE) particles at 77°C[123] through FTIR difference spectra. Initial exposure to water does not produce a significant change on the surface of the LDPE particles. The addition of $Al(CH_3)_3$ showed a marked increase in amount adsorbed on the LDPE as indicated by strong CH_3-related bands in the difference spectra. The addition of water vapor generated methanol and Al_2O_3 at the surface. Here the authors infer the growth of Al_2O_3 by the appearance of O–H stretching modes in the IR spectrum. The authors were unable to detect

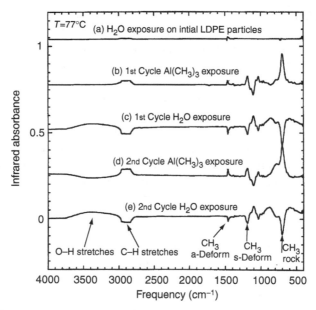

FIGURE 4.13 FTIR difference spectra of polyethylene particles recorded after (a) H_2O exposure on initial particles, (b) first cycle $Al(CH_3)_3$ exposure, (c) first cycle H_2O, (d) second $Al(CH_3)_3$ exposure, and (e) second H_2O exposure. (From Ferguson, J. D. et al., *Chemistry of Materials*, 16, 5602, 2004. With permission.)

bulk mode vibrations of Al_2O_3 until 10 ALD cycles. After 10 cycles, the Al_2O_3 modes grew steadily.

Returning to ALD growth of SiO_2 on SiO_2 powder, one drawback to using $SiCl_4$ as a precursor is that the reaction must be conducted at 300°C or higher. The constraint can be avoided using amines to catalyze the reaction of $SiCl_4$ with the SiO_2.[105,126–130] In this reaction scheme depicted in Figure 4.14, the first half reaction is catalyzed by an amine (often triethylamine, pyridine, or ammonia), allowing the reaction to take place near room temperature. The reaction is catalyzed by allowing pyridine to form a hydrogen bond with the surface silanol species. This is observed with FTIR by the disappearance of silanol species at 3747 cm^{-1} and the appearance of C–H modes in the 2900–2800 cm^{-1} region. As a result of the pyridine hydrogen bond, the Si–O group is rendered more nucleophillic and thus, able to react with the incoming chlorosilane at much lower temperatures (room temperature vs. 300°C). After chlorosilane reaction with the hydroxyl group, pyridine is no longer involved in the primary surface reaction.

The base-catalyzed ALD has been used to perform self limiting pore size reduction in porous silica gas separation membranes.[129,130] When a base

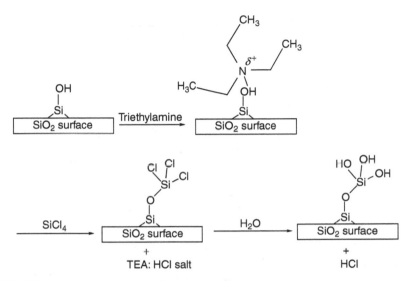

FIGURE 4.14 Reaction scheme for room temperature triethylamine catalyzed atomic layer deposition of silica using SiCl₄ and H₂O.

larger in kinetic diameter than the silane is used, the pore size reduction is directed by the size of the base. At room temperature, the traditional ALD reactions using $SiCl_4$ and H_2O will not occur. When the amine is excluded from the pore, there is no reaction of $SiCl_4$ in this region and, therefore, no additional reduction in pore diameter. This technique allows for continued reduction of larger pores after catalyst exclusion from smaller pores. The end result is a monodisperse pore size. Figure 4.15 presents nitrogen permeance vs. pressure data for a mesoporous silica membrane support with viscous flow defects and the same membrane after 10 and 20 AB modification cycles, using pyridine-catalyzed ALD of silica. The support used in this experiment showed a substantial contribution of 3.3% of the total nitrogen flux via viscous flow, that is, positive relation of gas flux with applied pressure drop due to pinholes or defects in the membrane layer. After 10 AB cycles, the viscous flow contribution of the defects was reduced to less than 1% of the total nitrogen flow. After 20 AB cycles, there was no evidence of any viscous flow contribution to the total N_2 permeance. The end result is a defect-free silica membrane with a near unimodal pore size distribution.

The technique for powder ALD presented above leads to a problem with regard to FTIR characterization. In order to process greater amounts of bulk powder in a vacuum process, a fluidization technique is generally employed in order to enhance the sorption steps between incoming reactants and the

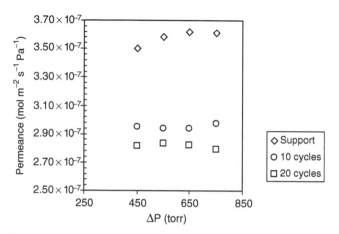

FIGURE 4.15 Nitrogen permeance as a function of average pressure for a defective mesoporous support and after 10 and 20 AB deposition cycles. (From McCool, B. A. and DeSisto, W. J., *Industrial and Engineering Chemistry Research*, 43, 2478, 2004. With permission.)

substrate particles.[121] To the best of our knowledge, FTIR has not been used to dynamically study the surface reactions on particles within a fluidized bed reactor. Using immobilized particles as a surrogate may not be suitable as the immobilized particles will not experience the same conditions as fluidized particles. For example, the fluidized bed will see the effects of convection playing a role through differences in the diffusion of precursors to the particle surface and temperature gradients. These changes may lead to differences in the growth rate and reaction mechanisms between bulk powders and powders immobilized for FTIR analysis.

Performing ALD on powders in supercritical fluid instead of a vacuum may provide some solutions to this problem. There is the obvious advantage of using an environmentally benign solvent that spontaneously separates from the oxide. In the case of silica, there is no caking of the silica as found with other nonaqueous solvents as the modified powder retains its original texture.[131] Moreover, supercritical carbon dioxide (SCF CO_2) has the unique ability to remove the adsorbed water layer on the silica,[131] and this is important for water-sensitive reactions such as silanization.[49] Additionally, it has been shown that by using supercritical fluid a greater extent of modification is possible and that even the inaccessible silanols are, in fact, accessible.[132] In particular, it has been shown that the inaccessible sites can exchange with D_2O and are reactive with an octadecylsilane when the reactions are conducted using SCF CO_2 as a solvent. This is not unique to fumed silica and is shown to also occur in mesoporous SiO_2 powders and films.

Figure 4.16 presents the IR absorbance spectra of fumed silica exposed to D_2O in a vacuum at 400°C and in supercritical CO_2 at 50°C and 200 atm. Both the SCF CO_2 and vacuum-exchanged SiO_2 exhibit strong bands near $2755\ cm^{-1}$ due to the deuterated, isolated silanols. However, the SCF-treated SiO_2 also exhibits a band located at $2680\ cm^{-1}$. This band falls directly where we would expect the deuterated inaccessible silanol species to be located and shows that D_2O accesses and exchanges with these sites using SCF CO_2 as a solvent.

We have also shown that supercritical fluid is a viable delivery medium for the base-catalyzed ALD of silica on silica powder.[133] Using the same reaction sequence presented in Figure 4.14, SiO_2 ALD was performed on a fumed silica powder spread as a thin film on a KBr window. The reactions were performed at 50°C and 200 atm CO_2. Each step was monitored by recording difference spectra versus the unmodified silica. Figure 4.17 shows the first three steps of the first ALD cycle. Spectrun A is the difference spectrum after addition of the triethylamine (TEA) base catalyst. The negative peaks 3747, 3650, and $3,550\ cm^{-1}$ indicate the formation of hydrogen bonds at the isolated, interparticle, and hydrogen-bonded silanols, respectively. Upon

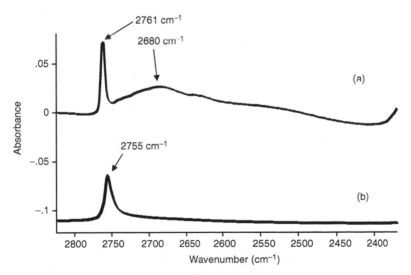

FIGURE 4.16 IR spectra of (a) fumed SiO_2 after D_2O exposure in 200 bar SCF CO_2 and (b) after exposure to D_2O in vacuum at 400°C. (From McCool, B. A. and Tripp, C. P., *Journal of Physical Chemistry B*, 109, 8914, 2005. With permission.)

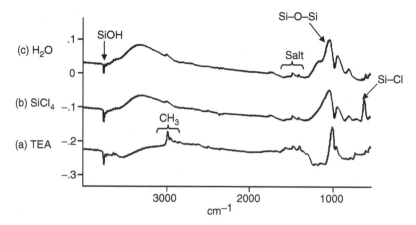

FIGURE 4.17 First three steps of TEA-catalyzed ALD of SiO$_2$ on fumed silica powder. Unmodified powder is used as the background.

addition of SiCl$_4$ (spectrum B), we see the appearance of SiCl modes around 500 cm^{-1} and the formation of Si–O–Si modes around 1080 cm^{-1} indicating the reaction of SiCl$_4$ at the TEA hydrogen-bonded silanols on the surface of the silica. The addition of water (spectrum C) shows an increase in the Si–O–Si species and the disappearance of the Si–Cl modes indicating successful hydrolysis of the Si–Cl surface species. Figure 4.18 shows the SiO$_2$ growth rate for three TEA-catalyzed cycles. The increase in silica bulk modes located at 1080–1150 cm^{-1} with each cycle shows that the growth rate is linear.

As with the section on FTIR application to sol–gel synthesis and characterization, in this section, we have presented an introduction to ALD and a review of how FTIR is being used to aid in the synthesis and characterization of ALD films and how ALD can be used to modify powders and fibers. We have also shown how supercritical fluid ALD may prove to be a powerful tool for dynamic monitoring of the film growth with IR, while bulk powders being modified are experiencing the same reaction conditions as the powder being monitored.

4.5 CONCLUSIONS

In this chapter we have provided information on the use of FTIR to interrogate the surface of powders and fibers. The emphasis was on the sampling considerations that enable the researcher to gain unique information about the gas/solid, nonaqueous liquid/solid and water/solid interfacial regions of

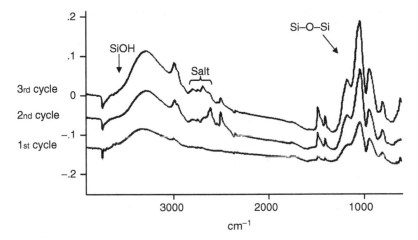

FIGURE 4.18 Three complete SiO_2 supercritical fluid ALD cycles on fumed silica powder.

powders and fibers. While FTIR is amenable to many techniques and sampling methods, our bias in selecting the choice of sampling approach is to keep the spectroscopy as simple as possible. In other words, transmission works best for powders and fibers, and the use of other infrared methods should be employed only when transmission fails.

Through a review of the literature, we have found that FTIR is typically used as a simple characterization tool to verify chemical composition and surface makeup of powders and fibers. In Section 4.3 and Section 4.4, we provide the reader with a more detailed analysis of the application of IR spectroscopic measurement to the synthesis and surface characterization of sol–gel oxide powders and their surface modification using ALD processes. Both of these topics represent areas that are current and have garnered much interest. It is noted, however, that the IR methods and information described for sol–gel oxide powders should provide a guide for their use with other types of powders and fibers. In Section 4.3 and Section 4.4, we provide not only a survey of FTIR's use in sol–gel and ALD processing, but also examples of how FTIR may be used to yield greater information content than is traditionally gained from using FTIR as a simple characterization tool.

REFERENCES

1. Kiselev, A. V. and Lygin, V. I., *Infrared Spectra of Surface Compounds*, Hallstead, New York, 1975.

2. Hair, M. L., *Infrared Spectroscopy in Surface Chemistry*, Marcel Dekker, New York, 1967.
3. Little, L. H., *Infrared Spectra of Adsorbed Species*, Academic Press, New York, 1966.
4. Bell, A. T. et al., *ACS Symposium Series, vol. 137: Vibrational Spectroscopies for Adsorbed Species*, 1980.
5. Blyholder, G. D. et al., *Special Issue in Honor of Robert Eischens. (A selection of papers presented at the 208th ACS meeting, 'Infrared Studies of Surface and Adsorbed Species' held in Washington, D.C., 21–23 August 1994.) [In: Colloids Surf., A, 1995; 105(1)]*, 1995.
6. Scheuing, D. R., Ed., *ACS Symposium Series, 447: Fourier Transform Infrared Spectroscopy in Colloid and Interface Science. Developed from a Symposium Sponsored by the Division of Colloid and Surface Chemistry of the American Chemical Society at the 199th National Meeting Boston, MA, April 22–27, 1990*, 1990.
7. Morrow, B. A., Surface groups on oxides, In *Spectroscopic Analysis of Heterogeneous Catalysts, Part A: Methods of Surface Analysis*, Fierro, J. L. G., Ed., Elsevier, Amsterdam, 1990.
8. Parfitt, G. D., Chemical characterization of pigment surfaces, *Journal of the Oil and Colour Chemists' Association*, 54 (8), 717, 1971.
9. Burneau, A. and Gallas, J.-P., Hydroxyl groups on silica surfaces. Vibrational spectroscopies, *Surface Properties of Silicas*, 145, 1998.
10. Sheppard, N., 50 years in vibrational spectroscopy at the gas/solid interface. Some personal and group endeavors, *Surface Chemistry and Catalysis*, 27, 2002.
11. Unger, K. K., Surface structure of amorphous and crystalline porous silicas. Status and prospects, *Advances in Chemistry Series*, 234 (Colloid Chemistry of Silica), 165, 1994.
12. Wachs, I. E., Raman and IR studies of surface metal oxide species on oxide supports: supported metal oxide catalysts, *Catalysis Today*, 27 (3–4), 437, 1996.
13. Zecchina, Z. et al., Infrared spectra of molecules adsorbed on oxide surfaces, *Applied Spectroscopy Reviews*, 21 (2), 259, 1985.
14. Hair, M. L., Transmission infrared spectroscopy for high surface area oxides, *ACS Symposium Series*, 137 (Vib. Spectrosc. Adsorbed Species), 1, 1980.
15. Morrow, B. A. and Gay, I. D., Infrared and NMR characterization of the silica surface, *Surfactant Science Series*, 90 (Adsorption on Silica Surfaces), 9, 2000.
16. Rochester, C. H., Infrared spectroscopic studies of powder surfaces and surface-adsorbate interactions including the solid/liquid interface. A review, *Powder Technology*, 13, 157, 1976.
17. Guiton, T. A. and Pantano, C. G., Infrared reflectance spectroscopy of porous silicas, *Colloids and Surfaces*, 74, 33, 1993.
18. Poston, P. E. et al., In situ detection of adsorbates at silica/solution interfaces by Fourier transform infrared attenuated total reflection spectroscopy using a silica-coated internal reflection element, *Applied Spectroscopy*, 52 (11), 1391, 1998.

19. Ninness, B. J. et al., An in situ infrared technique for studying adsorption onto particulate silica surfaces from aqueous solutions, *Applied Spectroscopy*, 55, 655, 2001.

20. Ninness, B. J. et al., The importance of adsorbed cationic surfactant structure in dicatating the subsequent interaction of anionic surfactants and polyelectrolytes with pigment surfaces, *Colloids and Surfaces A: Physicochemical and Engineering Aspects*, 203, 21, 2002.

21. Blitz, J. P. et al., Comparison of transmission and diffuse reflectance sampling techniques for FT-IR spectrometry of silane-modified silica gel, *Applied Spectroscopy*, 40, 829, 1986.

22. White, R. L. and Nair, A., Diffuse reflectance infrared spectroscopic characterization of silica dehydroxylation, *Applied Spectroscopy*, 44, 69, 1990.

23. Benziger, J. B. et al., IR photoacoustic spectroscopy of silica and aluminum oxide, *ACS Symposium Series*, 288, 449, 1985.

24. Gillis-D'Hamers, I. et al., Siloxane bridges as reactive sites on silica gel Fourier transform infrared-photoacoustic spectroscopic analysis of the chemisorption of diborane, *Journal of the Chemical Society, Faraday Transactions*, 86, 3747, 1990.

25. Kinney, J. B. and Staley, R. H., Reactions of titanium tetrachloride and trimethylaluminum at silica surfaces studied by using infrared photoacoustic spectroscopy, *Journal of Physical Chemistry*, 87, 3735, 1983.

26. Van Der Voort, P. et al., Reaction of NH3 with triclorosilatcd silica gel: a study of the reaction mechanism as a function of temperature, *Journal of the Chemical Society, Faraday Transactions*, 89, 2509, 1993.

27. Frost, R. L. and Vassallo, A. M., The dehyroxylation of the kaolinite clay minerals using infrared emission spectroscopy, *Clays and Clay Minerals*, 44, 635, 1996.

28. van der Vlies, A. J. et al., Chemical principles of the sulfidation of tungsten oxides, *Journal of Physical Chemistry B*, 106 (36), 9277, 2002.

29. Griffiths, P. R. and de Haseth, J. A., *Fourier Transform Infrared Spectroscopy*, Wiley, New York, 1986.

30. Bellamy, L. J., *The Infra-red Spectra of Complex Molecules*, 3rd Ed., Chapman and Hall, Thetford, U.K., 1975.

31. Smith, A. L., Infrared spectra-structure correlation for organosilicon compounds, *Spectrochimica Acta*, 16, 87, 1960.

32. Tripp, C. P. and Hair, M. L., Transmission infrared spectra of adsorbed polymers using a dual beam FTIR instrument, *Applied Spectroscopy*, 46, 100, 1992.

33. Parry, D. B. and Harris, J. M., Attenuated total reflection FTIR spectroscopy for measuring interfacial reaction kinetics at silica surfaces. *Proceedings of the Fourth Symposium on Chemical Modified Surface*, 3(Chem. Modif. Oxide Surf.), 127, 1990.

34. Neivandt, D. J. et al., Polarized infrared attenuated total reflection for the in situ determination of the orientation of surfactant adsorbed at the solid/solution interface, *Journal of Physical Chemistry B*, 102, 5107, 1998.

35. Kiselev, A. V. and Lygin, V. I., *Infrared Spectra of Surface Compounds*, Wiley, New York, 1975.
36. Gillis-D'Hamers, I. et al., Fourier-transform infrared photo-acoustic spectroscopy study of the free hydroxyl groups vibration, *Journal of the Chemical Society, Faraday Transactions*, 88, 2047, 1992.
37. Bertoluzza, A. et al., Raman and infrared-spectra on silica-gel evolving toward glass, *Journal of Non-Crystalline Solids*, 48 (1), 117, 1982.
38. Tripp, C. P. and Hair, M. L., The reaction of chloromethylsilanes with silica: a low frequency infrared study, *Langmuir*, 7, 923, 1991.
39. Murthy, R. S. S. et al., Quantitative variable-temperature diffuse reflectance infrared Fourier transform spectrometric studies of modified silica gel samples, *Analytical Chemistry*, 58, 3167, 1986.
40. Curthoys, G. et al., Hydrogen bonding in adsorption on silica, *Journal of Colloid and Interface Science*, 48, 58, 1974.
41. Kanan, S. M. and Tripp, C. P., Prefiltering strategies for metal oxide based sensors: use of chemical displacers to selectively cleave adsorbed organophosphonates from silica surfaces, *Langmuir*, 18, 722, 2002.
42. Allian, M. et al., Infrared spectroscopy study of the adsorption of carbonyl compounds on severely outgassed silica: spectroscopic and thermodynamic results, *Langmuir*, 11, 4811, 1995.
43. White, L. D. and Tripp, C. P., A low-frequency infrared study of the reaction of methoxymethylsilanes with silica, *Journal of Colloid and Interface Science*, 224, 417, 2000.
44. Morrow, B. A. et al., Infrared spectra of adsorbed molecules on thin silica films, *Journal of the Chemical Society, Chemical Communications*, 1282, 1984.
45. Tripp, C. P., Low frequency infrared studies of silica at the solid/gas and solid/liquid interface, *Journal of Molecular Structure*, 408/409, 133, 1997.
46. Ballinger, T. H. et al., Transmission infrared spectroscopy of high area solid surfaces. A useful method for sample preparation, *Langmuir*, 8, 1676, 1992.
47. Tripp, C. P., Unpublished work, 1994.
48. Tripp, C. P. and Hair, M. L., Direct observation of surface bonds between self-assembled monolayers of octadecyltrichlorosilane and silica surfaces: a low frequency IR study at the solid/liquid interface, *Langmuir*, 11, 1215, 1995.
49. Combes, J. R. et al., Chemical modification of metal oxide surfaces in supercritical CO_2: in situ infrared studies of the adsorption and reaction of organosilanes on silica, *Langmuir*, 15, 7870, 1999.
50. Griffiths, D. M. et al., Infra-red study of hydrogen-bonding interactions at the solid/liquid interface, *Journal of the Chemical Society, Faraday Transactions 1*, 70, 400, 1974.
51. Rochester, C. H., Infrared spectroscopic studies of adsorption behavior at the solid/liquid interface, *Advances in Colloid and Interface Science*, 12, 43, 1980.
52. Fleer, G. J. and Lyklema, J., Adsorption of polymers, In *Adsorption from Solution at the Solid/Liquid Interface*, Parfitt, G. D. and Rochester, C. H., Eds., Academic Press, New York, 1983.

53. Buckland, A. D. et al., Infrared study of adsorption on silica from two-component and three component liquid mixtures, *Journal of the Chemical Society, Faraday Transactions 1*, 74, 2393, 1978.
54. Vander Linden, C. and Van Leemput, R., Adsorption studies of polystyrene on silica 1. Monodisperse adsorbate, *Journal of Colloid and Interface Science*, 67, 48, 1978.
55. Kawaguchi, M. et al., Adsorption of polystyrene and poly(methy methacrylate) onto a silica surface studied by the infrared technique, *Journal of the Chemical Society, Faraday Transactions*, 86, 1383, 1990.
56. Thies, C., The adsorption of polystyrene-poly(methyl methacrylate) mixtures at the solid–liquid interface, *Journal of Physical Chemistry*, 70, 3783, 1966.
57. Eltekov, Y. A. and Kiselev, A. V., Adsorption of macromolecules on the solid adsorbents surfaces, *Journal of Polymer Science Polymer Symposium*, 61, 431, 1977.
58. Killmann, E. and Bergmann, M., Fraction of H-bonded segments of N-ethyl-pyrrolidone, oligomeric and polymeric vinylprrolidone adsorbed from $CHCl_3$ on silica measured by IR spectroscopy, *Colloid and Polymer Science*, 263, 372, 1985.
59. Patel, A. et al., Studies of cyclic and linear poly(dimethylsiloxanes):30. Adsorption studies on silica in solution, *Polymer*, 32, 1313, 1990.
60. Fontana, B. J. and Thomas, J. R., The configuration of adsorbed alkyl methyacrylate polymers by infrared and sedimentation studies, *Journal of Physical Chemistry*, 65, 480, 1961.
61. Tripp, C. P. and Hair, M. L., The reaction of alkylchlorosilanes with silica at the solid/liquid and solid/gas interface, *Langmuir*, 8, 1961, 1992.
62. Tripp, C. P. and Hair, M. L., Controlled flocculation–deflocculation behavior of adsorbed block copolymers in colloidal dispersions by modifying segment/surface interactions: the use of small displacer molecules to selectively cleasve interparticle bonds, *Langmuir*, 10, 4031, 1994.
63. Poston, P. E. et al., In situ detection of adsorbated at silica/solution interfaces by Fourier transform attenuated total reflection spectroscopy using a silica-coated internal reflection element, *Applied Spectroscopy*, 52, 1391, 1998.
64. Johnson, H. E. and Granick, S., Exchange kinetics between the adsorbed state and free solution: poly(methyl methacrylate) in carbon tetrachloride, *Macromolecules*, 23, 3367, 1990.
65. Li, H. and Tripp, C. P., Spectroscopic identification and dynamics of adsorbed cetyltrimethylammonium bromide (CTAB) structures on TiO_2 surfaces, *Langmuir*, 18, 9441, 2002.
66. Li, H. and Tripp, C. P., Use of infrared bands of the surfactant headgroup to identify mixed surfactant structures adsorbed on titania, *Journal of Physical Chemistry B*, 108 (47), 18318, 2004.
67. Li, H. and Tripp, C. P., Interaction of sodium polyacrylate adsorbed on TiO_2 with cationic and anionic surfactants, *Langmuir*, 20 (24), 10526, 2004.
68. Jiang, C. et al., Identification of lipid aggregate structures on TiO_2 surface using headgroup IR bands, *Journal of Physical Chemistry B*, 109 (10), 4539, 2005.

69. Jiang, C. et al., Infrared method for in situ studies of polymer/surfactant adsorption on silica powders from aqueous solution, *Applied Spectroscopy*, 57 (11), 1419, 2003.

70. Connor, P. A. et al., Infrared spectroscopy of the TiO_2/aqueous solution interface, *Langmuir*, 15, 2402, 1999.

71. Connor, P. A. et al., New sol–gel attenuated total reflection infrared spectroscopic method for analysis of adsorption at metal oxide surfaces in aqueous solutions. Chelation of TiO_2, ZrO_2, and Al_2O_3 surfaces by catechol, 8-quinolinol, and acetylacetone, *Langmuir*, 11, 4193, 1995.

72. Brinker, C. and Scherer, G., *Sol–Gel Science: The Physics and Chemistry of Sol–Gel Processing*, 1990.

73. Matos, M. C. et al., The evolution of Teos to silica-gel and glass by vibrational spectroscopy, *Journal of Non-Crystalline Solids*, 147, 232, 1992.

74. Almeida, R. M. et al., Characterization of silica-gels by infrared reflection spectroscopy, *Journal of Non-Crystalline Solids*, 121 (1–3), 193, 1990.

75. Gallardo, J. et al., Structure of inorganic and hybrid SiO_2 sol–gel coatings studied by variable incidence infrared spectroscopy, *Journal of Non-Crystalline Solids*, 298 (2–3), 219, 2002.

76. Gallardo, J. et al., Thermal evolution of hybrid sol–gel silica coatings: a structural analysis, *Journal of Sol–Gel Science and Technology*, 19 (1–3), 393, 2000.

77. Takada, S. et al., Mechanical property and network structure of porous silica films, *Japanese Journal of Applied Physics, Part 1: Regular Papers, Short Notes and Review Papers*, 43 (5A), 2453, 2004.

78. Tangwiwat, S. and Milne, S. J., Barium titanate sols prepared by a diol-based sol–gel route, *Journal of Non-Crystalline Solids*, 351 (12–13), 976, 2005.

79. Wang, Y. and Santiago-Aviles, J. J., Synthesis of lead zirconate titanate nanofibres and the Fourier-transform infrared characterization of their metallo-organic decomposition process, *Nanotechnology*, 15 (1), 32, 2004.

80. Khalil, K. M. S. et al., Formation and characterization of different ceria/silica composite materials via dispersion of ceria gel or soluble ceria precursors in silica sols, *Journal of Colloid and Interface Science*, 287 (2), 534, 2005.

81. Milt, V. G. et al., NO_x trapping and soot combustion on BaCoO3-y perovskite: LRS and FTIR characterization, *Applied Catalysis B-Environmental*, 57 (1), 13, 2005.

82. Khalil, K. M. S. et al., Preparation and characterization of thermally stable porous ceria aggregates formed via a sol–gel process of ultrasonically dispersed cerium(IV) isopropoxide, *Microporous and Mesoporous Materials*, 78 (1), 83, 2005.

83. Chen, J. Y. et al., Synthesis of (Y,Gd)(2)O-3: Eu nanopowder by a novel co-precipitation processing, *Journal of Materials Research*, 19 (12), 3586, 2004.

84. Zhao, Y. E. et al., FTIR spectra of the M(EDTA)(n-) complexes in the process of sol–gel technique, *Journal of Superconductivity*, 17 (3), 383, 2004.

85. Borges, C. et al., Structural state and redox behavior of framework Co(II) in CoIST-2: a novel cobalt-substituted aluminophosphate with AEN topology, *Journal of Physical Chemistry B*, 108 (24), 8344, 2004.

86. Albertazzi, S. et al., New Pd/Pt on Mg/Al basic mixed oxides for the hydrogenation and hydrogenolysis of naphthalene, *Journal of Catalysis*, 223 (2), 372, 2004.

87. Bernard, C. et al., Hydrothermal synthesis of LaMnO3 + delta: FTIR and WAXS investigations of the evolution from amorphous to crystallized powder, *Journal of Materials Science*, 39 (8), 2821, 2004.

88. Guo, J. J. et al., Novel synthesis of high surface area MgAl$_2$O$_4$ spinel as catalyst support, *Materials Letters*, 58 (12–13), 1920, 2004.

89. Vivekanandhan, S. et al., Glycerol-assisted gel combustion synthesis of nano-crystalline LiNiVO$_4$ powders for secondary lithium batteries, *Materials Letters*, 58 (7–8), 1218, 2004.

90. Catauro, M. et al., The influence of microwave drying on properties of sol–gel synthesized V-Nb catalysts, *Materials Engineering (Modena, Italy)*, 14 (3), 301, 2003.

91. Li, Z. et al., SEM, XPS, and FTIR studies of MoO$_3$ dispersion on mesoporous silicate MCM-41 by calcination, *Materials Letters*, 57 (29), 4605, 2003.

92. Andrade, A. L. et al., Surface modifications of alumina–silica glass fiber, *Journal of Biomedical Materials Research Part B-Applied Biomaterials*, 70B (2), 378, 2004.

93. Chatterjee, M. et al., Mullite fibre mats by a sol–gel spinning technique, *Journal of Sol–Gel Science and Technology*, 25 (2), 169, 2002.

94. Li, Z. F. and Ruckenstein, E., Intercalation of conductive polyaniline in the mesostructured V$_2$O$_5$, *Langmuir*, 18 (18), 6956, 2002.

95. Zhu, H. Y. et al., Gamma-alumina nanofibers prepared from aluminum hydrate with poly(ethylene oxide) surfactant, *Chemistry of Materials*, 14 (5), 2086, 2002.

96. Shim, J. W. et al., Effect of modification with HNO$_3$ and NaOH on metal adsorption by pitch-based activated carbon fibers, *Carbon*, 39 (11), 1635, 2001.

97. Kobelke, J. et al., Effects of carbon, hydrocarbon and hydroxide impurities on praseodymium doped arsenic sulfide based glasses, *Journal of Non-Crystalline Solids*, 284 (1–3), 123, 2001.

98. Zhuravlev, L. T., Concentration of hydroxyl groups on the surface of amorphous silicas, *Langmuir*, 3, 316, 1987.

99. Yang, R. T., Adsorbents: fundamentals and applications, *Book*, 131, 2003.

100. McCool, B. A. and Tripp, C. P., unpublished results, 2004.

101. Ritala, M. et al., Atomic layer deposition, In *Handbook of Thin Film Materials*, vol. 1, Academic Press, 2002.

102. George, S. M. et al., Surface chemistry for atomic layer growth, *Journal of Physical Chemistry*, 100 (31), 13121, 1996.

103. Ritala, M. et al., Atomic layer deposition of oxide thin films with metal alkoxides as oxygen sources, *Science (Washington, D.C.)*, 288 (5464), 319, 2000.

104. Cameron, M. A. et al., Atomic layer deposition of SiO_2 and TiO_2 in alumina tubular membranes. Pore reduction and effect of surface species on gas transport, *Langmuir*, 16 (19), 7435, 2000.

105. Klaus, J. W. et al., Growth of SiO_2 at room temperature with the use of catalyzed sequential half-reactions, *Science (Washington, D.C.)*, 278 (5345), 1934, 1997.

106. Klaus, J. W. et al., Atomic layer controlled growth of SiO_2 films using binary reaction sequence chemistry, *Applied Physics Letters*, 70 (9), 1092, 1997.

107. Sneh, O. et al., Atomic layer growth of SiO_2 on Si(100) using $SiCl_4$ and H_2O in a binary reaction sequence, *Surface Science*, 334 (1–3), 135, 1995.

108. Berland, B. S. et al., In situ monitoring of atomic layer controlled pore reduction in alumina tubular membranes using sequential surface reactions, *Chemistry of Materials*, 10 (12), 3941, 1998.

109. Ott, A. W. et al., Modification of porous alumina membranes using Al_2O_3 atomic layer controlled deposition, *Chemistry of Materials*, 9 (3), 707, 1997.

110. Dillon, A. C. et al., Surface chemistry of Al_2O_3 deposition using $Al(CH_3)_3$ and H_2O in a binary reaction sequence, *Surface Science*, 322 (1–3), 230, 1995.

111. Ott, A. W. et al., Al_2O_3 thin film growth on Si(100) using binary reaction sequence chemistry, *Thin Solid Films*, 292 (1–2), 135, 1997.

112. Ninness, B. J. et al., Formation of a thin TiO_2 layer on the surfaces of silica and kaolin pigments through atomic layer deposition, *Colloids and Surfaces, A: Physicochemical and Engineering Aspects*, 214 (1–3), 195, 2003.

113. Sammelselg, V. et al., TiO_2 thin films by atomic layer deposition: a case of uneven growth at low temperature, *Applied Surface Science*, 134 (1–4), 78, 1998.

114. Gu, W. and Tripp, C. P., Role of water in the atomic layer deposition of TiO_2 on SiO_2, *Langmuir*, 21 (1), 211, 2005.

115. Keranen, J. et al., Preparation, characterization and activity testing of vanadia catalysts deposited onto silica and alumina supports by atomic layer deposition, *Applied Catalysis A: General*, 228 (1–2), 213, 2002.

116. Bedair, S. M., Selective-area and sidewall growth by atomic layer epitaxy, *Semiconductor Science and Technology*, 8 (6), 1052, 1993.

117. Lee, J.-S. et al., Selective area growth at multi-atomic-height steps arranged on GaAs (111). A vicinal surfaces by atomic layer epitaxy, *Applied Surface Science*, 112, 132, 1997.

118. Leskela, M. and Ritala, M., Atomic layer deposition (ALD): from precursors to thin film structures, *Thin Solid Films*, 409 (1), 138, 2002.

119. Haukka, S. et al., Analytical and chemical techniques in the study of surface species in atomic layer epitaxy, *Thin Solid Films*, 225 (1–2), 280, 1993.

120. Puurunen, R. L. et al., Growth of aluminium nitride on porous silica by atomic layer chemical vapour deposition, *Applied Surface Science*, 165 (2–3), 193, 2000.

121. Wank, J. R. et al., Vibro-fluidization of fine boron nitride powder at low pressure, *Powder Technology*, 121 (2–3), 195, 2001.

122. Ferguson, J. D. et al., Atomic layer deposition of SiO_2 films on BN particles using sequential surface reactions, *Chemistry of Materials*, 12 (11), 3472, 2000.

123. Ferguson, J. D. et al., Atomic layer deposition of Al_2O_3 films on polyethylene particles, *Chemistry of Materials*, 16 (26), 5602, 2004.

124. Kang, S. W. et al., Infrared spectroscopic study of atomic layer deposition mechanism for hafnium silicate thin films using $HfCl_2$ $N(SiMe_3)(2)$ (2) and H_2O, *Journal of Vacuum Science and Technology A*, 22 (6), 2392, 2004.

125. Ferguson, J. D. et al., Surface chemistry and infrared absorbance changes during ZnO atomic layer deposition on ZrO_2 and $BaTiO_3$ particles, *Journal of Vacuum Science and Technology A*, 23 (1), 118, 2005.

126. Tripp, C. P. and Hair, M. L., Chemical attachment of chlorosilanes to silica: a two-step amine-promoted reaction, *Journal of Physical Chemistry*, 97 (21), 5693, 1993.

127. White, L. D. and Tripp, C. P., An infrared study of the amine-catalyzed reaction of methoxymethylsilanes with silica, *Journal of Colloid and Interface Science*, 227 (1), 237, 2000.

128. Klaus, J. W. and George, S. M., Atomic layer deposition of SiO_2 at room temperature using NH3-catalyzed sequential surface reactions, *Surface Science*, 447 (1–3), 81, 2000.

129. McCool, B. A. and DeSisto, W. J., Synthesis and characterization of silica membranes prepared by pyridine-catalyzed atomic layer deposition, *Industrial and Engineering Chemistry Research*, 43 (10), 2478, 2004.

130. McCool, B. A. and DeSisto, W. J., Self-limited pore size reduction of mesoporous silica membranes via pyridine-catalyzed silicon dioxide ALD, *Chemical Vapor Deposition*, 10 (4), 190, 2004.

131. Tripp, C. P. and Combes, J. R., The interaction of supercritical CO_2 with the adsorbed water and surface hydroxyl groups on silica, *Langmuir*, 14, 7348, 1998.

132. McCool, B. and Tripp, C. P., Inaccessible hydroxyl groups on silica are accessible in supercritical CO_2, *Journal of Physical Chemistry*, 109 (18), 8914, 2005.

133. McCool, B. A. and Tripp, C. P., Unpublished results, 2005.

5 Chemical Characterization of Silica Powders and Fibers: Application to Surface Modification Procedures

Valentin A. Tertykh

CONTENTS

5.1 INTRODUCTION

Chemical surface modification of silica powders (precipitated and pyrogenic silicas, silica gels, and aerogels of silica) renders new opportunities for their effective use as fillers of polymers, thickeners of dispersion media, adsorbents, silica-based bonded phases for different kinds of chromatographic separation, supports of catalysts, and active compounds [1–8]. Application of chemical modification procedures to the mineral fiber surfaces (especially glass and basalt fibers) is of primary importance in solving the problems of fiber glass-reinforced thermoplastic and thermosetting polymers [9].

The approaches employed for imparting desired functional properties to the silica surface under chemical modification depend on character of a practical or scientific task in question, and may involve such major methods as weakening of intermolecular interactions of surrounding molecules with sites of the surface, or, conversely, enhancement of these interactions with a view to lend specificity and selectivity to the surface, or else modification of the silica surface to transfer the compounds to their heterogeneous state (immobilization of active compounds).

Although modification of the silica surface can result through thermal treatment (dehydration and dehydroxylation), hydrothermal treatment (hydroxylation and hydration), adsorption of diverse substances (particularly substances with a large molecular weight), deposition of carbon or metals, the most ample opportunities for a purposeful change of silica properties are provided by attachment of one or another chemical compound as a result of various chemical reactions with the participation of surface sites. At present, mechanisms of such reactions, as well as the structures of the formed surface compounds, are studied by the spectroscopy in IR, UV, and visible regions, by the solid-state ^{29}Si, ^{13}C (and other nuclei) cross-polarization magic-angle spinning NMR method (CP–MAS NMR), isotope exchange combined with mass spectrometry, and by other contemporary physicochemical techniques [5,10–12]. Under favorable conditions (particularly in the situation with nonporous, highly disperse preparations of fumed silicas) the infrared spectroscopy enables one to exert a practically direct control over proceeding of chemical reactions in a surface layer.

5.2 TYPES OF HETEROLYTIC REACTIONS INVOLVING SILICA SURFACE

The majority of the reactions involving the silica surface sites concerns to the heterolytic processes of substitution, addition, elimination, and rearrangement [3]. Ideas advanced by Ingold [13], Eaborn, and Sommer [14,15] respectively, for chemical reactions of carbon and silicon compounds were applied [16–18] for the development of approaches to classification of the heterolytic conversions with the participation of the silica surface sites.

It is appropriate to consider just these surface groups as a center of the reagent attack [3]. In the general case, the directions of the potential attack will be determined by a nature of the surface groups, the character of electron density distribution on atoms of the site participating in a chemical reaction, and by the nature of a corresponding electrophilic or nucleophilic reagent. From this point of view it is possible to expect that the silicon atoms of the surface of the pristine silica (in structure of $\equiv$SiOH or $=$SiOSi$=$ groups) are the most preferable for an attack by nucleophilic reagents. Opposite, the atoms of oxygen in structure of the same groups are favorable to a greater degree for an attack by electrophilic reagents.

Strelko and Kanibolotskii [17] made the reasonable suggestion that in a case of solid surface, an access of the attacking reagent from rear is likely excluded, and as a consequence, reaction mechanism would be realized in much the same way as a process of the intramolecular substitution (S_Ei or S_Ni mechanism). There are strong grounds for believing [14,15] that the majority of reactions with participation of organosilicon compounds may proceed by such mechanism. Thus, at the stage determining the reaction rate, formation of the quasicyclic (mainly, four-centered) transient complexes, followed by disruption of the existing bonds and formation of the new bonds, is postulated [15].

Extending these concepts to reactions involving silica sites, for example, an interaction of surface $\equiv$SiOH groups with trimethylchlorosilane, is defined as the process of electrophilic substitution of proton in structural silanol group (S_Ei mechanism) and should proceed according to the following scheme:

$$\equiv SiOH + (CH_3)_3\,SiCl \;\rightleftharpoons\; \begin{array}{c} H \;\text{----}\; Cl \\ \backslash \quad\quad \backslash \\ O \;\text{----}\; Si(CH_3)_3 \\ | \\ -Si- \\ | \end{array} \;\longrightarrow\; \equiv SiOSi(CH_3)_3 + HCl$$

In a case of alcoholysis and ammonolysis of the preliminarily chlorinated silica surface, the nucleophilic reagent attacks the surface silicon atom and reaction of nucleophilic substitution is carried out (mechanism S_Ni), as, for instance, reaction of the surface $\equiv$SiCl groups with methanol:

When it comes to the probable structure of the transient complexes for the reactions involving silica surface sites, the following circumstances are usually taken into consideration. In a general case for silicon compounds, when the coordination number is equal to five, it should be expected that the structures, which are close to tetragonal pyramid or trigonal bipyramid, be formed. In accordance with the theoretical considerations [15] of processes of the nucleophilic substitution at silicon atom, attack from the side in relation to Si–X bond, where X, leaving substituent, corresponds to a transient complex with structure of the tetragonal pyramid. Attack with the same nucleophilic reagent, but from the face and along the edge of silicon-containing tetrahedron, should take place with transient state having the trigonal bipyramid structure (with dissimilar atoms arrangement in each particular case).

By now, among the studied chemical reactions involving functional groups on the silica surface, reactions of the nucleophilic and electrophilic addition (Ad_N, Ad_E, $Ad_{N,E}$), and processes of the elimination (E) were recognized [3,18].

Possible types of the heterolytic reactions involving silica surface sites may be represented by the following scheme:

Among the reactions used for chemical modification of a surface there is a large class of processes where an attack is made by an electrophilic reagent through oxygen atoms of silanol groups of the surface. These processes are referred to as reactions of electrophilic substitution of protons (S_Ei). They proceed during interactions with various chloro and alkoxysilanes, organosilazanes, organosiloxanes, numerous organoelemental compounds, and halogenides of various elements. The class includes also processes of electrophilic addition (Ad_E) to groups $\equiv SiOH$ (for example, in the case of isocyanates or ethylene imine). In the latter case, formation of grafted aminoethoxysilyl groups takes place:

$$\equiv SiOH + H_2C\!-\!CH_2 \ \ \overset{\longrightarrow}{\longleftarrow} \ \ \overset{H\ \text{----}\ NH\!-\!CH_2}{\underset{O\ \text{----}\ CH_2}{-\!Si\!-}} \ \ \longrightarrow \ \equiv SiOCH_2CH_2NH_2$$

Another class of reactions involves an attack of a nucleophilic reagent on silicon atoms of the silica surface. This class includes processes of nucleophilic substitution ($S_N i$) (e.g., in the situation with interactions between silanol groups of the surface and hydrohalogens or alcohols) and processes of nucleophilic addition (Ad_N), e.g., reactions of a solid-phase hydrosilylation of olefins [19] with participation of groups $\equiv SiH$ attached to the silica surface:

$$\equiv SiH + H_2C\!=\!CHR \ \ \overset{\longrightarrow}{\longleftarrow} \ \ \overset{R}{\underset{CH_2}{\overset{CH}{-\!Si\!-}}} \ \ \longrightarrow \ \equiv SiCH_2CH_2R$$

In a number of cases, the function of reactive sites on the silica surface is simultaneously performed by silicon atoms and oxygen atoms of siloxane bonds, so that the corresponding processes proceed by mechanism $Ad_{N,E}$, for example:

$$\equiv Si\!-\!O\!-\!Si\equiv \ + FH \ \ \overset{\longrightarrow}{\longleftarrow} \ \ \overset{H\!-\!F}{\equiv Si\!-\!O\!-\!Si \equiv} \ \ \longrightarrow$$

$$\longrightarrow \ \equiv Si\!-\!O\!-\!H + \ \equiv Si\!-\!F$$

It is also necessary to allow for the possibility of proceeding of elimination (E) and rearrangement processes with participation of chemical surface compounds [3,18]. These transformations of the surface chemical compounds will be briefly analyzed later on.

Introduction of new sites (for example, groups $\equiv$SiR, $\equiv$SiOR, $\equiv$SiNH$_2$, $\equiv$SiHal, $\equiv$SiH, etc., where R is an aliphatic or aromatic radical, Hal is a halogen atom) into a surface layer of silica leads to an increase in the number of potential heterolytic transformations with their participation. Grafted organic radicals can also participate in homolytic processes of substitution or addition (S$_H$ or Ad$_H$). These processes have been the subjects of experimental studies, but the amount of experience gained for the present is not large.

5.3 REGULARITIES OF REACTIONS INVOLVING SILICA SURFACE SITES

In studies of the peculiarities of chemical reactions involving the surface sites, two main approaches have been proposed [3,18], when the reagents with different electron structure and geometry attack the similar isolated site, or, conversely, the different fixed groups interact with the same reagent.

The interaction of the isolated silanol groups of dehydrated silica (thermal treatment at 400°C and more) with methylchlorosilanes of the $Cl_nSi(CH_3)_{4-n}$ series, where $n = 1$–4, proceeds predominantly monofunctionally (1:1) according to the scheme:

$$\equiv SiOH + Cl_nSi(CH_3)_{4-n} \rightarrow \ \equiv SiOSi(CH_3)_{4-n}Cl_{n-1} + HCl$$

As this takes place, the activation energies of the process of electrophilic substitution of proton in isolated silanol group of silica surface Q decrease with increasing, n and they are equal to 159, 126, 105, and 80 kJ/mol of the grafted groups for $n = 1$–4, respectively [20].

As to the given reaction series for the simplest case, a quantitative relationship between activity and structure may be described by the Hammet equation, taking advantage of the principle of linearity in the free energy changes (LFE-principle). And really, lgk/k_0 vs. $\Sigma_i\sigma_i^*$ dependence (Figure 5.1a) is linear (k_0 and k—the rate constants in the process of chemisorption of trimethylchlorosilane, taken as a standard in series under study, and of the given chlorosilane; $\Sigma_i\sigma_i^*$—the total inductive effect of the substituents at the silicon atom). It is essential that the activation energy vs. $\Sigma_i\sigma_i^*$ relationship is also consistent with the LFE-principle (Figure 5.1b), thus giving evidence of its applicability, not only for free energies, but also for the reaction activation energies. It is evident that a relationship between the activation energies and the relative rate constants is expressed as a linear

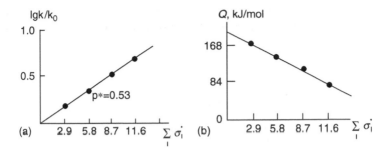

FIGURE 5.1 Dependence of the relative rate constants lgk/k_0 (a) and the activation energies Q (b) vs. the total inductive effect of the substituents in the reagent $\Sigma_i \sigma_i^*$ for chemisorption of methylchlorosilanes $Cl_nSi(CH_3)_{4-n}$ ($n = 0-4$) by isolated silanol groups of the silica surface.

dependence. The value of the reactivity constant ρ^*, as estimated based on the slope of lgk/k_0 vs. $\Sigma_i \sigma_i^*$ dependence, for the given reaction series, is equal to $+0.53$. It is known that the positive value of ρ^* is characteristic of the reactions characterized by an essential effect of the electronegative substituents. Really, as was noted above, the rate constants for the reactions of electrophilic substitution of proton in the silanol group exhibit a sequential rise in the series from $(CH_3)_3SiCl$ to $SiCl_4$ at the same temperature. Low absolute value of the Taft reaction constant, as compared to the ρ^* values for the similar reactions of the organosilicon compounds in solutions [15], resulted from the fact that the structural silanol group on the silica surface acts in this process as a more weak nucleophilic reagent, as compared to hydroxyls of water and alcohols molecules.

Chemisorption of methylchlorosilanes on the hydrated silica surface proceeds in mild conditions, beginning with room temperature. In this case, activation energies of the reaction are close to each other and are equal to about 40 kJ/mol of the grafted groups for all compounds of the series studied. It may be connected with participation in reaction more active nucleophilic reagent than isolated silanol group of the dehydrated silica surface. Such active sites of the hydrated surfaces are most likely molecules of the firmly adsorbed water.

A closer look at the data concerning a reactivity of trimethylsilylating reagents of the general formula $(CH_3)_3SiX$, where X is halogen, alkoxy or other group reactive with respect to silanol groups of the silica surface, shows that the most dense modifying layers were obtained by use of organosilicon compounds with Si–N(H,R) bonds, in particular hexamethyldisilazane or trimethyl (dimethylamino)silane. The observed order in the modifying reagents activity: organosilazanes > organoalkoxysilanes > organochlorosilanes, provides strong evidence in favor of the formation of four-center transient complexes in reaction of the electrophilic proton substitution in isolated silanol groups of the silica surface:

$X = Cl, OR, NR_2$ etc

As will readily be observed from the scheme, for a given type of reaction (mechanism $S_{E}i$), the proton-accepting properties of the substituent X at silicon atom of the attacking molecule have a great importance, because these properties characterize a degree of nucleophilic assistance of a leaving group. Really, when comparing the reactivities of chlorosilanes and alkoxysilanes, activities of the corresponding organosiloxanes and aminosilanes in the reaction of electrophilic substitution of proton in the isolated silanol groups of the silica surface, the following sequence was obtained:

$$Si-N(H, R) > Si-O(H, R) > Si-Cl > Si-O(Si) > Si-C(H, R)$$

For the first time, this sequence has been described in the paper [18]. It is characteristic that the similar sequence remains in the case of carrying out of reactions in liquid phase in the organic bases presence [21,22].

For series of compounds studied, there is a linear correlation between the activation energies of a chemisorption process and proton-accepting properties of the atoms bound to silicon within a reactive functional group of the modifying reagent (Figure 5.2a).

Actually, as was shown in [18], there is a quite satisfactory correlation between the calculated values of the activation energy for the process and those of proton affinity for the carbon, chlorine, oxygen, and nitrogen atoms. At the same time, direct experimental data, characterizing proton affinity of the same atoms within studied organosilicon compounds, are absent. The accepting properties of these atoms can be characterized to some degree by the values of the 3750 cm^{-1} band (it belongs to the stretching vibrations of O–H in the isolated silanol surface groups) shifts, $\Delta\nu_{OH}$, at formation of hydrogen bonds with the molecules of the organosilicon compounds studied. It was found that there is a linear relationship between the $\Delta\nu_{OH}$ values and those for activation energies of the process (Figure 5.2b).

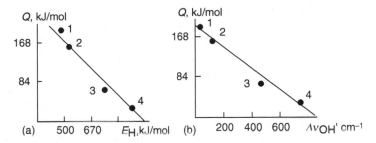

FIGURE 5.2 Dependence of the activation energies Q for reaction of electrophilic substitution of proton in surface silanol groups vs. the proton-accepting properties of atoms bound with silicon E_H (a) and the values of $3750\ cm^{-1}$ band shifts $\Delta\nu_{OH}$ (b) after adsorption of organosilicon compounds: (1) $Si(CH_3)_4$, (2) $(CH_3)_3SiCl$, (3) $(CH_3)_3SiOCH_3$, (4) $(CH_3)_3SiN(CH_3)_2$.

For the explanation of differences in reactivities of molecules, interacting with isolated silanol groups according to mechanism S_Ei, Gorlov, with co-authors [23–25], made the supposition that the activation barrier of such reactions is determinated only by the deformation energy of the attacking molecule, as its distance from the surface fragments, carrying the –OH group, is relatively large. On the contrary, for reactions, realizing according to mechanism S_Ni, it is supposed that the deformation energy of the corresponding surface fragment bring in the main contribution into the activation barrier. Supposed approach (a deformation model) to explanation of differences in reactivities of molecules, attacking the same surface site, was found to be the most suitable only for case when the leaving group and structure of the corresponding four-center transient complex are identical. Really, this model explained rather well the differences in reactivities of chlorosilanes in series from $SiCl_4$ to $(CH_3)_3SiCl$ [20]. At the same time, the application of deformation model to other methylhalosilanes, for which the following row of activities in relation to surface silanol groups was obtained [26,27]:

$$(CH_3)SiI > (CH_3)_3SiBr >> (CH_3)_3SiCl$$

results in less confident explanation. Calculations of the deformation energies for molecules $(CH_3)_3SiI$, $(CH_3)_3SiBr$, and $(CH_3)_3SiCl$ gave [25] not strongly differing values: 128, 144, and 165 kJ/mol correspondingly. At the same time, the reactivities of trimethylhalosilanes differ impressively [26,27]. Whereas a rather high temperature (above 300°C) is necessary for chemisorption of trimethylchlorosilane by isolated silanol groups of the silica surface, the reaction with trimethylbromosilane is proceeding at 50°C with a high rate, and the full substitution of protons of silanol groups in the reaction with trimethyliodosilane occurs already at room temperature. Maximum attained concentrations of the grafted trimethylsilyl groups are equal for all trimethyl-halosilanes studied, and they correspond to concentration of the isolated silanol groups on the surface of the pristine silica.

Interaction of trimethylhalosilanes with the silica surface may be realized in several ways, namely according to the mechanism of electrophilic proton substitution in the isolated silanol groups by the trimethylsilyl ones, or by addition of organosilanes to the siloxane bonds (mechanism $Ad_{N,E}$). Consideration must be given to the homolytic reactions to proceed, especially in the case of $(CH_3)_3SiI$, although such reactions are not distinctive for silicon compounds. Really, the results obtained show that the radical reactions and addition of $(CH_3)_3SiI$ and $(CH_3)_3SiBr$ molecules to the siloxane bonds do not occur at the given temperatures. First of all, the halogens (in the form $\equiv SiI$ or $\equiv SiBr$ groups) were not found in the surface compounds, suggesting that the chemisorption with the participation of siloxane bonds is absent. Secondly, the grafting of trimethylbromosilane and trimethyliodosilane does not occur on methoxylated silica at the moderate temperatures, and consequently, the radical processes do not proceed. The fact that the concentration of trimethylsilyl groups in the reaction products is equal to concentration of the hydroxyl groups on the surface of pristine fumed silica, gives the evidence for the reaction proceeding between surface sites and trimethylhalosilanes by the mechanism of electrophilic proton substitution in the structural silanol group ($S_E i$). Actually, elimination of a proton from the reaction site (due to reaction of methoxylation of silica surface $\equiv SiOH \rightarrow \equiv SiOCH_3$) results in the fact that the reaction with $(CH_3)_3SiI$ and $(CH_3)_3SiBr$ does not proceed under the usual conditions. It is important that the bond length decreases and the silicon–halogen bond energy increases in the same sequence as it takes place for the activities of trimethylhalosilanes in relation to the surface silanol groups. When explaining the differences in reactivity of trimethylhalosilanes, the easier polarizability of the large-sized atoms (and the appropriate silicon-halogen bonds), as well as the differences in the proton-accepting properties of the atoms related to the silicon atom, should be taken into consideration.

For trimethylsilyl-containing compounds with Si–N bond, for example, for trimethylpseudohalosilanes having general formula $(CH_3)_3SiX$, where $X = -N_3$, $-NCO$, $-NCS$, $-NCNSi(CH_3)_3$ groups, it was found that the given compounds are chemisorbed by the fumed silica surface (prepared at 600°C) even at room temperature. As it follows from the infrared spectra, the absence of distinct changes in the intensity of the surface silanol groups band (3750 cm^{-1}) together with simultaneous appearance of the absorption bands characteristic for the chemisorbed trimethylsilyl and the silicon–nitrogen-containing $\equiv SiN_3$, $\equiv SiNCO$, $\equiv SiNCS$, and $\equiv SiNCNSi(CH_3)_3$ groups, is attributed to the participation of the siloxane bridges in the surface reaction (mechanism $Ad_{N,E}$):

$$\equiv Si-O-Si \equiv + (CH_3)_3SiX \rightarrow \ \equiv SiOSi(CH_3)_3 + \ \equiv SiX$$

$$(X = -N_3, -NCO, -NCS, -NCNSi(CH_3)_3)$$

Registration of the IR absorption bands, characteristic for the chemisorbed $\equiv$SiX groups on the methoxylated silica surface after its contact with the vapors of N-containing trimethyl-substituted silanes, serve as an additional argument in a favor of the direct addition of $(CH_3)_3SiN_3$, $(CH_3)_3SiNCO$, $(CH_3)_3SiNCS$, $(CH_3)_3SiNCNSi(CH_3)_3$ molecules to siloxane bonds. The grafted $\equiv SiN_3$, $\equiv SiNCO$, $\equiv SiNCS$ or $\equiv SiNCNSi(CH_3)_3$ groups are easily subjected to the hydrolysis or methanolysis reactions [27].

As temperature increases, the process of electophilic proton substitution in the isolated silanol surface groups with formation of the grafted trimethylsilyl groups is also proceeding. However, as distinct from other Si–N-containing compounds (e.g., hexamethyldisilazane [28], trimethyl(dimethylamino)silane [20], dimethylaminotriethoxysilane [18], trimethylsilylpiperidine and trimethylsilylimidazole [29]), which react with the $\equiv$SiOH groups, even at room temperatures, the complete involvement of the structural silanols into reaction with trimethylazidosilane was possible after the heating up to 400°C (up to 500°C in a case of trimethylthioisocyanatesilane), and up to 550°C when using trimethylisocyanatesilane and bis(trimethlsilyl)carbodiimide [27]. Reactivity of trimethylpseudohalosilanes with Si–N bonds in respect to the surface $\equiv$SiOH groups decreases in the following row:

$$(CH_3)_3SiN_3 > (CH_3)_3SiNCS > (CH_3)_3SiNCO, \ (CH_3)_3SiNCNSi(CH_3)_3$$

Characteristically, that the activity of the same reagents in reactions of the siloxane bond splitting changes in the reverse order.

To explain the differences in reactivities of the trimethylpseudohalosilanes in the reaction with the same silica surface sites, the geometric structures and electronic configuration of the attacking molecules have been analyzed. Calculations have shown that the length of the Si–N bond depends on the composition of the pseudohalogenide group and decreases in the next row, $SiN_3 > SiNCS > SiNCO$. Stability of the molecular system, as it is determined by an enthalpy of formation of the corresponding trimethylpseudohalosilane, increases with the decrease of the Si–N bond length. Thus, the decline in activity of the $(CH_3)_3SiX$ compounds correlates with calculation data, showing the growth of the stability of the reagent molecules and the decrease of the equilibrium length of the Si–N bonds. The calculations testify also about the distinct charge disjunction on the atoms of the X-group of the trimethylpseudohalosilane molecules. This circumstance can favor to dissociation of the reagent molecules on the corresponding surface sites. As was shown in the paper [30], in the presence of organic bases vapors, the chemisorption of the nitrogen-containing trimethylpseudohalosilanes is carried out at room temperature. This result testifies that the electron-donating properties and the proton affinity of nitrogen atoms in the studied trimethylpseudohalosilanes are expressed much more poorly, than it takes place in the triethylamine, diethylamine or pyridine molecules.

As regards to reactions of the heterolytic splitting, the Si–C bonds usually exhibit a rather low activity. This assertion concerns also to the reactions involving the silica surface sites. For instance, according to our data [20], tetramethylsilane reacts with the isolated hydroxyl groups of the dehydrated silica surface at temperatures above 550°C. Activation energy of the process, which is executed according to the $S_E i$ mechanism, is equal to about 185 kJ/mol of the grafted trimethylsilyl groups. However, the reaction with trimethylcyanosilane is the exception. As was shown in [31], the $(CH_3)_3SiCN$ molecules can rather easily enter into the reaction with the ≡SiOH-groups of the silica surface. Even at room temperature, about 70% of the isolated silanol groups are substituted by the trimethylsilyl ones for one hour. Reaction with trimethylcyanosilane does not proceed, if a proton is excluded from the reaction site due to methoxylation of the surface. These experimental data testify that the $(CH_3)_3SiCN$ molecules chemisorption is a process of the electrophilic proton substitution in the silanol groups. When analyzing the reasons of the sharp differences in an activity of the $(CH_3)_3SiCH_3$ and $(CH_3)_3SiCN$ molecules in the reaction with the surface ≡SiOH groups, account should be taken of the higher electronegativity of the CN-group, as compared to the methyl one. It may result in the increase in the effective positive charge on the silicon atom in the trimethylcyanosilane molecule. It is significant also that the proton-accepting properties of the CN-groups are higher than those of the methyl ones. Thus, according to [31] shift of the 3750 cm^{-1} band at adsorption of $(CH_3)_3SiCN$ vapors is equal to 330 cm^{-1}, whereas $\Delta\nu_{OH}$ at adsorption of tetramethylsilane amounts to 40 cm^{-1} only.

Chemisorption of bis(trimethylsilyl)sulfate $[(CH_3)_3SiO]_2SO_2$ on the silica surface begins even at room temperature, and the complete involvement of the OH-groups into reaction occurs at 100°C [32]. Reaction is realized according to $S_E i$ mechanism without any sulfonation of the silica surface. When comparing the reactivities of different trimethylsilyl-containing compounds with the Si–O bonds at interaction with the surface ≡SiOH groups, it was noted that activity of the modifying reagent exhibits the strong dependence on the nature of oxygen atom of the substituent:

$$(CH_3)_3 SiOSO_2 OSi(CH_3)_3 > (CH_3)_3 SiOCH_3 > (CH_3)_3 SiOSi(CH_3)_3$$

Actually the trimethylmethoxysilane molecules react with the surface silanol groups at 180°C, but the hexamethyldisiloxane molecules above 380°C. Characteristically, that high activity in respect to the surface silanol groups, is typical also for the other sulfur-containing organosilicon compounds, for example for trimethylsilylfluoromethylsolfonate $(CH_3)_3 SiOSO_2 CF_3$ and hexamethyldisylthione $(CH_3)_3 SiSSi(CH_3)_3$ [29].

At analysis of interaction trimethylhalosilanes, trimethylpseudohalosilanes, and other organic and organosilicon compounds with silica surface, it is important to allocate the kinetic model of process, to describe a limiting stage and the most probable mechanism of reaction. As is known, owing to geometrical and chemical heterogeneity of dispersed solids, kinetics of chemical reactions involving surface sites are not described by models developed for homogeneous processes proceeding in a liquid or gas phase. In this connection, the simulation methods for estimation of the activation parameters of chemisorption processes on silica and other oxides, taking into consideration the various types of their surface heterogeneity, deserve attention. Important significance, and certain prognosticated force, can have also correlations established between various quantum-chemical and thermodynamic characteristics of the modifying reagents structure and these activation parameters, enabling to evaluate reactivity of the surface sites in chemisorption processes with their participation.

Models of chemical reaction involving sites of nonuniform surfaces, approximately taking into consideration the discrete and exponential distribution of chemisorption activation energies with constant and changing preexponential factors, were offered in the works [40–43]. In the latter case, the possibility exists of calculating isokinetic temperature and rate constant of the appropriate chemical reaction, as it was determined for case of chemisorption of a number organosilicon compounds on the silica surface. Authors [33–36] pointed out, also, that for the more complete description of the chemisorption kinetics, consideration must be given to the contribution into observed conversion rate of the lateral interactions, both between chemisorbed species and these species and activated transient complexes. Solutions of the integrated kinetic equations were obtained using the methods of a perturbations theory. For finding of the main route of the chemisorption process, it was offered to compare parameters of various types of sites of the nonuniform surface (or distribution on parameters of these sites) with activation parameters of the reaction. An important problem of a chemisorption theory is a calculation of such activation parameters on the basis of physicochemical properties of the reagent molecules and active surface sites. Bogillo [33] has offered the approach to evaluation of parameters of distributions of the silica surface on the reactionary constants of chemisorption for a number methyl-chlorosilanes and trimethylpseudohalosilanes, using the experimental kinetic data obtained in our works [20,26–28,30–32,37–43]. The signs of average reactionary constants found in the paper [33] correspond to the mechanism of electrophilic proton substitution in surface silanol groups, as it was formerly supposed for these compounds. The high deviations of these magnitudes testify about significant heterogeneity of the reactive sites.

Formation of transient complexes with hydrogen or donor–acceptor bonds involving surface site and reagent molecule, can make a contribution into the height of chemisorption activation barrier. The possible methods for the calculation of such complexes stability, chemisorption activation energies, and their connection with parameters of the reagent electronic structure have been analyzed by the authors of works [44–46]. Using quantum-chemistry methods, the ionization potential and electron affinity, positive and negative charges on atoms of the modifying reagent molecule were calculated. Relying on such data, it is possible to evaluate stability of transient complexes between the reagent molecules and hydroxyl groups of the silica surface.

As is shown, in the majority of cases, chemisorption activation energies of a number monofunctional organosilicon compounds, containing trimethylsilyl group $(CH_3)_3SiX$, are increased with decreasing Si–X bond energy, increasing inductive constant of the substituent X and decreasing reaction enthalpies. The majority linear correlations was found to be observed between reaction activation energies of the appropriate organosilicon compounds with hydroxyl groups of the silica surface and electron affinity of these reagents, or parameters, including this quantity.

Gun'ko, with co-authors [44–47], within the framework of theory of a transient state and statistical theory of absolute reaction rates using adiabatic or dynamic coordinate of reaction, have calculated rate constants of surface reactions of various types. They had analyzed the approaches to description of nature of the activation barriers, dynamic changes in local electronic states of attacking molecules, and surface sites. It was demonstrated that the contribution to energy of activation for the majority of reactions involving surface silanol groups on 50%–70%, is governed by the expenditures on H^+-transfer in cyclic transient complexes.

5.4 EFFECT OF ELECTRON- AND PROTON-DONATING COMPOUNDS ON CHEMISORPTION PROCESSES

Earlier we [3,48] developed the concepts about catalytic influence of the electron- and proton-donating compounds on proceeding of reactions involving silica surface sites. Thus, interaction of isolated silanol groups of dehydrated silica surface with trimethylchlorosilane is proceeding at 300°C, but the same reaction is completed at room temperature in the presence of triethylamine vapors. Chemical reaction of the surface $\equiv$SiOH groups with alkoxyorganosilanes begins about at 180°C, but the same reaction is effectively executed at room temperature after introducing hydrogen chloride into the reaction volume. There is good reason to believe that the possible directions of such catalytic influence for the reactions of electrophilic substitution of proton in the surface silanol groups consist in the manifestation of proton-accepting properties of the catalyst, as it takes place in the case of $N(C_2H_5)_3$,

or else this effect emerges in the construction of cyclic transient complexes involving the surface $\equiv$SiOH group, the reagent molecule and the catalyst:

It is reasonable to expect [48,49], that due to donor–acceptor interactions in the cyclic transient structures, the charge separation is reduced to a minimum, and thus the rupture and formation of the bonds is proceeding in the mild conditions.

It was supposed also [48,49] that the catalytic effect of ammonia and triethylamine, as well as other organic bases, on chemisorption of organosilicon compounds consists in the hydrogen bond formation with protons of the surface silanol groups, as a result of which enhancement of the nucleophilicity of oxygen atom takes place. Really, the experimental data were obtained [48–61] that this scheme is more preferable in comparison with the mechanism in which formation of the more reactive intermediate between the modifying organosilicon reagent and organic base molecule takes place [21,22,62,63].

Gun'ko, with co-authors [64–67], believe that similar catalytic processes proceed in accordance with the Langmuir–Hinshelwood mechanism, and an exponential growth of a rate constant is determined by the enthalpic factor. In response to the essential downturn of the reaction temperature, an influence tunnel H^+-transfers grows. Thus, the more adequate results were obtained by the method of dynamic coordinate of the reaction. The theoretical analysis predicts [64,66] that at the consistent vibrations of the surface atoms, reagent and organic base molecules the activation parameters, and the rate constant of the reaction, can change on some orders. It is suggested that such reduction of the activation energy magnitude occurs at the expense of respective alterations of density of the local states (orbital control) and atom charges (enthalpic contribution).

5.5 REACTIONS OF ELIMINATION AND ADDITION

Formerly, the data were obtained [68] that thionyl chloride interacts with the hydroxyl groups of the silica surface according to the mechanism of electrophilic proton substitution ($S_E i$), and after that, the formed surface compounds transform owing to elimination process (E), and the ultimate result of such transformation corresponds to the product of the reaction of nucleophilic substitution on the silicon atom:

We have suggested that such transformations of the surface chemical compound, which are reminiscent in a certain degree of some catalytic transformations on the oxide surfaces, can be considered as a more general class of reactions uniting the processes proceeding in accordance with the $S_E i$ and $S_N i$ mechanisms. As is shown in [3,69], this type of reaction, involving the surface silanol groups, is realized also in the other cases. In particular, at chemisorption of WCl_6 [70] by the isolated silanol groups, the formed surface chemical compounds ($\equiv SiOWCl_5$) are decomposed at heating with elimination of tungsten oxochloride:

$$\equiv SiOWCl_5 \rightarrow\ \equiv SiCl + WOCl_4$$

The surface compounds $\equiv SiOWOCl_3$ formed at reaction of the silanol groups with $WOCl_4$, undergo transformations with formation of $\equiv SiCl$ groups and WO_2Cl_2 elimination [71]. In the case of the WCl_5, chemisorption was detected the $WOCl_3$ release [72]. Elimination processes (E) were detected also at chemisorption of some halogenides and oxohalogenides of other elements (CCl_4, $TiCl_4$, $MoOCl_4$, $MoCl_5$, PCl_5) [68,73–76]. In all these reactions, the formation of the surface $\equiv SiCl$ groups, and elimination of the appropriate relatively volatile oxochlorides, was detected. In many cases, the factors determining the stability of the formed surface compounds remain unknown.

Reaction of phenols with the $\equiv SiNH_2$ groups, which were previously introduced into the surface layer of chlorinated silica, was used for the preparation of grafted phenol derivatives [77]. It was shown that the quaternary ammonium compounds formed on the silica surface at the initial stage (the process of electrophilic addition Ad_E) destruct at 100°C–140°C, with formation of the grafted ethereal groups (reaction of elimination):

$$\equiv Si\text{–}NH_2 + C_6H_5OH \xrightarrow{(Ad_E)} \equiv Si\text{–}NH_3^+ \cdots {}^- OC_6H_5 \xrightarrow{(E)} \equiv Si\text{–}OC_6H_5 + NH_3$$

Analogous route of the surface reactions $Ad_E \rightarrow E \rightarrow S_Ni$, with formation of groups $\equiv SiI$, was detected at chemisorption of hydrogen iodide on the silica, previously aminated in a similar manner [78].

Recently, the considerable possibilities of the solid-phase hydro-silylation reactions for the formation of dense modifying coatings has been established (see in Refs. [19,79]). It was shown that $\equiv SiH$ groups grafted to the silica surface may be practically quantitatively involved into chemical reactions with alk-1-enes, both in the presence of the Speier catalyst, and in the absence of catalyst at high temperature. Modified silicas with grafted hydrocarbon radicals containing 6, 8, 10, 14, 16 or 18 carbon atoms were synthesized by such method. The achieved concentrations of grafted groups decrease from 3.70 µmol/m^2 for $n = 6$–0.71 µmol/m^2 for $n = 18$. However, in spite of the decrease of grafted hydrocarbon group concentration with increasing n, the hydrophobic properties of modified silicas are enhanced. It is obvious that hydrocarbon radicals with the high content of carbon atoms do not have a brush-like structure and most likely these groups are inclined at some angle to the silica matrix surface. In the case of styrene chemisorption via hydrosilylation, reaction the content of grafted phenylethyl groups amounts to 1.53 µmol/m^2 [80]. Catalytic addition of functional olefins (acetyl acetone, vinyl acetate, acrylamide) to $\equiv SiH$ groups on the silica surface proceeds with the lesser yield (from 75 to 30%). Opposite, catalytic addition of 2-hydroxyethylmethacrylate to the surface $\equiv SiH$ groups is carried out without difficulty [81], according to the following scheme:

$$\equiv SiH + H_2C{=}C{-}COOCH_2CH_2OH \underset{CH_3}{\overset{}{|}}$$

$$\begin{array}{c} \equiv Si{-}CH_2{-}CH{-}COOCH_2CH_2OH \\ \underset{CH_3}{|} \quad \underset{CH_3}{|} \\[2mm] \equiv Si{-}\underset{CH_3}{\overset{CH_3}{|}}{-}COOCH_2CH_2OH \end{array}$$

As is shown in [82], the surface $\equiv SiH$ groups, resulting by thermal destruction of methoxylated silica, can be introduced in the solid-phase hydrosilylation reaction. In such a manner, the proposed method may be used for the synthesis of organosilicas containing hydrolytically stable Si–C bonds between the silica surface and organic functional groups without using any organosilicon reagents.

The attachment of one of the participants of the hydrosilylation process to solid surface provides more convenient and favorable position into the reaction mechanism studies. The experimental data [83] give evidence that solid-phase catalytic hydrosilylation reactions involve the formation of intermediate complexes containing both of the reagents and the Speier catalyst. The possibility has been also established of execution of the noncatalytic reaction of solid-phase hydrosilylation of olefins at high pressures and temperatures. The modifying coatings obtained using solid-phase hydrosilylation reactions have a rather high hydrolytic stability and reproducible properties. The reaction of solid-phase hydrosilylation can be viewed as one of promising methods of synthesis of surface chemical compounds with Si–C bonds [19,79].

5.6 APPLICATIONS OF CHEMICALLY-MODIFIED SILICAS

Substitution of structural silanol groups by methylsilyl groups, through reactions with appropriate organosilicon compounds, moderates substantially the energy of adsorption of polar adsorbates (one can observe a decrease in values of adsorption and heat of adsorption), and the silica surface acquire stable hydrophobic properties. At the same time, such a chemical modification of surface prevents aggregation of disperse silica particles, and facilitates their more uniform distribution in hydrocarbons and polymeric media. Moreover, attachment of sufficiently long hydrocarbon radicals (C_{16}–C_{18}) makes it possible to use such modified silicas as a basis for creation of adsorbents intended to concentrate organic compounds, which is of importance for designing methods of analysis for contents of pesticides and other organic impurities when effecting monitoring of environment [84].

Attachment of a layer of chemically grafted organic radicals with various functional groups ($-NH_2$, $-COOH$, $-CH=CH_2$, $-CN$, $-SH$, etc.) that differ in their physicochemical properties and reactivities, enables one to enhance desired characteristics of the silica surface, which offers ample scope and novel potentiality for practical application of modified silicas. Similar functional silicas find much use in the normal-phase variant of high-performance liquid chromatography (HPLC). Besides, they may have much promise in development of chemically active fillers of polymers, in designing of selective adsorbents, including highly specific affinity and immune affinity adsorbents for separation and purification of biopolymers.

Further, chemically modified silicas acquired a great importance in the sphere of production of heterogeneous metal complex catalysts, and immobilization of enzymes, other biologically active compounds, antibodies, cells, sections of tissues, and microorganisms on an inorganic matrix. This approach is promising for development of chemical and biological sensors whose application makes it possible to exert a pulsed or continuous control over corresponding components in a gas phase or liquid medium. For example, chemical modification of the surface of electrodes with functional organosilicon compounds (sol–gel transformations), is rather widely used for attachment of electroactive compounds in a surface layer (electrochemical sensors). From this standpoint, the most promising is chemical binding of an active compound, because construction of sensing elements of control systems by the polymer encapsulation methods does not provide a perpetual accessibility of an active compound and uniformity of its distribution in a film, as well as a necessary strength of bonding with the electrode surface.

Purposeful variation of properties of surface of silicas is of great importance for carrying out numerous scientific and practical tasks in adsorption, chromatography, catalysis, and chemistry of filled polymers. Presently, functionalized silica surfaces find much use for production of immobilized reagents, which are successfully employed in synthetic organic chemistry [85]. Especially significant is the ever-growing role of chemically modified silicas in modern materials science and nanodimensional engineering. Worthy of note is, also, wide use of polysilsesquioxanes produced by the sol–gel method (co-hydrolysis of organofunctional trialkoxysilanes and tetraalkoxysilane); organically modified silicates (ormosils), including silicates doped with diverse dyes; and organically modified ceramics (ormocers) for producing novel optical materials and lasers, sensitive elements of sensing devices, and biocompatible ceramics [87–90]. New vistas in production of modified silicas are furnished by chemical modification of surface and core of ordered mesoporous siliceous materials (such as MCM-41 and MCM-48) [91–93], which possess an exceptionally extended surface and unique structure of pores accessible to large molecules. It is not ruled out that such modified silicas will form the basis for development of adsorbents and catalysts of a new generation. Of significance is, also, formulation and establishment of principles of mosaic nanodimensional modification of siliceous supports, because under favorable conditions on the surface of such supports, it is possible to construct models of active centers of multi-component metal complex catalysts and biological catalysts, as well as to create ultraselective and affinity sorbents. Functionalized silica surfaces find more and more much use and comprise novel dispersion media, polymeric compositions, and modern materials.

5.6.1 Modified Silicas in Chromatography and Solid-Phase Extraction

Modified silica surfaces provide a considerable expansion of capabilities of chromatography and sorption techniques, which find much use for separation, purification, and concentration of substances [2,84,94–104]. It is appropriate to mention here that, in the situation with various types of chromatographic separation (gas-adsorption chromatography, HPLC, affinity chromatography), a surface layer of silica should contain sites that are able to participate in diverse kinds of intermolecular interactions. Thus, gas-adsorption chromatography and reversed-phase HPLC make use of siliceous supports with attached nonspecific methyl, phenyl, octyl, and octadecyl groups, while, in the case of normal-phase HPLC, it is necessary to use modified silicas with quaternary ammonium compounds or with aminopropyl, cyanopropyl, carboxyl, propylsulfo, alkyldiol, and other groups capable of participating in donor–acceptor interactions with molecules of compounds, which are separated. In the situation with affinity chromatography, which depends on formation of biospecific complexes (enzyme–substrate, enzyme–inhibitor, lectin–glycoproteid, antigen–antibody, etc.), one of the constituents of such a complex should be chemically bonded to the surface of a silica matrix. The function of a biospecific ligand is most often performed by lectins, protein A, certain dyes, polysaccharides, amino acids, and monoclonal and polyclonal antibodies, which are attached to functional silica surface with the help of diverse cross-linking agents. Determination of contents of medicinal preparations in biologic fluids, containing substantial amounts of albuminous compounds, calls for application of modified porous silica matrices of diphilic nature, i.e., there should be a hydrophilic external surface (accessible for proteins) and organophilic internal surface (inaccessible to proteins). Such a structure of modifying coatings makes it possible to retain macromolecules of proteins on the external surface of a sorbent and to effect separation of medicinal preparation molecules (which have smaller sizes and which interact with the hydrophobic internal surface of pores) by the reversed-phase mechanism. It should also be noted that functionalized silica surfaces find use for covalent (chemospecific) chromatography of biopolymers.

Functionalized silica surfaces made the basis for designing effective chiral stationary phases for separation of optical isomers. With the view of producing enantioselective sorbents of this kind, the chiral selectors of various electronic and geometric structures are attached to the silica surface. Such grafted selectors must afford to provide two- or three-point interactions with separated optical isomers [105].

Of significance is, also, the fact that chemically modified silicas are employed for adsorption concentration of ions and molecules in analytical practice (solid phase extraction, SPE) [2,84,95–107]. For instance, columns, which contain silicas with grafted octadecylsilyl groups, are used for preconcentration of phenols, pesticides, and herbicides. Octadecylsilica, with adsorbed derivatives of dithiocarbamates, crown esters, and other compounds, is utilized for SPE of metals. In particular, C_{18}-columns with adsorbed tri-n-octylphosphine oxide are used for extraction and determination of uranium and thorium in water [108]. In the case of a repeated use of C_{18}-cartridges with adsorbed bonding or complexing compounds, there proceeds a gradual elution of such compounds from the surface of a sorbent. Therefore, many researchers put forth their efforts to develop methods for chemical bonding of most important analytical reagents on silica and other appropriate matrices. Such modified silicas, with covalently bonded complexing compounds, are referred to as complexing silicas. By now, the efforts have yielded a number of rather versatile methods, suitable for immobilization of a large majority of analytical reagents [3,8,19,79,95–102]. Characteristically, the researchers of such methods derived benefit from the experience, which was acquired when designing procedures for immobilization of enzymes and other biologically active compounds on inorganic matrices [109]. For example, in order to attach 4-(2-pyridylazo)resorcinol, 1-(2-pyridylazo)-2-naphthol, and 8-hydroxyquinoline to the silica surface, a use was made of the single-stage Mannich reaction [110].

Although complex-forming adsorbents on the basis of organic polymers possess an increased capacity with respect to extracted compounds in comparison with mineral supports, the developed modified silicas with grafted analytical reagents do not swell in water and organic solvents. Moreover, they are distinguished for a high rate of mass exchange with a surrounding medium and, hence, for better kinetic characteristics. Complexing silicas may be classified according to the nature of an attached functional group (nitrogen-, oxygen-, sulfur-, phosphorus-containing silicas, modified silicas with functional groups involving unsaturated carbon-carbon bonds, and polyfunctional silica surfaces) [3], or according to the nature of a bond, between an ion and a grafted ligand (donor–acceptor bonds, ion exchange, charge-transfer complexes, and inclusion complexes of the "host-guest" type).

5.6.2 METAL COMPLEX CATALYSTS ANCHORED ON SILICAS

It is known that complexes of transition metals are attached to support surfaces for the purpose of producing catalysts which preserve specificity and selectivity of homogeneous metal complexes, and gain technological advantages afforded by heterogeneous contacts [111–114]. Attachment of metal complexes to a surface prevents agglomeration of coordination unsaturated sites and enhances thermostability of complexes. At the same time, one cannot foretell in advance, a possible influence of the surface of a support on catalytic activity and selectivity of attached complexes. Although production of heterogeneous metal complex catalysts on siliceous matrices involves adsorption (both of metal complexes prepared preliminarily using suitable solvents, and of metal complexes synthesized concurrently in a surface layer with participation of metal ions and of adsorbed ligands) or ion exchange, the most commonly employed methods consist in formation of coordination bonds between metal ions and ligands chemically bonded to surface. In this case, modified silicas perform a role of a polymeric macroligand. Anchoring of organic ligands on surfaces of silicas is also often effected using functional organoalkoxysilanes, whose functional groups which contain ligands are bound to surface through hydrolytically stable bonds Si–C, while alkoxy groups and silanol groups form bonds Si–O–Si.

Introduction of oxygen- and nitrogen-containing ligands, and ions of various transition metals, made it possible to obtain heterogeneous metal complex catalysts that possess a high activity and selectivity in processes of linear and cyclic oligomerization, hydroformylation, and hydrosilylation of olefins, and in some other reactions. Attachment of chiral ligands and complexes on the silica surface made it possible to expand substantially the potentialities of asymmetric heterogeneous catalysis, which is employed for synthesis of optically active compounds. Besides, of interest is also application of metal complexes anchored on modified silicas in reactions of transfer and activation of oxygen.

5.6.3 IMMOBILIZATION OF ENZYMES ON MODIFIED SILICA MATRICES

The main aim of immobilization is production of heterogenized enzymes, which are readily separated from reaction media, and which can be used repeatedly under flow reactor conditions. Immobilization of enzymes on silica matrices is effected by both the adsorption method and method of covalent binding on surface [115,116]. Chemical bonding permits one to create more stable immobilized preparations; in particular preparations that are elution-resistant in varying surrounding medium factors (ionic strength, pH, concentration of a substrate, and content of substances, which may lead to desorption of enzymes).

At present, the available gamut of methods for covalent bonding of enzymes with silica surface is rather broad. As a rule, all of them involve a preliminary modification of surface by bifunctional silane, and subsequent introduction of active groupings that are able to form covalent bonds with side residues of amino acids of enzymes. The most commonly employed is a preliminary modification of silica by 3-aminopropyltriethoxysilane, and subsequent activation of the surface by various reagents (such as glutaraldehyde, diazonium salts, isothiocyanates, carbodiimides, acylazides, haloalkyls, and acid chlorides) [117]. A good performance is also shown [109] by the methods for activating the amino-containing silica surface with 2,4-toluene diisocyanate:

$$\equiv SiCH_2CH_2CH_2NHC(O)NH \quad - \bigcirc - CH_3$$
$$NHC(O)NH \quad - Enzyme$$

and with cyanuric chloride:

$$\equiv SiCH_2CH_2CH_2NH \quad - \underset{N-}{\overset{N=}{\bigvee}} \overset{Cl}{\underset{N}{\bigvee}}$$
$$NH - Enzyme$$

The function of starting organosilicas used for production of activated matrices can be performed, not only by aminoderivatives, but also by organosilicas with other functional groups (such as $\equiv SiCH=CH_2$, $\equiv SiH$, $\equiv SiRSH$). In particular, activated silica matrices for immobilization of enzymes were prepared through reactions between grafted vinyl groups and maleic anhydride, addition of acrolein to surface silicon hydride groups, and interaction of grafted sulfhydryl groups with Ellman's reagent [109].

Since properties of an immobilized enzyme are determined by a package of properties of a biocatalyst and support as well as are dependent on a method for binding the biocatalyst to the support surface, when carrying out a concrete task it proves useful to test various methods for binding, with a view to choose the procedure that is optimal from the standpoint of activity and stability of products. In the case of porous silica matrices, the geometric characteristics of a support exert a substantial effect on rate of binding of enzymes, on the capacity of a matrix with respect to a protein, and on manifestation of activity of immobilized preparations, which is due to inside-diffusion retardation. Activity of immobilized preparations is, to a great extent, dependent on pH of a medium, where binding of an enzyme takes place. It has been found that the maximum degree of binding of a protein to the silica surface is observed at pH values close to the isoelectric point of the protein.

Although practical application of immobilized enzymes has much potentiality in fine organic synthesis and in various fields of biotechnology, it is in the sphere of creation of biosensors that their application is the most impressive. In biosensors, a use is made either of flow minireactors with immobilized enzymes, or of a biocatalyst that is attached directly to the surface of ion-selective electrodes. In the first case, the function of supports is performed by silica-based materials, which possess a rigid structure and, therefore, set good hydrodynamic conditions in a flow of a buffer and of a solution that is analyzed. Placing an appropriate transducer at the outlet of such a minireactor with an immobilized enzyme makes it possible to create analyzers for a broad spectrum of substances, which are substrates or effectors of the enzyme. The role of transducers is most often played by an oxygen electrode, flow spectrophotometer, H^+, and other ion-selective electrodes, as well as by a multipurpose enthalpimetric transducer. In the second case, enzymes are immobilized directly on the surface of a sensitive element (such as ion-selective field-effect transistor, piezoelectric quartz element, and optical fiber element). So long as the supports used contain silicon dioxide, the chemical attachment of enzymes to their surface should be effected with allowance for the experience acquired in the sphere of immobilization of active compounds on modified silicas.

In recent years there is a great interest in effecting enzyme-catalyzed reactions in nonaqueous solvents, with a view to increase yields of products or to carry out novel syntheses. Owing to their stability in organic solvent silica matrices may have much promise in the capacity of supports for appropriate enzymes. For example, it is possible to perform synthesis of various peptides with the help of immobilized proteases, and with allowance for reversibility of hydrolysis of peptide bonds [109]. In this case, as a consequence of the absence of racemization, the products prepared possess a higher optical purity in comparison with the classical peptide synthesis.

5.6.4 CHEMICALLY MODIFIED SILICAS IN POLYMERS AND DISPERSION MEDIA

In a lot of cases, efficiency of application of functionalized silica surfaces in polymers and dispersion media is determined by ratio of hydrophilic sites to hydrophobic sites in a surface layer. This index is of critical importance for thickeners of lubricants and fillers of polymeric materials. Substitution of hydroxyl groups by grafted organic radicals favors a better distribution of silica particles in organic media. At the same time, the complete screening of the silica surface (e.g., by alkyl groups) may impair structural and mechanical characteristics of thickened systems, and filling of polymeric materials may worsen their physicomechanical properties.

A substantial reinforcement of filled systems can be attained in the situation when active sites of modified silicas ($-NH_2$, $-COOH$, $-SH$, $\equiv SiH$, etc.) are capable of chemical interaction with functional groups of polymeric macromolecules. For the first time, such a chemical mechanism of reinforcement was employed in glass-fiber composites [9]. From this standpoint, functionalized silica surfaces containing grafted amine groups may have much promise, as chemically active fillers of chlorine- and bromine-containing elastomers and copolymers, epoxy resins, polyurethane rubbers, melamino-formaldehyde and resorcinol–formaldehyde latex and resins, sulfonated ethylene–propylene–diene terpolymers, and carboxyl-containing polymeric systems. Specifically, in the case of introduction of modified silicas with amine groups into carboxyl-containing polymers, formation of salt-type cross bonds between the filler and polymer is observed [3,81]:

On the contrary, modified silicas with carboxyl groups may considerably reinforce butadiene–vinylpyridine copolymer, elastomeric terpolymer of vinylpyridine, styrene, and butadiene, etc. Silicas with grafted unsaturated groups can be advantageously employed for reinforcement of a broad range of olefin-containing copolymers and elastomers. Chemisorption of sterically hindered phenols on the surface of highly disperse silicas makes it possible to produce fillers and thickeners that have antioxidant properties [3].

Applications of highly disperse silica in varnishes and paints are based on thixotropic properties of this thickener and on its ability to increase viscosity of systems and to decrease sedimentation of pigments of various types. One of the salient features of such systems with siliceous thickeners consists in the fact, that, at an infinitely small shearing stress, their behavior is similar to that of solids, and at high shearing stresses, to that of liquid compositions. The thixotropic properties of highly disperse silica lie in the fact that the deformation caused by shear, stirring, or shaking leads to breaking of bonds at points of contact of particles (of aggregates of particles) or of solvate shells of particles, and to subsequent restoration of them at the state of rest. Application of functionalized silica surfaces enables one to exert control over viscosity of a system and to enhance substantially the water-resisting property of a coating.

5.7 CHEMICAL MODIFICATION OF MINERAL FIBERS

Chemical modification of surfaces of glass and basalt fibers by bifunctional organosilicon compounds of the general formulae X_3SiRY, where X is alkoxy, halogen, or other hydrolysable, and capable of reacting with surface silanols group, RY-organofunctional group such as alkyl, aryl, alkene, amino, methacrylate, epoxy, mercapto etc, is mainly used for adhesion improvement in mineral fiber reinforced polymeric composites [9,118]. In dependence with method and conditions of fiber treatment (modification from dilute aqueous or organic solutions, drying conditions, state of the hydrated layer, and hydroxylation degree of the fiber surface) organosilane coupling agents are condensed into polysiloxane structures, or they are chemically bonded to the surface. As the result of polycondensation processes and reactions with surface silanol groups, such organosilanes form adsorbed or chemisorbed layers located at the interface between inorganic fiber and polymer. These layers enhance compatibility of the materials, and improve wetability at the fiber/matrix interface. Owing to interdiffusion and formation of interpenetrating polymeric networks [119], the significant change of the physicomechanical properties of the filled system are observed. Chemical modification of fiber surfaces give rise in durability of the composites and their resistance to corrosive attack of moisture and salt water. Especially strong interfacial bonding takes place in a case of chemical interaction between RY-groups of the modified fiber surface and functional groups of the thermosetting resin (crosslinking).

Chemical modification of mineral fiber surfaces and the nature of fiber–polymeric matrix interactions have been intensively studied with Fourier transform infrared spectroscopy, x-ray photoelectron spectroscopy, Auger electron spectroscopy, ellipsometry, secondary ion mass spectroscopy, nuclear magnetic resonance spectroscopy on various nuclei, scanning electron microscopy, radioisotope labeling, and other currently available methods [120].

Among the modifying reagents, organosilanes with amino-, methacryl-, epoxy- or vinyl group have found widespread application and their fiber/polymeric matrix interface structures have been widely studied. Functional organosilanes with amine groups are effective coupling agents for epoxy resins and polyurethanes. Really, the chemistry of amine-containing coupling agent alone allows it to react with both the glass fiber surface and the epoxy matrix to increase the fiber/matrix adhesion [121–125]. For amine-crosslinked epoxy resin and glass fiber composites, glycidylpropyltrialkoxysilane was an effective modifying reagent, in which the coupling function is glycidyl group; this group is expected to react with an amine group belonging to the epoxy network [126].

The papers [127–129] describe surface grafting of hyperbranched dendritic polyamidoamine onto glass fiber, or disperse silica having surface amino groups (prepared by treatment of surface with 3-aminopropyltriethoxy-silane or by introduction of polymer with terminal NH_2-groups). Grafting reaction was executed by repeating Michael addition of methyl acrylate to surface amino groups, followed by amidation of end groups (reaction of the resulting ester surface compounds with ethylenediamine or hexamethylene-diamine). In a case of silica, the amino group content increased from 0.40 to 8.30 mmol/g after 10th generation [128]. Obtained by this means, modified glass fibers [127] were applied for postgrafting of poly(isobutyl vinyl ether) and poly(2-methyl-2-oxazoline). A somewhat different route for preparation of grafting of polyamidoamine dendrimer hybrids was proposed in the paper [130]. Michael addition reaction was also used [131] for surface fuctionaliza-tion of aminated glass, silicon crystals, and silica microspheres with vinylic monomers (methyl vinyl ketone, methyl acrylate, methacrolein, and acrolein). In a case of aldehydes, Schiff base bond formation was also observed.

In composites based on mineral fibers and thermoplastic polymers, role of modifying reagent may consist in an improvement of the interdiffusion processes, when incorporation of glass fibers into the thermoplastic matrix leads to alteration of the mechanical properties of the filled systems. For example, as was shown in [132], the composite material based on commercial-grade polypropylenes, filled with different contents of basalt fibers, presented deterioration of both mechanical characteristics, for example, stress and strain at yield with increasing of the fiber content. On the other hand, the impact strength was four-fold higher than that of unfilled polymer. A poor adhesion between the polypropylene matrix and the basalt fibers was detected. But when polypropylene grafted with maleic anhydride (acting as a coupling agent) was added into the blend, the tensile properties of the obtained materials, and their impact strengths, increased significantly. The adhesion improvement was confirmed by scanning electron microscopy, as well. Thus, the modification of the polymeric matrix led to higher mechanical charac-teristics of the filled system. Improvement of the physicomechanical properties of polypropylene can be also executed by chemical modification of glass fibers with 3-methacryloxypropyltrimethoxysilane [133]. Introduction of amine-containing fillers in maleic anhydride-containing copolymers may result in formation of the amic acid and imide structures at the fiber/matrix interface [134].

Is known that the performance of thermoplastic composites depends on the intrinsic properties of the composite components, the quality of the fiber–matrix interface, and the crystalline properties of polymeric matrix. It was found [135,136] that copolymer polypropylene with maleic anhydride did indeed improve surface adhesion between fibers and polymeric matrix (adhesion promoter) and that, as a result, various mechanical properties were markedly enhanced. In our opinion, chemical interactions in such filled system have a dominant role. Really, introducing into composite the copolymer polypropylene with acrylic acid (instead of polypropylene with grafted maleic anhydride) showed lower mechanical properties, in spite of good adhesion between fibers and polymeric matrix [137]. In line with the assumption of authors [138], strong interaction between the filler surface and the elastomer with maleic anhydride leads to formation of a soft-engineered interphase layer, which has been achieved, via grafting. Authors of the paper [139] are also inclined to believe that chemical bonds are formed between maleated polyolefin and fiber surface. In a case of isotactic polypropylene, an essential effect on its crystallization behavior was detected at introducing into the glass fiber-filled system of copolymer polypropylene with maleic anhydride [140].

ACKNOWLEDGMENT

The work was financially supported by the Foundation of Fundamental Investigations of the Ministry for Education and Science of Ukraine (Project No. 03.07/00099).

REFERENCES

1. Iler, R. K., *The Chemistry of Silica: Solubility, Polymerization, Colloid and Surface Properties and Biochemistry of Silica*, Wiley, New York, 1979.
2. Unger, K. K., *Porous Silica. Its Properties and Use as a Support in Column Liquid Chromatography*, Elsevier, Amsterdam, 1979.
3. Tertykh, V. A. and Belyakova, L. A., *Chemical Reactions Involving Silica Surface*, Naukova Dumka, Kiev, 1991, (in Russian).
4. Bergna, H. E., Ed., *The Colloid Chemistry of Silica*, Advances in Chemistry Series 234, American Chemical Society, Washington, DC, 1994.
5. Vansant, E. F., Van Der Voort, P., and Vrancken, K. C., *Characterization and Chemical Modification of the Silica Surface*, Elsevier, Amsterdam, 1995.
6. Legrand, A. P., Ed., *The Surface Properties of Silicas*, Wiley, New York, 1998.
7. Papirer, E., Ed., *Adsorption on Silica Surfaces*, Marcel Dekker, New York, 2000.

8. Chuiko, A. A., Ed., *Silica Surface Chemistry, Vol. 1*, UkrINTEI, Kiev, pp. 1–2, 2001, (in Russian).
9. Plueddemann, E. P., Ed., *Interfaces in Polymer Matrix Composites*, Composite Materials 6, Academic Press, New York, 1974.
10. Little, L. H., *Infrared Spectra of Adsorbed Species*, Academic Press, London, 1966.
11. Hair, M. L., *Infrared Spectroscopy in Surface Chemistry*, Marcel Dekker, New York, 1967.
12. Kiselev, A. V. and Lygin, V. I., *Infrared Spectra of Surface Compounds and Adsorbed Species*, Nauka, Moscow, 1972, (in Russian).
13. Ingold, C. K., *Structure and Mechanism in Organic Chemistry*, Cornell University Press, Ithaca, NY, 1969.
14. Eaborn, C., *Organosilicon Compounds*, Butterworth, London, 1960.
15. Sommer, L. H., *Stereochemistry, Mechanism and Silicon. An Introduction to the Dynamic Stereochemistry and Reaction Mechanisms of Silicon Centers*, McGraw-Hill Book Company, New York, 1965.
16. Budd, S. M., *Phys. Chem. Glasses*, 2, 111–114, 1961.
17. Strelko, V. V. and Kanibolotskii, V. A., *Kolloid Zhurn.*, 33, 750–756, 1971.
18. Tertykh, V. A. and Pavlov, V. V., *Adsorbtsiya i Adsorbenty*, 6, 67–75, 1978.
19. Tertykh, V. A. and Belyakova, L. A., In *Adsorption on New and Modified Inorganic Sorbents*, Dabrowski, A. and Tertykh, V. A., Eds., Elsevier, Amsterdam, pp. 147–189, 1996.
20. Tertykh, V. A., Pavlov, V. V., Tkachenko, K. I., and Chuiko, A. A., *Teor. Eksp. Khim.*, 11, 174–181, 1975.
21. Lork, K. D., Unger, K. K., and Kinkel, J. N., *J. Chromatogr.*, 352, 199–211, 1986.
22. Buszewski, B., Nondek, l., Jurasek, A., and Berek, D., *Chromatographia*, 23, 442–446, 1987.
23. Gorlov, Yu. I., Zayats, V. A., and Chuiko, A. A., *Teor. Eksp. Khim.*, 25, 756–757, 1989.
24. Chuiko, A. A. and Gorlov, Yu. I., *Silica Surface Chemistry. Surface Structure, Active Sites, Sorption Mechanisms*, Naukova Dumka, Kiev, 1992, (in Russian).
25. Gorlov, Yu. I., *React. Kinet. Catal. Lett.*, 50, 89–96, 1993.
26. Tertykh, V. A., Belyakova, L. A., Varvarin, A. M., Lazukina, L. A., and Kukhar', V. P., *Teor. Eksp. Khim.*, 18, 717–722, 1982.
27. Tertykh, V. A., Belyakova, L. A., and Varvarin, A. M., *React. Kinet. Catal. Lett.*, 40, 151–156, 1989.
28. Tertykh, V. A., Chuiko, A. A., Mashchenko, V. M., and Pavlov, V. V., *Zhurn. Fiz. Khim.*, 47, 158–163, 1973.
29. Chmielowiec, J. and Morrow, B. A., *J. Colloid Interface Sci.*, 94, 319–327, 1983.
30. Varvarin, A. M., Belyakova, L. A., Tertykh, V. A., Lazukina, L. A., and Kukhar', V. P., *Teor. Eksp. Khim.*, 21, 739–745, 1985.
31. Varvarin, A. M., Belyakova, L. A., Tertykh, V. A., Lazukina, L. A., and Kukhar', V. P., *Teor. Eksp. Khim.*, 23, 117–120, 1987.

32. Varvarin, A. M., Belyakova, L. A., Tertykh, V. A., Lazukina, L. A., and Kukhar', V. P., *Teor. Eksp. Khim.*, 25, 377–380, 1989.
33. Bogillo, V. I., In *Adsorption on New and Modified Inorganic Sorbents*, Dabrowski, A. and Tertykh, V. A., Eds., Elsevier, Amsterdam, pp. 237–284, 1996.
34. Bogillo, V. I., *React. Kinet. Catal. Lett.*, 50, 75–81, 1993.
35. Bogillo, V. I. and Shkilev, V. P., *Langmuir*, 12, 109–114, 1996.
36. Dabrowski, A., Bogillo, V. I., and Shkilev, V. P., *Langmuir*, 13, 936–944, 1997.
37. Tertykh, V. A., Pavlov, V. V., Tkachenko, K. I., and Chuiko, A. A., *Teor. Eksp. Khim.*, 11, 415–417, 1975.
38. Tertykh, V. A., Mashchenko, V. M., and Chuiko, A. A., *Dokl. AN SSSR*, 200, 865–868, 1971.
39. Tertykh, V. A., Belyakova, L. A., Varvarin, A. M., Lazukina, L. A., and Kukhar', V. P., *Dokl. AN UkrSSR*, N5, 58–61, 1983.
40. Varvarin, A. M., Belyakova, L. A., Tertykh, V. A., Lazukina, L. A., and Kukhar', V. P., *Ukr. Khim. Zhurn.*, 50, 849–851, 1984.
41. Varvarin, A. M., Belyakova, L. A., Tertykh, V. A., Lazukina, L. A., and Kukhar', V. P., *Dokl. AN SSSR*, 293, 1390–1393, 1987.
42. Varvarin, A. M., Belyakova, L. A., Tertykh, V. A., Lazukina, L. A., and Kukhar', V. P., *Teor. Eksp. Khim.*, 24, 496–500, 1988.
43. Varvarin, A. M., Belyakova, L. A., Tertykh, V. A., Lazukina, L. A., and Kukhar', V. P., *Ukr. Khim. Zhurn.*, 54, 829–833, 1988.
44. Gun'ko, V. M., Bogillo, V. I., and Chuiko, A. A., *ACH-Models Chem.*, 131, 561–570, 1994.
45. Bogillo, V. I. and Gun'ko, V. M., *Langmuir*, 12, 115–124, 1996.
46. Gun'ko, V. M., *React. Kinet. Catal. Lett.*, 50, 97–102, 1993.
47. Gun'ko, V. M., *Colloids Surf. A*, 101, 279–286, 1995.
48. Tertykh, V. A., Pavlov, V. V., and Vatamanyuk, V. I., *Adsorbtsiya i Adsorbenty*, 4, 57–62, 1976.
49. Tertykh, V. A., *Macromol. Symp.*, 108, 55–61, 1996.
50. Kaas, R. L. and Kardos, J. L., *Amer. Chem. Soc. Polym. Prepr.*, 11, 258–265, 1970.
51. Boksanyi, L., Liardon, O., and Kovatz, E., *Adv. Colloid Interface Sci.*, 6, 95–137, 1976.
52. Kiselev, A. V., Kuznetsov, B. V., and Lanin, S. N., *J. Colloid Interface Sci.*, 69, 148–156, 1979.
53. Tertykh, V. A., *Adsorbtsiya i Adsorbenty*, 11, 3–11, 1983.
54. Engelhardt, H. and Orth, P., *J. Liq. Chromatogr.*, 10, 1999–2022, 1987.
55. Blitz, J. P., Murthy, R. S. S., and Leyden, D. E., *J. Am. Chem. Soc.*, 109, 7141–7145, 1987.
56. Blitz, J. P., Murthy, R. S. S., and Leyden, D. E., *J. Colloid Interface Sci.*, 126, 387–392, 1988.
57. Blitz, J. P., Murthy, R. S. S., and Leyden, D. E., *J. Colloid Interface Sci.*, 121, 63–69, 1988.
58. Tripp, C. P. and Hair, M. L., *J. Phys. Chem.*, 97, 5693–5698, 1993.

59. Hair, M. L. and Tripp, C. P., *Colloid Surf. A*, 105, 95–103, 1995.
60. Tripp, C. P., Kazmaier, P., and Hair, M. L., *Langmuir*, 12, 6407–6409, 1996.
61. White, L. D. and Tripp, C. P., *J. Colloid Interface Sci.*, 227, 237–243, 2000.
62. Kinkel, J. N. and Unger, K. K., *J. Chromatogr.*, 316, 193–200, 1984.
63. Buszewski, B., Jarasek, A., and Garaj, I., *J. Liq. Chromatogr.*, 10, 2325–2336, 1987.
64. Gun'ko, V. M., Voronin, E. F., Pakhlov, E. M., and Chuiko, A. A., *Langmuir*, 9, 716–722, 1993.
65. Pakhlov, E. M., Gun'ko, V. M., and Voronin, E. F., *React. Kinet. Catal. Lett.*, 50, 305–310, 1993.
66. Cun'ko, V. M., In *Fundamental and Applied Aspects of Chemically Modified Surfaces*, Blitz, J. P. and Little, C. H., Eds., Royal Society of Chemistry, Cambridge, pp. 270–279, 1999.
67. Cun'ko, V. M., Vedamuthu, M. S., Henderson, G. L., and Blitz, J. P., *J. Colloid Interface Sci.*, 228, 157–170, 2000.
68. Pavlov, V. V., Tertykh, V. A., Chuiko, A. A., and Kazakov, K. P., *Adsorbtsiya i Adsorbenty*, 4, 62–69, 1976.
69. Tertykh, V. A., In *Organosilicon Chemistry III. From Molecules to Materials*, Auner, N. and Weis, J., Eds., VCH, Weinheim, pp. 670–681, 1998.
70. Babich, I. V., Plyuto, Yu. V., and Chuiko, A. A., *Zhurn. Fiz. Khim.*, 62, 516–519, 1988.
71. Babich, I. V., Plyuto, Yu. V., and Chuiko, A. A., *Dokl. AN UkrSSR*, N4, 39–41, 1987.
72. Mutovkin, P. A., Babich, I. V., Plyuto, Yu. V., and Chuiko, A. A., *Ukr. Khim. Zhurn.*, 59, 727–730, 1993.
73. Shimizu, M. and Low, M. J. D., *J. Am. Ceram. Soc.*, 54, 271–272, 1971.
74. Gomenyuk, A. A., Babich, I. V., Plyuto, Yu. V., and Chuiko, A. A., *Zhurn. Fiz. Khim.*, 64, 1662–1664, 1990.
75. Plyuto, Yu. V., Gomenyuk, A. A., Babich, I. V., and Chuiko, A. A., *Kolloid Zhurn.*, 55 (N6), 85–89, 1993.
76. Mutovkin, P. A., Plyuto, Yu. V., Babich, I. V., and Chuiko, A. A., *Ukr. Khim. Zhurn.*, 57, 367–370, 1991.
77. Voronin, E. F., Bogomaz, V. I., Ogenko, V. M., and Chuiko, A. A., *Teor. Eksp. Khim.*, 16, 801–807, 1980.
78. Voronin, E. F., Tertykh, V. A., Ogenko, V. M., and Bogomaz, V. I., *Zhurn. Fiz. Khim.*, 55, 234–236, 1981.
79. Tertykh, V. A., Yanishpolskii, V. V., Bereza-Kindzerska, L. V., Pesek, J. J., and Matyska, M. T., In *Chemistry, Physics and Technology of Surfaces*, Chuiko, A. A., Ed., N4-6, KM Academia, Kiev, pp. 69–90, 2001.
80. Tertykh, V. A. and Tomachinsky, S. N., *Funct. Mater.*, 2, 58–63, 1995.
81. Tertykh, V. A., Yanishpolskii, V. V., and Bolbukh, Yu. N., *Macromol. Symp.*, 194, 141–146, 2003.
82. Tertykh, V. A., Belyakova, L. A., and Simurov, A. V., *Mendeleev Commun.*, N2, 46–47, 1992.

83. Belyakova, L. A. and Simurov, A. V., *Ukr. Khim. Zhurn.*, 61 (N4), 24–29, 1995.
84. Johnson, W. E., Fendinger, N. J., and Plimmer, J. R., *Anal. Chem.*, 63, 1510–1513, 1991.
85. Hodge, P. and Sherrington, D. C., Eds., *Polymer-Supported Reactions in Organic Synthesis*, Wiley, Chichester, 1980.
86. Mackenzie, J. D. and Bescher, E. P., *J. Sol–Gel Sci. Tech.*, 13, 371–377, 1998.
87. Makote, R. and Collinson, M. M., *Anal. Chim. Acta*, 394, 195–200, 1999.
88. Walcarius, A., *Electroanalysis*, 10, 1217–1235, 1998.
89. Levy, D., *Chem. Mater.*, 9, 2666–2670, 1997.
90. Haas, K. H., *Adv. Eng. Mater.*, 2, 571–582, 2000.
91. Zhao, X. S., Lu, G. Q., and Millar, G. J., *Ind. Eng. Chem. Res.*, 35, 2075–2090, 1996.
92. Moller, K. and Bein, T., *Chem. Mater.*, 10, 2950–2963, 1998.
93. Yang, C.-M. and Chao, K.-J., *J. Chin. Chem. Soc.*, 49, 883–893, 2002.
94. Kiselev, A. V. and Yashin, Ya. I., *Gas-Adsorption Chromatography*, Plenum Press, New York, 1969.
95. Lisichkin, G. V., Ed., *Modified Silicas in Adsorption, Catalysis and Chromatography*, Khimiya, Moscow, 1986, (in Russian).
96. Unger, K. K., *Packing and Stationary Phases in Chromatographic Techniques*, Marcel Dekker, New York, 1990.
97. Scott, R. P. W., *Silica Gel and Bonded Phases*, Wiley, New York, 1993.
98. Hetem, M. J. J., *Chemically Modified Silica Surfaces in Chromatography, a Fundamental Study*, Hüthig Verlag, Heidelberg, 1993.
99. Buszewski, B., Jezierska, M., Welniak, M., and Berek, D., *J. High Res. Chromatogr.*, 21, 267–281, 1998.
100. Biernat, J. F., Konieczka, P., Tarbet, B. J., Bradshaw, J. S., and Izatt, R. M., *Separ. Purif. Method*, 23, 77–348, 1994.
101. Laskorin, B. N., Strelko, V. V., Strazhesko, D. N., and Denisov, V. I., *Sorbents Based on Silica Gel in Radiochemistry*, Atomizdat, Moscow, 1977, (in Russian).
102. Zaitsev, V. N., *Complexing Silicas: Syntheses, Structure of Bonded Layer and Surface Chemistry*, Folio, Khar'kov, 1997, (in Russian).
103. Moors, M., Massart, D. L., and Mcdowall, R. D., *Pure. Appl. Chem.*, 66, 277–304, 1994.
104. Stevenson, D., *J. Chromatogr. B*, 745, 39–48, 2000.
105. Davankov, V. A., Navratil, J. D., and Walton, H. F., *Ligand Exchange Chromatography*, CRC Press, Boca Raton, FL, 1988.
106. Moors, M., Massart, D. L., and Mcdowall, R. D., *Pure. Appl. Chem.*, 66, 277–304, 1994.
107. Stevenson, D., *J. Chromatogr. B*, 745, 39–48, 2000.
108. Shamsipur, M., Yamini, Y., Ashtari, P., Khanchi, A., and Ghannadimarageh, M., *Separ. Sci. Technol.*, 35, 1011–1019, 2000.
109. Tertykh, V. A. and Yanishpolskii, V. V., In *Adsorption on Silica Surfaces*, Papirer, E., Ed., Marcel Dekker, New York, pp. 523–564, 2000.

110. Tertykh, V. A., Yanishpolskii, V. V., and Panova, O. Yu., *J. Therm. Anal. Calorim.*, 62, 545–549, 2000.
111. Bailar, J. C. Jr., *Catal. Rev.–Sci. Eng.*, 10, 17–36, 1974.
112. Ermakov, Yu. I., Zakharov, V. A., and Kuznetsov, B. N., *Anchored Complexes on Oxide Carriers in Catalysis*, Nauka, Novosibirsk, 1980, (in Russian).
113. Lisichkin, G. V. and Yuffa, A. Ya., *Heterogeneous Metal Complex Catalysts*, Khimiya, Moscow, 1981, (in Russian).
114. Price, P. M., Clark, J. H., and Macquarrie, D. J., *J. Chem. Soc. Dalton Trans.*, N2, 101–110, 2000.
115. Weetall, H. H., *Nature*, 223, 959–960, 1969.
116. Weetall, H. H., *Science*, 166, 615–617, 1969.
117. Weetall, H. H., *Appl. Biochem. Biotechnol.*, 41, 157–188, 1993.
118. Plueddemann, E. P., *Silane Coupling Agents*, Plenum Press, New York, 1982.
119. Sperling, L. H., *Interpenetrating Polymer Networks and Related Materials*, Plenum Press, New York, 1981.
120. Suzuki, N. and Ishida, H., *Macromol. Symp.*, 108, 19–53, 1996.
121. Almoussawi, H., Drown, E. K., and Drzal, L. T., *Polym. Compos.*, 14, 195–200, 1993.
122. Wang, T. W. H., Blum, F. D., and Dharani, L. R., *J. Mater. Sci.*, 34, 4873–4882, 1999.
123. Amdouni, N., Sautereau, H., and Gerard, J. F., *J. Appl. Polym. Sci.*, 45, 1799–1810, 1992.
124. Hamada, H., Ikuta, N., Nishida, N., and Maekawa, Z., *Composites*, 25, 512–515, 1994.
125. Connell, M. E., Cross, W. M., Snyder, T. G., Winter, R. M., and Kellar, J. J., *Compos., A.–Appl. Sci. Manuf.*, 29, 495–502, 1998.
126. Salmon, L., Thominette, F., Pays, M. F., and Verdu, J., *Polym. Compos.*, 20, 715–724, 1999.
127. Fujiki, K., Sakamoto, M., Yoshikawa, S., Sato, T., and Tsubokawa, N., *Compos. Interface*, 6, 215–226, 1999.
128. Tsubokawa, N., Ichioka, H., Satoh, T., Hayashi, S., and Fujiki, K., *React. Funct. Polym.*, 37, 75–82, 1998.
129. Fujiki, K., Sakamoto, M., Sato, T., and Tsubokawa, N., *J. Macromol. Sci.-Pure Appl. Chem.*, 37, 357–377, 2000.
130. Ruckenstein, E. and Yin, W. S., *J. Polym. Sci. A-Polym. Chem.*, 38, 1443–1449, 2000.
131. Nitzan, B. and Margel, S., *J. Polym. Sci. A.–Polym. Chem.*, 35, 171–181, 1997.
132. Botev, M., Betchev, H., Bikiaris, D., and Panayiotou, C., *J. Appl. Polym. Sci.*, 74, 523–531, 1999.
133. Tselios, C., Bikiaris, D., Savidis, P., Panayiotou, C., and Larena, A., *J. Mater. Sci.*, 34, 385–394, 1999.
134. Bayer, T., Eichhorn, K. J., Grundke, K., and Jacobasch, H. J., *Macromol. Chem. Phys.*, 200, 852–857, 1999.
135. Yoon, B. S., Shin, D. C., Suh, M. H., and Lee, S. H., *Polymer-Korea*, 22, 633–641, 1998.

136. Gassan, J. and Biedzki, A. K., *J. Thermoplast. Compos. Mater.*, 12, 388–398, 1999.
137. Roux, C., Denault, J., and Chapagne, M. F., *J. Appl. Polym. Sci.*, 78, 2047–2060, 2000.
138. Jancar, J., Dibenedetto, A. T., and Dianselmo, A., *Chem. Pap.-Chem. Zvesti.*, 50, 228–232, 1996.
139. Zhou, X. D., Dai, G. C., Guo, W. J., and Lin, Q. F., *J. Appl. Polym. Sci.*, 76, 1359–1365, 2000.
140. Bogoevagaceva, G., Janevski, A., and Mader, E., *J. Adhes. Sci. Technol.*, 14, 363–380, 2000.

6 On the Border of ..., A Magnetic Resonance Point of View

André Pierre Legrand and Hubert Hommel

CONTENTS

6.1 MAGNETIC RESONANCE SPECTROSCOPIES

Among characterization spectroscopies, magnetic resonances are liable to produce specific analyses of bulk and interface of solid powders, fibers, or composites. Such methods need the presence of atoms with a magnetic moment or electrons belonging to a probe molecule (spin label). The localization of those one in the material informs on the chemical bounding or mobility of the probe. The material investigated can be studied during the different steps of the synthesis and through the treatments. This enables

245

adjustment of the material for specific applications. Nevertheless, this approach must be associated with some other spectroscopies as infrared or electronic ones (Auger, LEED, x-ray, etc.).

6.1.1 THE NUCLEAR MAGNETIC RESONANCE OF CONSTITUTIVE NUCLEI

The importance of nuclear magnetic resonance (NMR) is well known in liquids for the determination of the structure and of the dynamic of molecules. Owing to the considerable progress of the technique, this powerful method is also applicable to solids.

The basic experience in NMR is to put the sample containing atoms with spins in a strong magnetic field, B_0. The Zeeman Hamiltonian analogous to the classical energy of a magnetic momentum in a magnetic field $-\mu B$ is

$$\hat{H} = \gamma \vec{I} B_0$$

where γ is the gyromagnetic factor, $\vec{I}$ is the spin operator, and B_0 the magnetic field, which is generally oriented along the Oz axis. The eigenvalues or energy levels are then given by

$$E_m = \gamma m \hbar B_0$$

where m is the magnetic quantum number which has $2I+1$ values between $+I$ and $-I$ by integer steps. By adding a small harmonic perturbation, namely an oscillating magnetic field at the Larmor frequency along a direction perpendicular to the Oz axis, transitions between the energy levels given above are induced following the selection rule, $\Delta m = \pm 1$ between the initial and final state, if $h\nu = \Delta E$ (Abragam 1961; Slichter 1990).

The peculiarity of these spins is that they are sensitive mainly to magnetic interactions which are relatively weak, like the deformation of the electronic cloud under the action of B_0. Therefore, they can act as local probes, giving information about the local microscopic environment at a distance of a few angstroms around the atom.

Now, as the spectrometers are pulsed apparatus, the transient behavior in the time domain is experimentally easily detectable and gives important new data. As an example, the silicon nucleus (^{29}Si), contrary to ^{1}H, is an isotope with a low natural abundance of 4.7 percent. The same is true for ^{13}C (1.1 percent), which is a component of the organic molecules adsorbed or grafted on the surface. These diluted species are nevertheless in dipolar interaction with the abundant proton (^{1}H) species. To observe the nature of the different chemical species at the interface or in the bulk of silica, it is necessary to suppress this dipolar interaction. Through well-designed manipulation either in the real space (magic angle spinning MAS, Andrew, Bradbury, and Eades 1958) or in the spin space (decoupling (Bloch 1958), cross-polarization (Pines, Gibby, and Waugh 1973), different pulse sequences (Mehring 1976; Haeberlen 1976; Jeener 1982)) the different interactions can be selectively averaged and the different components of an overall environment can be studied separately. Such improvements were obtained through digitization techniques and mathematical algorithms such as Fast Fourier Transform.

6.1.2 THE ELECTRON PARAMAGNETIC RESONANCE OF SPIN LABELS

Most materials have complete electronic orbitals (achieved for example, by covalent bonding) and therefore do not exhibit any paramagnetic center nor electron paramagnetic resonance (EPR) signals. One of the techniques used is to add a reporter group to the system, whose signal gives information about its environment possibly precise enough to overcome the lack of direct signals. This supplementary molecule can be simply mixed with the system, in which case it is called a spin probe, or it can be covalently bonded to a definite point, in which case it is called a spin label. Probably the most frequently used spin labels are different nitroxide free radicals that can even be purchased commercially, and whose EPR signals have been thoroughly studied for their own sake and are now very well documented and understood. Generally, they are composed by the nitroxide part that gives the EPR signal and by another functional group able to react with the molecule to be labeled. Two examples are given in Figure 6.1.

Unlike most free radicals, the unpaired electron is relatively unreactive because it is well screened by the methyl groups and the molecule can be stable for several years, at least when stored in a refrigerator. The spin

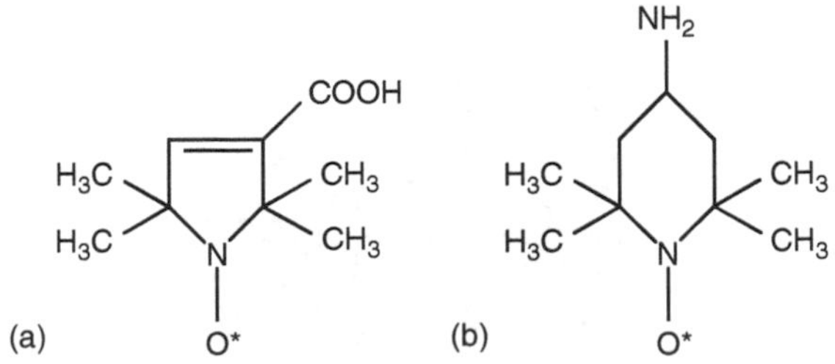

FIGURE 6.1 Example of two nitroxide free radicals: (a) the 2-2-5-5-tetramethyl-3-pyrroline-1-àxyl-3-carboxylic acid and (b) the 4-amino-tempo.

Hamiltonian of the nitroxide free radical is known (Berliner 1976). It can be written

$$\hat{H} = \beta \vec{S} \overline{\overline{g}} \vec{B} + \vec{I} \overline{\overline{A}} \vec{S}$$

where $S = 1/2$ is the spin of the unpaired electron, $I = 1$ the spin of the nitrogen nucleus, $\overline{\overline{g}}$ is the gyromagnetic tensor, and $\overline{\overline{A}}$ is the hyperfine tensor. The shape of the EPR spectrum of nitroxide free radicals has indeed been extensively studied and shown to be sensitive to the molecular Brownian motion of the label. For example, if the motion is very fast, the spectrum appears as three well-resolved Lorentzian lines which can be explained by the Kivelson theory (Kivelson 1960). The rotational correlation time falls in the range $3 \times 10^{-11} - 3 \times 10^{-9}$ s. In the slow tumbling region, the shape is influenced by the anisotropic part of the spin Hamiltonian and can be explained by the more comprehensive Freed theory (Schneider and Freed 1989). Extra low and high field peaks appear which can easily be distinguished from the preceding ones. In our case of polymers at interfaces, the key observation is that in a certain temperature range the spectra appear as a superposition of the two kinds of reference spectra distinguished previously, namely the fast and the slow motion spectra. The physical meaning of this observation is that labels belonging to trains are hindered by the contact with the solid surface and experience a slow motion whereas labels belonging to loops and tails experience a fast motion in solution (Figure 6.2).

By computer simulation the two contributions can be separated, and by a double integration of the experimental spectra, which are the first derivatives of the absorption, it is possible to evaluate the two fractions—the population of labels in solution (P_{sol}) and the population of labels adsorbed on the solid (P_{ads}) (Figure 6.3). This information is statistical in nature and can be related to the concentration profile of monomers from the surface (Hommel 1995).

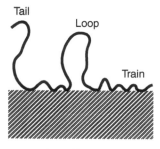

FIGURE 6.2 Schematic representation of loops, tails, and trains of a chain at the solid-liquid interface.

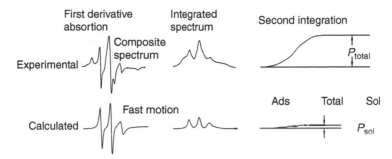

FIGURE 6.3 Method of analysis of the spectra; the fast motion spectrum is calculated using the theoretical expression of Kivelson (1960) and adjusted to the experimental one and after two integrations the two fractions can be evaluated. (From Kivelson, D., *J. Chem. Phys.*, 33, 1094, 1960.)

6.2 MAIN MATERIALS AS POWDERS OR FIBERS

6.2.1 SILICA

Silica exists in various forms which can be classified on the basis of four main features: crystal structure, dispersity, surface composition, and porosity. The silicon atoms exposed at the surface will tend to maintain their tetrahedral coordination with oxygen. They complete their coordination at room temperature by attachment to monovalent hydroxyl groups, forming silanol groups Si–OH. With regard to the surface of crystalline silica, all of the hydroxyls are considered as free or isolated. In contrast, the surface structure of amorphous silica is highly disordered and we cannot expect such a regular arrangement of hydroxyl groups. It can be assumed that some of them are adjacent to each other and are possibly capable of interacting by hydrogen bonding. Such hydroxyl groups are termed vicinal (Figure 6.4). Hence the surface of amorphous silica may be covered by isolated and vicinal hydroxyl groups. Sindorf and Maciel (1980), using ^{29}Si solid state NMR, clearly demonstrated the presence of geminal $=Si(OH)_2$ and single $\equiv Si(OH)$ on silica gels at different degrees of hydration. They considered that the surface is quite heterogeneous and may contain segments of surface resembling both the (111) and (100) faces of cristobalite. Adsorption of a water molecule is assumed to break the Si–O–Si bridge, giving rise to two single Si–OH groups. Breaking two bonds on the same silicon produces the formation of a geminal Si–$(OH)_2$ and two single Si–OH groups. It is these surface characteristics which determine the applications of the silica, be it for chromatography, dehydration, polymer reinforcement, or other processes.

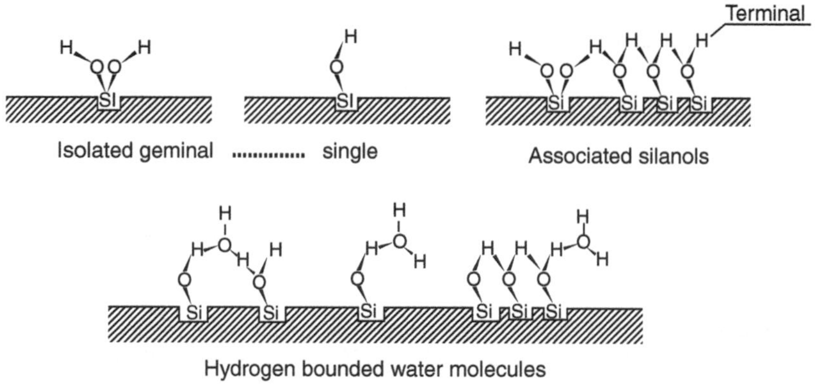

FIGURE 6.4 Different hydroxyl groups.

The silanol can react with different organic species. The chemist considering the yield of a given reaction can be satisfied by a schematic description like

$$\equiv Si-OH + OH-R \rightarrow\ \equiv Si-O-R + H_2O$$

However, with the developed analytical tools existing now, it is desirable to characterize exactly the formed species which can be different from the assumed formula especially for such heterogeneous reactions.

6.2.1.1 High Resolution ^{29}Si Spectroscopies

Different silicas have been investigated by ^{29}Si NMR. The pioneering breakthrough in this field was initiated in particular by Maciel, and it has since been extensively reviewed (Maciel 1994; Maciel and Ellis 1994; Maciel et al. 1994). The most popular and generally successful surface-selective polarization strategies to date use ^{1}H$\rightarrow^{29}$Si cross-polarization (CP/MAS). The dipolar coupling varies as the inverse cube of the ^{1}H–^{29}Si internuclear distance and the cross-polarization rate constant has essentially an r^{-6} dependence. Hence the dynamics of cross-polarization can be used to discriminate in favor of the surface nuclei. For this kind of amorphous solid, it is interesting to know the arrangement of the atoms at relatively short distances, the angles and lengths of the chemical bonding, and also the arrangements between several Si(–O–)$_4$ tetraedra. Now the NMR signals give precisely such kind of information. Generally, three lines are observed: one at -90 ppm/TMS, characteristic of the geminal silanols $=Si(OH)_2$ (Q$_2$), another at -100 ppm/TMS, characteristic of the single silanols $\equiv Si(OH)$ (Q$_3$), and a

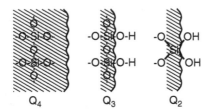

FIGURE 6.5 ^{29}Si chemical shifts. Q_4: -109.3 (Maciel and Sindorf 1980); -109 (Albert and Bayer 1991); -108.7 (Apperley, Hay, and Raval 2002). Q_3: -99.8 (Maciel and Sindorf 1980); -100 (Albert and Bayer 1991); -101.2 (Apperley, Hay, and Raval 2002). Q_2: -90.6 (Maciel and Sindorf 1980); -91 (Albert and Bayer 1991); -92.7 (Apperley, Hay, and Raval 2002). Remark: Q_1 had been never reported in the literature.

third one at -110 ppm/TMS, corresponding to the siloxane bridges (Q_4) (Maciel and Sindorf (1980)) (Figure 6.5). Nevertheless, this last value depends on the method of preparation of the silica. For fused quartz, the chemical shift of the Q_4 is spread from -85 to -120 ppm due to the distortion of the $Si(-O-)_4$ tetraedra (Dupree and Pettifer 1984; Devine et al. 1987). This phenomenon has been observed on some fume silica, causing a false determination of the relative proportions of silanols (Taïbi, Hommel, and Legrand 1992).

Performing successively two experiments, one with magic angle spinning only and the other one by coupling magic angle spinning and cross-polarization with different contact times, it is possible to estimate two parameters that correspond approximately to the relative intensities of the lines. On one hand, f_g is the fraction of geminal silanols over all silanols. On the other hand, f_s is the fraction of silanols over all species, silanols, or siloxane bridges. By a systematic comparison of the values of f_g and f_s for different silicas and different treatments, valuable information concerning the structure of amorphous silicas can be obtained (Legrand et al. 1990, 1993, 1998, 1999, 2002; Luhmer et al. 1996; Piquemal et al. 1999) (Figure 6.6).

A self-consistent model was proposed to interpret the distribution of experimental points (Figure 6.6a) (Legrand et al. 1999). One can assume a square flat array representing the surface silica, where at each intersection is located a silicon, bonded to four oxygen atoms, themselves bonded to silicons. Using a random model of adsorption of water molecules, which are susceptible to break the $Si-O-Si$ bridge when adsorbed, it is possible to estimate the relative proportions of Q_2 and Q_3 through this Monte Carlo method. The different proportions so obtained constitute a cloud into which the main experimental points set. In Figure 6.6, the line ($\varepsilon = 1$) shows the average value along with the cloud spreads. Some experimental results are situated below this line, so a more formal approach can be used. Single SiO_4 tetrahedron

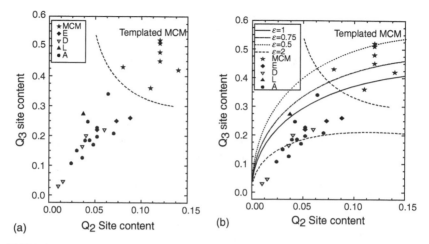

FIGURE 6.6 (a) Q_2 and Q_3 site contents of different silicas (Jelinek et al. 1992; Leonardelli et al. 1992; Legrand et al. 1993) and different MCM silicas with their templates (Piquemal et al. 1999). (b) Self-consistent model (Legrand et al. 1999).

is considered the basic building blocks of the silica framework, without referring to any subunit formed by several SiO_4 tetrahedra. x_i denotes the fraction of silicon atoms bearing i OH usually denominated Q_{4-i}. Let p be the probability of formation of an OH group and $(1-p)$ be that of disappearance of an OH (e.g., by condensation). To take into account a large hydroxyl site content observed in MCM silicas, a weighting factor ε is introduced. This has to be understood so that after a breaking of the first Si–O–Si bridge, another breaking at the same Si site has the probability εp instead of p. Such probability signifies an easier ($\varepsilon > 1$) or more difficult ($\varepsilon < 1$) breaking. (Nevertheless, one has to take into account that $\varepsilon p \leq 1$). Consequently, each experimental result corresponds to an "ε" value. Figure 6.7 shows that the process of formation depends of the average amount of hydroxyl/silicon. It would be easier to increase the hydroxylation when this amount is low.

Silica dissolution in alkaline solutions had been studied using ^{29}Si NMR (Wijnen et al. 1989). Fume silica dissolution rates depend on the alkali metal hydroxide type. This rate increases in the order (LiOH=CsOH)< (RbOH=NaOH)< KOH. Silica gel dissolution involves formation of monomeric silicic acid, Q_0. The monomeric anions oligomerize into dimer species, which in turn form cyclic and linear trimer species. The structure of highly polymerized silicate species depends on the alkali metal cation, i.e., low pH silicate solutions have structurally different silicate species as a function of alkali metal hydroxide. Incorporation of fume silica, as residue of silicon

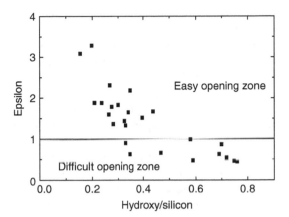

FIGURE 6.7 Values of e determined for different silica owing to the model. (From Legrand, A. P., Sfihi, H., and Bouler, J. M., *Bone*, 25, 103, 1999a.)

preparation, into cement is used for special concrete preparation. Characterization and dissolution was examined using the same method (Taïbi 1991; Lajnef et al. 1994).

6.2.1.1.1 Hydroxylation of Amorphous Fume Silicas

The reinforcing effect of different grades of fumed silicas on silicone elastomers will be influenced by the surface roughness that increases with the specific surface area. These variations can be shown by calling on NMR, infrared spectroscopic methods, and adsorption methods including IGC analysis (Barthel et al. 2004). A series of different silica were prepared, controlling the conditions of preparation during the process, to evaluate the surface roughness. Silica sample Wacker HDK-S13 having a specific surface area ≤ 200 m^2/g could be considered as flat at a molecular level, but above this critical specific surface area of 200 m^2/g, such as Wacker HDK-T30 and T40, and moreover microporous HDK-T30f1 and HDK-T30f2 samples, the surface roughness increases readily. Because the global surface silanol density does not change significantly with the specific surface area, the local silanol density increases strongly with increasing surface roughness or microporosity as it is clearly evidenced by both NMR and IR spectroscopies. They show correlatively a strong decrease of the number of isolated silanol groups.

One observes clearly on the ^{29}Si NMR (CP/MAS) spectra that for both silica samples, if the Q_2/Q_3 ratio remains relatively constant, the number of Q_4 silicon atoms is significantly lower for the HDK-T30 silica than for the HDK-S13 silica. Fewer Q_4 atoms are found in the vicinity of the HDK-T30 surface protons than in the vicinity of the HDK-S13 ones. Because of the relative stability of the global silanol density of the silica samples studied (about 1.8

OH/nm^2), this must correspond to a highest local density of silanol groups on the HDK-T30 silica surface.

^{1}H MAS NMR spectroscopy is used to distinguish physically bonded water molecules to isolated and associated hydroxyls (Bronnimann, Zeigler, and Maciel 1988; d'Espinose de la Caillerie, Aimeur, and Legrand 1997). ^{1}H spectra demonstrate that the relative concentration of isolated silanol groups corresponding to the sharp peak component of the spectra decreases with increasing surface area and is no longer observed for the microporous HDK-T30 sample (Figure 6.8).

These samples were examined by IR spectroscopy (Figure 6.9) in the characteristic region of the OH vibrations. The relative concentration of isolated silanol groups decreases when the specific surface area increases and the microporous HDK-T30f2 sample no longer exhibits isolated silanol, showing an augmentation of the local silanol surface density.

6.2.1.1.2 Hydride Si–H Formation on Pyrogenic Silica

Properties of silicas mainly depend on their specific area and, above all, the nature and concentration of their hydroxyl groups. In order to modify the surface properties of silicas, two approaches can be suggested: the preparation of covalently attached bonded-phase materials by the reaction of the sililating agent with hydroxyl groups of the silica surface or the direct modification of the silica surface itself. The conversion of silanol groups to hydrides constitutes one of the possible ways to apply this second approach.

The possibility of hybridization of silica had been shown in particular by Chu et al. (1993) and for industrial application by Heeribout et al. (1999). Using ^{1}H MAS and ^{29}Si CP/MAS identification of H–Si bonds had been attributed to specific peaks, as shown in Figure 6.10.

6.2.1.1.3 Grafted Silicas

Another approach for studying the surface of silicas is to use their reactivity toward different organic molecules. It is possible to represent the reaction schematically as follows:

$$R–OH + HO–Silica \rightarrow R–O–Silica + H_2O$$

or using equivalent relations for other reagents. This is a useful method to determine the amount of hydroxyls with this chemical titration, but the chemistry is certainly more complex because of the different site occupancies and the type of association the hydroxyls have. Snyder and Ward (1966) assume that vicinal silanols would be more reactive than isolated single silanols with chlorosilanes, for example. A comparative analysis was done on series of different silicas by chemical titration and NMR. Underestimation of the hydroxyl amount by chemical method is observed (Figure 6.11). Nevertheless, the NMR method is susceptible of limitation for some fume silicas.

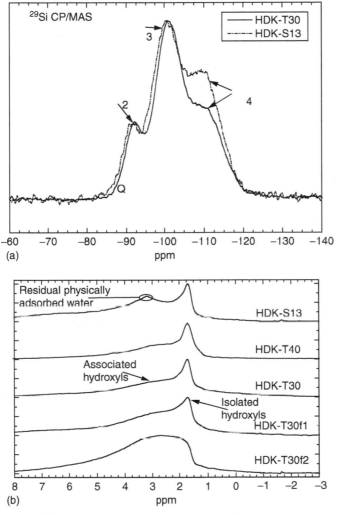

FIGURE 6.8 (a) ^{29}Si MAS-only spectra of HDK-S13 (1356 repetitions; 90 s repetition delay) and HDK-T30 (1157 repetitions; 5 ms CP contact time; 5 s repetition delay); (b) ^{1}H MAS NMR spectra of different silica (400, 32, 128, 400, 128 repetitions; 8, 20, 8, 8, 4 s repetition delay; 12 kHz spinning speed; 500 MHz). Spectra are normalized with the same maximum.

Qualitative analysis. Organofunctional groups bonded to solid surfaces are indispensable as column packing materials for chromatography. Proper methods for characterizing them have long been looked for. Among them, conventional ^{13}C NMR for liquids may be used (Snyder and Ward 1966;

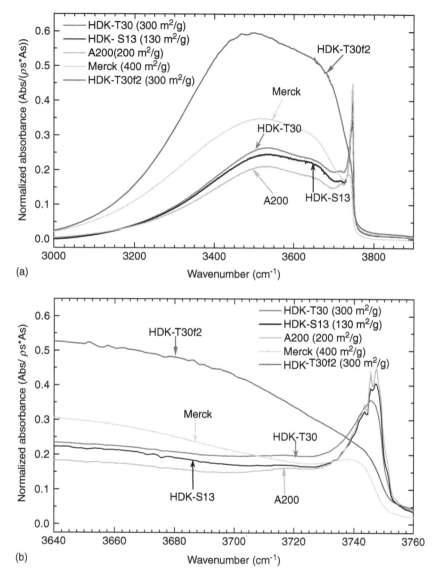

FIGURE 6.9 Infrared spectra of HDK silica samples. The narrow band around 3747 cm^{-1} are assignated to isolated single silanols. (a) Spectrum range: 3000–3900 cm^{-1}. (b) Extended spectrum range: $3640–3760 \text{ cm}^{-1}$ of the same spectra. (From Burnau, A. and Gallas, J. -P., *The Surface Properties of Silicas*, Legrand, A. P., Ed., Wiley, New York, pp. 147–234, 1998.)

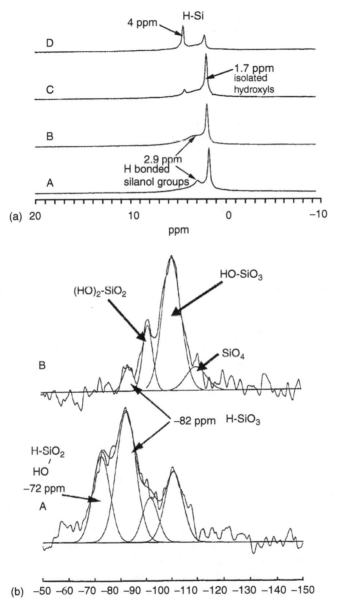

FIGURE 6.10 (a) ^{1}H MAS NMR spectra of treated silicas. (A) Silica treated at 1073 K with nitrogen flow rate, (B) silica treated at 1273 K with nitrogen flow rate, (C) silica treated at 1073 K with hydrogen flow rate, (D) silica treated at 1273 K with hydrogen flow rate. (b) Experimental and simulated ^{29}Si CP/MAS spectra of hydrogen treated silicas. (A) 1273 K treated silica, 27,500 scans; (B) 1073 K treated silica, 10,000 scans.

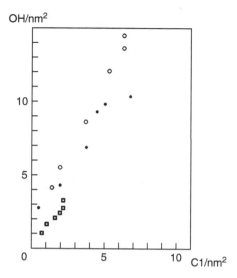

FIGURE 6.11 Comparison of hydroxyl group contents of silica as measured by calorimetry (OH/nm^2) and derivatization (C_1/nm^2). □ precipitated silica, ● silica gel, ○ fume silica. (From Legrand, A. P., Hommel, H., Tuel, A., Vidal, A., Balard, H., Papirer, E., Levitz, P., Czernichowski, M., Erre, R., Van Damme, H., Gallas, J. P., Hemidy, J. F., Lavalley, J. C., Barres, O., Burneau, A., and Grillet, Y., *Adv. Colloid Interface Sci.*, 33, 91, 1990b.)

Tanaka, Shinodo, and Saito 1979; Gilpin and Gangoda 1984; Gilpin 1984). In this case, it is necessary to add a solvent in such a way that the $^1H-^{13}C$ dipolar interaction be averaged.

Since the early 1980s, solid state NMR spectroscopy of ^{13}C and ^{29}Si has been of use for the analysis of grafts on silica. It is interesting to compare results obtained by ^{13}C CP/MAS (Chiang et al. 1982) and conventional liquid-state techniques (Tanaka, Shinodo, and Saito 1979).

An improvement of the analysis can be done if ^{13}C and ^{29}Si measurements are done together. The first example may be the paper of Bayer et al. (1983). About ten silanes were grafted and six commercial reversed-phase materials were investigated from a qualitative point of view. The authors have shown that a polymerization of bifunctional reagents may occur along the surface of silica. At the same time, Sindorf and Maciel (1983, 1983a) observed the same phenomenon with polyfunctional chloromethylsilanes. Figure 6.12 and Figure 6.13 summarize series of experimental determinations and plausible attributions of grafts on the silica surface.

As an example, the grafting of γ-aminopropyltriethoxysilane on silica can be analyzed using such methods. Although the amount of grafting could be identical, the organization of the grafts onto the surface could be preparation-dependent, giving significant differences of efficiency of the chromatographic columns. The origin of such differences is due to the formation aminopropylsilyl grafted brush polymer on the surface (Legrand et al. 1990).

Characterization of silica–dimethylsiloxane hybrids using nonhydrolytic sol–gel synthesis was done by ^{1}H–^{29}Si 2D NMR spectroscopy (Apperley, Hay, and Raval 2002). This method enables evidence for the structures of these composites. The siloxane component consists of short D units, which show a good compatibility with the silica on a molecular level, probably as a result of copolymerization between Q and D units.

Quantitative analysis. The interest in a quantitative determination is mainly directed to that of ^{29}Si, because these are the active sites for the phenomenon of hydrolysis and/or polymerization.

For qualitative analysis utilizing CP/MAS, it is necessary to follow the evolution of each peak of a spectrum as a function of the contact time. A simplified procedure was proposed by Sindorf and Maciel (1982), choosing an intermediate contact time CT in such a way that $T_{1\rho H} \gg CT \gg T_{SiH}$, which enabled quantitative determination through the respective peak areas. Although this procedure is sometimes questionable, it is very useful.

Using this method, Sindorf and Maciel (1982, 1983a) studied the reactivity of the different silanols to hexamethyldisilazane. This reaction was done on different types of silica, and it gives a trimethylsilyl graft. In that case, no polymerization can occur because of the monofunctionality of the reagent. It has been observed that the geminal hydroxyl silanol sites are proportionally more reactive than lone hydroxyl silanol sites.

An inverse phenomenon was observed on a Cab–O–Sil M5 silica, carefully prepared following the procedure of Gobet and Kovats (1977). The silylating agents were dimethyl [3, (3,3-dimethylbutyl) dimethylsilyl]-amine, dimethyl(trimethylsilyl[JH19])-amine, and dimethyl (dimethylsilyl)-amine, for the preparation of DM.B, DM.M, and DM.H, respectively (Tuel et al. 1990) (Figure 6.14). Figure 6.15 shows that the reactivity of single or associated silanols is higher than that of geminal silanols. This different behavior certainly has its origin in the nature of the silica, and also on the way of preparing its surface before grafting.

To interpret those results, some hypotheses have been proposed and discussed (Linton et al. 1985; Miller et al. 1985). It seems that the most reactive subset of the geminal and single silanols are those that are hydrogen-bonded. The two hydrogen-bonded hydroxyls belonging to the same geminal site or to geminal–vicinal or purely vicinal may not occur to the same extent.

Dynamic behavior of the graft. Classical measurements of ^{1}H and/or ^{13}C spectra and their peculiar relaxation times give information on the restricted

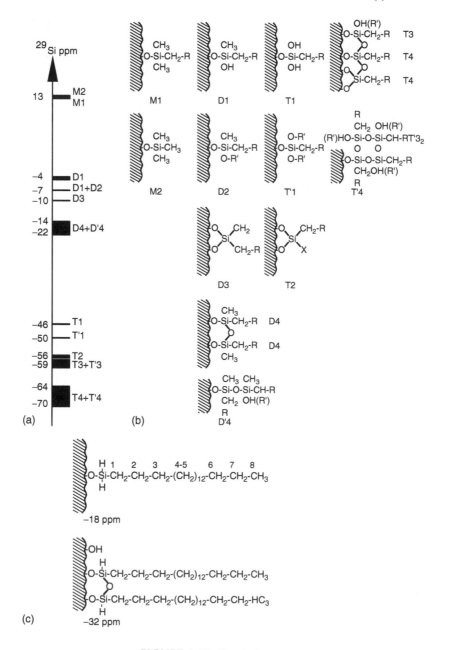

FIGURE 6.12 (See facing page.)

mobility of adsorbed molecules (Pfeifer (1972), Ellingsen and Resing (1980)). For alkyl chains, an interesting approach is given by ^{13}C spin labeling at different places along the chain. The line width of the abundant ^{13}C evolves as a function of the grafting ratio and nature of solvents (Gilpin and Gangoda 1984; Gilpin 1984). This method confirms the "brush" model, with an increase of the mobility from the anchored to the free part of the chain.

With solid state NMR, the characteristic time of cross polarization was analyzed as a function of the position of the carbon along an alkyl chain (Sindorf and Maciel 1983b). Significant differences were observed between anchored and free end carbons of the chain. The determination of this time, which is shorter when the mobility is restricted, confirms the fact that there is a distribution of mobility all along the chain, even in the absence of solvent.

Measurements of ^{13}C relaxation time T_1 for alkyl chains in the presence of solvent show a dependence on the degree of coating and an influence of the

---

FIGURE 6.12 ^{29}Si Chemical shifts (ppm) of bonded phases. Owing Pfleiderer et al. 1990: "D'_4, T'_3 and T_4 are species polymerized perpendicular to the surface; D_4, T_3, and T_4 are polymerized along the surface. T_2 and T_3 were used to differentiate between a silicon having two $(-O-Si-O-)_n$ units as neighbors (T_2) and a silicon having one $(-O-Si-O-)_n$ unit and one $(-O-Si-R-)$ unit as neighbors (T_3), because of different chemical shifts of the T_2 and T_3 groups due to the different chemical surroundings." Assignment depends on the authors, although spectra are similar. (From Pfleiderer, B., Albert, K., and Bayer, E., *J. Chromatgr.*, 506, 343, 1990.) M1: + 12.3 (Bayer et al. 1983); + 12 (Hetem et al. 1989); + 12 (Albert and Bayer 1991); 13 (Meiouet et al. 1991); M2: + 13.2 (Bayer et al. 1983); D1: −3 (Hetem et al. 1989); −4 (Pfleiderer, Albert, and Bayer 1990); −4 (Albert and Bayer 1991); D2: −11 (Hetem et al. 1989); −7.2 (Pfleiderer, Albert, and Bayer 1990); D1+D2: −9.9, −8.4, −7.9 (Bayer et al. 1983); D3: −16 (Bayer et al. 1983); −11 (Hetem et al. 1989); −10 (Pfleiderer, Albert, and Bayer 1990); −10 (Albert and Bayer 1991); −16 (Pursch et al. 1997); D4: −16 (Hetem et al. 1989); D4+D'4: −14 to −21 (Pfleiderer, Albert, and Bayer 1990); −14 to −22 (Albert and Bayer 1991); D3+D'4: −16, −17, −18 (Bayer et al. 1983); T1: −48 (Bayer et al. 1983); −48 (Hetem et al. 1989); −46 (Pfleiderer, Albert, and Bayer 1990); −46 (Albert and Bayer 1991); −46 (Meiouet et al. 1991); −49 (Pursch et al. 1997); T'1: −48 (Bayer et al. 1983); (X=OH) −55 (Hetem et al. 1989); −50 (Pfleiderer, Albert, and Bayer 1990); −50 (Albert and Bayer 1991); T2: −57 (Bayer et al. 1983); −55 (Hetem et al. 1989); −56 ($R \geq CH_3$) (Pfleiderer, Albert, and Bayer 1990); −56 (Albert and Bayer 1991); −57 (Tuel et al. 1992); T3+T'3: −53.2, −56.6 (Bayer et al. 1983); −59 (Pfleiderer, Albert, and Bayer 1990); −59 (Albert and Bayer 1991); −56 (Meiouet et al. 1991); −56 (Pursch et al. 1997); T4+T'4: −66.2 & −66 (Bayer et al. 1983); −64 to −70 (Pfleiderer, Albert, and Bayer 1990); −64 to −70 (Albert and Bayer 1991); −65 (Meiouet et al. 1991); −66 (Tuel et al. 1992); −65 (Pursch et al. 1997); −68 (Tsuji, Jones, and Davis 1999).

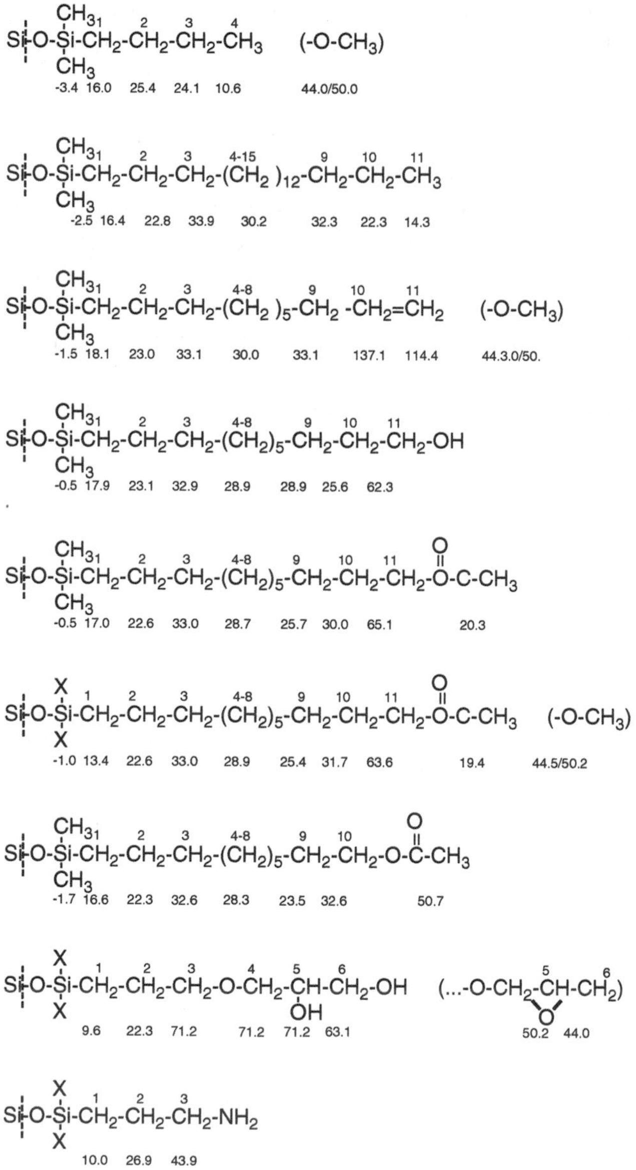

CH₃
|
Si–O–Si–CH₂–CH₂–CH₂–CH₃   (–O–CH₃)
|
CH₃
-3.4 16.0 25.4 24.1 10.6   44.0/50.0

CH₃
|
Si–O–Si–CH₂–CH₂–CH₂–(CH₂)₁₂–CH₂–CH₂–CH₃
|
CH₃
-2.5 16.4 22.8 33.9   30.2   32.3 22.3 14.3

CH₃
|
Si–O–Si–CH₂–CH₂–CH₂–(CH₂)₅–CH₂–CH₂=CH₂   (–O–CH₃)
|
CH₃
-1.5 18.1 23.0 33.1   30.0   33.1   137.1 114.4   44.3.0/50.

CH₃
|
Si–O–Si–CH₂–CH₂–CH₂–(CH₂)₅–CH₂–CH₂–CH₂–OH
|
CH₃
-0.5 17.9 23.1 32.9   28.9   28.9 25.6 62.3

CH₃   O
| ‖
Si–O–Si–CH₂–CH₂–CH₂–(CH₂)₅–CH₂–CH₂–CH₂–O–C–CH₃
|
CH₃
-0.5 17.0 22.6 33.0   28.7 25.7 30.0 65.1   20.3

X   O
| ‖
Si–O–Si–CH₂–CH₂–CH₂–(CH₂)₅–CH₂–CH₂–CH₂–O–C–CH₃   (–O–CH₃)
|
X
-1.0 13.4 22.6 33.0   28.9 25.4 31.7 63.6   19.4   44.5/50.2

CH₃   O
| ‖
Si–O–Si–CH₂–CH₂–CH₂–(CH₂)₅–CH₂–CH₂–O–C–CH₃
|
CH₃
-1.7 16.6 22.3 32.6   28.3 23.5 32.6   50.7

X
|
Si–O–Si–CH₂–CH₂–CH₂–O–CH₂–CH–CH₂–OH   (...–O–CH₂–CH–CH₂)
| | \O/
X   OH
9.6 22.3 71.2   71.2 71.2 63.1   50.2 44.0

X
|
Si–O–Si–CH₂–CH₂–CH₂–NH₂
|
X
10.0 26.9 43.9

FIGURE 6.13 ¹³C Chemical shifts (ppm) of bonded phases. (From Bayer, E., Albert, K., Reiners, J., Nieder, M., and Müller, D., *J. Chromatogr.*, 264, 197, 1983; Albert, K. and Bayer, E., *J. Chromatogr.*, 544, 345, 1991.)

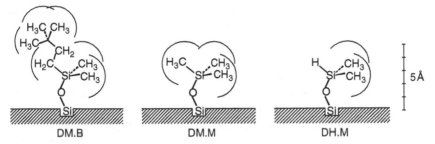

FIGURE 6.14 Structure of the triorganylsiloxy substituents, (3,3-dimethylbutyl) dimethylsiloxy, DM.B, trimethylsiloxy, DM.M, and dimethylsiloxy, DM.H.

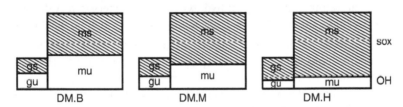

FIGURE 6.15 Surface site occupancy fractions for different grafts for geminal and single silanols, assuming that geminal silanols are not disubstituted, for a Cab–O–Sil M5 silica with 5 OH nm^{-2} (g: geminal, m: single silanol, u: unreacted, s: reacted). (From Tuel, A., Hommel, H., Legrand, A. P., and Kovats, E. Sz., *Langmuir*, 6, 770, 1990.)

end capping treatment (Albert, Evers, and Bayer 1985). This suggested to the authors that a bending of the chains is possible. Nevertheless, the problem is much more complicated. The behavior is not homogeneous—the tail and anchoring ends may have different mobility and conformations. This is indeed strongly dependent on the nature of the chain itself. For example, if instead of an alkyl chain a flexible poly(ethylene oxide) chain is grafted, this inhomogeneity can be clearly demonstrated. Using relaxation times of ^{1}H and ^{13}C as a function of the frequency of measurement, it has been shown that, from a dynamical point of view, a partitioning of the monomer units occurs. One part, in the vicinity of the surface, has restricted mobility and another part, more free, consists of loops and trains of the chain. There is also an anisotropic motion of the monomer units. This behavior is very dependent on the porous structure of the silica (Facchini and Legrand 1984; Tajouri et al. 1985).

Studies of stationary phases with long *n*-alkyl chains by ^{13}C solid state NMR not only informs on the chemical moieties present at the interface with silica but is able to show the chain order (Pursch et al. 1997). Variation of the mobility all along the chain *via* the relaxation time $T_{1\rho H}$ on 18-alkyl chains was observed by Sindorf and Maciel (1983b). With long 34-alkyl chains, the study as a function of the temperature shows a trans/gauche conformation change (Figure 6.16). This phenomenon is of importance for the chromatographic selectivity of some compounds).

Remarks. It has been shown that amorphous silicas have surface properties that depend on their conditions of preparation. NMR procedures, although apparently simple, oblige one to take heed of different characteristic times T_{1H}, $T_{1\rho H}$, CT, and Hartmann–Hahn matching, all relative to the duty cycle. In effect, this last time has to be longer than the T_1s. The consequence is that measurements can be time consuming, particularly for MAS. This difficulty can be overcome in some cases by adding Fe_2O_3 (Dobson et al. 1988). Nevertheless, measurements can be done if the amount of hydroxyl per unit mass is not too small; as an example, the study of hydroxyls in glass-fibers is very difficult.

To conclude, other actual or potential applications may be mentioned:

- Incorporation of industrial fumed silica into Portland cement has interesting consequences for the properties of the concrete (Dobson et al. 1988).
- Strong natural glassy rods secreted by a sponge (Monorhaphis) are several millimeters in diameter and may be up to one meter in length. They are remarkable for their flexibility, better than glass-fibers of the same size (Levi et al. 1989). Such sponge skeletons contain amount of silanols comparable to that of precipitated silica (Taïbi 1991).

6.2.1.1.4 Biochromatography

The importance of the separation of natural macromolecules like proteins is ever increasing both in research and in industry, particularly in the biomedical, agro-alimental, and biotechnology sectors. Pure products or at least mixtures of well-defined composition from natural or recombinant mixtures are needed. In liquid chromatography, solid minerals like silica, which have very good mechanical properties and which can be prepared in a well-controlled way, are used as a stationary phase wherein the exchange of solute molecules takes place. All the physical properties like granulometry, porosity, and surface chemistry are relevant for explaining the mechanism of separation. However, such materials present a major drawback, which is their ability to adsorb irreversibly many macromolecules of interest in biology, mainly in an aqueous medium where charged electronegative groups appear on the surface (Andrade and Hlady 1986; Norde 1986; Horbett and Brash 1987; Andrade 1995) and to denature proteins.

One of the methods used to minimize this effect is the coating of the silica beads with a polymer such as a polysaccharide like dextran substituted with a calculated amount of positively charged groups, namely DEAE (Touhami

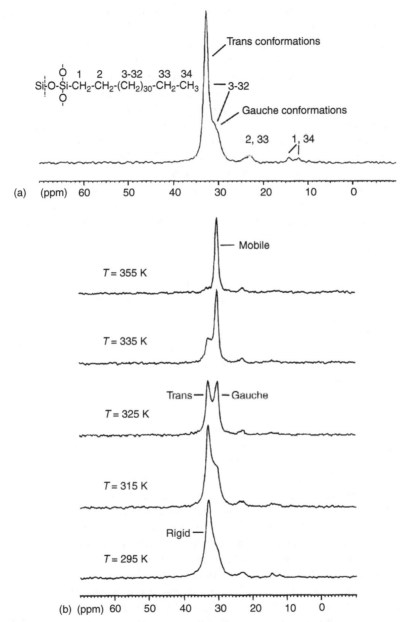

FIGURE 6.16 (a) ^{13}C CP/MAS NMR structural assignment of a surface polymerized C_{34} phase; (b) evolution as a function of the temperature. (From Pursch, M., Brindle, R., Ellwanger, A., Sander, L. C., Bell, C. M., Händel, H., and Albert, K., *Sol. Stat. Nucl. Magn. Res.*, 9(2–4), 191, 1997.)

et al. 1993b, 1993a). However, the mechanism of chromatography becomes now influenced by the conformations and mobility of the chains at the interface, and these new parameters must also be investigated. In particular, here the modification induced in the surface layer when the dextran is functionalized with *para*-aminobenzamidine (*p*-ABA) is investigated in a group that has been shown to be useful in affinity chromatography in the separation of the serine proteases like trypsin, thrombin, and plasmin. Indeed, it is claimed that the amidine group mimics the cationic lateral chain of the arginyl and eventually of the lusyl binding sites of the protein substrates.

The polysaccharide used was dextran T70 M_w = 70,000 g mol^{-1}. Dextran was substituted by positively charged diethylaminoethane (DEAE) groups. The silica was "polygosil" of porosity 300 Å, granulometry 15–25 (m, and specific surface area 160 m^{2}g^{-1}. The silica was passivated with the DEAE-dextran. The dextran was also functionalized with the *p*-ABA. A typical chemical formula for the compound is given in Figure 6.5 (Figure 6.17).

When these samples are put in a phosphate buffer (0.05 M, 0.1 M NaCl, pH 7.4), which is the usual condition for the protein separation, the segmental motions are increased compared to dry samples and the EPR spectra appear as the superposition of two populations as explained above. In Figure 6.18, the rotational correlation times for the fast fraction are shown as a function of the inverse of absolute temperature. The motion of the chains bearing *p*-ABA groups, or more precisely, of the loops and tails bearing the ligand, are somewhat slower than without this side arm.

In Figure 6.19, the ratio of the population of labels in solution over that of labels adsorbed on the surface is given as a function of the inverse of absolute temperature. This parameter gives information on the conformation of the chains, flat on the surface, or extended in solution. The results show that the chains with the *p*-ABA ligand have a more abundant fast fraction of labels or are more extended into solution than the chains without this side arm.

In summarizing these results, it appears that the functional species covalently fixed on the dextran chains have a marked effect on the microscopic behavior of the coated layer. First, the dynamics are altered and the local Brownian motion is restricted. Secondly, however, due to steric hindrances, the loops and tails are also relatively more abundant. The mobility and accessibility of the sites can be investigated in this way, providing a microscopic description of the separation mechanism (Hommel, Van Damme, and Legrand 2004). The coating by dextran is also interesting for drug delivery by coating nanoparticles with it (Fattal and Vauthier 2002; Vauthier and Couvreur 2004).

Remarks: Some perspectives for pharmacology. An exciting application of colloidal particles based on biodegradable polymers is their development as

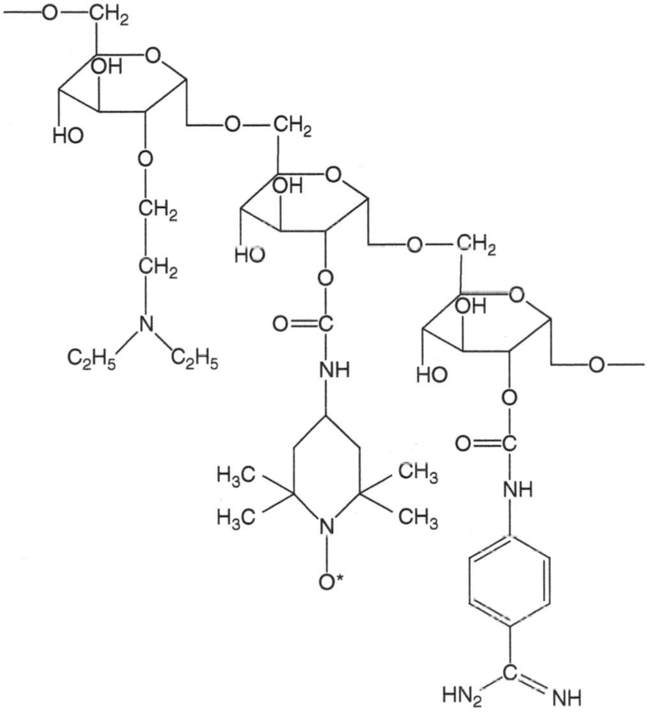

FIGURE 6.17 Typical formula for the DEAE-dextran eventually substituted with *p*-ABA and spin-labeled with 4-amino-TEMPO (From Touhami et al. *Colloid Surf B. Biointerfaces*, 1, 189, 1993a).

carriers for the in vivo delivery of drugs. Generally, after intravenous injection, these nanoparticles are rapidly removed from the blood stream due to a massive uptake by macrophages of the Mononuclear Phagocytes System. Thus they mainly concentrate in the liver, where interesting therapeutic applications could be proposed. This specific distribution of the nanoparticles could be related to their recognition as foreign bodies by macrophages. A major challenge is to design nanoparticles able to carry a drug to a specific site in the body by escaping macrophage capture and enabling specific recognition of a biological target at a molecular level. To reduce their uptake by macrophage, the most popular approach was to create a steric barrier made of poly(ethylene oxide) (PEO) on the nanoparticle surface to repel blood proteins. Other hydrophilic polymers, including dextran, were shown to produce a steric repulsive effect. Taking into account the different kinds of dextran coated particles, it can be expected that the conformation of

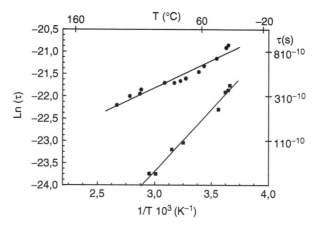

FIGURE 6.18 Evolution with temperature of the rotational correlation time of the fast fraction of labels in a phosphate buffer: (v) without a p-ABA ligand; (λ) with a p-ABA ligand. (From Touhami, A., Hommel, H., Legrand, A. P., Serres, A., Muller, D., and Jozefonvicz, J., *Colloid Surf. B: Biointerfaces*, 1, 189, 1993. Reprinted with permission from Elsevier.)

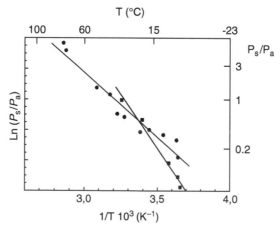

FIGURE 6.19 Evolution with temperature of the ratio of the two populations of labels (desorbed or adsorbed), characteristics of the conformation (flat or extended) of the chains, in a phosphate buffer: (■) without a p-ABA ligand; (●) with a p-ABA ligand. (From Touhami, A., Hommel, H., Legrand, A. P., Serres, A., Muller, D., and Jozefonvicz, J., *Colloid Surf. B: Biointerfaces*, 1, 189, 1993.)

dextran chains at the nanoparticles can be different leading to the observation of different in vivo fates (Barrat et al. 2001, Fatal et al. 2002, Vauthier et al. 2004). A major obstacle to investigate this hypothesis is to find a suitable and relevant method to highlight possible differences in the conformation of dextran chains on the surface of spherical particles of a few hundreds of nanometers. Precisely, it has been demonstrated the usefulness of different Magnetic Resonance Spectroscopy to achieve such a goal (Chauvierre et al. 2004). This is the next step in progress.

6.2.1.1.5 Polymer Composites: Silica-Filled Tire Rubbers

The addition of fillers into elastomers is of significant commercial importance due to the improvement of various technical properties of the final materials. The incorporation of reinforcing particles in an elastomeric matrix increases its elastic modulus and fracture properties, such as the stress at breakage. It is well known that the reinforcing potential of filler is also dramatically dependent on the quality of its dispersion in the rubber matrix. These effects vary with the amount of filler, the surface characteristics of the silica particles, and with temperature, thus highlighting the importance of the interface between the filler and the polymer.

Since the appearance of the "GREEN" or "ENERGY TIRE" concept by Michelin Company in the 1990s, by the utilization of silica as complementary reinforcing filler, this process is now more and more used. Owing its polar hydrophilic surface, silica has to be modified, in particular, by the use of coupling agent to shield its surface and make reaction with the rubber matrix.

Incorporation of fillers (silica, carbon black) in elastomers (BR, SBR) is associated with the formation of bound rubber, which results from interactions taking place between the surface of the solid and the polymer. These interactions are expected to be dependent on the surface reactivity of the solid and thus should affect the mobility of the polymer chains in the vicinity of the solid particles. Solid-state NMR spectroscopy appears in this respect to be well suited for monitoring the polymer/filler interphase and to follow its evolution vs. solid-surface reactivity (Legrand et al. 1992). Using ^{1}H spin–spin relaxation time T_2 measurement, it was shown that bound rubber has two components: a tightly and a loosely held phase. The modification of the filler surface reactivity by grafting of alkyl chains (methyl, hexadecyl) is associated with a change in the distribution of the two bound rubber components, which is particularly evidenced in the silica polybutadiene composites. In such a case, the influence of the grafting reaction can be understood in terms of efficiency, i.e., the number of surface active sites still available after modification. The smaller this number, the smaller is the tightly held bound rubber component. Carbon black/SBR composites appear to be less sensitive to the effect of the surface deactivation. A more recent study (Brinke et al. 2002) of natural rubber filled with Stöber silica pure or grafted with trifunctional organosilanes, using the $^1H\,T_2$ measurement, shows the existence of three domains of mobility. The first domain originates from a layer of rubber chains and coupling agents tightly bound to the silica surface. This domain depends on the chemical structure of the coupling agent. An intermediate region is related to the network chains, with physically adsorbed chain portions acting as network junctions. A last one domain with high mobility originates from free extractable rubber chains in the presence of a good solvent. Such conclusions are comparable to those usually proposed for rubber filled with carbon black.

Local information about silica-rubber interface can equally be obtained using ^1H-^{29}Si CP/MAS NMR (Simonutti et al. 1999). So, larger or reduced tightness of adhesion of 1,4-*cis*-polysioprene to precipitated silica surface can be evaluated. Consequently effects of incorporation of a bidentate coupling agent and of vulcanization are shown.

6.2.2 Ceramic Powders

Physical properties of solids are greatly influenced by the arrangements of atoms present in the network. Nanostructured systems can offer improved and new properties compared to those of bulk materials. Consequently, in such materials, the influence of interfaces and interphases are of noticeable importance. Such interactions between the grains can lead to new physical and chemical properties. Ceramic composites, obtained by dispersing nano-scale particles of one constituent into larger particles of a second constituent, have shown such improvements in mechanical properties (micro/nano composites). In particular, superplasticity has been obtained by mixing two equally fine constituents (nano/nano composites). Different modes of preparation have been used to make nanosized particles in large enough quantities. Among them, laser ablation of solid targets and plasma or laser induced reactions of gaseous mixtures have been developed.

Such nanosized particles present spherical particles, more or less aggregated, with a mean diameter in the range 10–100 nm. The characterization of such material needs the utilization of complementary methods, as presented in Table 6.1. In particular, solid state NMR is able to provide useful information depending of the chemical elements contained in the material. As an example, the characterization of nanostructured Si_3N_4/SiC composites, using ^{13}C, ^{29}Si and ^{27}Al for additives, is able to control the feedback between synthesis and sintering of high temperature resisting ceramics (Legrand and Sénémaud 2003).

TABLE 6.1
Available Information Depending on the Method Used

Method	Surface	Bulk	Crystalline	Amorphous	Local or Short Order	Average or Long Order	Sample Type
CA		x				x	Powder
EPR		x				x	Powder
EXAFS		x			x	x	Powder
IR		x			x	x	Powder
ND		x	x			x	Powder
NMR	x	x	x	x	x	x	Powder
SXS		x				x	Any
TEM		x	x	x			Nanopowder
XD		x	x			x	Powder
XPS	x		x	x		x	Any
SA	x					x	Any

The x indicates the dominant domain of expertise of the method. Chemical analysis (CA), electron paramagnetic resonance (EPR), extended x-ray absorption fine structure (EXAFS), infrared spectroscopy (IR), soft x-ray spectroscopy (SXS), solid state nuclear magnetic resonance (NMR), specific surface area determination (SA), transmission electron microscopy and electron diffraction (TEM), x-ray (XD) and neutron (ND) diffraction, x-ray photoelectron spectroscopy (XPS). (From Legrand, A. P. and Sénémaud, C., Eds., *Nanostructured Silicon-Based Powders and Composites.* Taylor & Francis, 2003.)

6.2.2.1 Nanosized Silicon Carbide and Nitride

Silicon carbide powders have been prepared by an infrared laser pyrolysis process of SiH_4 and C_2H_2 gases of different average size from 15 to 50 nm (SiC 163–177). Cubic (3C) to polytype structures are partially detected by x-ray, although NMR is able to do it easily (Figure 6.20) (Tougne et al. 1993).

In Si_3N_4/SiC nanoparticles, a composite structure, although evidenced by TEM, NMR, detects well-crystallized Si_3N_4 islands surrounded by semi amorphous SiC medium containing free radicals as observed by taking into account the difference of ^{29}Si relaxation times (Figure 6.21) and EPR observation (Curie law behavior of the magnetization).

6.2.3 BIOLOGICAL INTERFACES

Biological systems present a large variety of interfaces. Their behavior implicates complex mechanisms of which few have been completely elucidated. A restricted domain, although of importance for application, is

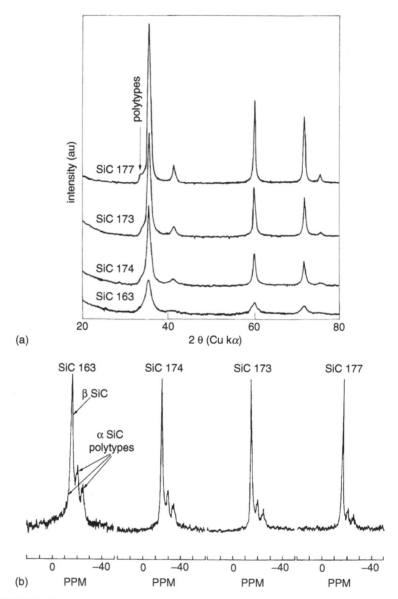

(a)

(b)

FIGURE 6.20 (a) X-ray diffraction patterns of the as-synthesized SiC powders (a.u., arbitrary units). (b) Solid state ^{29}Si MAS-NMR spectra at 59.6 MHz of as-formed silicon carbide (duty cycle, 500 s; 90° pulse, 5 μs; number of scans, 161, 100, 164, and 60). Spectra are classified as function of decreasing residence time into the laser irradiation cell.

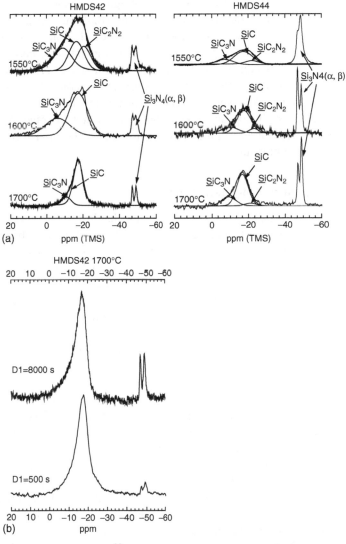

FIGURE 6.21 (a) Solid state ^{29}Si MAS–NMR spectra at 99.36 MHz with a rotation frequency of 5 kHz, of two nanosized Si/C/N powders, referenced HMDS42 and HMDS44, heat-treated at 1550, 1600, and 1700°C during 1 hour in nitrogen atmosphere and decomposition, using a least square method, into different assumed silicon environments. (b) Influence of the length of the duty cycle (D1) on the relative intensities of the two components of each spectrum. Spectra are adjusted so that the maximum of each spectrum is similar to the others. Signal over noise depends on the number of scans, and consequently, of the total time of accumulation, which is different for some spectra.

that of the mineralization in the vertebrate kingdom. Such a mechanism is analyzed fundamentally through model systems (Ball and Voegel 2003) to control and to mimic it for applications. From medical applications, such as bone repair in human surgery or bioprosthesis implantation, contradictory approaches have to be considered. For the first one the objective is to reinforce the inorganic mineral constituent of bones, on the contrary the second one intends its inhibition to prevent dysfunction of the prosthesis.

Examples of such opposite approaches (Senna and Legrand 2004) are presented here with analysis of the mineral implanted and its transformation or appearance after implantation.

6.2.3.1 Programmed Mineralization

Calcification processes implicated in the mechanisms of biodegradation, osteoconduction, and osteoinduction is the subject of many observations and analyses. Particular attention is given to the obtainment of materials usable for bone repair. Among them, resorbable phosphate-based ceramics are nowadays commonly used for different applications. Characterization of such materials in orthopedic and maxillofacial surgery through osteoformation (Marchandise, Belgrand, and Legrand 1992), selection of the better characteristics of such materials (Miquel et al. 1990; Bohner et al. 1996), and evaluation of the influence of drug delivery into the bone structure (Legrand et al. 2000; Bohic et al. 2000) requires a variety of evaluation methods.

As such materials are more or less crystallized, constituted of powders or porous ceramics to permit the osteointegration, solid state NMR is efficient for the characterization of such materials. Moreover, it can be used to analyze the explants or the inorganic part of bone mineral.

An example (Legrand 2002) of such material is macroporous biphasic ceramic phosphate (MBCP), a mixture of hydroxyapatite (HAP: $Ca_{10}(PO_4)_6(OH)_2$) and β-tricalcium phosphate (β-TCP: $Ca_3(PO_4)_2$), the composition of which is being optimized for medical utilization (Toth et al. 1991; AFNOR 1993; Ishikawa, Ducheyne, and Radin 1993).

^{31}P MAS/NMR spectra of two compositions HAP/β-TCP MBCP show that such ceramics are neither a pure nor a simple mixture (Figure 6.22).

After 78 weeks, the material was explanted and analyzed by ^{31}P MAS. Figure 6.23 (left) shows an important evolution of the material. A bone-like peak around 2.8 ppm is observed. ^{31}P MAS saturation recovery shows that the spectrum is constituted of two types of materials (Figure 6.23 right). The peak observed at 2.8 ppm is attributed to newly formed bone and the residual broad line to unresorbed MBCP.

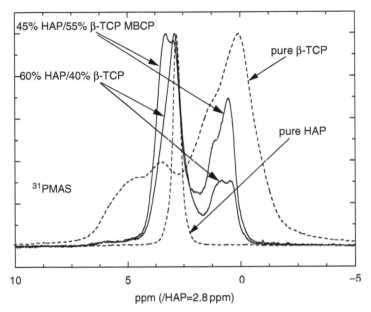

FIGURE 6.22 ^{31}P MAS/NMR spectra of pure HAP, pure β-TCP, and MBCP. 45/55 or 60/40% relative compositions were determined by x-ray diffraction.

The influence of macro and micro porosity of similar ceramics was demonstrated similarly (Marchandise, Belgrand, and Legrand 1992), owing development of new efficient resorbable material for human chirurgical applications.

Other kinds of similar biomaterials have been proposed. Experiments done in vitro have demonstrated possible applications.

Studies of in vitro growing process of calcium phosphates in presence of collagen fibers and modified hyaluronate (Liu et al. 1999) or hydroxyapatite-collagen-hyaluronic acid composite (Bakos, Solda, and Hernandez-Fuentes 1999) show an interesting approach—using composite systems mimicking the biological environment. Similarly, hydroxyapatite and tricalcium phosphate have been incorporated into polyhydroxybutyrate in different proportions (Wang and Ni 2004). In vitro studies were conducted using an acellular simulated body fluid and they have shown formation of mineral crystals of apatitic structure on the surface of such composites. Polydimethylsiloxane–CaO–SiO$_2$–TiO$_2$ and poly(tetramethylene oxide)–CaO–SiO$_2$–TiO$_2$ hybrids showed apatite-forming ability in simulated body fluid and interesting mechanical properties (Kawashita et al. 2004).

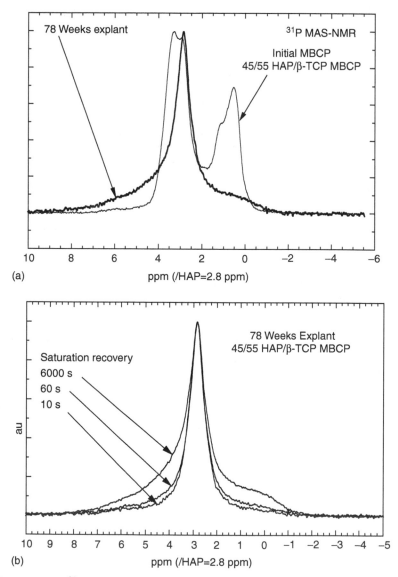

FIGURE 6.23 ^{31}P MAS/NMR of: (a) initial 45/55 MBCP and 78 weeks explant; (b) saturation recovery of explant showing the composite structure of the spectrum.

6.2.3.2 Pathological Mineralization

After implantation of breast silicone prostheses or stented bioprosthetic heart valves, a calcification process usually develops. Analysis of the inorganic part of such deposits is of interest to determine its nature and origin (Figure 6.24).

6.2.3.2.1 Breast Implants Silicone Polymers

In spite of continual advancements in surgical techniques and the development of new devices, late complications remain a significant factor of breast implant surgery (Collis and Sharpe 2000). The durability of these implants does not

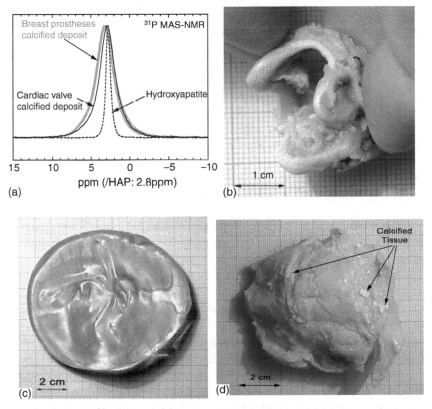

FIGURE 6.24 (a) ^{31}P MAS NMR spectra of hydroxyapatite compared to cardiac valve and breast prosthesis calcified mineralized deposits. (b) Calcified deposit onto cardiac valve, observed after explantation. (c) Explanted breast prosthesis. (d) Internal explanted capsule which covered the breast prosthesis, with calcified deposits.

match the life expectancy of the patients. The reoperation rates of patients are very high and they are due to

- Excessive fibrous capsules (Janowsky, Kupper, and Hilka 2000; Kuhn et al. 2000)
- Percolation of silicone gel through the silicone shell
- Change in shape
- Rupture of prosthesis (Brown et al. 2000)
- Mineralization (Peters et al. 2001)

Breast prostheses harvested in different hospitals in Canada and in the United States were participating in the retrieval program of explanted devices by the Quebec Biomaterials Institute (Laval University, Quebec, Canada). Different models of devices implanted for 10 years or more were selected, as they were highly mineralized (Guidoin et al. 1991).

6.2.3.2.2 Heart Repair Natural Porcine Valve and Pericardium Valves

Such bioprostheses also harvested from different hospitals in Canada and the US have two origins:

- Porcine: aortic leaflets and/or valves of pigs treated with glutaraldehyde are mounted on stents
- Pericardium: pericardium of the heart of calf that is trimmed, mounted on a stent, and fixed with glutaraldehyde

The replacement of human valves is different from aortic and mitral to tricuspid ones. It is frequent for the aortic site (diameter 20–23 mm) and mitral (larger diameter, 23–31 mm). It is less frequent for tricuspid (larger diameter, 22–30 mm) valves. Reoperations are very frequent after 7 and 10 years in the aortic and mitral valves.

The origin of the progressive complications, in particular for the left side of the heart, are the age degradation and calcification, which reduce the opening of the leaflets and permit blood leaks at closure (Herijgers et al. 1999; Grabenwoger et al. 2000; Vasudev et al., 2000).

The objective would be to prevent this deposit formation. One possible approach could be to modify the surface of the biomaterial by adsorption of inhibitors. Koutsopoulos and Dalas (2000) have shown the inhibition of hydroxyapatite formation in aqueous solutions by amino acids with hydrophobic side groups. Such experiments are far from medical utilization, in particular for long-term behavior, but are of interest for future applications.

Finally, the comparison between programmed and pathological mineralization, not only done through NMR experimental results but observed by x-ray diffraction, shows the bone-like structure of the inorganic part of the regenerated parts or of the deposits. Such observations, although the complexity of the systems studied to distinguish the interface from bulk, demonstrate the ability of NMR to contribute to a better knowledge and improvements of such materials.

6.2.4 Microporous Carbons and Others

Granular activated carbons are prepared from naturally occurring materials such as coal, petroleum, coconut husks, peat, polymers, etc. A high internal surface area developed during the preparation permits the adsorption of molecules (Neely and Isacoff 1982). Exchange processes occurring in the adsorbate–adsorbent systems had been studied using line shape and relaxation times measurements not only in such material (Estrade-Szwarckopf, Auvray, and Legrand 1970), but in a large class of adsorbents and catalysts (Fraissard et al. 1985; Fraissard and Lapina 2002).

Activated carbon adsorbents prepared at relative high temperatures are more or less conductive materials and consequently disturbing NMR measurements. Charcoal contains localized paramagnetic free radicals which can be observed by EPR (Uebersfeld, Etienne, and Combrisson 1954). After heat treatment at a temperature lower than 600°C of a charcoal, porosity is obtained and adsorption of different molecules can be accomplished. The adsorbed molecule moves on the surface during its sticking time but exchange occurs with the external liquid or gas medium. To obtain information on the behavior of the adsorbed phase, special NMR measurements have been developed (Motchane 1962; Barszczewski-Jacubowicz 1963; Legrand et al. 1999). Owing the presence of paramagnetic centers inside the material, dipolar coupling between the protons of the adsorbed molecule and the paramagnetic centers was considered. Using the cross-polarization method between these two spin species, a physical coupling was demonstrated. The type of effect observed (Solid State Effect, Abragam 1961 p. 293) showed that the sticking time of the adsorbed molecule is long enough relative to the relaxation times. The interest of the method is its ability to identify the nature of the physical interaction between the spin species. Is the free radical on the surface of the adsorbent and is it able to interact with an adsorbed molecule? The answer is more subtle than the question because paramagnetic centers are able to:

- Directly polarize the constitutive nuclei of the adsorbed molecule (Grivet 1964).

• Indirectly polarize, through protons of the skeleton of the charcoal, the constitutive nuclei of the adsorbed molecule. This effect is restricted to protonated or partially protonated adsorbed molecule (Legrand et al. 1999).

Such effects depend of the accessibility of the adsorbed molecule to the vicinity of the paramagnetic free radical. This accessibility depends on the temperature at which measurements are done. At low temperature, around 20 K, experiments done with liquid hydrogen molecule of different isotopic composition (HH, HD, or DD) demonstrate that only an indirect coupling exists. Consequently, HH, HD (Figure 6.25), and not DD show the effect. This phenomenon is comforted by the following remark: HH, owing to quantification rules, has two rotational states—para and ortho (Motizuki and Nagamiya 1956). Only the ortho state is detectable by NMR. The thermodynamic equilibrium state at 20 K is the para state. Starting with a room temperature percentage of 25% para, ortho-para conversion must occur at 20 K through the same physical coupling mentioned above (Legrand and Heslot 1997). The observation confirms that no conversion occurs—this is a complementary proof of an indirect coupling. Moreover, in the liquid HD contacting the adsorbent charcoal, an internal Overhauser effect inside the freely rotating molecule permits to observe an apparent solid-state effect on the deuterium nucleus. This shows, although the mechanisms are complex to identify, that the answer to the above question can be obtained: free radicals are located inside cavities and bulk of the charcoal is not accessible (distance between free radical and molecule <3 Å) to hydrogen molecules around 20 K, but if experiments are done at higher temperatures with protonated or fluorinated molecules, accessibility is obtained. This shows the influence

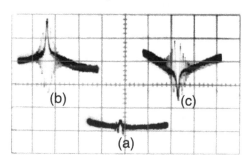

FIGURE 6.25 ^{1}H NMR signal of HD: (a) natural signal of the bulk liquid, (b) enhancement of the signal by dynamic polarization, (c) inverted signal. Such effects are obtained by saturation of forbidden transitions between electronic and nuclear spins coupled by static dipolar interaction (From Legrand A.P. and Uebersfeld J., *CR. Acad. Sci. B*, 264, 337, 1967.)

of activation energy in the adsorption process of the molecule to the surface of the charcoal.

Such results show the influence of the porosity and its accessibility to the adsorbates. New developments in NMR are progressing for the study of such porous materials. In particular, the use of xenon as a NMR probe is of a great interest (Raftery et al. 1994).

6.3 GENERAL CONCLUSION

Nowadays, nanoscience is a tremendous domain of expansion for research and applications. Although it could seem to be in fashion, this domain was developing for a long time under different names such as grinding, powder preparation, adsorption on divided solids, solid catalysts, clusters, etc. Nevertheless, new materials are prepared and new applications are developed, needing previous methods of study and characterization. Among them, magnetic resonances show their utility through the examples selected here. This restricted presentation must be considered as an introduction, or, referring to the title of the chapter, a border of a domain of utilization.

REFERENCES

Abragam, A., *The Principles of Nuclear Magnetism*, Clarendon Press, Oxford, 1961.

AFNOR (Association Française de normalisation), Norme S94-066, AFNOR, Paris, 1993.

Albert, K. and Bayer, E., *J. Chromatogr.*, 544, 345, 1991.

Albert, K., Evers, B., and Bayer, E., *J. Magn. Reson.*, 62, 428, 1985.

Andrade, J. D., Ed, *Surface and Interfacial Aspects of Biomedical Polymers*, Plenum, New York, 1995.

Andrade, J. D. and Hlady, V., *Adv. Polym. Sci.*, 79, 1, 1986.

Andrew, E. R., Bradbury, A., and Eades, R. G., *Nature*, 182, 1659, 1958.

Apperley, D., Hay, J. N., and Raval, H. M., *Chem. Mater.*, 14, 983, 2002.

Bakos, D., Solda, M., and Hernandez-Fuentes, I., *Biomaterials*, 20, 191, 1999.

Ball, V. and Voegel, J.-C., *l'Actualité Chimique Octobre*, 11–26, (in French) 2003.

Barrat, G., Couarraze, G., Couvreur, P., Dubernet, C., Fattal, E., Gref, R., Labarre, D., Legrand, P., Ponchel, G., and Vauthier, C., Polymeric micro and nanoparticles as drug carriers, In *Polymeric Biomaterials*, Dumitriu Ed., 2nd ed., Vol. 28, Marcel Dekker Inc., New York, p. 753, 2001.

Barszczewski-Jacubowiez, M., *These Ann. Phys.*, 13, 8, 1963.

Barthel, H., Balard, H., Bresson, B., Legrand, A. P., Burneau, A., and Carteret, C., In *Organosilicons IV*, Auner, N. and Weis, J., Eds., Wiley, Weinheim, 2004.

Bayer, E., Albert, K., Reiners, J., Nieder, M., and Müller, D., *J. Chromatogr.*, 264, 197, 1983.

Berliner, L. J., Ed, *Spin Labeling Theory and Applications*, Academic Press, New York, p. 592, 1976.

Bloch, F., *Phys. Rev.*, 111, 841, 1958.

Bohic, S., Rey, C., Legrand, A. P., Sfihi, H., Rohamizadech, R., Martel, C., Barbier, A., and Daculsi, G., *Bone*, 2, 341, 2000.

Bohner, M., Lemaitre, J., Legrand, A. P., d'Espinose de la Caillerie, J.-B., and Belgrand, P., *J. Mater. Sci. Mater. Med.*, 7, 457, 1996.

Brinke, J. W., Litvinov, V. M., Wijnhoven, J. E. G., and Noordermeer, J. W. M., *Macromolecules*, 35, 10026, 2002.

Bronnimann, C. E., Zeigler, R. C., and Maciel, G. E., *J. Am. Chem. Soc.*, 110, 2023, 1988.

Brown, S. L., Middleton, M. S., Berg, W. A., Soo, M. S., and Pennello, G., *J. Am. Roentgenol.*, 175, 1057, 2000.

Burnau, A. and Gallas, J.-P., Vibrational spectroscopies, In *The Surface Properties of Silicas*, Legrand, A. P., Ed., Wiley, New York, pp. 147–234, 1998.

Chauvierre, C., Vauthier, C., Labarre, D., and Hommel, H., *Colloid. Polym. Sci.*, 282, 1016, 2004.

Chiang, Chwan-Hwa, Liu, Nan-I, and Koenig, J. L., *J. Colloid Interface Sci.*, 86, 26, 1982.

Chu, C.-H., Jonsson, E., Auvinen, M., Pesek, J. J., and Sandoval, J. E., *Anal. Chem.*, 65, 808, 1993.

Collis, N. and Sharpe, D. T., *Plast. Reconstr. Surg.*, 105, 1979, 2000.

d'Espinose de la Caillerie, J.-B., Aimeur, M. R., and Legrand, A. P., *J. Colloid Interface Sci.*, 194, 434, 1997.

Devine, R. A. B., Dupree, R., Farnan, I., and Capponi, J., *J. Phys. Rev. B*, 5, 2560, 1987.

Dobson, C., Goberdhan, D. G. C., Ramsay, J. D. F., and Rodger, S. A., *J. Mater. Sci.*, 23, 4108, 1988.

Dupree, E. and Pettifer, R. F., *Nature*, 308, 523, 1984.

Ellingsen, D. S. and Resing, H. A., *J. Phys. Chem.*, 84, 2204, 1980.

Estrade-Szwarckopf, H., Auvray, J., and Legrand, A. P., *J. Chim. Phys.*, 7–8, 1970 see also p. 1292

Facchini, L. and Legrand, A. P., *Macromolecules*, 17, 2406, 1984.

Fattal, E. and Vauthier, C., In *Encyclopedia of Pharmaceutical Technology*, Swarbrick, J. and Boylan, J. C., Eds., Marcel Dekker Inc., New York, p. 1874, 2002.

Fraissard, J. and Lapina, O., *Magnetic Resonance in Colloid and Interface Science*, NATO Science series II Mathematics, Physics and Chemistry, 76.

Fraissard, J., Gutsze, A., Michel, D., Pfeifer, H., and Winkler, H., *Adv. Colloid Interface Sci.*, 23, 1985.

Gilpin, R. K. and Gangoda, M. E., *Anal. Chem.*, 56, 1470, 1984.

Gilpin, R. K. J., *Chromatogr. Sci.*, 22, 371, 1984.

Gobet, F. and Kovats, E. Sz., *Adsorption Sci. Technol.*, 1, 77, 1977.

Grabenwoger, M., Fitzal, F., Gross, C., Hutschala, D., Bock, P., Brucke, P., and Wolner, E. J., *Heart Valve Dis.*, 9, 104, 2000.

Grivet, J. P., *C.R. Acad. Sci.*, 259, 776, 1964.

Guidoin, R., Rolland, C., King, M. W., Roy, P. E., and Therrien, M., *Silicone in Medical Devices.* FDA Conference of Proceedings, Baltimore, February 1–2, 1991.

Haeberlen, U., *High Resolution NMR Spectroscopy in Solids: Selective Averaging. Supplement No. 1 to Advances in Magnetic Resonance*, Academic Press, New York, 1976.

Heeribout, L., d'Espinose, J. B., Legrand, A. P., and Mignani, G., *J. Colloid Interface Sci.*, 215, 296, 1999.

Herijgers, P., Ozaki, S., Verbeken, E., Van Lommel, A., Jashari, R., Nishida, T., Lennens, V., and Flameng, W., *Thorac. Cardiovasc. Surg.*, 11 (4), 171, 1999.

Hetem, M., Van de Ven, L., De Haan, J., Cramers, C., Albert, K., and Bayer, E., *J. Chromatogr.*, 479, 269, 1989.

Hommel, H., *Adv. Colloid Interface Sci.*, 54, 209, 1995.

Hommel, H., Van Damme, H., and Legrand, A. P., *Ann. Chim. Sci. Mat.*, 29 (1), 67, 2004.

Horbett, T. A. and Brash, J. L., Eds., *Proteins at Interfaces ACS Symp. Ser. 343*, American Chemical Society, Washington, D.C., 1987.

Ishikawa, K., Ducheyne, P., and Radin, S., *J. Mater. Sci. Mater. Med.*, 4, 168, 1993.

Janowsky, E., Kupper, L. L., and Hilka, B. S., *Neur. Engl. J. Med.*, 342, 781, 2000.

Jeener, J., In *Advances in Magnetic Resonance*, Waugh, J. S., Ed., Vol. 10, Academic Press, New York, 1982.

Jelinek, L., Dong, P., Rojas-Pazos, C., Taïbi, H., and Kovatz, E. S. Z., *Langmuir*, 8, 2152, 1992.

Kawashita, M., Kamitakahara, M., and Kokubo, T., *Ann. Chim. Sci. Mat.*, 29 (1), 7, 2004.

Kivelson, D., *J. Chem. Phys.*, 33, 1094, 1960.

Koutsopoulos, S. and Dalas, E., *Langmuir*, 16, 6739, 2000.

Kuhn, A., Singh, S., Smith, P. D., Ko, F., Falcone, R., Lyle, W. G., Maggi, S. P., Wells, K. E., and Robson, M. C., *Ann. Plast. Surg.*, 44, 387, 2000.

Lajnef, M., Taïbi, H., Legrand, A. P., and Hommel, H., In *Application of NMR Spectroscopy to Cement Science*, Colombet, P. and Grimmer, A. R., Eds., Gordon & Breach Science Publishers, London, p. 181, 1994.

Legrand, A. P., Ed, *The Surface Properties of Silicas*, Wiley, New York, 1998.

Legrand, A. P., *Encyclopedia of Surface and Colloid Science*, Marcel Dekker, Inc., New York, 2002.

Legrand, A. P. and Heslot, A., *Pratique de la Physique et de la Chimie Quantiques*, Ellipses, 158, 1997.

Legrand, A. P. and Sénémaud, C., Eds., *Nanostructured Silicon-Based Powders and Composites*. Taylor & Francis, 2003.

Legrand, A. P. and Uebersfeld, J., *CR. Acad. Sci. B*, 264, 337, 1967.

Legrand, A. P., Auvray, J., and Uebersfeld, J., *J. Chim. Phys.*, 210, 1964.

Legrand, A. P., Hommel, H., Taïbi, H., Miquel, J. L., and Tougne, P., *Colloids Surf.*, 45, 391, 1990a.

Legrand, A. P., Hommel, H., Tuel, A., Vidal, A., Balard, H., Papirer, E., Levitz, P., Czernichowski, M., Erre, R., Van Damme, H., Gallas, J. P., Hemidy, J. F., Lavalley, J. C., Barres, O., Burneau, A., and Grillet, Y., *Adv. Colloid Interface Sci.*, 33, 91, 1990b.

Legrand, A. P., Lecomte, N., Vidal, A., Haidar, B., and Papirer, E., *J. Appl. Polym. Sci.*, 46, 2223, 1992.

Legrand, A. P., Taïbi, H., Hommel, H., Tougne, P., and Leonardelli, S. J., *Non-Cryst. Solids*, 155, 122, 1993.

Legrand, A. P., Hommel, H., and d'Espinose de la Caillerie, J.-B., *Colloids Surf. A: Physicochem. Eng. Aspects*, 158, 157, 1999.

Legrand, A. P., Sfihi, H., and Bouler, J. M., *Bone*, 25, 103, 1999a.

Legrand, A. P., Bresson, B., and Bouler, J. M., *Bioceramics*, Giannini, S. and Moroni, A., Eds., 13, Trans Tech Publications Inc., Ütikon-Zürich. p. 759, 2000.

Legrand, A. P., Bresson, B., Guidoin, R., and Famery, R., *J. Biomed. Mater. Res. (Appl. Biomater.)*, 63, 390, 2002.

Leonardelli, S., Facchini, L., Frétigny, C., Tougne, P., and Legrand, A. P., *J. Am. Chem. Soc.*, 114, 6412, 1992.

Levi, C., Barton, J. L., Guillemet, C., Le Bras, E., and Lehuede, P., *J. Mater. Sci. Lett.*, 8, 337, 1989.

Linton, R. W., Miller, M. L., Maciel, G. E., and Hawkins, B. L., *Surf. Interface Anal.*, 7, 196, 1985.

Liu, L. S., Thompson, A. Y., Heidaran, M. A., Poser, J. W., and Spiro, R. C., *Biomaterials*, 20, 1097, 1999.

Luhmer, M., d'Espinose, J.-B., Hommel, H., and Legrand, A. P., *Magn. Reson. Imaging*, 14 (7:8), 911, 1996.

Maciel, G. E., In *Nuclear Magnetic Resonance in Modern Technology*, Maciel, G. E., Ed., Vol. 447, Kluwer Academic Press, Dordrecht, Netherlands, p. 225, 1994.

Maciel, G. E. and Ellis, P. D., In *NMR Techniques in Catalysis*, Bell, A. T. and Pines, A., Eds., 231, Marcel Dekker, New York, 1994.

Maciel, G. E. and Sindorf, D. W., *J. Am. Chem. Soc.*, 102, 7607, 1980.

Maciel, G. E., Bronnimann, C. E., Zeigler, R. C., Chuang, I-S. , Kinney, D. R., and Keiter, E. A., In *The Colloid Chemistry of Silica*, Bergna, H. E., Ed. *Adv. Chem. Ser.*, ACS, Washington, D.C., 1994 and references therein.

Marchandise, X., Belgrand, P., and Legrand, A. P., *Magn. Reson. Med.*, 28, 1, 1992.

Mehring, M., High resolution NMR spectroscopy in solids, In *NMR: Basic Principles and Progress*, Diehl, P., Fluck, E., and Kosfeld, R., Eds., Vol. 11, Springer, Berlin, 1976.

Meiouet, F., Felix, G., Taïbi, H., Hommel, H., and Legrand, A. P., *Chromatographia*, 31 (7/8), 335, 1991.

Miller, M., Linton, R. W., Maciel, G. E., and Hawkins, B. L., *J. Chromatogr.*, 319, 9, 1985.

Miquel, J. L., Facchini, L., Legrand, A. P., Marchandise, X., Lecouffe, P., Chanavaz, M., Donnazan, M., Rey, C., and Lemaitre, J., *Clin. Mater.*, 5, 115, 1990.

Motchane, J.-L., *Thèse Ann. Phys.*, 13, 7, 1962.

Motizuki, K. and Nagamiya, T., *J. Phys. Soc. Jpn*, 11 (2), 93, 1956.

Neely, J. W. and Isacoff, E. G., *Carbonaceous Adsorbents for the Treatment of Ground and Surface Waters*, Marcel Dekker Inc., New York, 1982.

Norde, W., *Adv. Colloid Interface Sci.*, 25, 267, 1986.

Peters, W., Smith, D., Lugowski, S., Pritzker, K., and Holmyard, D., *Plast. Reconstr. Surg.*, 107, 356, 2001.

Pfeifer, H., *NMR Basic Principles and Progress*, Vol. 7, Springer, Berlin, 1972.

Pfleiderer, B., Albert, K., and Bayer, E., *J. Chromatgr.*, 506, 343, 1990.

Pines, A., Gibby, M. G., and Waugh, J. S., *J. Chem. Phys.*, 59, 569, 1973.

Piquemal, J.-Y., Manoli, J.-M., Beaunier, P., Ensuque, A., Tougne, P., Legrand, A. P., and Brégeault, J.-M., *Micr. Meso. Mat.*, 29, 291, 1999.

Pursch, M., Brindle, R., Ellwanger, A., Sander, L. C., Bell, C. M., Händel, H., and Albert, K., *Sol. Stat. Nucl. Magn. Res.*, 9(2 4), 191, 1997.

Raftery, D. and Chmelka, B. F., *Xenon NMR Spectroscopy in NMR Basic Principles and Progress*, Vol. 30, Springer, Berlin, p. 111, 1994.

Schneider, D. J. and Freed, J. H., In *Biological Magnetic Resonance*, Berliner, L. J. and Reuben, J., Eds., Vol. 8, Plenum Publishing Corporation, New York, p. 1, 1989.

Senna, M. and Legrand, A. P., *Ann. Chim. Sci. Mat.*, 29 (1), 1, 2004.

Simonutti, R., Comotti, A., Negroni, F., and Sozzani, P., *Chem. Mater.*, 11, 822, 1999.

Sindorf, D. W. and Maciel, G. E., *J. Am. Chem. Soc.*, 102, 7606, 1980.

Sindorf, D. W. and Maciel, G. E., *J. Phys. Chem.*, 86, 5208, 1982.

Sindorf, D. W. and Maciel, G. E., *J. Am. Chem. Soc.*, 105, 3767, 1983.

Sindorf, D. W. and Maciel, G. E., *J. Phys. Chem.*, 87, 5516, 1983a.

Sindorf, D. W. and Maciel, G. E., *J. Am. Chem. Soc.*, 105, 1848, 1983b.

Slichter, C. P., *Principles of Magnetic Resonance*, Springer, Berlin, 1990.

Snyder, L. R. and Ward, J. W., *J. Phys. Chem.*, 70, 3941, 1966.

Taïbi, H., *Thèse. Spécialité Chimie-Physique*, Université P.M. Curie, Paris, France, 1991.

Taïbi, H., Hommel, H., and Legrand, A. P., *J. Chim. Phys.*, 89, 445, 1992.

Tajouri, T., Facchini, L., Legrand, A. P., Balard, H., and Papirer, E., *Bull. Soc. Chim. Fr.*, 6, 1143, 1985.

Tanaka, K., Shinodo, S., and Saito, Y., *Chem. Lett.*, 179, 1979.

Toth, J. M., Hirthe, W. M., Hubbard, W. G., Brantley, W. A., and Lynch, K. L., *J. Appl. Biomater.*, 2, 40, 1991.

Tougne, P., Hommel, H., Legrand, A. P., Herlin, N., Luce, M., and Cauchetier, M., *Diamond Relat. Mat.*, 2, 486, 1993.

Touhami, A., Hommel, H., Legrand, A. P., Serres, A., Muller, D., and Jozefonvicz, J., *Colloid Surf. B: Biointerfaces*, 1, 189, 1993a.

Touhami, A., Hommel, H., Legrand, A. P., Serres, A., Muller, D., and Jozefonvicz, J., *Colloid Surf. A: Physicochem. Eng. Aspects*, 75, 57, 1993b.

Tsuji, K., Jones, C. W., and Davis, M. E., *Micro. Meso. Mat.*, 29, 339, 1999.

Tuel, A., Hommel, H., Legrand, A. P., and Kovats, E. Sz., *Langmuir*, 6, 770, 1990.

Tuel, A., Hommel, H., Legrand, A. P., Gonnord, M. F., Mincsovics, E., and Siouffi, A. M., *J. Chim. Phys.*, 89, 477, 1992.

Uebersfeld, J., Etienne, A., and Combrisson, J., *Nature*, 174, 614, 1954.

Vasudev, S. C., Chandy, T., Sharma, C. P., Mohanty, M., and Umasankar, P. R., *J. Biomater. Appl.*, 14, 273, 2000.

Vauthier, C. and Couvreur, P., Miscellaneous biopolymers and biodegradation of synthetic polymers, in *Handbook of Biopolymers*, Matsura, J. P. and Steinbuchel, A., Eds., Wiley-V/C, New York, 2004, Vol. 9 (in press).

Wang, M. and Ni, J., *Ann. Chim. Sci. Mat.*, 29 (1), 17, 2004.

Wijnen, P. W. J. G., Beelen, T. P. M., de Haan, J. W., Rummens, C. P. J., Van de Ven, L. J. M., and Van Santen, R. A., *J. Non-Cryst. Sol.*, 109, 85, 1989.

7 Vapors, Liquids, and Polymers in Solid–Solid Confinements: Application to the Agglomeration and Ageing of Powders and to the Reinforcement of Polymers

Henri Van Damme

CONTENTS

7.1 INTRODUCTION: SOLID–SOLID CONTACTS AND SOLID–SOLID CONFINEMENTS

Interparticle "contacts" in powders or, more generally, in dispersed materials are seldom true solid–solid contacts over the whole apparent interface area. Due to surface roughness, which is a quasi-universal feature of powders [1–4], and/or to the presence of a third medium, adsorbed or not, what is called a "contact" is most of the time a complex micro- or nanoporous geometry which is better described in terms of confinement than in terms of true solid–solid interface (such as found in ceramics for instance). Real atom-to-atom contact is achieved at very tiny spots only and, in most places, the contact is an interparticle gap in the nm range (Figure 7.1 and Figure 7.2). It is the purpose of this chapter to analyze the properties of matter in such narrow and often geometrically complex gaps. More specifically, we will concentrate on two phase equilibria or phase transitions: the vapor–liquid transition (capillary transition) in interparticle "contacts" in powders on one hand, and the liquid-glass transition in the confined regions between filler particles in polymer nanocomposites on the other hand. The vapor–liquid transition in confined geometries (capillary condensation) is an old subject [5–7], but recent theoretical, simulation, and experimental work has shed a new light on it, in particular on mechanical [8–12], hysteresis [13–19], and kinetic [20–25] aspects. This is directly applicable to the agglomeration and ageing of powders in a wet atmosphere. On the other hand, the glass transition of thin polymer films has also been intensively investigated in recent years, both in self-supporting and in supported films [26–29]. Dramatic shifts of the glass transition temperature have been measured, which provide a strong basis for an often invoked mechanism of rubber reinforcement by fillers (see Chapter 9 in this volume). As will be shown, the wetting properties of the solid surfaces prove to be key, both for agglomeration and ageing and for reinforcement.

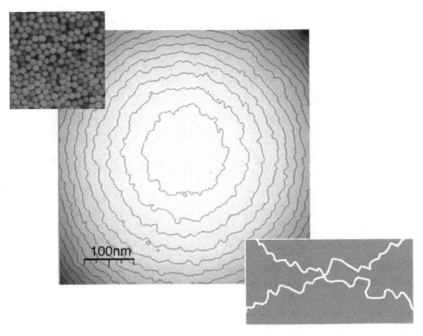

FIGURE 7.1 Top left: Laser microscopy image of 1.5 μm silica particles prepared by the Stöber method. (From Lootens, D., PhD thesis, Université Pierre et Marie Curie, Paris, 2004.) Central: AFM picture of one of the particles. The iso-level contour lines reveal the surface roughness, which is typically of the order of 10 nm. Bottom right: sketch of the contact between two such particles. The "contact" zone is a geometrically complex confined space.

7.2 EXPERIMENTAL EVIDENCE FOR CONFINEMENT EFFECTS

Before entering into the thermodynamics of phase transitions in confined geometries, we will present typical examples of the experimental behavior— the understanding of which is sought after—taken from recent literature. Section 7.2.1 will be devoted to simple experiments aiming at studying the avalanche and failure behavior of granular materials and powders in humid atmosphere [20,21,23]. It illustrates clearly the relationship between the kinetics of capillary condensation and the well-known deterioration of powder flowability in wet environment. Section 7.2.2 will be devoted to two examples of polymer/filler nanocomposites. The first example [30] illustrates a case where confinement of the polymer leads to a remarkable enhancement of its mechanical properties whereas the second example [31] illustrates a case where, in spite of the exceptional properties of the filler

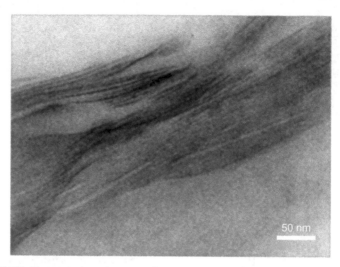

FIGURE 7.2 Transmission electron microscopy image of an ultrathin section of a montmorillonite-polyurethane nanocomposite sample (scale bar: 50 nm). In most places, the polymer is confined between individual nm-thick flexible clay platelets. Due to the enormous aspect ratio and surface area of the clay particles, not much clay is needed to have *the whole* polymer in such slab-type confinements with an average thickness which is not much larger than the gyration radius of the polymer chains.

particles, a poor reinforcement is obtained, presumably due to the bad quality of the interface.

7.2.1 THE COHESION OF POWDERS IN WET CONDITIONS

Experiment 1. The first experiment is as follows (Figure 7.3) [21]. The powder is made of small glass beads of uniform size ($\sim$200 μm). It is introduced in a drum in which the temperature, T, and the relative vapor pressure of a condensable vapor (heptane or water), $P_V/P_{V,sat}$, are controlled. After a given "equilibration" time, the drum is slowly rotated and the maximum stability angle θ_m (i.e., the angle at which the first avalanche starts) is measured. According to classical granular materials mechanics (see below), θ_m is determined by inter-granular friction and cohesion. Any increase of θ_m beyond its value in totally dry conditions ($P_V/P_{V,sat} = 0$) may be interpreted as an increase of cohesion. As illustrated in Figure 7.4, a dramatic increase of θ_m is observed beyond some relative pressure ($\sim$0.70 for water and $\sim$0.90 for heptane). This is a clear signature of capillary condensation.

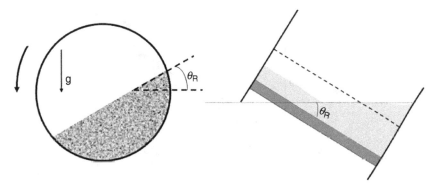

FIGURE 7.3 Cartoon of the experimental set-up used by Bocquet, L., Charlaix, E., Ciliberto, S., and Crassous, J., *Nature*, 396, 735–739, 1998; Restagno, F., Bocquet, L., Biben, T., and Charlaix, E., *J. Phys. Condens. Matter.*, 12, A419–A424, 2000; Fraysse, N., Thomé, H., and Petit, L., *Eur. Phys. J.*, B11, 615–619, 1999 (left), and by Valverde, J. M., Castellanos, A., and Ramos, A., *Phys. Rev.*, E62, 6851–6860, 2000 (right) for studying the ageing of granular media or powders (see text). The dashed line is the initial height of the powder bed.

Experiment 2. In a second type of experiment [23], the powder of glass beads (~ 50 µm) was first dried before introduction in the drum, and the weight of the drum, in which the relative humidity was maintained at 68%, was recorded. As illustrated in Figure 7.5, the weight does not increase as one would intuitively expect with, say, a $t^{1/2}$-dependent diffusion-limited regime

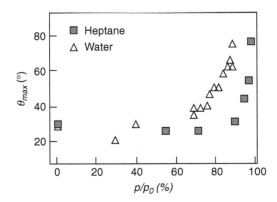

FIGURE 7.4 Evolution of the maximum stability angle θ_m of a bed of 200 µm glass beads, as a function of the condensable vapor relative pressure, in a rotating drum. (Adapted from Fraysse, N., Thomé, H., Petit, L., *Eur. Phys. J.*, B11, 615–619, 1999.)

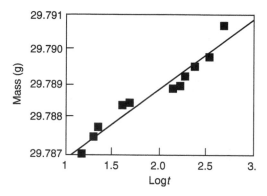

FIGURE 7.5 Time evolution of the mass of a bed of initially dry 50 μm glass beads, in a close container in which the relative humidity is kept at 68%. (Adapted from Restagno, F., Bocquet, L., Biben, T., Charlaix, E., *J. Phys. Condens. Matter.*, 12, A419–A424, 2000.)

and a final plateau. Instead, the observed kinetics is logarithmic, extending over indefinite times, without any sign of saturation.

Experiment 3. In a third type of experiment [20,23], θ_m was measured as a function of a waiting or "ageing" time, t_w. However, t_w is defined in a more subtle way than in the previous experiment. The beads (~ 50 or ~ 200 μm) were first "equilibrated" in the drum at a given temperature and relative humidity for a long time, and the drum was rotated for a few turns, leading to several avalanches. The system was then kept at rest for a time t_w, after which the experiment started. Thus, t_w is a waiting time well after a long "equilibration" with the humid atmosphere but immediately after perturbation of the contacts network in the granular bed. In spite of that, θ_m, or, more exactly, $\tan \theta_m$ was increasing logarithmically with the waiting time (Figure 7.6), according to:

$$\tan \theta_m \propto \ln \frac{(t_w)}{\cos \theta_m} \tag{7.1}$$

Thus, the cohesion of the bed is "endlessly" increasing, according to an ever slower kinetics. Furthermore, the higher the relative humidity, the faster the ageing process is. The effect is strong: a $\sim 10^{-4}$ relative mass uptake (corresponding to an average 5 nm thick water layer on the particles) is enough to increase the coefficient of friction by a factor of ~ 3.

Experiment 4. Finally, in a fourth type of experiment [12], a rectangular bed of powder (styrene butadiene) was first dried by passing dry nitrogen through it and then it was kept in this dry environment. The bed was then

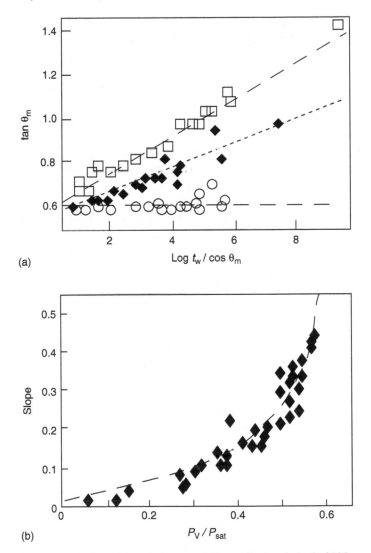

(a)

(b)

FIGURE 7.6 (a) Evolution of the maximum stability angle θ_m of a bed of 200 μm glass beads, as a function of the waiting time t_w (see text) from a few minutes to more than one week, at RH = 3% (○), 23 (◆), and 43% (□), in a rotating drum. (Adapted from Bocquet, L., Charlaix, E., Restagno, F., *C. R. Physique*, 3, 207–215, 2002.) The same behavior was observed with smaller beads. (Restagno, F., Bacqnet, L., Biben, T., Charlain, E., *J.phys. condeno. matter*, 12, A419–A424, 2000). (b) Variation of the slopes of the previous graph with water vapor relative pressure. The dashed line is the theoretical prediction (see Section 7.3).

inclined until an avalanche started. The free surface profile after the avalanche was recorded and the depth of the slip plane with respect to the initial powder surface (Figure 7.3) was measured as a function of several parameters, such as the bed width and the particle size. It was observed that, as the average particle size was decreased from ~350 μm to 8.5 μm, the depth of the slip plane was strongly increasing. With the coarser particles, the avalanche affected only the surface layer of grains, whereas with the finer powders, slip occurred in depth.

All together, what these experiments demonstrate is the long-known importance of inter-particle cohesion in the flow and rupture behavior of powders [32] and, in particular, the importance of capillary forces. In an ideally noncohesive powder at rest (van der Waals and other attractive forces much smaller than the grain weight), stability is totally determined by friction. In 1773, Coulomb showed that the surface layer of grains in a noncohesive heap remains stable as long as the component T of its weight, parallel to the slope, does not exceed a fraction k of the component N perpendicular to the slope (Figure 7.7): $T \leq kN$. The parameter $k = \tan \varphi$ is the solid friction coefficient and φ is the internal friction angle. A simple geometrical argument shows that, for a surface layer, the maximum stability angle is simply $\theta_m = \varphi$ (Figure 7.7).

Coulomb's criterion may be generalized to determine the stability conditions of any noncohesive granular medium. According to this criterion, the material remains stable as long as it does not contain any plane for which $\tau \geq k\sigma$, τ and σ being the tangential (shear) and normal (compressive) stress to this plane, respectively. The relationship $\tau \geq k\sigma$ is a failure criterion. A noncohesive granular medium in which there would be no compressive stress would be unable to sustain even the smallest shear stress.

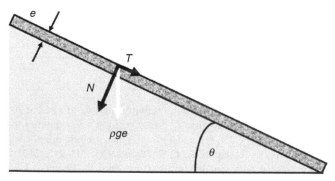

FIGURE 7.7 Equilibrium of a heap. A layer of thickness e and density ρ, making an angle α with the horizontal plane, will start avalanching when the parallel ($T = e\rho g \sin \alpha$) and perpendicular ($N = \rho g \cos \alpha$) components of its weight are related by $T = kN$, where k is the internal friction coefficient.

In a noncohesive heap, all stresses are due to the material weight and they increase linearly with the depth under the free surface. It can be shown by a classical Mohr–Coulomb analysis [33,34] that, in spite of this, the maximum stability angle is independent of the heap height and reads simply $\tan \theta_m = \tan \varphi = k$. The onset of flow is characterized by a limiting stress function, $\tau = k\sigma$, the yield locus (Figure 7.8). Stresses lower than this limit will cause negligible deformation, whereas in the limiting conditions there is a combination of shear and compressive stresses that will cause failure (an avalanche).

In the case of materials with inter-granular cohesive forces, it was proposed by Coulomb to consider friction and cohesion independently and to add them to each other. Assuming, in addition, that cohesive stresses are isotropic, the failure criterion becomes

$$\tau \geq k(\sigma + c) \quad \text{or} \quad \tau \geq k\sigma + C \tag{7.2}$$

A Mohr–Coulomb analysis shows that the stability conditions become cohesion-dependent. The yield locus writes $\tau = k\sigma + C$ (Figure 7.8) and the maximum stability angle for a heap is given by the following

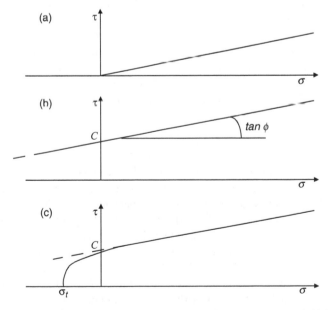

FIGURE 7.8 (a) Yield locus of a noncohesive granular medium. (b) Yield locus of a cohesive granular medium. (c) Yield locus of a cohesive fine powder.

expression:

$$\tan \theta_m = \tan \varphi \left(1 + \frac{C}{\rho g z \cos \theta_m} \right) \qquad (7.3)$$

ρ is the material density, g the acceleration of gravity, and z the depth under the heap surface. Thus, the maximum stability angle becomes a function of the heap height and failure occurs at depth. For a heap of finite height D, failure, if any, must necessarily happen at a depth smaller than D. Hence, the critical angle is decreasing with the heap height with a small heap sustaining steeper slopes than a large heap. For $z \rightarrow \infty$, the criterion for noncohesive media is recovered: $\tan \theta_m = k$ Cohesion is no longer involved. However, this is true only as long as C does not depend on σ, hence of z and of the heap size.

The previous expressions are generally well applicable to coarse and dense cohesive materials, such as wet sand, at high stresses. Fine cohesive powders contain highly porous agglomerates and are often much less dense. As a consequence, the yield locus shows a marked inwards curvature at negative σ values, corresponding to tensile stresses (Figure 7.8). This is due to the large pores, which are flaws that initiate fracture. As expected, this curvature decreases when the powder is consolidated under pressure. The crossing point of the yield locus and the negative σ axis is the tensile strength, σ_t, which is the maximum tensile stress that the powder can withstand. σ_t and C are of the same order of magnitude.

In light of the previous discussion, the four experiments that were summarized deliver a few simple messages. The first (experiment 4) is that, in fine dry powders, van der Waals cohesion is important for fine (~ 10 μm) powders, but no longer for coarse (~ 300 μm) powders. The attractive van der Waals force between two spherical particles of diameter d in contact is given by [35]

$$F_{vdW} = \frac{Ad}{24D^2} \qquad (7.4)$$

where A is the Hamaker constant and D the distance between the particles (center-to-center distance between the closest atoms). For two particles in contact, D is of the order of 0.3 nm. For perfectly smooth particles with a Hamaker constant of 5×10^{-20} J and a density of 10^3 kg/m^3, the van der Waals attraction would overcome the particle weight for particles smaller than 3 mm. This is not in agreement with everyday experience. The world would be very sticky if this were true! The missing element is the surface roughness, which, as pointed out above, limits the quality of contact. The diameter which has to be considered in Equation 7.4 is the typical diameter of the asperities, d_{asp}. With $d_{asp} \cong 0.2$ μm, which is the roughness size for the asperities of the xerographic toners used in experiment 4, the van der Waals attraction would overcome the particle weight for particles smaller than $\cong 100$ μm. This would also be the size below which cohesion is expected to interfere with the avalanche behavior, in agreement with the experiment.

The second message from the experiments is that a condensable vapor—water vapor in particular—may modify the flow behavior not only of fine powders, but also of coarse powders. This is no real surprise, knowing the capillary condensation phenomenon. What is a surprise though, are the low relative pressures at which the influence starts (Figure 7.6, bottom) and the very small amounts of condensate involved. In terms of IUPAC nomenclature, this would almost correspond to condensation in micropores or, at least, of nm size. Once more, this suggests that condensation in very narrow gaps between asperities is involved at the start. The other surprising fact is the extraordinary slow kinetics of condensation. This is true the first time the powder is put in a humid atmosphere (Figure 7.5), and also each time the powder bed is perturbed (Figure 7.6, top). In the later case, old contacts are "broken" and new contacts are formed, leading to a transfer of liquid bridge from one spot to another. All these thermodynamic and kinetic aspects of cohesion will be analyzed in terms of surface roughness and confinement in Section 7.3.

7.2.2 POLYMER NANOCOMPOSITES

The last decade has seen the development of new classes of nanocomposites in which the reinforcing filler particles are no longer complex particles, such as silica or carbon black aggregates, but geometrically simple particles with at least one dimension truly in the nanometer range and a large aspect ratio [36–39]. Among those nanoscopic filler particles are smectite clays and carbon nanotubes. As reported hereafter, dramatically different results have been obtained with each of those filler materials, which exemplify the effect of confinement and good or bad wetting.

7.2.2.1 Clay–Polymer Nanocomposites

Smectite clays are made of large (1 μm lateral dimension is common) 1 nm thick silicate sheets. Montmorillonite, an aluminum–silicate, is the most common mineral of this family. After adequate ion exchange with organophilic cations, smectite clays may be dispersed in organic monomers (which are later polymerized) or directly into molten polymers. When successful, this procedure leads to total delamination of the clay sheets into the polymer matrix. Thanks to the small thickness and the large lateral extension of the individual clay sheets, a huge filler–polymer interface area is generated, of the order of 800 m^2/g of clay. Simultaneously, with only a few percentages of

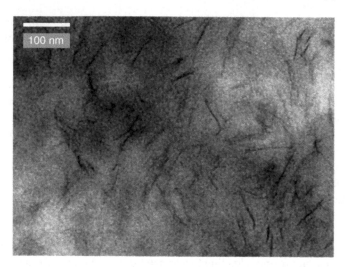

FIGURE 7.9 A successful example of total and homogeneous dispersion ("delamination") of a montmorillonite clay in a polymer matrix. The average distance between clay platelets is of the order of 50 nm.

delaminated clay in the polymer matrix, the average thickness of polymer between clay sheets drops rapidly—below a few tens of nanometers. Thus, a few percent of delaminated clay is enough to convert a bulk polymer into a totally interfacial or confined material (Figure 7.9).

A remarkably successful example is provided by nylon 6/montmorillonite nanocomposites [30]. With only 4 volume percent montmorillonite, a relative increase of the tensile modulus larger than 2.5 was obtained. What is even more remarkable is that the reinforcement so obtained is larger than the most optimistic predictions from effective medium theories for composites. In effective medium theories, the effective modulus of the composite is obtained from the general relation between an averaged stress and an averaged strain [40]

$$\langle \sigma \rangle = M_{\mathrm{eff}} \langle \varepsilon \rangle \tag{7.5}$$

In the simplest form of the theories, the averages are obtained by averaging over the volume fractions of each phase, ϕ_1 and ϕ_2

$$\langle \sigma \rangle = \phi_1 \sigma_1 + \phi_2 \sigma_2 \quad \text{and} \quad \langle \varepsilon \rangle = \phi_1 \varepsilon_1 + \phi_2 \varepsilon_2 \tag{7.6}$$

with $\phi_2 = 1 - \phi_2$ and a lower and an upper bound for the modulus are predicted by arranging the two phases in two extreme configurations. The lower bound is obtained by arranging the filler phase and the matrix phase in

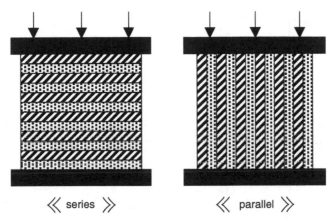

FIGURE 7.10 Sketch of the series and parallel configurations used to calculate the lower and upper bounds, respectively, of the compressive or tensile modulus of a two-phase composite.

series, whereas the upper bound is obtained by arranging them in parallel (Figure 7.10). In the series configuration, the matrix and the filler experience the same stress ($\sigma_1 = \sigma_2$). In the parallel configuration, they experience the same strain ($\varepsilon_1 = \varepsilon_2$). This leads to the following results:

$$\frac{1}{M_{\text{series}}} = \phi_1 \frac{1}{M_1} + (1 - \phi_1) \frac{1}{M_2} \tag{7.7}$$

$$M_{\text{parallel}} = \phi_1 M_1 + (1 - \phi_1) M_2 \tag{7.8}$$

It can be shown rigorously that, if the properties of each phase are not modified by the other phase, the properties of the composite are necessarily between those bounds:

$$M_{\text{series}} \leq M_{\text{eff}} \leq M_{\text{parallel}} \tag{7.9}$$

In the case considered here, with a continuous nylon 6 matrix and a discontinuous silicate filler phase, the effective modulus is expected to be close to the lower bound. The upper bound would correspond to a situation where the reinforcing filler would percolate through the matrix and directly support the load applied to the sample in the mechanical test. This is far from being the case here (Figure 7.9). In spite of this, as shown in Figure 7.11, the

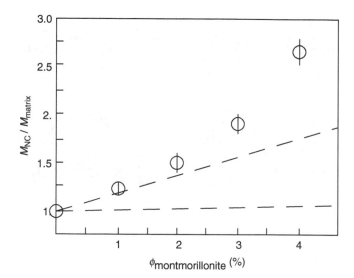

FIGURE 7.11 Tensile modulus of a delaminated nylon 6/montmorillonite nanocomposite (Figure 7.9): (○) experimental results from Ji, X. L., Jing, J. K., Jiang, W., and Jiang, B. Z., *Polym. Eng. Sci.*, 42, 983–993, 2002; the bottom line is the predicted lower (series) bound for the effective modulus whereas the upper line is the predicted upper (parallel) bound.

measured modulus is not only close to the upper bound but it is even *larger* than the upper bound.

What is shown by this result is that the composite cannot be considered as a simple mixture of two components, even in an optimized configuration. The properties of the nylon matrix have been deeply modified by the clay platelets. Thus, a three-phase model was developed by the authors, in which a gradient of modulus was introduced around the clay particles [30]. The polymer modulus was assumed to decrease continuously from the clay surface, reaching the bulk value at a distance of 7 nm. This is of the same order as half the average distance between clay platelets in the composite. Hence, it implies that most of the polymer is in confined situation, with different properties from those of the bulk. Good agreement with the experimental data was obtained using this model. It is not the aim of this chapter to discuss the choice of parameters used in this model. The important point, which will be discussed in Section 7.4, is the physical basis for the assumption of a gradient of modulus in confined geometry.

7.2.2.2 Carbon Nanotubes–Polymer Nanocomposites

The best known nanofibers are so far carbon nanotubes [41,42]. Basically, a carbon NT is a rolled graphene sheet. Carbon NTs exist in two forms: single wall tubes and multiple wall tubes (SWNTs and MWNTs, respectively). SWNTs tend to form aligned bundles in which the NTs are held together by weak van der Waals forces. Intrinsically, carbon SWNTs are the most attractive reinforcing filler particles, due to their exceptional mechanical properties, combining strength, and flexibility. Thanks to the sp^2 character of the carbon–carbon bonds in a graphene sheet, the intra-plane stiffness is remarkably high (C_{11} of the order of 1 TPa). The graphene sheet is also very flexible, thanks to the ability of the carbon atoms to change their hybridization when the sheet is bent. Hence, a SWNT is expected to be a remarkably strong and flexible nanofiber. Molecular dynamics computations predict a Young modulus for axial deformations larger than 1 TPa as soon as the NT diameter becomes larger than 1 nm (the bonds are weakened by the curvature in NTs with a smaller diameter). This rigidity and also the remarkable flexibility have been confirmed by direct measurements and observations. In addition, tensile strengths and strains of the order of 50 GPa and 10%, respectively, have been measured. All of this is way above the properties of the best classical carbon fibers.

In spite of these remarkable properties, reinforcement of polymer matrices by SWNTs has so far proven to be disappointing. In a recent study [31,41], an epoxy matrix was reinforced with SWNTs bundles, up to 35 weight percent, which corresponds to a volume fraction of 20% (Figure 7.12). At such a high loading, the increase of Young modulus was less than a factor of 2. This is 15 times less than what is expected from an effective medium calculation for a homogeneous and isotropic dispersion of short fibers with a high aspect ratio. The following relation should apply [43]

$$M = M_m \phi_m + \frac{1}{6} M_f \phi_f \qquad (7.10)$$

The subscript m and f applies to the matrix and filler, respectively. Note that this relationship corresponds to the upper bound described above (Equation 7.8), in which the factor 1/6 has been introduced to consider the disorder of fiber orientations. This relation applies well for polymers reinforced by short classical carbon fibers. The results with SWNT are well below this. Actually, they are scarcely larger than the lower bound (Equation 7.7; Figure 7.13).

A first possible reason for the (so far) disappointing impact of high loadings of SWNT on the modulus of the nanocomposite is the fact that, in spite of the excellent homogeneity of the dispersion (Figure 7.12), the NTs are

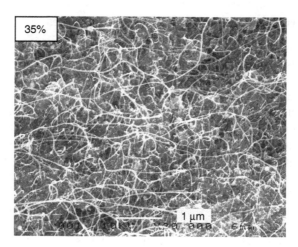

FIGURE 7.12 TEM micrograph of an SWNT-epoxy nanocomposite. (Adapted from Vaccarini, L., Goze, C., Bernier, P., and Rubio, A., to be published. Vaccarini, L., Thesis, Université de Montpellier II, France, 2000.)

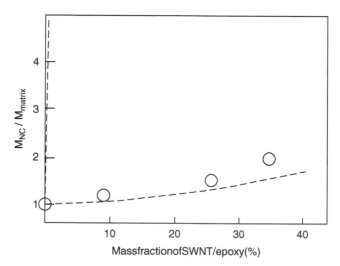

FIGURE 7.13 Young's modulus of an epoxy/carbon nanotube nanocomposite: ($\bigcirc$), experimental values from three point bending test. (From Vaccarini, L., Goze, C., Bernier, P., and Rubio, A., to be published. Vaccarini, L., Thesis, Université de Montpellier II, France, 2000); also shown are the upper (parallel, dashed curve) and lower (series, dotted curve) bounds predicted by effective medium theories.

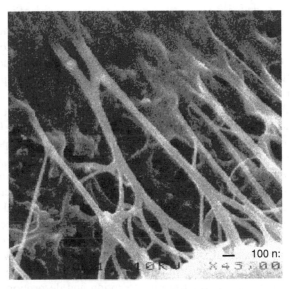

FIGURE 7.14 Electron micrograph of the fracture surface of an epoxy/carbon NT nanocomposite. (From Vaccarini, L., Goze, C., Bernier, P., and Rubio, A., to be published. Vaccarini, L., Thesis, Université de Montpellier II, France, 2000.) The nanofibers (NT bundles) are pulled out of the matrix.

not dispersed as individual NTs, but as bundles. Thus, as tensile load is applied to the sample, the NTs may slide with respect to each other within their bundle. A second possible reason is the poor quality of the epoxy–NT interface. NTs are known to have a very inert surface, like graphene sheets, with a very low out-of-plane surface energy (basically, only dispersion forces may be exchanged). As shown by the SEM micrograph of fracture surfaces (Figure 7.14), the NT bundles are pulled out but do not break, which means that no or little stress has been transmitted to the nanofibers. This is a clear sign of poor wetting of the NT surface by the epoxy polymer. Thus, in spite of the remarkable mechanical properties of carbon NTs, poor reinforcement is obtained, which may be assigned to the total absence of beneficial effect on the polymer in the interfacial or confined regions. A negative effect may even be considered! Why this is not unreasonable will be discussed in Section 7.4.

7.3 THE AGEING OF POWDERS AND THE VAPOR–LIQUID TRANSITION IN CONFINED GEOMETRY

When molecules are in contact with any porous material, they are submitted to two effects. The first is an attraction to the wall or, more specifically, to some sites on the wall. This leads to adsorption, strictly speaking. The second effect is simply that they are confined in a finite box, with well-defined boundary (curvature) conditions. It is usually considered that this leads to a bulk first order phase transition which is capillary condensation, i.e., liquefaction at a chemical potential which is lower than that of the saturating vapor, μ_{sat}.

It is not easy to separate clearly the effect of confinement from that of adsorption. A clear reason is that there are always some attractive interactions with the solid surface, at least the London dispersion forces. So far, the most detailed information on the effect of confinement in disordered porous media comes from molecular simulation studies, the only ones able to take into account the effect of morphological disorder of the walls. Analytical theories are unable to cope with morphological complexity and are restricted to simple shapes. Important phenomena have nevertheless been predicted [13,14] and some of them have been verified by experiment [44]. For instance, the critical temperature for the vapor–liquid transition in a slit-shaped pore or an infinite cylinder with no external surface is lower than in bulk. Below this critical capillary temperature T_{cc}, adsorption is characterized by a discontinuity, the capillary condensation. Furthermore, adsorption and desorption branches do not coincide so that a hysteresis loop is predicted. However, this hysteresis is not related to any pore blocking in constrictions or necks since the theory considers a single and morphologically simple pore. Nor is it related to variations in wetting conditions (contact angle). Actually, during adsorption, at pressures higher than the hysteresis closure point, the adsorbed film remains in a metastable state, being unable to overcome the free energy barrier for condensation. The system does not have fluctuations of sufficient amplitude to find the equilibrium state [16]. On the contrary, during evaporation, the fluid is always at equilibrium. At temperatures higher than T_{cc}, the discontinuity and the hysteresis loop disappear and the adsorption/desorption isotherm exhibit monotonic evolution. The stronger the confinement, the lower the critical temperature is (for a given adsorbate). Section 7.3.1 is devoted to a brief summary of the stability problem, following the approach of Evans [14]. It will be followed by an analysis of dynamical aspects and its application to ageing. Throughout, we will use a macroscopic approach, using concepts such as surface energies and curvature radii. Furthermore, we will neglect adsorption, i.e., evolution of the solid/vapor surface energy with vapor pressure or that of the solid/liquid/vapor surface energy with film thickness.

7.3.1 CAPILLARY CONDENSATION REVISITED

Let us consider a simple slit-shaped pore of width L between two smooth walls of area A (Figure 7.15). The pore is in equilibrium with a vapor at a pressure P_v and temperature T. The volume AL of the system is fixed. The system is open

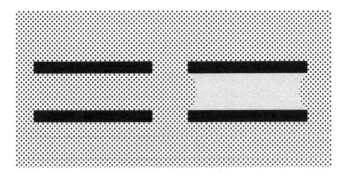

FIGURE 7.15 A slit-shaped pore with parallel walls embedded in a vapor reservoir at a given chemical potential, in the two situations considered here: full of vapor (left) or full of liquid (right).

and the chemical potential of the molecules in the pores is imposed by the pressure of an external reservoir which can exchange molecules with the pore. The thermodynamic potential, which is minimum at equilibrium in such conditions, is the grand potential, Ω

$$\Omega = F - G = U - TS - \sum \mu_i N_i \tag{7.11}$$

where F, G, U, and S are the free energy, free enthalpy, internal energy, and total entropy of the system (i.e., the pore), respectively. The variable N_i is the total number of molecules, vapor or liquid, of species i.

We consider a case where there is only one species present (water molecules). For a purely volumic system, $\Omega = -PV$. For a system with an interface, an interface energy term has to be added:

$$\Omega = \gamma A - PV \tag{7.12}$$

The question is whether or not there is a critical width, L_c, under which the pore might be filled with the liquid, at an equilibrium vapor pressure lower than $P_{V,sat}(T)$. It is known that for $L \to \infty$ (no confinement, i.e., two isolated walls), no condensation is possible in undersaturation conditions ($P_V < P_{V,sat}(T)$) A quasi-liquid adsorbed film may be present on each wall, but this film is very thin compared to the pore width. In order to know whether the situation is different or not in confined situations, i.e., in pores with a finite width, we have to compare the two extreme situations: the pore full of vapor on one hand and the pore full of liquid on the other hand (Figure 7.15).

$$\Omega_{wVw} = -P_V AL + 2\gamma_{wV}A \qquad (7.13)$$

$$\Omega_{wLw} = -P_L AL + 2\gamma_{wL}A \qquad (7.14)$$

where the subscript w is used for wall. It is assumed that the phases are homogeneous, i.e., that the density profile $\rho(z)$ is constant in the direction perpendicular to the walls. By difference, one obtains:

$$\Omega_{wLw} - \Omega_{wVw} = -(P_L - P_V)AL + 2(\gamma_{wL} - \gamma_{wV})A \qquad (7.15)$$

The condition for existence of a liquid phase crossing the pore is $\Omega_{wLw} - \Omega_{wVw} \leq 0$ and the condition for coexistence is $\Omega_{wLw} - \Omega_{wVw} = 0$.

Equation 7.15 may be put in a more useful form by using the Young-Dupré relationship for wetting (see the Introduction by J. Berg and Chapter 9 by M. Nardin in this volume). One gets, in coexistence conditions:

$$(P_L - P_V) = -\frac{2}{L}\gamma_{LV}\cos\theta \qquad (7.16)$$

which is nothing but Laplace relationship for the pressure jump a liquid–vapor interface with mean curvature $H = -\cos\theta/L$. Equation 7.16 may be further modified by expanding the pressures as functions of the chemical potential around μ_{sat}:

$$-\frac{2}{L}\gamma_{LV}\cos\theta = \int_{\mu_{sat}}^{\mu}\left(\frac{\partial P_L}{\partial\mu} - \frac{\partial P_V}{\partial\mu}\right)d\mu \qquad (7.17)$$

Since $(\partial\mu/\partial P)_{T,N,A} = \bar{V} = 1/\rho$, from the total differential of the free enthalpy ($\bar{V}$ is the molar volume $[\mathrm{m}^3\,\mathrm{mol}^{-1}]$ and ρ the bulk density $[\mathrm{mol}\,\mathrm{m}^{-3}]$), one obtains

$$\frac{2}{L}\gamma_{LV}\cos\theta = \Delta\mu\Delta\rho = P_V(\mu) - P_L(\mu) \qquad (7.18)$$

$\Delta\mu = \mu_{sat} - \mu$ is the (positive) undersaturation in chemical potential and $\Delta\rho = \rho_L - \rho_V$ is the difference between the bulk densities of the liquid and the vapor. Equation 7.18 shows that a liquid-filled pore will be the stable configuration of the system in undersaturation conditions ($\Delta\mu > 0$) as soon as the width of the pore will be smaller than a critical value:

$$L_c = \frac{2\gamma_{LV}\cos\theta}{\Delta\mu\Delta\rho} \qquad (7.19)$$

For Equation 7.19 to be satisfied in physically meaningful conditions ($L_c > 0$ for $\mu \leq \mu_{sat}$), good or at least partial wetting conditions have to be encountered, $\gamma_{wL} \leq \gamma_{wV}$, so that the wall/liquid interface is preferred over the wall/vapor interface and $\cos \theta \geq 0$. Those are the conditions for capillary condensation at $P_V < P_{V,sat}(T)$. In bad wetting conditions ($\gamma_{wV} < \gamma_{wL}$), no capillary condensation is possible. In order to fill the pore with liquid in those conditions, the chemical potential has to be higher than that of the saturating vapour pressure. This cannot be achieved by increasing the vapour pressure above $P_{V, sat}(T)$ because condensation would occur outside the pore at $P_V = P_{V, sat}(T)$. However, applying an excess pressure p to the liquid so obtained, with an inert gas for instance, will increase its chemical potential by $\bar{V}p$ and force the liquid to enter. This is the so-called forced intrusion process, applied in mercury porosimetry.

Assuming that the vapor is a perfect gas, $\Delta\mu$ may be replaced by $RT\ln(P_{V, sat}/P_V)$. As another approximation, $\Delta\rho$ may be replaced by $\rho_v L$, since the density of the vapor is much smaller than that of the liquid. This leads to:

$$\ln\frac{P_{V, sat}}{P_V} = \frac{2\gamma_{LV}\cos\theta}{RT\rho_L l_c} \tag{7.20}$$

or

$$\ln\frac{P_V}{P_{V, sat}} = -\frac{2\gamma_{LV}\bar{V}\cos\theta}{RTL_c} \tag{7.21}$$

which is Kelvin equation, relating the vapor pressure of a liquid filling a pore to the size of the pore and to the contact angle. It tells us that if we increase the pressure of the vapor in which the pore is embedded, the pore will be filled with liquid when the pressure reaches the value given by the equation. However, it does not tell us anything on the *kinetics* of this condensation.

7.3.2 NUCLEATION, METASTABILITY, AND HYSTERESIS

The typical signature of capillary condensation on adsorption isotherms is a stepwise increase of the amount of adsorbed molecules. A priori (Kelvin equation or, more generally, Equation 7.19 for instance), the process is expected to be reversible. In other words, one expects the step to occur at the same relative pressure during condensation and evaporation. It is well known that this is not the case. Emptying of the pores always occurs at lower pressure than filling of the pores. This hysteresis is classically assigned to a contact angle difference in adsorption and desorption and/or to geometrical porous network effects [7]. This cannot be ruled out and happens most probably in a number of situations. However, the thermodynamic approach outlined in the previous sections leads to the conclusion that there is another general reason for the widespread occurrence of a hysteresis in capillary condensation–evaporation, that is, the metastability of the non (or less) wetting phase (the vapor) during condensation.

Let us consider again the mechanism of condensation in slab geometry. Capillary condensation occurs because the wetting phase (the liquid) decreases the overall surface energy of the system. However, this decrease occurs only when the pore is *totally* filled with the liquid, i.e., when the energy-consuming liquid/vapour interfaces have disappeared. Several routes are possible to reach this final state. The simplest (in theoretical terms) route would be the formation of a wetting film on each wall and the growth of each film until they coalesce when their thickness becomes equal to $L_c/2$ (Figure 7.16). In fact, the energy balance of this process is extremely unfavorable. The final state is, of course, the state with the lowest grand potential, but before reaching this state the system has gone through intermediate states with a higher potential. Indeed, as long as the films do not merge, each of them develops a liquid/vapor interface which impedes the energy balance.

Neglecting specific molecule-surface interactions (i.e., neglecting adsorption or, in thermodynamic terms, the disjoining pressure[*]), the grand potential of a slit-shaped pore with a perfectly wetting ($\gamma_{wV} = \gamma_{wL} + \gamma_{LV}$) liquid film of thickness e on each side is given by

$$\Omega_{wLVLw} = -2P_L eA - (L-2e)AP_V + 2\gamma_{LV}A + 2\gamma_{wL}A \qquad (7.22)$$

The "excess" grand potential with respect to the pore filled with the vapour phase only is obtained by subtracting Equation 7.13 from Equation 7.22:

$$\Delta\Omega = \Omega_{wLVLw} - \Omega_{wVw} = 2eA(P_V - P_L) = 2eA\Delta\rho\Delta\mu \qquad (7.23)$$

Thus, $\Delta\Omega$ increases with e and drops to zero when condensation occurs (Figure 7.17). The maximum of this excess term is reached when $e = L_c/2$, right before condensation

$$\Delta\Omega_{max} = \Delta\rho\Delta\mu LA = v_p\Delta\rho\Delta\mu \qquad (7.24)$$

---

[*] The full expression, including the e-dependence of surface tensions, would write:
$$\Omega_{wLVLw} = -2P_L eA - (L-2e)AP_V + 2\gamma_{LV}A + 2\gamma_{wL}A + 2AW_{wLv}(e)$$
in which the last term takes adsorption into account, i.e., interactions of the films with the solid walls. The disjoining pressure is defined as:
$$\Pi_d = P_L - P_V = dW_{wLV}(e)/de.$$

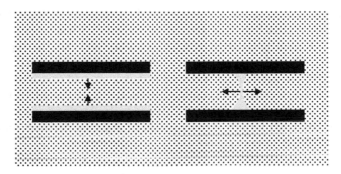

FIGURE 7.16 Two virtual mechanisms of capillary condensation in a slit-shaped pore. The first one is much more free energy costly than the second one.

$v_p = LA$ is the volume of the pore. $\Delta\Omega_{max}$ corresponds to the free energy barrier which has to be overcome for condensation to occur.[*]

The path just described is less favorable. As shown by Restagno et al. [24], a more realistic and less energy costly mechanism would involve the lateral growth of a bridge with a radius of curvature equal to Kelvin radius (Figure 7.16). It can be shown that, in this case,

$$\Delta\Omega_{max} = \Delta\rho\Delta\mu V_{nucleus} + \gamma_{LV}A_{LV} + \gamma_{wL}A_{wL} \qquad (7.25)$$

which depends on the volume of the liquid nucleus and on its free surface and on its interface area with the solid walls.

Whatever the mechanism, there is always—as long as the temperature is below the critical temperature—an energy barrier to be overcome for the formation of a liquid nucleus. As long as this barrier has not been passed, the vapor is in a metastable state. This means that pressures higher than Kelvin pressure have to be reached before condensation starts. Thus, in adsorption experiments, the condensation branch is always shifted towards higher pressures. Similarly, in experiments performed at constant pressure and decreasing temperature, there will be a shift towards lower temperatures. Hysteresis behavior has also been evidenced in recent experiments using a surface force apparatus, in which the separation between two surfaces is first decreased and then increased. A strong metastability of the vapor phase is observed for $L < L_c$ during approach.

---

[*] If adsorption is taken into account, Equation 7.24 becomes $\Delta\Omega max = \Delta\rho\Delta\mu(L - 2e_{eq})$ A where e_{eq} is the equilibrium thickness of the adsorbed layers at P_V.

It should be pointed out that nucleation does *not* occur during evaporation and in the desorption branch of adsorption isotherms. When the pore is full of liquid and Kelvin pressure is reached, nothing prevents the liquid from evaporating. Thus, the pressure, temperature, and critical distance of decondensation are the equilibrium pressure, temperature, and critical distance of the process. In agreement with common practice, it is the desorption branch of adsorption–desorption isotherms which has to be used for pore size determination (and the reason for this is independent of any considerations on contact angle, pore shape, or pore connectivity).

7.3.3 KINETIC ASPECTS AND THE AGEING OF POWDERS

The existence of a free energy barrier for nucleation of the liquid has direct kinetic consequences. Statistically, the probability for condensation to occur within a pore is now determined by Maxwell–Boltzmann statistics and the average waiting time before condensation obeys Arrhenius law

$$\tau = \tau_0 \exp\left(\frac{\Delta\Omega}{RT}\right) \qquad (7.26)$$

Surprisingly, this thermally activated character of capillary condensation has been neglected for a long time. It was only recently that it was pointed out [23].

Although the slab geometry is interesting for understanding the fundamentals of capillary condensation, it is too idealized to be a good model of

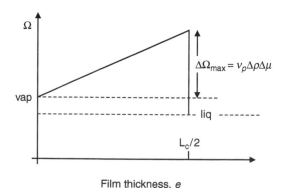

Film thickness, *e*

FIGURE 7.17 Evolution of the grand potential during the first virtual capillary condensation process of Figure 7.16, left in a slit-shaped pore of width L_c, at a vapor pressure given by the Kelvin equation, in a perfect wetting situation.

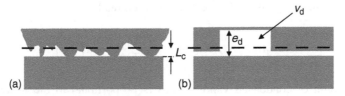

FIGURE 7.18 Model used by Restagno, F., Bocquet, L., Biben, T., and Charlaix, E., *J. Phys. Condens. Matter.*, 12, A419–A424, 2000 in their analysis of the thermally activated dependence of capillary condensation: (a) a rough surface facing a smooth surface, and (b) its model, with defects of excess volume v_d and area a_d separating the asperities and limiting the lateral growth of the liquid bridge.

moisture uptake and ageing in powders. A more realistic model would be a pore made of two rough surfaces facing each other, or more simply, a rough surface facing a smooth surface (Figure 7.18). This is the model taken by Restagno et al. [23]. It has the disorder required for modeling correctly the contact regions between grains in a powder, with a broad distribution of local "pore" sizes.

When macroscopic contact is established between the two surfaces, real contact is established only on a few very localized spots. In such regions, condensation is expected to occur quasi-instantaneously, leading to a set of wetted islands around the true contacts. Once these islands have formed, they should grow laterally up to the point where the distance between the surfaces is equal to L_c. However, in so doing, the meniscus encounters regions where the distance is larger than L_c. In order to tackle the problem analytically, the authors replaced the unfavorable regions by "defects" with an average gap e_d and an area a_d. The average time to overcome the defect is then given by Equation 7.26, with

$$\Delta \Omega = a_d (e_d - L_c) \Delta \rho \Delta \mu \tag{7.27}$$

$a_d(e_d - L_c) = v_d$ is the "excess" volume of the defect. After a time t, only the defects with activation time τ smaller than t have been overcome. Using Equation 7.26 and Equation 7.27, those defects may be identified as those for which their excess volume v_d obeys the following relationship

$$v_d < v_{d,max}(t) = RT(\Delta \rho \Delta \mu)^{-1} \ln(t/\tau_0) \tag{7.28}$$

The number of filled defects at time t, $N(t)$ is of the order of $N(t) = v_{d,\,max}(t)/v_0$, where v_0 is the width of the distribution of excess volumes. Once the liquid bridge has passed a defect, it wets the surrounding area until new defects are encountered. Assuming that this wetted area has a typical size δA_0, depending on the surface roughness, the total wetted area reads

$$A_{\text{wet}}(t) \cong N(t)\delta A_0 = \frac{\delta A_0}{[\Delta\mu/RT]\Delta\rho v_0}\ln\left(\frac{t}{\tau_0}\right)$$ (7.29)

The amount of condensed vapor, which is proportional to the wetted area, follows the same evolution. The adhesion force, which is proportional to the Laplace pressure jump across the meniscus and to the wetted area, follows also the same time dependence. Ultimately, the same trend is also expected for the cohesion C of a packing of grains, provided each contact can be modeled by the type of configuration shown in Figure 7.18. Using Equation 7.3, the predicted time dependence of the maximum stability angle θ_m is in good agreement with experiment (Figure 7.6b).

7.4 POLYMER REINFORCEMENT AND THE LIQUID–GLASS TRANSITION IN CONFINED GEOMETRY

In Section 7.2.2, we have seen that in nanocomposites, depending on the nature of the "walls" (the filler particles) confining the polymer, dramatically different results may be obtained. In the case of a good polymer–solid interface (montmorillonite/nylon 6), a remarkable improvement of the polymer modulus was obtained over a significant distance from the solid surface. On the contrary, in the case of poorly wettable solid (carbon NT/epoxy), the polymer properties were apparently depressed. This is what we will try to put (qualitatively) on a thermodynamic basis now. We will restrict the discussion on the situation where the thickness of the regions in which the polymer is confined (the average filler–filler distance) is relatively homogeneous and larger than the radius of gyration of the polymer coil. This is the case in the two examples considered above (Figure 7.9 and Figure 7.12). It is not the case in intercalated materials, where the thickness of the polymer layers may be so small that the conformation of the macromolecules has no relationship anymore with that of a macromolecule in the bulk polymer.

7.4.1 THE GLASS TRANSITION IN THIN POLYMER FILMS: EXPERIMENTAL RESULTS

Why should the mechanical properties of a confined polymer be different from those of a bulk polymer? Direct and local measurements of profiles of mechanical properties in composites and nanocomposites at the (sub)nanometer scale are still in their infancy, but intense investigations have been made in recent years on the properties of thin polymer films. The measurements were focused on the glass transition temperature, T_g. As discussed elsewhere in this volume (Chapter 9 by H. Haidar), an increase of T_g has direct and obvious consequences on the mechanical properties of rubber. The consequences are less important in the case of amorphous polymers which are already in the glassy state.

In thin polymer films, it has now been verified by a number of different techniques, including ellipsometry, x-ray reflectivity, Brillouin light scattering, and positron annihilation spectroscopy, that the glass transition temperature may be significantly modified with respect to that of the bulk material [26,27]. In the case of freely-standing (or "self-supporting") films, a large reduction of T_g is observed by as much as 80 K. Remarkably, the film thickness over which these effects are observed is well beyond the nanometer scale and may be as large as 100 nm. In polymer films supported on a solid substrate (which is the case of direct interest for our purposes), the effect may be either a decrease or an increase of T_g, depending on the interaction with the substrate. In cases where there are strong specific interactions between the polymer and the substrate, that is, in good wetting conditions such as those encountered with poly(2-vinyl pyridine) [45] or PMMA [46,47] on silica for instance, the glass transition temperature increases by as much as several tens of K with decreasing film thickness. On the contrary, in bad wetting conditions, such as polystyrene on a variety of substrates [26,27], the trend is opposite and of equivalent magnitude. A particularly illustrative case is that of PMMA, a relatively polar polymer, on SiO_x surfaces (oxidized silicon wafers) and on the same surfaces after silanization with HDMS [28]. On the polar surface, T_g increased by up to 7°C above the bulk value for 18 nm thick films, whereas on the nonpolar silanized surface, T_g decreased by 10°C below the bulk value for films 21 nm thick (Figure 7.19).

The general interpretation of these changes is that the mobility of the polymer chains is either enhanced—close to a free surface or in bad wetting situations, or depressed—close to a strongly adsorbing or in good wetting situations. Roughly speaking, the first case corresponds to a situation where the solid support is less polarizable than the polymer and the second to a situation where it is more polarizable than the polymer [48]. The theoretical explanation for these effects is still the matter of some debate. In general, explanations fall in either one of two categories: finite size effects on one hand and surface or interface effects on the other hand. Before discussing them, it may be useful to be reminded of some basic aspects of the glass transition in terms of free volume.

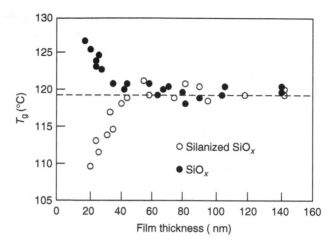

FIGURE 7.19 Evolution of the glass transition temperature in thin PMMA films as a function of film thickness. On a substrate interacting weakly with the polymer (silanized SiO_x surface), T_g decreases when the film becomes thinner. On a strongly interacting substrate (SiO_x surface), T_g increases. (Adapted from Fryer, D. S., Nealey, P. F., and de Pablo, J. J., *Macromolecules*, 33, 6439–6447, 2000.)

7.4.2 THE FREE VOLUME PICTURE OF THE GLASS TRANSITION: A REMINDER

At high temperature (say, at a temperature higher than the melting temperature of the crystalline phase, if it exists), an amorphous polymer is but a (very) viscous liquid in which the segmental mobility is high and where the whole polymer chains are undergoing large reptation motions. The temperature dependence of the viscosity follows generally a classical Arrhenius law

$$\eta = \eta_0 \exp\left(\frac{\Delta G_a}{RT}\right) \tag{7.30}$$

ΔG_a is an activation Gibbs free energy. At $T \cong 0{,}7$ or $0{,}8 \ T_m$, the melting temperature, the viscosity enters into a regime where it increases faster than what is predicted from the Arrhenius law and where the melt exhibits increasingly clear viscoelastic properties. The viscosity behavior is generally correctly described by the empirical Vogel–Fulcher–Tammann (VFT) law:

$$\eta = \eta_0 \exp\left(\frac{Cst}{T - T_0}\right) \tag{7.31}$$

T_0 is a temperature at which the viscosity would diverge. Simultaneously, elastic properties are observed at increasingly low frequencies.

In fact, the system does not remain fluid until T_0. Repetition and large segment movements freeze in before T_0 is reached and the material enters into the glassy state at a temperature $T_g > T_0$. It behaves as a hard, elastic, and brittle material. The shear modulus is typically in the MPa range above T_g and in the GPa range below T_g.

A simple model of the glass transition is the free volume model first proposed by Cohen, Turnbull, and Crest [49–51]. It relies on the assumption that in polymer melts, there is a void space called free volume, v_f, which may be redistributed without changing the energy of the system. This redistribution allows for the generation of local voids large enough for a chain segment to accomplish a local motion of amplitude comparable to its own size. This confers a fluid character to the system. The important point is that the free volume is strongly temperature-dependent and vanishes at some finite temperature. Hence, there is a temperature at which redistribution even of the total free volume of the system is no longer able to generate locally the void space for a significant segment movement. This is the glass temperature.

The concept of free volume is a rather subtle concept. Each polymer segment is considered to be in a cage, the wall of which being the neighboring segments from the same chain or from another chain. The interaction potential of a segment in its cage is of the Lennard–Jones type, with an inflexion point at some distance. This inflexion point separates a short range domain (compact cage) from a longer range domain (expanded cage). Let us consider two neighboring cages trying to exchange void volume. This may be achieved by moving the separating wall in such a way that the volume lost by one cage is gained by the other. If the two cages are initially in the compact configuration, some energy will have to be injected into the system because the energy recovered by contracting one cage is less than the energy required to expand the other. On the contrary, if the two cages are in the expanded state, the void rearrangement may be achieved without net injection of energy and even with a gain of energy. Cohen and Turnbull called *free volume* that part of the void volume which may be redistributed without energy cost. Since the void volume increases with temperature with a coefficient α_{glass} and α_{liq} in the glassy and liquid state, respectively, there is a temperature T_0 such that, for $T \leq T_0$, the void volume will no longer be free volume but merely expanded solid volume whereas for $T \geq T_0$, a fraction of the void volume is really free volume. One has

$$v_f = v_0(\alpha_{liq} - \alpha_{verre})(T - T_0) \qquad (7.32)$$

v_0 is the minimum free volume of a cage.

At temperatures higher than T_0, the width of the free volume distribution increases rapidly in parallel with the configuration entropy since this does not require any energy. The most probable free volume distribution (the distribution which maximizes the number of possible configurations) may be computed by classical statistical techniques, leading to

$$p(v) = \left(\frac{\gamma}{v_f}\right) \exp\left(\frac{\gamma v}{v_f}\right) \tag{7.33}$$

$p(v)$ is the probability for having a free volume cage between v and $v + dv$, and v_f is the average free volume per cage. Assuming that a minimum free volume, v^*, is necessary for a diffusive motion to occur, the following expression is obtained by integration of Equation 7.33 for the diffusion coefficient and for the fluidity (reciprocal viscosity):

$$D \text{ or } \phi = \frac{1}{\eta} \propto \exp\left(-\frac{\gamma v^*}{v_f}\right) \tag{7.34}$$

Considering the temperature dependence of the average free volume (Equation 7.32), Equation 7.34 is in agreement with the empirical VFT equation.

Although the free volume theory does not explain everything, it captures the essential features of the glass transition. The key point which has been used in the models proposed so far to explain the shift of T_g in thin polymer films is the density dependence of relaxation processes. In fact, one may equally well speak of a glass transition density instead of a glass transition temperature.

7.4.3 THE GLASS TRANSITION IN THIN POLYMER FILMS: MODELS

The first explanation which comes to mind to explain the decrease of T_g in freely-standing films is that their average density is lower than that of the bulk polymer due to a finite size effect [52]. Indeed, starting from simple free volume calculations, it may be predicted that a decrease in the room temperature density of only $\sim 1\%$ should lead to an increase in free volume large enough to depress T_g by as much as 40 K in polystyrene, for instance. However, no density differences have been reported so far (but one should admit that density differences so small are extremely difficult to measure) [26,27]. In addition, calculation of the loss of cohesive energy due to the finite thickness of a film in which the attractive forces are van der Waals forces shows that the predicted loss is far not enough to lead to the *permanent* density decrease required to explain the observed decrease of T_g [53].

Another explanation has been proposed in which it is considered that the free interface introduce new degrees of freedom related to the sliding motion of each chain along its own path [54,55]. In the bulk, this sliding is blocked at the end points, but in a thin film it would be allowed for chain arcs touching the surface. However, this explanation only applies to the case of long polymer chains with radius of gyration larger than the film thickness. Thus, it cannot be extrapolated to the cases of polymer/filler nanocomposites in which we are interested.

The most attractive model so far has been proposed by Long and Lequeux [53]. Actually, the authors proposed two models which apply to weak and to strong interfaces, respectively.

7.4.3.1 The Weak Interface Case (Bad Wetting)

The model is based on the idea that in thin freely-standing films and in thin films weakly bound to a support, the small size of the system allows for large density *fluctuations*. As long as the amplitude of these fluctuations is large enough to *temporarily* bring the density of the film below the density of the glass, they would prevent the film from being glassy. The predicted decrease of T_g is of the right order of magnitude compared to the observed values. The rationale is as follows.

Any elastic body of bulk modulus K and volume V undergoes volume fluctuations of amplitude $\Delta V / V$, such that

$$\left(\frac{\Delta V}{V}\right)^2 = \left(\frac{\delta\rho}{\rho}\right)^2 = \frac{T}{KV} \qquad (7.35)$$

In a freely-standing film, each half of the film of thickness $D = h/2$ fluctuates independently of the other and undergoes thickness fluctuations of amplitude

$$\frac{\Delta D}{D} = \left(\frac{T}{KD^3}\right)^{1/2} \qquad (7.36)$$

Let us consider the influence of these fluctuations on the glass transition temperature. At temperatures higher than T_g, the density is lower than ρ_g. This density increases when the temperature is decreased and, at temperatures below T_g, it becomes equal to ρ_{bulk}, larger than ρ_g. However, it fluctuates between two typical values which are $\rho^+ = \rho_{vol} + \delta\rho$ and $\rho = {}^-\rho_{vol} - \delta\rho$.

In particular, the lower limit is

$$\rho^- = \rho_{\text{vol}} - \left(\frac{8T}{Kh^3}\right)^{1/2} \qquad (7.37)$$

As long as these fluctuations bring the film temporarily in a state of density lower than ρ_{g}, the film cannot be glassy. To be glassy, the lower value of the fluctuating density must remain higher than the bulk glass density. One may define a new glass transition temperature, $T'_{\text{g}} \le T_{\text{g}}$, such that

$$\rho_{\text{vol}}(T'_{\text{g}}) - \delta\rho = \rho_{\text{g}} \qquad (7.38)$$

The relative change of glass transition temperature as compared to the bulk is given by

$$\frac{\Delta T_{\text{g}}}{T_{\text{g}}} \propto \frac{-1}{\alpha_{\text{T}}} \frac{\delta\rho}{\rho_{\text{vol}}} = \frac{-1}{\alpha_{\text{T}}} \left(\frac{8T}{Kh^3}\right)^{1/2} \qquad (7.39)$$

The parameters in this relationship (thermal expansion coefficient α_{T}, modulus K) are those of the bulk polymer. Equation 7.39 predicts $\Delta T_{\text{g}}/T_{\text{g}} \propto h^{-3/2}$ in reasonable agreement with experience.

7.4.3.2 The Strong Interface Case (Good Wetting)

The previous model is unable to explain the *increase* of T_{g}, which is observed in films in strong interaction with a support. A different approach has to be used. A possible clue, followed by Long and Lequeux, is that a strongly adsorbed polymer film with a thickness smaller than the radius of gyration of the chain behaves as a nonhomogeneous end-grafted polymer brush. In such layers, the chains have stretched configuration which, due to entropic elasticity, tend to compress the layer. This leads to a density higher than that of the bulk. The higher the grafting density, Σ, the stronger the effect is. The variable Σ is the number of monomer segments which are connected to the surface per unit surface area. This leads to an extra contribution δK to the bulk modulus, which reads

$$\delta K = 3T\Sigma^2 a \qquad (7.40)$$

a is the size of a monomer segment. In the very strong adsorption limit, it may be assumed that at the substrate/polymer interface, every monomer segment in contact with the surface is adsorbed, i.e., $\Sigma \cong a^{-2}$. However, due to loops, this density decreases as one goes away from the interface. It may be shown that $\Sigma(z) \propto a/z$. The average extra modulus calculated over the total thickness h of the adsorbed layer is then

$$\delta K = \frac{3k_{\mathrm{B}}T}{a^2 h} \tag{7.41}$$

This leads to a new equilibrium density, $\rho = \rho_{\mathrm{bulk}} + \delta\rho$, such that

$$\frac{\delta\rho}{\rho_{\mathrm{vol}}} = \frac{\delta K}{K} = \frac{3k_{\mathrm{B}}T}{a^2 h K} \tag{7.42}$$

The corresponding change in glass transition temperature reads

$$\frac{\Delta T_{\mathrm{g}}}{T_{\mathrm{g}}} = \frac{1}{\alpha_{\mathrm{T}}}\frac{\delta K}{K} = \frac{1}{\alpha_{\mathrm{T}}}\frac{3k_{\mathrm{B}}T}{a^2 h K} \tag{7.43}$$

With typical numerical values for the parameters, this equation predicts an increase of T_{g} of the order of 0.1 T_{g}, i.e., several tens of Kelvins for a 10 nm thick strongly bound film of a polymer with a bulk T_{g} around 400 K (PMMA for instance). However, the predicted increase of modulus is relatively modest, of the order of 10^7 Pa.

7.4.4 Back to Nanocomposites and Confined Polymer Layers

Whatever the theoretical explanation for the meso- to long range modifications of molecular mobility in thin films, indirectly revealed by a ΔT_{g}, they have far reaching consequences for our understanding of the properties of confined polymer layers in nanocomposites. Let us come back to the two examples of nanocomposites described in Section 7.2.2. The montmorillonite/nylon 6 composite is clearly a case of good wetting and strong interface, in which the modulus of the matrix is improved in an interfacial zone extending over distances of the order of 10 nm. The link between reinforcement and the strength of the polymer–surface interaction has been directly evidence by ^{15}N NMR [56] (Figure 7.20). The larger the chemical shift, the larger the modulus is. With an ultimate interface area of the order of 800 m^2/g, a flexible plate-like morphology, a huge aspect ratio, and a tunable surface chemistry, smectite clays are particularly able to modify deeply the properties of the polymer matrix in which they are embedded, due to a combination of interfacial and confinement effects. Simple geometrical arguments show that with interfacial regions 50 nm wide, a volume fraction of 1% of delaminated clay in the nanocomposite would be enough to convert the total mass of polymer into interfacial polymer.

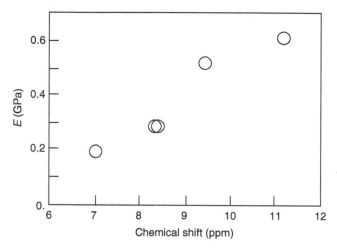

FIGURE 7.20 Relation between the tensile Young modulus of a nylon/montmorillonite nanocomposite and the intensity of the polymer–clay interaction, as revealed by the ^{15}N NMR chemical shift of the polymer molecules. (Adapted from Usuki, A., Koiwai, A., Kojima, Y., Kawasumi, M., Okada, A., Kurauchi, T., and Kamigaito, O., *J. Appl. Polym. Sci.*, 55, 119–124, 1995.)

On the other hand, the carbon nanotube/epoxy composite is most probably a case of bad wetting. This is not a surprise, considering the very low surface energy of graphite. The lesson is clear: using a nanomaterial as reinforcing filler is not a guarantee for improved properties. It all depends on wetting.

Not everything is clear, though. The previous argument is a good clue to explain the reinforcement of rubbery materials because there is a strong increase of modulus when the glass transition is crossed from above. However, it is much more questionable in the case of amorphous polymer matrices which are *already* in the glassy state. Shifting the glass transition temperature upwards in this type of material leads but to a minor increase of modulus at temperatures well below T_g (room temperature). The reason why, in spite of that, remarkable reinforcements have been observed remains to be understood.

7.5 CONCLUSION

In this chapter, it was shown how confinement (the presence of solid walls), wetting conditions (the interactions with the walls), and disorder (roughness) may modify the thermodynamics and the kinetics of the vapor–liquid and the liquid–glass transition. These are old problems, but it was only recently that a consistent framework involving all aspects was developed. This sheds new light on the two topics covered in this chapter: the ageing of powders in moist environments and their action on the mechanical properties of polymer matrices. There is no doubt that many more examples could be found in living matter, where the ubiquitous presence of flexible interfaces makes confinement and disorder not an exception, but a rule.

REFERENCES

1. Avnir, D., Farin, D., and Pfeifer, P., *Nature*, 308, 261–264, 1984.
2. Farin, D. and Avnir, D., In *The Fractal Approach to Heterogeneous Chemistry*, Avnir, D., Ed., Wiley, Chichester, 271–293, 1989.
3. Van Damme, H., Levitz, P., Gatineau, L., Alcover, J. F., and Fripiat, J. J., *J. Colloid Interface Sci.*, 122, 1–8, 1988.
4. Van Damme, H., In *Adsorption on Silica Surfaces*, Papirer, E., Ed., Marcel Dekker, New York, 119–166, 2000.
5. Thomson (Lord Kelvin), W., *Phil. Mag.*, 42, 448–452, 1871.
6. Zsigmondy, R., *Z. Anorg. Allgem. Chem.*, 71, 356–377, 1911.
7. Defay, R. and Progogine, I., *Tension Superficielle et Adsorption*, Desoer, Liège, 1951.
8. Hornbaker, D. J., Albert, R., Albert, I., Barabasi, A.-L., and Schiffer, P., *Nature*, 387, 765, 1997.
9. Albert, R., Albert, I., Hornbaker, D., Schiffer, P., and Barabasi, A.-L., *Phys. Rev.*, E56, R6271–R6274, 1997.
10. Halsey, T. C. and Levine, A. J., *Phys. Rev. Lett.*, 80, 3141–3144, 1998.
11. Tegzes, P., Albert, R., Paskvan, M., Barabasi, A.-L., Vicsek, T., and Schiffer, P., *Phys. Rev.*, E60, 5823–5826, 1999.
12. Valverde, J. M., Castellanos, A., and Ramos, A., *Phys. Rev.*, E62, 6851–6860, 2000.
13. Ball, P. C. and Evans, R., *Langmuir*, 5, 714–723, 1989.
14. Evans, R., *J. Phys. Condens. Matter*, 2, 8989–9007, 1990.
15. Page, K. S. and Monson, P. A., *Phys. Rev. E.*, 54, 6557–6564, 1996.
16. Celestini, F., *Phy. Lett.*, A28, 84–87, 1997.
17. Pellenq, R. J.-M. and Denoyel, R. P. O., In *Fundamentals of Adsorption 7*, Kaneko, K., Kanoh, H., and Hanzawa, Y., Eds., IK International Publisher, New Delhi, 352–359, 2000.
18. Gelb, L. D., Gubbins, K. E., Radhakrishnan, R., and Sliwinska-Bartkowiak, M., *Rep. Prog. Phys.*, 62, 1573–1659, 1999.
19. Pellenq, R. J.-M. and Levitz, P. E., *Mol. Phy.*, 100, 2059–2077, 2002.
20. Bocquet, L., Charlaix, E., Ciliberto, S., and Crassous, J., *Nature*, 396, 735–739, 1998.
21. Fraysse, N., Thomé, H., and Petit, L., *Eur. Phys. J.*, B11, 615–619, 1999.
22. Crassous, J., Bocquet, L., Ciliberto, S., and Laroche, C., *Europhys. Lett.*, 47, 562–565, 1999.

23. Restagno, F., Bocquet, L., Biben, T., and Charlaix, E., *J. Phys. Condens. Matter.*, 12, A419–A424, 2000.
24. Restagno, F., Bocquet, L., and Biben, T., *Phys. Rev. Lett.*, 84, 2433–2436, 2000.
25. Bocquet, L., Charlaix, E., and Restagno, F., *C. R. Physique*, 3, 207–215, 2002.
26. Forrest, J. A. and Jones, R. A. L., In *Polymer Surfaces, Interfaces, and Thin Films*, Karim, A. and Kumar, S., Eds., World Scientific, Singapore, 251–294, 2000.
27. Forrest, J. A. and Dalnoki-Veress, K., *Adv. Colloid Interface Sci.*, 94, 167–196, 2001.
28. Fryer, D. S., Nealey, P. F., and de Pablo, J. J., *Macromolecules*, 33, 6439–6447, 2000.
29. Bansal, A., Yang, H., Li, C., Cho, K., Beniceuriz, B.C., and Kumar, S.K., Schadler, L.S., *Nature Mater.*, 4, 693–698, 2004.
30. Ji, X. L., Jing, J. K., Jiang, W., and Jiang, B. Z., *Polym. Eng. Sci.*, 42, 983–993, 2002.
31. Vaccarini, L., Goze, C., Bernier, P., and Rubio, A., to be published.Vaccarini, L., Thesis, Université de Montpellier II, France, 2000.
32. Rietema, K., *The Dynamics of Fine Powders*, Elsevier, London, 1991.
33. Schofield, A. N. and Wroth, C. P., *Critical State Soil Mechanics*, MacGraw-Hill, London, 1968.
34. Nederman, R. M., *Statics and Kinematics of Granular Materials*, Cambridge University Press, New York, 1992.
35. Israelachvili, J, *Intermolecular and Surface Forces*, 2nd ed., Academic Press, London, 176–212, 1992.
36. Giannelis, E. P., Krishnamoorti, R., and Manias, E., *Adv. Polym. Sci.*, 18, 107–147, 1999.
37. Alexandre, M. and Dubois, P., *Mater. Sci. Eng.*, 28, 1–63, 2000.
38. Biswas, M. and Ray, S. S., *Adv. Polym. Sci.*, 155, 167–221, 2001.
39. Ajayan, P. M., Schadler, L. S., and Braun, P. V., *Nanocomposite Science & Technology*, Wiley-VCH, Weinheim, 2003.
40. Beran, M. J. and McCoy, J. J., *Int. J. Solid Struct.*, 6, 1033–1054, 1970.
41. Vaccarini, L., and Bernier, P., In *Nanomatériaiux* O.F.T.A-TEC&DOC, Paris, 221–252 (2001).
42. Smith, B. W. and Luzzi, D. E., In *Introduction to Nanoscale Science and Technology*, Di Ventura, M., Evoy, S., and Heflin, J. R. Jr., Eds., Kluwer, Boston, MA, 137–182, 2004.
43. Cox, H. L., *British J. Appl. Phys.*, 3, 72–81, 1952.
44. Morishige, K., Fujii, H., Uga, M., and Kinakawa, D., *Langmuir*, 13, 3494–3498, 1997.
45. van Zanten, J. H., Wallace, W. E., and Wu, W., *Phys. Rev. E*, 53, R2053–R205X, 1996.
46. Keddie, J. L., Jones, R. A. L., and Cory, R. A., *Faraday Discuss*, 98, 219–230, 1994.
47. Grohens, Y., Brogly, M., Labbe, C., David, M. O., and Schultz, J., *Langmuir*, 14, 2929, 1998.

48. de Gennes, P. G., *Rev. Mod. Phys.*, 57, 827–863, 1985.
49. Cohen, M. H. and Turnbull, D., *J. Chem. Phys.*, 31, 1164–1169, 1959.
50. Cohen, M. H. and Turnbull, D., *J. Chem. Phys.*, 34, 120–125, 1961.
51. Cohen, M. H. and Grest, G. S., *Phys. Rev.*, B20, 1077–1098, 1979.
52. Reiter, G., *Europhys. Lett.*, 23, 579–583, 1993.
53. Long, D. and Lequeux, F., *Eur. Phys. J.*, E4, 371–387, 2001.
54. de Gennes, P. G., *Eur. Phys. J.*, E2, 201–205, 2000.
55. de Gennes, P. G., *C. R. Acad. Sci. Paris Ser. IV*, 1179–1186, 2000.
56. Usuki, A., Koiwai, A., Kojima, Y., Kawasumi, M., Okada, A., Kurauchi, T., and Kamigaito, O., *J. Appl. Polym. Sci.*, 55, 119, 1995.

8 Fiber Surfaces in Textile Industry: Application for the Characterization of Wear or Comfort Properties of Modern Fabrics

Marc Renner and Marie-Ange Bueno

CONTENTS

8.1 FIBERS FOR TEXTILE INDUSTRY AND THEIR SURFACE PROPERTIES

Textile industry is based on the use of fibers—typical materials whose length is very important in comparison to their thickness. The morphological

325

property is the aim of the mechanical behavior of fibers and their ability to be transformed in supple and soft structures and products. For more than 5000 years, fibers have been manufactured by humans into strings, ropes, yarns, and fabrics. Until the end of the nineteenth century, only natural fibers have been used. We can easily classify these natural fibers in following categories:

- Vegetal seminal fibers like cotton are fine, quite strong, and easy to grow. Cotton by itself represents nearly half of the world fiber production. Nevertheless, the production of this low cost and very popular textile fiber is limited by the unsustainable way of growing, due to the high demand of water and the need of pesticides to fight against flies and aphids which bring on sticky fibers. The cotton fiber (Figure 8.1) displays usually a smooth surface and is well known for its soft touch. Chemical treatments like sodium hydroxide under mechanical tension improve the brightness of the cotton fibrous structure.
- Other vegetal fibers, like flax (Figure 8.2), hemp, or sisal are usually coarser than cotton. Their mechanical properties are quite interesting, and these types of fibers can be taken in count for technical applications.
- Animal hairs show scales on their surface. This particular morphology induces interesting physical properties of the fibrous structure like high thermal insulation properties. In the case of a contact between fibers, the scales can interact as an irreversible

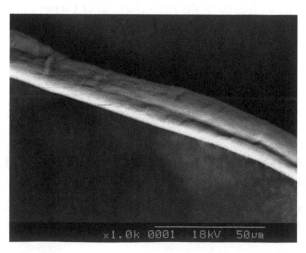

FIGURE 8.1 Cotton fiber.

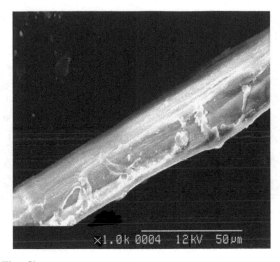

FIGURE 8.2 Flax fiber.

click and ratchet mechanism cold felting. This effect is well known in the area of wool fibers (Figure 8.3) and products.
- Silk is produced by some insects (e.g., worms, spiders, etc.) to built shells, traps, or draglines. The mechanical properties of these very long fibers, called "filaments," are fascinating and encourage humans to imitate them.

FIGURE 8.3 Wool fiber.

FIGURE 8.4 Viscose fiber.

Thanks to the works of chemists and physicists like Thomas Edison and De Chardonnet. Artificial fibers were developed by cellulose regeneration at the end of the nineteenth century. The way to man-made fibers was open and the challenge was fineness and strength. Today, artificial fibers are still used and helpful. The possibility to produce cellulose fibers with regeneration of the solvent, i.e., without major pollution, is a new trend. The well-known viscose fiber (Figure 8.4) shows a specific multilobal morphology due to its coagulation during the wet spinning process. Other natural polymers like chitin can also be spun into fibrous structures.

Based on petroleum transformation, synthetic fibers became very popular during and after the Second World War. At first, polyamide (Figure 8.5) and later polyester put into light typical physical properties. The amorphous

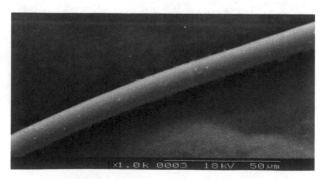

FIGURE 8.5 Polyamide fiber.

and crystalline aspects of the fibrous structure are the result of the conjunction of the basic properties of the polymeric material and the drawing effect during the spinning of the fiber.

8.2 FIBERS AND FIBROUS STRUCTURES

8.2.1 APPROACH AT DIFFERENT SCALES

There are different ways to obtain a fibrous structure with staple fibers. The first one consists of manufacturing an intermediate linear structure, which is called yarn, in order to transform it into a surface by using processes such as weaving, knitting, or braiding. In a second possible way, individual fibers can be directly assembled into a fiber web bound together as a so-called "nonwoven" structure. Mechanical aspects and surface properties of the fibrous structures are strongly related to the type of the fiber assembling at different scales. We consider 3 typical scales of fibrous structures:

- The microscopic scale: between 0.1 and 10 μ. This scale gives a good idea of the morphology of individual fibers and of the fiber surface.
- The mesoscopic scale: between 10 μ and 1 mm. This scale illustrates typically the intermediate fibrous structures—such as yarns or nonwovens—bundling effects.
- The macroscopic scale: between 1 and 100 mm, visible to the naked eye, the fibrous product shows its structural morphology.

Figure 8.6 illustrates the 3 scales in the case of cotton fibers at 3 stages—individual fiber, single yarn, and "single jersey" knitted fabric (classical T-shirt fabric).

For a better comprehension of some physical properties of the fibers, an investigation into this so-called nanoscopic scale (between 100 nm and 10 μ) can also be required.

Microscopic scale 0.1–10μ   Mesoscopic scale10–1000μ Macroscopic scale 1–100mm

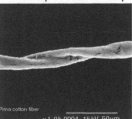 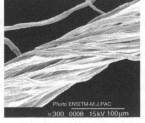 

FIGURE 8.6 Cotton fibers at 3 scales.

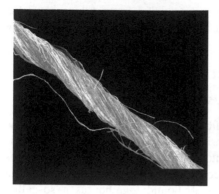 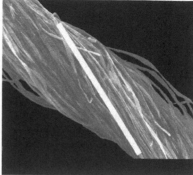

FIGURE 8.7 Cotton single yarn: twisted structure.

8.2.2 YARNS

In most cases, yarns are single bundles of staple fibers, obtained by ring or rotor processes show parallel fibers, oriented with an angle relative to the yarn axis. The twist gives cohesion to the yarn (Figure 8.7). It induces normal forces between fibers, increases interfiber tangential friction forces, and gives to the yarn its mechanical cohesion. The yarn is characterized by its mass per unit length (yarn count: tex = mg/m) and its twist value (rev/m). The twist angle is an intrinsic parameter of one type of yarn and independent of the yarn count. This specific arrangement gives to the yarn a low bending rigidity in comparison to a unique fiber of the same count (in the case of monofilament). It is also usual to twist two single yarns together, thus a two-plied yarn is obtained. If the two-plied yarn is twisted in the opposite direction of the initial twist of each single yarn, both single yarns loose twist to give it to the assembly. In a balanced situation (Figure 8.8), the individual fibers are parallel to the two-plied yarn axis.

8.2.3 FABRICS

Based on yarns, fabrics are usually woven or knitted. When they are woven (Figure 8.9), warp and weft yarns are perpendicularly intertwined. The intertwining is systematically achieved at each contact point for plain woven fabrics. It is not the case for other woven structures like twill weave or sateen which show less bending and shear rigidity. Woven structures are used in all the fields of textile industry, for garments or technical products, and at any time when dimensional stability and or specific mechanical resistance and behavior is required. Nevertheless, weaving technology is expensive and

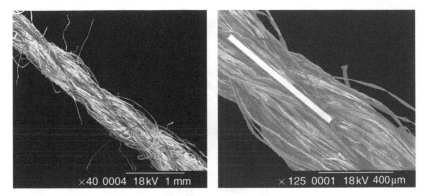

FIGURE 8.8 Cotton two-plied yarn: balanced structure.

is replaced by nonwoven technology when it is possible, particularly for technical applications.

In the case of knitted fabrics, two kinds of technologies are available: weft-knitted fabrics (Figure 8.10), which are essentially based on the interlacing of a single yarn, and warp-knitted fabrics (Figure 8.11), which are the result of the interlacing between a great number of parallel warp yarns. Usually yarns have a greater mobility in weft-knitted fabrics than in warp knitted or woven structures where the bending rigidity of the fabric is directly related to the bending rigidity of the yarn. Weft knitting is highly productive and the products allow great deformation under low stress. This technology is particularly adapted for tights fitting garments. Warp knitting allows specific

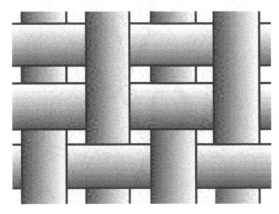

FIGURE 8.9 Woven fabric.

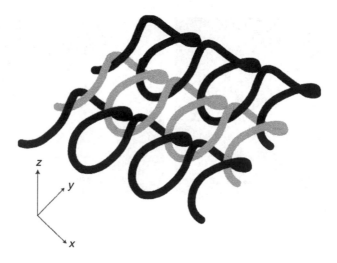

FIGURE 8.10 Weft-knitted fabric.

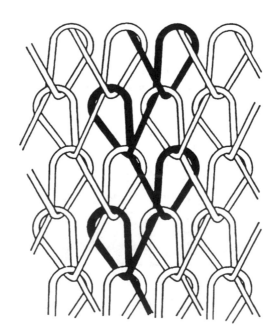

FIGURE 8.11 Warp-knitted fabric.

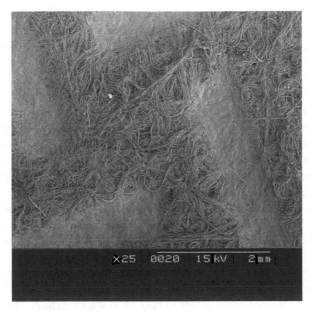

FIGURE 8.12 Nonwoven fabric.

design to obtain given mechanical properties with higher productivity than weaving (home furnishing, automotive industry, sport garments, and articles).

Nonwovens fabrics (Figure 8.12) are processed without yarn construction. Nevertheless, they can have a specific structure, for example, holes (high pressure water jet or needling) or bonding points. Since this way to transform fibers in fibrous surfaces can be done at a low cost, nonwovens technologies have been for many years a good answer to the needs in the areas of hygiene and industrial products. Because of the random distribution as well as of the low level of organization of the fibers at the mesoscopic scale, nonwovens products are not yet suitable for the garment industry uses.

8.3 TRIBOLOGICAL ASPECTS OF THE HAND (HANDLE) OF FABRICS

8.3.1 PHYSIOLOGICAL ASPECTS

The state of fabric surface, i.e., structure (yarn arrangement) and surface hairiness, is very important for the tactile feeling of fabrics. The couple hand-brain is obviously the center of the human tactile sense. The human hand has four kinds of receptors that are sensitive to mechanical stimulation

(Figure 8.13): the fast adapting receptors, Meissner and Pacinian corpuscles, have a reaction in the beginning and the end of the stimulus, the slowly adapting receptors, Merkel disk and Ruffini corpuscle, keep being excited during the whole stimulus period. These four kinds of mechanoreceptors have different sensitivity thresholds, are in different depth in the dermis, and are active in perceiving of a surface contact. A movement between the surface and the fingers is necessary to have a sensitive touch. In fact, the human hand is not so accurate for a surface state evaluation without a rubbing movement, i.e., just compression [1]. The first theory about the texture perception was that mechanoreceptors fibers are sensitive to the contact frequency between the skin and the surface asperities [2,3]. With this theory, the faster the hand span is, the rougher seems to be the surface—but it is not the case because the perception of a texture does not change with the speed of the frictional movement (when the rubbing speed does not exceed 25 cm s^{-1}). The second theory takes this further remark into consideration: the human hand is sensitive to the energy of the skin strain [4–7].

Furthermore, the handle of fabrics is also influenced by thermal transfer and cool or warm feeling, particularly during the first seconds of the contact. Nervous ends are dedicated either to cool or to warm feeling. The sensitivity of the human hand is important in terms of relative comparison but not very efficient for absolute evaluation of temperature.

8.3.2 MECHANICAL ASPECTS AND CHARACTERIZATION

The KES-F (Kawabata Evaluation System for Fabrics) [8] is well known for the characterization of tensile, shear, compression, friction, and roughness

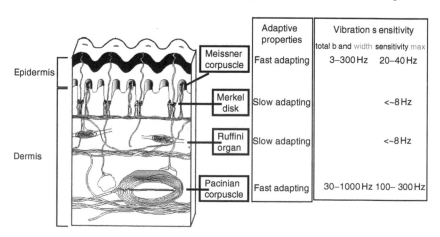

FIGURE 8.13 Human skin structure: mechanical receptors.

properties of fibrous structures which typically show high transformation under low stress. In the case of a tribological approach, this method allows the measurement of a coefficient of friction, the mean deviation of that coefficient, and the mean deviation of the profile. Nevertheless, if these parameters are sufficient enough to characterize material differences, they are not sensitive enough to identify fine differences of fibrous surfaces as well as their finishing treatments [9].

By hypothesis that human hand is sensitive, in a given frequency range, to the energy of the skin strain during the contact, we suggest a mechanical investigation of the fibrous surface based on spectrum analysis of the vibration signal occurred during the contact between the surface and a sensor in motion.

A multi-directional tribometer [10] is used for this measurement. It consists in three parts: drive of the sample, sensor, and a signal-processing unit (Figure 8.14). The sample carrier is a 140 mm diameter rotary disk. The sensor is positioned at one end of a balance arm with counterweight at the other. This arm is fixed on the frame that holds the sample carrier. The sensor is a piezoelectric accelerometer. The probe attached to the sensor is a steel wire (0.5 mm in diameter and 5 mm in length) with its axis radial to the sample carrier. The surface to be scanned is a rotary disk of 110 mm diameter. Measurements are performed at a chosen rotary speed (linear speed range from 20 to 100 mm s^{-1}).

The Fourier analysis of the electrical signal from the sensor consists of computing the autospectrum relative to frequency by a spectrum analyzer. The autospectrum is the average of several instantaneous spectra during sample

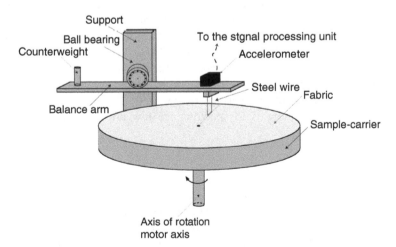

FIGURE 8.14 Multi-directional tribometer.

carrier rotation. Each spectrum is expressed in power spectral density (PSD) relative to frequency. The PSD is obtained as follows:

$$\text{PSD}(f) = \frac{|X(f)|^2}{K \Delta f}$$

where f, frequency (Hertz, Hz); $X(f)$, Fourier transform of the temporal signal $x(t)$, which corresponds here to the signal from the sensor (m s^{-2}); PSD, power spectrum density $[(\text{m s}^{-2})^2 \text{ Hz}^{-1}]$; Δf, step in frequency domain; $\Delta f = 1$ Hz; K, coefficient relative to the windowing (dimensionless). In our case, we used a Hanning window, then $K = 1.5$.

Most woven or knitted fabrics have a periodic structure based on the basic pattern (the kind of weave or knit). Thus, the PSD shows one or several peaks corresponding to the periodicity of the fabric structure. The number of peaks is equal to the number of periodicities scanned by the probe during sample rotation. For example, in the case of a woven fabric, the PSD has three peaks which correspond to warp yarns, weft yarns, and to the diagonal effect (Figure 8.15). An illustration of this kind of investigation is given for typical surface modification of cotton fabrics for a better comfort and hand: raising and sanding [11].

8.3.2.1 Raising

The objective of the raising process is to modify the heat insulation ability of fabrics. Raising consists of brushing a fabric with wire-covered material. Therefore, during the raising operation, the needles pull out some fibers from the yarn and so a fiber mat is formed on the fabric surface (Figure 8.16). This mat is quite thick and covers the fabric structure. Raising produces a typical hairiness with random distribution of length [11].

The fiber mat created on the fabric surface during the raising operation covers the fabric structure, therefore the probe has a very small contact with the ground elements of the surface (warp, weft, ... for a woven fabric). Therefore, the peak heights in spectrum are low after raising than before (Figure 8.17).

8.3.2.2 Sanding

The objective of the sanding process is to improve the fabric touch. Sanding is an abrasive wear, so it changes the state of fabric surface. During the sanding operation, yarns are worn (weft or warp yarns, depending on the kind of weave), i.e., peripheral fibers are broken but not pulled out [11]. For one worn fiber, two hairs are formed. The length of hairs depends on the kind of weave

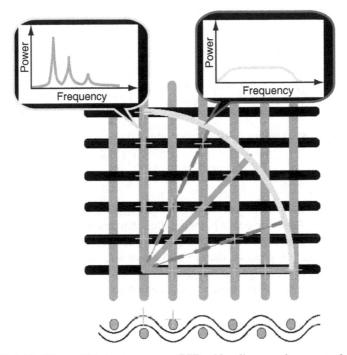

FIGURE 8.15 Effect of fabric structure on PSD with a linear and a punctual probe.

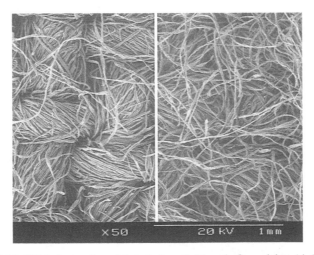

FIGURE 8.16 SEM picture for a fabric before (left) and after raising (right).

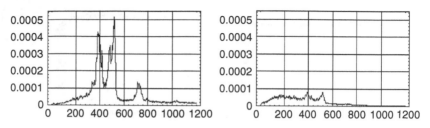

FIGURE 8.17 PSD before (left) and after raising (right).

and on the yarn density. Sanding gives short, but dense, regular hairs (Figure 8.18).

Figure 8.19 shows the spectrum for a cotton plain woven fabric before and after sanding. The three peaks point out the main directions of the plain woven fabric. These peaks are lower after sanding. In fact, the asperities are worn during the sanding process, therefore they are less visible and this phenomenon appears in the height of the peaks.

The frequency value of each peak is equal to

$$f = \frac{\pi D \omega}{l}$$

where D, diameter of the scanned surface (m); ω, rotation speed (rps); f, frequency (Hz); l, length of a spatial period (m).

FIGURE 8.18 SEM picture for a fabric before (left) and after raising (right).

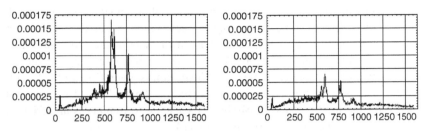

FIGURE 8.19 PSD before (left) and after (right).

A spatial period shows the distance between two asperities (for example, two wales in the case of a knitted fabric).

The height of the frequency peak corresponding to this spatial period evolves in the same direction as the friction force due to these asperities. This friction force depends on the height of the asperities and the material. Hence, for a given material, the maximum peak is obtained by the asperity height. That defines the fabric surface roughness. The asperity height is governed by the yarn undulation in the fabric. Indeed, the rougher the structure is, the higher the maximum frequency peak is. This parameter is called roughness-friction criterion.

In the case of nonstructured fabrics, such as nonwovens, the previous described method is not optimal. A new measurement method [12,13] has been developed and patented (Figure 8.20). The sample is clamped on a rotary disk, and a very thin metallic plate (50 μm thin) rubs the tested surface. During the contact, the plate vibrates according to eigenvalues of frequencies. Strain gauges are fixed on this plate and measure its vibrations (eigenvalues). Like in

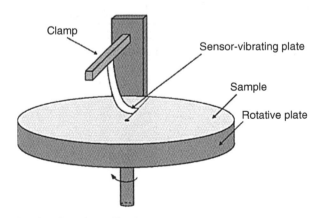

FIGURE 8.20 Vibrating plate tribometer.

the previous method, a spectral analysis of the vibration signal is performed. Since the frequencies of peaks are eigenvalues (Figure 8.21), they depend on the material, on the geometry, as well as on the size of the vibrating plate. Further, the magnitude of the peaks and then their energy depend on the surface state of the tested samples. An illustration is given (Figure 8.22) with the comparison of 4 nonwoven tissues, based on the analysis of the spectral energy for the modes 1 and 3. Mode 1 is sensitive to roughness and friction whereas mode 3 is sensitive to roughness [14].

8.3.3 THERMAL ASPECTS AND CHARACTERIZATION

The warm or cool effect during the first seconds of contact between the hand and the fabric is directly related to the real area of contact, which is the area of heat conduction, as well as to the presence of air, which acts as an insulator. For a given fiber material, a great area of contact allows a more important heat transfer than a smaller one, therefore the handle feels cooler [15].

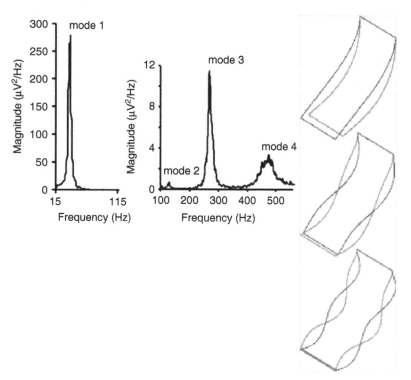

FIGURE 8.21 Eigenvalves and mode shapes of the vibrating plate.

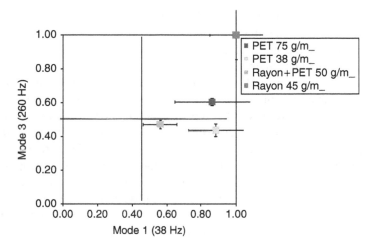

FIGURE 8.22 Comparative analysis of nonwoven fabrics based on spectral energy.

An apparatus has been developed [16] to assess the real contact area by an indirect technique. The method measures the transient heat conduction in a vacuum. Transient heat conduction is converted into thermal energy with the help of molecular interactions under a non-homogeneous temperature distribution. Heat conduction requires contact between surfaces. For identical materials, when the area of contact increases, the conduction phenomenon raises. Furthermore, when the real contact area is large, the material absorbs more energy. The energy E absorbed by the system during the time t is proportional to the real area of contact, as shown below. Fourier's equation:

$$\frac{E}{\Delta t} = \lambda A \frac{\Delta T}{e}$$

where λ, thermal conductivity of the material in the heat flow direction (W m^{-1} K^{-1}); A, real area of contact (m^2); ΔT, difference of temperature (K) between the hot plate and the ambient; e, fabric thickness (m).

For homogeneous materials, the Fourier's equation indicates that for a given temperature gradient, heat flow increases with thermal conductivity of the material. Therefore, the more a material absorbs thermal energy during contact by conduction, the higher are its conductive properties with heat flow in the normal direction to the surface.

$$\vec{\phi} = -\lambda \, \vec{\nabla} T$$

where, considering the heat flow in the normal direction of the surface, $\vec{\varphi}$, heat flow density in the normal direction of the surface (W m^{-2}); λ, thermal conductivity of the material in the heat flow direction $(\text{W m}^{-1}\,\text{K}^{-1})$; T, temperature field depending on the time (K).

For a fibrous material, the thermal conductivity is a combination of thermal conductivity of the air and of the fiber (weighted respectively by the fraction of the volume taken up by each component). However, if the fabric is in a vacuum, the contribution of air contained in the fabric is negligible and the most significant effect is the conductivity of fibers.

In order to measure the thermal energy absorbed by a fabric, an apparatus has been developed. It consists of a guarded hot plate shown in Figure 8.23. It has a square, thin, central aluminum test plate (100 mm² area) surrounded by a coplanar ring aluminum plate (292 mm² area). The central plate and the guard ring are separated by a thin polystyrene band (2 mm width). They are heated by two independent cemented resistances heater wires. The ring plate is at a temperature of 0.5°C higher than the test plate, i.e., 33.5°C, to avoid lateral heat loss from the central test plate and prevent any heat transfer from the ring plate to the test plate.

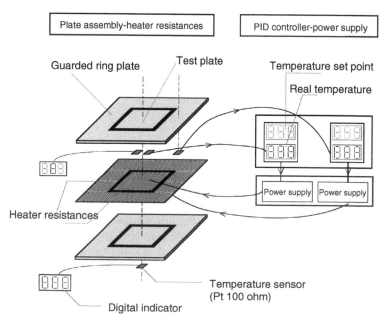

FIGURE 8.23 Thermal transfer measurement on fabric (guarded hot plate).

After bringing the test plate and the guard ring plate to the required operating temperature, 33°C and 33.5°C, respectively, it is necessary to wait until equilibrium is reached between the system and the ambient environment. At the balanced state, the electrical power supplied by the resistances remains constant. At this point, the fabric sample is placed on the surface of the test plate. There is an instantaneous heat loss by the plate and a temperature gradient exists through the fabric. The electrical power considered as a function of time required by the system (plate plus sample) to reach 33°C again is recorded. Since the guard ring avoids any lateral heat loss, the measurement reflects the vertical heat flow between the test plate and the sample. The thermal energy lost by the test plate is equal to the energy absorbed by the fabric plus a constant C_1. The heat flow generated by the test plate is equal to the electrical power supplied by the heater wire resistances plus a constant C_2. The constants C_1 and C_2 are independent from the fabric. They only depend on the imposed temperature gradient ΔT. In our study, this gradient is constant and equal to 13°C, i.e., the temperature difference between the atmosphere temperature (20°C) and the operating temperature (33°C).

Under these conditions, the thermal energy transferred to the fabric is the difference between the power measured at the balanced state without the sample and the power measured with a sample on the test plate. The thermal energy absorbed by the sample is calculated by multiplying instantaneous power absorbed by the fabric during the time interval. Hence, it is very important to have a short time interval (meaning less than one second—in our case, 0.8 s). At a steady state (meaning a longer time interval), the energy absorbed is mainly related to raw material rather than to the structure of the fibrous material.

An example is given with comparative heat transfer and friction/roughness measurements for weft-knitted fabrics. The influence of material and structure is characterized at 3 different scales [15].

- At microscopic scale: the difference between two kinds of cotton (cotton 1 fine, cotton 2 coarse)
- At mesoscopic scale: the difference between single yarn and plied yarn (same global count)
- At macroscopic scale: the influence of the stitch length (from fine to coarse)

We can observe in Figure 8.24 and Figure 8.25 that fine fibers, plied yarns, and reduced stitch length lead to smooth contact (reduced friction/roughness) and to higher heat transfer (cooler feeling). On the contrary, coarse fibers, single yarns, and important stitch length lead to rough contact and low heat

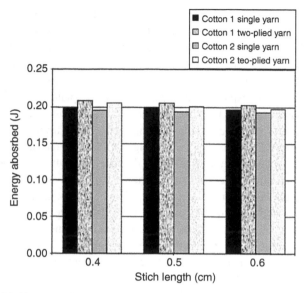

FIGURE 8.24 Heat transfer: influence at different scales.

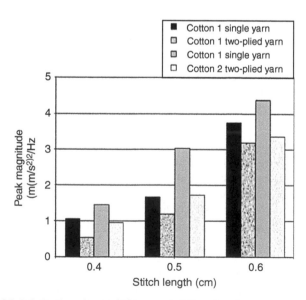

FIGURE 8.25 Friction/roughness: influence at different scales.

transfer. Regarding the influence at microscopic and mesoscopic scale (fiber and yarn), the effect on tribological and thermal aspect is essentially due to the bending properties; low-bending moment induces high compliance and therefore smooth contact and high contact area. The influence of stitch length on thermal transfer is essentially due to the presence of air. This influence is reduced when the test is done under vacuum.

REFERENCES

1. Morley, J. W., Goodwin, A. W., and Darian-Smith, I., Tactile discrimination of gratings, *Experimental Brain Research*, 49, 291–299, 1983.
2. Krueger, L. E., David Katz's Der Aufbau der Tastwelt: a synopsis, *Perception and Psychophysics*, 7, 337–341, 1970.
3. Zigler, M. J., Rewiew of Katz Der Aufbau der Tastwelt, *Psychological Bulletin*, 23, 326–336, 1926.
4. Lederman, S. J., Tactile roughness of grooved surfaces: the touching process and effects of macro- and microsurface structure, *Perception and Psychophysics*, 16, No. 2, 385–395, 1974.
5. Kudoh, N., Tactile perception of textured surfaces: effects of temporal frequency on perceived roughness by passive touch, *Tohoku Psychologica Folia*, 47, No. 1–4, 21–28, 1988.
6. Taylor, M. M. and Lederman, S. J., Tactile roughness of grooved surface: a model and the effect of friction, *Perception and Psychophysics*, 17, No. 1, 23–36, 1975.
7. Lederman, S. J., Tactual roughness perception: spatial and temporal determinants, *Canadian Journal of Psychology*, 37, No. 4, 498–511, 1983.
8. Kawabata, S., *The Standardisation and Analysis of Hand Evaluation*, The Textile Machinery Society of Japan, Osaka, 1980.
9. Bueno, M.-A. and Renner, M., Comparison of a new tribological method for the evaluation of the state of the fabric surface with the KES-F surface tester, *Journal of the Textile Institute*, 92, Part 1. No. 3, 212–227, 2002.
10. Bueno, M.-A., Lamy, B., Renner, M., and Viallier, P., Tribological investigation of textile fabrics, *Wear*, 195, 192–200, 1996.
11. Bueno, M.-A., Viallier, P., Durand, B., Renner, M., and Lamy, B., Instrumental measurement and macroscopical study of sanding and raising, *Textile Research Journal*, 67, No. 11, 779–787, 1997.
12. Bueno, M. -A., Fontaine, S., and Renner, M., Dispositif pour mesurer l'état de surface d'un matériau et procédé de mise en oeuvre, Brevet France, No. 00 07490/FR, 2000.
13. Bueno, M. -A., Fontaine, S., and Renner, M., Method and device for assessing the surface condition of a material, World Patent No. WO 01/94878 A1, US Patent 6,810,744 B2, 2001.

14. Fontaine, S., Marsiquet, C., Nicoletti, N., Renner, M., and Bueno, M.-A., Development of a sensor for surface state measurements using experimental and numerical modal analysis, *Sensors and Actuators A: Physical*, 120, 507–517, 2005.

15. Pac, M.-J., Bueno, M.-A., Renner, M., and El Kasmi, S., Warm-cool feeling relative to tribological properties of fabrics, *Textile Research Journal*, 71 (9), 806–812, 2001.

16. Bueno, M. A., Renner, M., and Nicoletti, N., Influence of fiber morphology and yarn spinning process on the 3D loop Shape of weft knitted fabrics, *Textile Research Journal*, 74, No. 4, 297–304, 2004.

Section III

The Solid-Polymer Interface

9 Wettability of Fibers and Powders: Application to Reinforced Polymeric Materials

Michel Nardin

CONTENTS

9.1 INTRODUCTION

Multicomponent materials and in particular, reinforced materials, have known a huge industrial development in the last decades, whatever the nature of both the reinforcement (i.e., powders, nanoparticles, fillers, fibers, and so on) and the matrix (metals, polymers, ceramics, etc.). There has been also an increasing interest in the scientific community to analyze and understand the physical and chemical nature of the interface between the reinforcing entities and the matrix, as well as its mechanical behavior. More precisely, most work in this field considered that good final mechanical performance or use properties of the resulting reinforced materials depend significantly on the quality of the interface that is formed between both solids. In other words, the adhesion established at this interface became one of the most important parameters in controlling the interfacial behavior, and consequently, the mechanical properties of reinforced materials. Therefore, it is understandable that a better knowledge of the adhesion phenomena is required for practical applications of multicomponent materials. The main difficulty in the scientific analysis of adhesion mechanisms stems from the fact that adhesion is at the boundary of several scientific fields, including physical chemistry of surfaces and interfaces, materials and macromolecular sciences, mechanics and micromechanics of fracture, rheology, etc. Consequently, the study of adhesion uses various concepts, depending on different special fields of expertise. This variety of approaches is emphasized by the fact that many theories of adhesion have been proposed which together are both complementary and contradictory, i.e., mechanical interlocking, electronic or electrostatic theory, theory of weak boundary layers, thermodynamic or adsorption model (also referring to wettability), diffusion or interdiffusion theory, and finally, chemical bonding theory.[1–3]

However, the thermodynamic model of adhesion, generally attributed to Sharpe and Schonhorn,[4] is certainly the most widely used approach in adhesion science at present. In this theory, it is considered that adhesion between two solids or between a solid and a liquid (more generally, between an adhesive and a substrate) is due to interatomic and intermolecular forces established at the interface, provided that an intimate contact is achieved. The most common interfacial forces result from van der Waals and Lewis acid–base interactions, as described below. The magnitude of these forces can generally be related to fundamental thermodynamic quantities, such as surface-free energies of both entities in contact. Generally, the formation of an assembly is obtained through a liquid–solid contact step, therefore, criteria for good adhesion becomes, essentially, criteria for good wetting, although this is a necessary but not a sufficient condition. It is clear that when an intimate contact has been established at an interface, other adhesion phenomena can therefore take place, such as, in particular, the creation of interfacial chemical bonds, the molecular reorganization and reorientation near the interface, the interdiffusion of molecular or macromolecular chains across this interface in the case of contact involving polymers, etc., each of these phenomena being able to enhance the level of adhesion.

In the first part of this chapter, wetting criteria based on the determination of a contact angle between a drop of liquid deposited onto a solid surface will be defined. Theoretical considerations concerning the determination of surface and interface free energies from wettability measurements will be examined. The estimation of a reversible work of adhesion from the surface properties of materials in contact will therefore be considered.

Experimental approaches concerning wettability measurements onto single fibers will be analyzed in the second part of this chapter. A following section will deal with the analysis of capillary impregnation of "plugs" constituted by divided solids, like powders or fibers, as well as of fiber fabrics. This imbibition can lead to the estimation of the surface energy of such divided solids using particular experimental conditions.

Finally, the role of the surface properties of fibers and fillers on the micromechanical behavior of the interface as well as the mechanical performance of reinforced materials will be briefly described.

9.2 WETTING CRITERIA, SURFACE AND INTERFACE FREE ENERGIES, AND WORK OF ADHESION

9.2.1 GENERAL CONSIDERATIONS

In a solid–liquid system, wetting equilibrium is defined from the profile of a sessile drop on a planar solid surface. Young's equation,[5] relating the surface tension γ of materials at the three-phase contact point to the equilibrium contact angle θ (Figure 9.1), is written as:

$$\gamma_{SV} = \gamma_{SL} + \gamma_{LV} \cos \theta \tag{9.1}$$

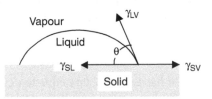

FIGURE 9.1 Schematic profile of a liquid drop onto a planar solid surface.

Subscripts S, L, and V refer, respectively, to solid, liquid, and vapor phases such that a combination of two of these subscripts corresponds to the given interface (i.e., SV corresponds to the solid–vapor interface, etc.).

The term γ_{SV} represents the surface free energy of the substrate after equilibrium adsorption of vapor from the liquid and is sometimes lower than the surface free energy γ_S of the solid in vacuum. This decrease is defined as the spreading pressure π ($\pi = \gamma_S - \gamma_{SV}$) of the vapor onto the solid surface. In most cases, in particular when dealing with polymer materials, π could be neglected and, to a first approximation, γ_S is used in place of γ_{SV} in wetting analyses. Similarly, it is considered that $\gamma_L = \gamma_{LV}$. In the case of high energy materials, it is possible to remove the effect of this spreading pressure by using the two liquid phase method described below.

The difference of energy dU stemming from the covering of a unit surface area dA of the solid by the liquid (Figure 9.1) can be expressed as

$$S = dU/dA = \gamma_{SV} - \gamma_{SL} - \gamma_{LV} \tag{9.2}$$

the quantity S, describing the spreading ability of the liquid, is called the spreading coefficient. Therefore, the inequality $S \geq 0$ constitutes a wetting criterion. It is worth noting that geometrical aspects or processing conditions, such as the surface roughness of the solid and applied external pressure, can restrict the applicability of this criterion.

However, a more fundamental approach leading to the definition of other wetting criteria is based on the analysis of the nature of forces involved at the interface and allows the calculation of the free energy of interactions between two materials in contact.

For solid–liquid systems, according to Dupré's relationship,[6] the adhesion energy W_{SL} is defined as

$$W_{SL} = \gamma_S + \gamma_L - \gamma_{SL} = \gamma_L(1 + \cos\theta) \tag{9.3}$$

in agreement with Equation 9.1 and neglecting the spreading pressure. Fowkes[7] proposed that the surface free energy γ of a given entity could be represented by the sum of the contributions of different types of interactions and, particularly by only two terms, i.e., a dispersive (London's interactions, or in other words, purely nonpolar interactions) and a nondispersive or polar component (superscripts D and ND respectively), as follows:

$$\gamma = \gamma^D + \gamma^{ND} \tag{9.4}$$

The last term in the right-hand side of this equation corresponds to all the nondispersion forces, including Debye and Keesom interactions, as well as hydrogen bonding. Intermolecular interactions are fully described and analyzed by Israelachvili.[8] Similarly, the reversible work of adhesion W_{12} between two entities 1 and 2 can be expressed as the sum of two components corresponding to dispersive and polar interactions, respectively:

$$W_{12} = W_{12}^D + W_{12}^{ND} \qquad (9.5)$$

Fowkes[9] has also considered that the dispersive part of these interactions can be well quantified as twice the geometric mean of the dispersive components of the surface energy of the two entities:

$$W_{12} = 2(\gamma_1^D \gamma_2^D)^{1/2} \qquad (9.6)$$

By analogy with the work of Fowkes, Owens and Wendt[10] and then Kaelble and Uy,[11] have suggested that the nondispersive part of interactions W^{ND} between two materials can be expressed as the geometric mean of the nondispersive components of their surface energy, although there is no theoretical reason to represent all the nondispersive interactions by this type of expression. Hence, the work of adhesion W_{12} becomes

$$W_{12} = 2(\gamma_1^D \gamma_2^D)^{1/2} + 2(\gamma_1^{ND} \gamma_2^{ND})^{1/2} \qquad (9.7)$$

For solid–liquid equilibrium (subscripts S and L, respectively), a direct relationship between the contact angle θ of the drop of a liquid on a solid surface and the surface properties of both liquid and solid is obtained from Equation 9.3, Equation 9.5 and Equation 9.6:

$$\cos \theta = 2 \frac{(\gamma_S^D \gamma_L^D)^{1/2}}{\gamma_L} + 2 \frac{W_{SL}^{ND}}{\gamma_L} - 1 \qquad (9.8)$$

By contact-angle measurements of droplets of different liquids of known surface properties, the components γ_S^D and γ_S^{ND} of the surface energy of the substrate can be then determined according to the following approach (Figure 9.2). By using nonpolar liquids ($\gamma_L = \gamma_L^D$), no polar interactions are then established between the liquid and the solid, and then W_{SL}^{ND} is equal to zero. Thus, Equation 9.8 reduces to

$$\cos \theta = 2\sqrt{\gamma_S^D} \frac{\sqrt{\gamma_L^D}}{\gamma_L} - 1 \qquad (9.9)$$

The variation of $\cos \theta$ versus $\sqrt{\gamma_L^D}/\gamma_L$ is a straight line with an intercept at the origin equal to (-1) and a slope equal to $2\sqrt{\gamma_S^D}$. It is therefore possible to

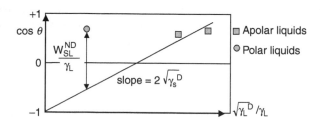

FIGURE 9.2 Schematic representation of the variation of the cosine of the contact angle θ as a function of the components of the surface energy of apolar and polar liquids for a given solid surface.

determine experimentally the dispersive component of the surface energy of the solid.

The experimental points corresponding to polar liquids are lying above this straight line (Figure 9.2) if polar interactions are established at liquid–solid interface. The difference of ordinates between them and the straight line is equal to W_{SL}^{ND}/γ_L and allows the determination of the magnitude of nondispersive (polar) interactions between the liquid and the solid.

Previous equations are only valid when an equilibrium contact angle is obtained. In practice, when a liquid drop is deposited onto a solid surface, the experimental values of the contact angle are usually ranging between two limit values, i.e., the advancing θ_a (maximum value) and the receding θ_r (minimum value) contact angles. Effectively, by means of a syringe, it is possible either to increase or to decrease the liquid drop volume (Figure 9.3). Therefore, the drop volume first increases but the solid–liquid interfacial area remains constant, leading to an increase of the contact angle. At a maximum critical value θ_a, this area suddenly increases. Vice versa, when the drop volume is decreased by

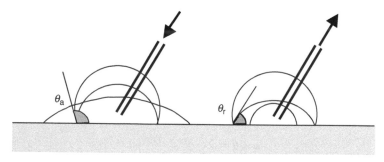

FIGURE 9.3 Schematic representation of advancing θ_a and receding θ_r contact angles obtained respectively by increasing or decreasing the liquid drop volume by means of a syringe.

means of the syringe, the contact angle first decreases to reach a minimum critical value θ_r, where the liquid–solid interfacial area suddenly decreases.

The wetting hysteresis is defined from these contact angles as $(\theta_a - \theta_r)$ or $(\cos\theta_a - \cos\theta_r)$. It depends mainly on the roughness and/or the chemical heterogeneity of the solid surface. It could also be related to the molecular reorganization of the solid–liquid interface, which can take place during advancing and receding wetting experiments.[12] It has to be mentioned that two types of hysteresis are observed:[13] a kinetic one, which depends on the wetting rate, and a thermodynamic one, which is not time or frequency dependent.

It is worth mentioning that the method described above and schematically represented in Figure 9.2 is largely used nowadays. However, other approaches based for example on a work of adhesion obtained from a harmonic mean[14] of dispersive and nondispersive components of the surface energy, instead of Fowkes' geometric means given in Equation 9.6, were used in the literature. In some other methods, the contact angles of only two given liquids onto a solid surface are measured and lead to two equations (Equation 9.7) where the dispersive and nondispersive components of the surface energy of this solids are the two unknowns. The resolution of these two equations allows therefore the determination of both components.

9.2.2 Two Liquid Phase Method

As seen before, the spreading pressure has been neglected in the previous approach. This pressure cannot be considered as negligible in the case of solids exhibiting high surface energies, like metals, ceramics, or glasses. It is usually difficult and often impossible to measure such a spreading pressure on a flat substrate. Moreover, for such surfaces, the wetting is generally complete and determination of surface energy from simple contact angle measurements becomes impossible. To cast off such a problem, a method called the "two liquid phase method," consisting of the measurement of contact angle of drops of a liquid deposited on high surface energy solids previously immersed in another liquid, has been developed.[15–18] Obviously, two immiscible liquids have to be used and, usually, systems involving water (subscript W) in hydrocarbon media (n-alkanes in particular, subscript A) are chosen (or vice versa). From a theoretical point of view, in Figure 9.1, the vapor phase is being replaced by an alkane, and then Young's and Dupré's equations become, respectively:

$$\gamma_{SA} = \gamma_{SW} + \gamma_{WA}\cos\theta \qquad (9.10)$$

$$W_{SW} = \gamma_S + \gamma_W - \gamma_{SW} = 2\sqrt{\gamma_S^D \gamma_W^D} + W_{SW}^{ND} \qquad (9.11)$$

$$W_{SA} = \gamma_S + \gamma_A - \gamma_{SA} = 2\sqrt{\gamma_S^D \gamma_A^D} \qquad (9.12)$$

Finally, substituting Equation 9.11 and Equation 9.12 into Equation 9.10 leads to:

$$\gamma_W - \gamma_A + \gamma_{WA}\cos\theta = 2\sqrt{\gamma_S^D}\left(\sqrt{\gamma_W^D} - \sqrt{\gamma_A^D}\right) + W_{SW}^{ND} \qquad (9.13)$$

In Equation 9.13, the contact angle θ is measured, whereas γ_S^D and W_{SW}^{ND} are the unknowns. The variation of the quantity $(\gamma_W - \gamma_A + \gamma_{WA}\cos\theta)$ versus $\left(\sqrt{\gamma_W^D} - \sqrt{\gamma_A^D}\right)$ is a straight line with an intercept at the origin equal to W_{SW}^{ND} and a slope equal to $2\sqrt{\gamma_S^D}$ (Figure 9.4). It is thus possible to determine the surface properties (dispersive component of the surface energy γ_S^D and nondispersive interactions with water W_{SW}^{ND}) of a high surface energy solid.

Such an approach is valid insofar as water can displace n-alkanes, so that an intermediate layer of alkanes will not exist to prevent contact between water and the solid surface. Such a criterion for displacement of alkanes by water has been analyzed by Shanahan et al.,[19] who proposed the following inequality:

$$W_{SW}^{ND} > 2\left(\sqrt{\gamma_S^D} - \sqrt{\gamma_A^D}\right)\left(\sqrt{\gamma_A^D} - \sqrt{\gamma_W^D}\right) \qquad (9.14)$$

Provided the nondispersive interaction term between the solid and water satisfies relation (Equation 9.14), water is able to displace the n-alkanes.

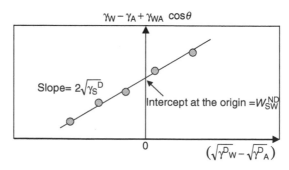

FIGURE 9.4 Schematic representation of the analysis of experimental data related to the cosine of the contact angle θ in the case of two liquid phase method. Determination of the surface energy characteristics of a high surface energy solid.

Such a criterion, however, can only be verified at the end of the experimental investigation.

A lot of solids that exhibit a high surface energy were characterized with this two liquid phase method in the literature. This is particularly true for glasses, metallic oxides, metals, ceramics, etc.[18]

9.2.3 ACID–BASE APPROACHES

At this step, it has to be mentioned that even if the notion of dispersive and nondispersive components of the surface energy constitutes a significant advance, it remains controversial from a theoretical point of view. Other approaches were investigated, particularly the generalized Lewis acid–base concept.[20] This concept describes the acid–base interactions as well as hydrogen bonds in terms of molecular frontier orbitals and allows for the analyzation of all the different degrees of electronic sharing between two entities. For this approach, a base and an acid are defined according to the energy level of molecular orbitals, the well-known "highest occupied molecular orbital" (HOMO) and "lowest unoccupied molecular orbital" (LUMO), respectively.[21] The magnitude of acid–base interactions is directly related to the difference of energy between HOMO and LUMO levels of the interacting entities. Finally, the equilibrium intermolecular or interatomic distance between these entities decreases when the energy of interactions increases. Effectively, short intermolecular distances, i.e., about 0.2 nm, close to those corresponding to covalent or ionic bonds, are related to the strongest acid–base interactions (hydrogen bonds for water molecules association, for example), whereas interatomic distances of about 0.5 nm, similar to distances currently observed for van der Waals interactions, are related to the weakest acid–base interactions.[22]

Fowkes and coworkers[23–26] have shown that these electron acceptor and donor interactions constitute a major type of interfacial forces—in other words, that the contribution of the polar (dipole–dipole) interactions to the thermodynamic work of adhesion could generally be neglected compared to both dispersive and acid–base contributions. They have also considered that the acid–base component $W^{ab}(=W^P)$ of the adhesion energy can be related to the variation of enthalpy $-\Delta H^{ab}$, corresponding to the establishment of acid–base interactions at the interface, as follows:

$$W^{ab} = f(-\Delta H^{ab})n^{ab} \qquad (9.15)$$

where f is a factor which converts enthalpy into free energy and is taken equal to unity, and n^{ab} is the number of acid–base bonds per unit interfacial area, close to about 6 μmoles/m^2. Therefore, from Equation 9.5, Equation 9.6, and

Equation 9.15, the total adhesion energy W_{12} becomes[27]

$$W_{12} = 2(\gamma_1^D \gamma_2^D)^{1/2} + f(-\Delta H^{ab})n^{ab} \tag{9.16}$$

The experimental values of the enthalpy $(-\Delta H^{ab})$ can be estimated from the work of Drago and coworkers,[28,29] who proposed the following relationship:

$$-\Delta H^{ab} = C^A C^B + E^A E^B \tag{9.17}$$

where C^A and E^A are two quantities which characterize the acidic material at the interface, whereas similarly, C^B and E^B characterize the basic material. The validity of Equation 9.17 was clearly evidenced for polymer adsorption on various substrates.[25] Another estimation of $(-\Delta H^{ab})$ can be carried out from the semi-empirical approach defined by Gutmann,[30] who has proposed that liquids may be characterized by two constants: an electron acceptor number AN and an electron donor number DN. For solid surfaces, electron acceptor and donor numbers, K_A and K_D, respectively, have also been defined and measured by inverse gas chromatography.[31–33] In this approach, the enthalpy of formation $(-\Delta H^{ab})$ of acid–base interactions at the interface between two solids 1 and 2 is now given by[30,31]

$$-\Delta H^{ab} = K_{A1}K_{D2} + K_{A2}K_{D1} \tag{9.18}$$

This expression was successfully applied to describe fiber–matrix adhesion in the field of composite materials by Schultz et al.[33] (see also Section 9.4 of this chapter). In these studies and as shown below, the practical adhesion, defined as normalized interfacial shear strength, was linearly related to the work of adhesion W between the fiber and the matrix, estimated from Equation 9.16 and Equation 9.18.

It must be mentioned that other approaches have been proposed to quantify acid–base interactions, in particular that of van Oss et al.[34–37] who define surface energy components γ^+ and γ^- to describe, respectively, the electron acceptor and electron donor characters of liquid and solid surfaces. According to these authors, the acid–base (or polar) component $\gamma^{ab}(=\gamma^P)$ of the surface energy of these liquids and solids is given by

$$\gamma^{ab} = 2(\gamma^+ \gamma^-)^{1/2} \tag{9.19}$$

and the work of adhesion between two entities 1 and 2 becomes

$$W_{12} = 2(\gamma_1^D \gamma_2^D)^{1/2} + 2(\gamma_1^- \gamma_2^+)^{1/2} + 2(\gamma_1^+ \gamma_2^-)^{1/2} \tag{9.20}$$

Acid–base components of the surface energy of any liquid or solid surfaces are determined from a set of given liquids, in particular, water, for which the magnitude of electron acceptor γ^+ and electron donor γ^- abilities are known or estimated. Such an approach is largely used in the field of materials and biological science.

At present, it is worth mentioning that a considerable amount of work is devoted to the acid–base concept and its applicability in the field of wettability. It is now well established that the acid–base interactions play a major role in adhesion science and are very useful to quantify the magnitude of hydrogen bonds at solid–solid interfaces.

9.2.4 CONCLUSION

It is now well known that the wettability of solid surfaces plays an important role as a basic physical phenomenon in numerous practical applications, like dissolution, dispersion, fabrics impregnation, cleaning, washing, coating, drying, etc. Moreover, as shown in the present section of this chapter, wettability measurements can be used as a tool for determining the surface energy of a solid, and particularly its dispersive and nondispersive components. These surface energy parameters are of major importance in the field of adhesion science since they can lead to the calculation of a thermodynamic reversible work of adhesion. It is relatively easy to carry out wettability experiments on well-defined and planar surfaces. On the contrary, the experimental determination of surface energy parameters through wettability measurements for divided solids is not a trivial task. Quite a lot of theoretical and experimental approaches have been therefore developed in the past several decades to assess such a determination. These approaches are briefly described in the next section.

9.3 WETTABILITY MEASUREMENTS ONTO DIVIDED SOLIDS

Measurements of contact angle by methods described in the previous section and which work well with flat surfaces cannot be directly applied to divided solids, such as fibers and powders. New approaches have necessarily been developed to obtain accurate and reproducible results of contact angle for these types of solids. The description of such methods is the object of the present section.

9.3.1 DROP-ON-FIBER SYSTEM

First, the contact angle θ of a liquid axisymmetrical drop deposited onto a cylindrical fiber can be measured by determining the shape of this drop following the original work of Carroll[38] and Yamaki and Katayama.[39] The schematic shape of the liquid drop and the definition of the parameters measured are shown in Figure 9.5.

The drop profile, called an undoloïd, can be represented by the following equation:

$$L = 2[aF(\phi,k) + nE(\phi,k)] \tag{9.21}$$

where

$$L = \ell/x_1, \quad n = x_2/x_1,$$

$$a = (x_2\cos\theta - x_1)/(x_2 - x_1\cos\theta),$$

$$k = [1 - (a^2/n^2)]^{1/2}$$

and

$$\phi = \sin^{-1}[(n^2 - 1)/(n^2 - a^2)]^{1/2},$$

and $F(\phi,k)$ and $E(\phi,k)$ are elliptic integrals of the first and second type, respectively.

Theoretically, it is possible to calculate the contact angle θ from Equation 9.21 knowing only the easily measurable parameters ℓ (drop length), x_1 (fiber radius), and x_2 (maximum drop radius). Unfortunately, however, Equation 9.21 has no analytical solution and θ has to be estimated by using numerical approaches. The results are generally reproducible, but the scatter on θ is usually in the order of $\pm 5\%$. Finally, this method can be applied when θ does not exceed about 60°. In fact, for larger values of θ, non-axisymmetrical drops (clamshell-like shape) are formed for which the measurement of contact angle remains impossible.

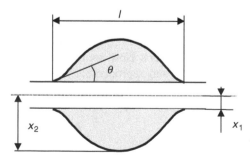

FIGURE 9.5 Shape and definition of parameters of drop-on-fiber system: ℓ and x_2 are respectively the length and the maximum radius of the drop; x_1 is the fiber radius and θ the contact angle.

The theoretical model proposed by Carroll and Yamaki and Katayama was revisited by different authors and particularly, by Song et al.[40] who introduced a generalized drop length–height method for the determination of contact angle in the drop-on-fiber systems. With this method, a large part of the drop profile is used in the calculation, which reduces the experimental error of the determination and improves the accuracy of the obtained results.

As an example, Carroll[38] has fully described that kind of measurement in two liquid phase system by calculating the contact angle of paraffin oil onto different types of polymeric fibers (polyester, nylon 6.6, fluorinated ethylene–propylene copolymer) totally immersed in water. He has also analyzed the effect of roughness of the fiber surface in the same way.[41] Such an approach has also been applied to untreated and plasma treated polyethylene fibers[42] in order to correlate their thermodynamic surface properties to the adhesion energy established at the interface between these fibers and an epoxy resin. About ten years ago, Walliser[43] developed a semiautomatic system (microwetting® system) allowing the measurement of contact angle of axisymmetrical liquid drops onto all types of fibers, but particularly polymeric fibers, such as aramid fibers. He has also fully described the experimental and solid–liquid interfacial thermodynamic conditions leading to a transition between axisymmetrical and non-axisymmetrical drops onto fibers (Figure 9.6). Very recently, Baley et al.[44] have used the approach of Song et al.[40] to determine the surface energy of plant fibers, i.e., untreated and surface treated flax fibers, in relation with the interfacial properties in flax fiber–polyester resin composites.

9.3.2 WILHELMY'S (TENSIOMETRIC) METHODS

Another experimental approach allowing us to determine the static and dynamic contact angles of liquid onto fibers (as well as onto thin films or

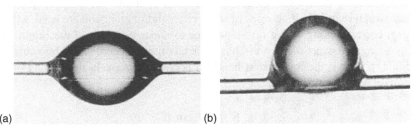

(a)   (b)

FIGURE 9.6 Photographs of liquid drops on polyethylene fibers: (a) axisymmetrical drop of α-bromonaphthalene, (b) non-axisymmetrical drop of polyethylene glycol. (From Walliser, A., *Caractérisation Des Interactions Liquid–Fibre élémentaire Par Mouillage*, PhD Thesis, Mulhouse (France), 1992.)

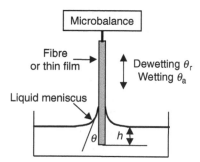

FIGURE 9.7 Schematic representation of the tensiometric method. θ_a and θ_r are respectively the advancing and receding contact angles whereas θ is the equilibrium contact angle; h is the immersion height.

plates) is the tensiometric or Wilhelmy's method (Figure 9.7). The fiber is suspended from the pan of an electrobalance above a beaker containing the liquid. By means of a motorized support, the beaker can be raised or lowered. During immersion and emersion cycles, the force F due to the meniscus of the liquid raised or lowered at the liquid–air interface and other phenomena described below is recorded.

The general equation describing the tensiometric measurement relates the variation of force F (the sample weight is considered as compensated by tarring prior to measurement) to the geometrical and thermodynamic parameters as follows:

$$F = p\gamma_L \cos \theta - \rho h A g \qquad (9.22)$$

The first term of this equation is the wetting force (p: fiber perimeter), whereas the second corresponds to the buoyancy (A: fiber cross-section area, ρ: solid–liquid density difference, g: gravitational acceleration). Sometimes, particularly when relatively high rates of immersion and/or emersion are used, a third term corresponding to the resistance due to shear viscosity of the liquid and involving a viscous shear stress has to be taken into account. The buoyancy is equal to zero if the immersion height h is equal to zero. In these conditions, Equation 9.22 reduces to

$$F = p\gamma_L \cos \theta \qquad (9.23)$$

During immersion and emersion, different contact angles are generally measured, the advancing θ_a and the receding θ_r dynamic contact angles respectively. Experimentally, Figure 9.8 describes schematically the variation

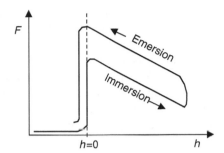

FIGURE 9.8 Schematic representation of the variation of the tensiometric force versus immersion depth h according to Equation 9.22 and taking into account different advancing and receding contact angles.

of the measured force versus the immersion length h of a fiber into the given liquid. Two straight lines with a negative slope are obtained in agreement with Equation 9.22 and taking into account both advancing and receding contact angles.

The wetting hysteresis, defined from these dynamic contact angles as $(\theta_a - \theta_r)$ or $(\cos \theta_a - \cos \theta_r)$, is a useful parameter related to either surface roughness, surface chemical heterogeneity, or surface molecular reorganization of fibers and thin films. As already mentioned, when high immersion and/or emersion rates are used, the values of the dynamic angles can be affected and only an apparent wetting hysteresis is obtained.

Fibers but also films or thin plates can be characterized by this tensiometric method. For example, Wei et al.[45] have studied the chemical stability in different media (air and liquids) of the surface of glass plates, which have been modified by grafting a silane coupling agent. Moreover, this method can lead to the estimation of the fiber diameter according to Equation 9.23, and its eventual variation along the fiber by using a liquid, such as an alkane, which totally wets the fiber surface ($\cos \theta = 1$). For example, Bascom[46] has applied this method for studying the wettability of carbon fibers. By immersing the fibers into hexadecane, he has determined a fiber diameter of 6.8 μm in good agreement with the actual value of 7 μm.

The previous approach cannot be directly used in the case of too viscous liquids.[47] It is therefore necessary to determine the wetting force acting on the fiber at zero immersion length. This has been used by Sauer[48] to determine the contact angle of melted polyethylene terephthalate at a temperature of 320°C on glass fibers.

In most cases, a two liquid phase tensiometric method is employed to determine the dispersive and acid–base components of the surface energy of fibers. This method is similar to the measurement of contact angle described in Figure 9.4. Therefore, the reduction of experimental data is performed according to Equation 9.13. In a few cases, formamide, which is also nonmiscible with alkanes, is used in place of water in order to avoid some problems related to the nonperpendicular penetration of the fiber across the liquid–liquid interface.[18] The principle is similar to this concerning the tensiometric method with one liquid phase, except that now two forces are successively measured (Figure 9.9); i.e., first a force $F_{a/air}$ at the alkane/air interface and, secondly, an additional force $F_{a/w}$ at the alkane/water interface.

For fibers of small diameter, the force due to buoyancy can be neglected. Moreover, alkanes totally wet the fiber surface, then $\cos \theta_f^{a/air} = 1$ (subscript f for fiber). Therefore, the force $F_{a/air}$ at the alkane/air interface is given by

$$F_{a/air} = p\gamma_A \qquad (9.24)$$

On the other hand, the force $F_{a/w}$ at the alkane/water interface can be written as

$$F_{a/w} = p\gamma_{WA}\cos \theta_f^{a/w} = p\xi \qquad (9.25)$$

where ξ can be considered as the tension of the adhesion.[49] Combining Equation 9.24 and Equation 9.25, ξ is given by

$$\xi = \frac{F_{a/w}}{F_{a/air}}\gamma_A \qquad (9.26)$$

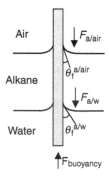

FIGURE 9.9 Schematic representation of the forces acting on a fiber immersed in two immiscible liquid phases: $F_{a/air}$ at the alkane/air interface and depending on the contact angle on the fiber $\theta_f^{a/air}$; $F_{a/w}$ at the alkane/water interface and depending on the contact angle $\theta_f^{a/w}$; $F_{buoyancy}$.

Finally, it is not necessary to measure or to calculate the contact angle at water/alkane interface to determine the surface energy of a fiber. Effectively, according to Equation 9.13, it is just necessary to plot the quantity $[\gamma_W - \gamma_A + \xi]$ versus $[\sqrt{\gamma_W^D} - \sqrt{\gamma_A^D}]$ to determine the dispersive component γ_f^D of the surface energy of the fiber from the slope of such a straight line and the nondispersive interactions with water W_{fW}^{ND} from its intercept at the origin.

As for planar surfaces, a criterion of displacement (Equation 9.14) of n-alkanes by water has to be fulfilled.

The tensiometric method, in one or two liquid phases, is largely used nowadays to determine the thermodynamic surface properties of fibers (and also of course thin films), whatever the nature of these latter. This is particularly the case for cylindrical and rigid fibers, like glass and carbon but also polymeric fibers, which can easily penetrate into the liquid phases without bending. As a matter of fact, it is generally observed that flexible fibers, like natural and more particularly plant fibers, can take a pronounced curved shape in contact with liquid interfaces and sometimes the penetration of such fibers into high surface tension liquids, water in particular, cannot be possible. However, a lot of results are available in the literature for different kinds of materials. For example, considering the most recent studies, the characterization of carbon fiber surfaces in terms of acidic and basic functional groups for untreated and surface treated fibers has been performed by dynamic contact angles.[50,51] Large differences concerning the surface energy of carbon fibers are observed. In most cases, the results obtained by wettability techniques are used to enhance interfacial adhesion behavior of such fibers in composites materials (see Section 9.4 of the present chapter) and, consequently, to improve the mechanical properties of these latter. Similar studies were performed on glass fibers. For example, surface treatments by silanization[52,53] are largely analyzed, since it is well known that the use of coupling agents, and in particular silanes, can greatly improved the level of adhesion at fiber–matrix interface in composite materials. The Wilhelmy method can also be used to analyze the effect of a sizing on glass fiber surface properties,[54] since such a sizing can play a major role on the thermodynamic and mechanical behavior of the fiber–matrix interface. Obviously, due to the importance of fiber surface properties in the field of textile and clothing industries (see Renner and Bueno[55] in this book), dynamic contact angle measurements are currently used to determine the surface energy of untreated and surface-treated natural fibers as well as polymeric fibers, such as poly(ethylene terephthalate) fibers,[56] polypropylene fibers,[57] and so on. Finally, even if as mention above the experiments are more difficult to carry out, the tensiometric method is largely used nowadays to characterize natural or plant fibers, for example, wood fibers,[58] cellulose,[59] flax fibers,[60,61] fiber plant straw fraction,[62] etc., mostly for application in the field of paper and/or composite materials.

9.3.3 WETTABILITY MEASUREMENTS ONTO POWDERS BY CAPILLARY RISE

The interest in determining contact angles of liquid onto powders and surface-free energy of these latter is now clearly demonstrated. For example, determination of surface energy of powders is widely used in many scientific areas such as pharmacology, reinforced materials, the textile industry, etc. However, the direct measurement of contact angles of a liquid drop on a solid is not applicable to small powder particles. Thus, a different way, based on the determination of the kinetics of liquid rise in a column was developed almost simultaneously by Lucas,[63] Washburn,[64] and Rideal[65] early in the 20th century.

In practice, the powder to be studied is precisely packed into a glass tube which is closed at its lower end by a filter (sintered glass, usually). This tube is brought into contact with a liquid, which rises into powder column due to capillarity. The rising rate of the liquid into the powder is measured either visually or by weight gain with an electrobalance (Figure 9.10 and Figure 9.11). This kinetics enables us to calculate contact angles, which are

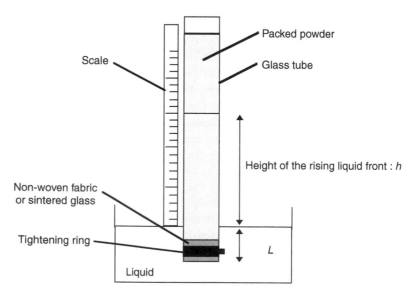

FIGURE 9.10 Schematic representation of an experimental setup for capillary height measurement.

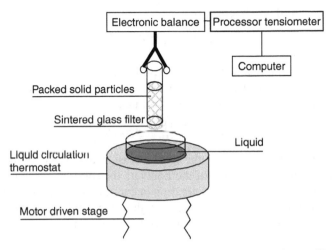

FIGURE 9.11 Schematic representation of an experimental setup for capillary weight measurement.

related to surface energy, through the general wettability relationships described previously.

The porous media can be represented (Figure 9.12) by an array of parallel but tortuous capillaries.

According to Poiseuille's law,[66] expressing the balance between viscous forces and capillary and hydrostatic forces (neglecting inertial effects), the rate

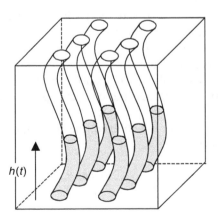

FIGURE 9.12 Schematic representation of capillary rise in porous media. $h(t)$ is the liquid height as a function of time.

of liquid penetration v is given by

$$v = \frac{dh}{dt} = \frac{(\chi R_D)^2}{8\eta} \frac{\Delta P}{h} \qquad (9.27)$$

with h the height reached by the liquid front at time t, R_D the mean hydrodynamic radius of pores, χ the tortuosity of pores, η the viscosity of the liquid, and ΔP the pressure difference, i.e.,

$$\Delta P = \frac{2\gamma_L \cos \theta}{R_S} - \rho g h \qquad (9.28)$$

where γ_L and ρ are respectively the surface tension and density of the liquid, θ is the advancing contact angle of the liquid on the solid, R_S is the mean static radius of pores, and g is the acceleration due to gravity. The tortuosity of the pores, defined as the ratio of the real length of the capillary to its projective onto one axis (one dimension), is obtained from both static and hydrodynamic radii according to $\chi = \sqrt{R_S/R_D}$.

Equation 9.27 is valid for the following restrictions: (i) steady-state laminar flow is considered, (ii) Newtonian liquids are used, and (iii) friction effect and inertial phenomena are negligible.

In the early stages of the progression (when h is largely lower that the maximum liquid height h_{eq}, see below), if the hydrostatic pressure can be neglected, integration of Equation 9.27 for the boundary condition $h = 0$ when $t = 0$ leads to the well-known Washburn's equation, which relates linearly the square of the height h^2 with the time t.

$$h^2 = \frac{R_D \gamma_L \cos \theta}{2\eta} t \qquad (9.29)$$

Two unknown terms are involved in relationship (9.29): R_D and θ. The common way to overcome this problem is to use first a total wetting liquid ($\theta = 0$, then $\cos \theta = 1$) to calculate R_D. In most cases, alkanes are used and more particularly, octane (with respect to its viscosity and vapor pressure). Knowing R_D and using columns of constant packing characteristics, it is therefore very easy to determine the contact angle θ of nonwetting liquids ($0° < \theta < 90°$).

If the equilibrium state is achieved, the maximum liquid height h_{eq} is given by (Jurin's law).

$$h_{eq} = \frac{2\gamma_L \cos \theta_{eq}}{\rho g R_S} \qquad (9.30)$$

with θ_{eq} the equilibrium (static) contact angle. Equation 9.30 leads to the determination of the static pore radius R_S. Thus, from the dynamic measurement and for a total wetting liquid ($\cos\theta = 1$), the tortuosity χ can be determined.

To overcome some difficulties, when the liquid front is not visible or when it is not certain that visible front reflects accurately the real progression of the liquid in the powder column, it can be of interest to perform weight gain measurements[67,68] instead of direct height measurements. It is also evident that an automatic weight measurement leads to a greater precision than an optical height determination.

The weight w of the rising liquid is related to the height h reached in the powder column according to

$$w = \varepsilon\rho\pi r^2 h \tag{9.31}$$

where ε is the porosity of the packed powder column, ρ the density of the liquid, and r is the inner radius of the tube. As previously, if the early stages of the progression are considered, and taking into account Equation 9.29, the following relationship is obtained:

$$w^2 = (R_D(\varepsilon\pi r^2)^2)\frac{\rho^2\gamma_L\cos\theta}{2\eta}t \tag{9.32}$$

which leads to

$$w^2 = c\frac{\rho^2\gamma_L\cos\theta}{2\eta}t \tag{9.33}$$

The term c is a geometric factor that is kept constant insofar as the packing density of the powder into the column remains constant. As previously, two unknown terms, c and θ, are involved in Equation 9.33 and one has to use first a total wetting liquid such as octane to determine the constant c, then θ is measured for other liquids which do not totally wet the powder surface.

The comparison between height and weight methods was carefully analyzed from a practical point of view by studying the capillary rise of polar and nonpolar liquids into different types of untreated and surface treated silica particles (silica flour, sand, and limestone).[69,70] In both cases, linear relationships relating the square of the liquid height h^2 (Figure 9.13) or weight w^2 versus time t were obtained, which lead to remarkably similar values of the components of surface energy (dispersive component γ_S^D and the nondispersive interaction term W_{SL}^{ND} with different polar liquids) of silica particles. Moreover, for a silica flour previously modified by grafting of

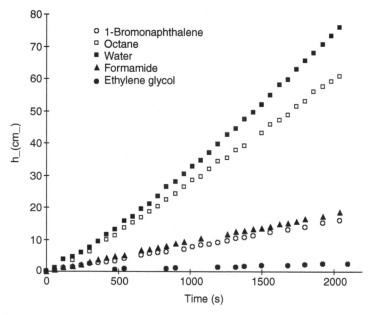

FIGURE 9.13 Experimental variations versus time of the square of the capillary height h^2 of different nonpolar and polar liquids into a packed column of untreated silica flour. (From Siebold, A., Walliser, A., Nardin, M., Oppliger, M., and Schultz J., *J. Colloid Interface Sci.*, 186, 60, 1997.)

trimethoxypropylsilane—in other words, a very hydrophobic surface mainly populated by $-CH_3$ groups—a value of about 23 mJ/m^2 for the surface energy of this powder was obtained in good agreement with literature data.[71]

However, several experimental problems have to be taken into account concerning the capillary method based on weight measurements, which greatly limit its applicability, even if it appears *a priori* easier and more accurate than height measurements. In fact, with this method, one measures a total weight, which includes in particular the weight of the meniscus formed by the liquid at the outer side of the column. This meniscus can easily be taken into account for a well-defined system like a glass tube. Obviously, it is more difficult to solve this experimental problem in the case of the irregular shape of a porous materials or a fiber fabrics, for example.[72,73] Moreover, as it will be shown in the Section 9.3.4, very complex phenomena related to the evaporation of the liquid and/or the existence of two apparent rising liquid fronts could strongly affect weight measurements and seem to indicate that the weight method cannot be applied to nonisotropic systems like fiber fabrics.

FIGURE 9.14 Micrographs of the shape of the meniscus of dodecane in a capillary glass tube: (a) at large velocity ($\theta \approx 33°$); (b) at equilibrium or zero velocity ($\theta \approx 0°$). (From Siebold, A., Walliser, A., Nardin, M., Oppliger, M., and Schultz, J., *Colloids Surf. A*, 161, 81, 2000.)

The dynamic advancing contact angle is generally larger than the static one, even for a total wetting liquid (Figure 9.14), and this fact is not considered in the literature for capillary rise in powder column in most cases. Hoffman[74] proposed an empirical relationship between the contact angle and the liquid penetration rate v and more precisely the capillary number $C_a = v\eta/\gamma$. This relationship takes the form of a power law $\theta = kC_a^{1/3}$ when C_a tends towards zero. Ngan and Dussan[75] found that this relationship does not hold for any fluid. Other authors have recently proposed another relationship.[76] Due to this problem of variation of contact angle during capillary progression, the validity of Washburn's equation is therefore controversial for determining the thermodynamic characteristics of powder surfaces.

Nevertheless, Washburn's approach is largely used to estimate the thermodynamic surface properties of powders packed into a column or deposited as a rather thin layer (a few hundreds of micrometers) on a flat surface (thin layer wicking technique). In this latter technique, the powder is deposited from a suspension, usually aqueous ones, onto a (glass) slide to form—after thorough drying and an eventual thermal treatment—a uniform thin layer of powder firmly adhering to the substrate. The capillary progression is carried out into such a thin layer and analyzed through theoretical approaches quite similar to these presented above. In particular, Chibowski and coworkers have largely analyzed fundamental and practical aspects of such a technique.[77–80] The thin layer wicking method for measuring the contact angles of liquids onto powdered materials was applied to a lot of minerals (see, for example).[81]

9.3.4 CAPILLARY RISE INTO FIBROUS SYSTEMS AND FABRICS

Washburn's approach can also be applied to fibrous systems—in particular, woven fabrics—whatever the nature of the fibers. In that case, the experiment is often carried out in a noncontrolled environment, i.e., in room atmosphere and temperature where the evaporation of the rising liquid occurs. A new model, based on the study of Bico,[82] describes the capillary rise into fabrics, taking into account the evaporation of the liquid has recently been proposed.[83,84] The new equation describing the capillary rise kinetics is the following one:

$$\frac{dh}{dt} = \frac{(\chi R_\mathrm{D})^2}{8\eta h}\left(\frac{2\gamma_\mathrm{L}\cos\theta}{R_\mathrm{S}} - \rho g h\right) - \frac{h}{\kappa} \tag{9.34}$$

where $\kappa = \varepsilon A/\alpha p$, with α the specific evaporation rate (m/s) of the liquid considered, p the perimeter of evaporation, A the cross-section area of the porous medium, and ε the porosity of the medium. In the early stages, when the hydrostatic pressure can be neglected, integration of Equation 9.34 for the boundary condition $h = 0$ at $t = 0$ leads to

$$h^2 = h_\mathrm{eq}^{*2}\cdot(1 - e^{-2t/\kappa}) \tag{9.35}$$

The height at equilibrium h_eq^*, obtained when the evaporation of the rising liquid occurs, is generally largely smaller than this corresponding to Jurin's law (Equation 9.30) due to such an evaporation. It is therefore easier to measure this height. Experimentally, κ can be determined from the slope of the straight line obtained by plotting $\ln(1 - h^2/h_\mathrm{eq}^{*2})$ versus $(-2t/\kappa)$, avoiding therefore the difficult determination of the parameter α. Afterwards, both R_S (from a total wetting liquid for which $\cos\theta = 1$) and, subsequently, the value of the contact angle θ of a given nonwetting liquid onto the fibers, are obtained from

$$R_\mathrm{S}\cos\theta = \frac{4\eta h_\mathrm{eq}^{*2}}{\gamma_\mathrm{L}\kappa} \tag{9.36}$$

Experimental results show that this model seems to work very well for the glass woven fabrics. They also show that the evaporation of the rising liquid, depending on external conditions (ventilation, temperature, ...) plays a major role on the capillary progression. This is illustrated in Figure 9.15 for example, concerning the capillary progression of an-alkane (n-nonane for the present case) in a glass fabric.[83] For a saturated atmosphere, a linear relationship h^2 versus time is obtained in good agreement with the standard Equation 9.29 based on Washburn's analysis. However, when the atmosphere is not controlled, it appears that the values of the capillary heights are located below the previous straight line and tend towards a maximum height at equilibrium h_eq^* lower than this given by Equation 9.30.

As described above, the problem of liquid evaporation during capillary rise can be easily analyzed by taking into account some experimental parameters related to such an evaporation, κ and h_eq^* in particular. It

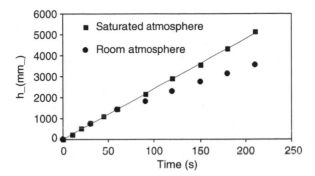

FIGURE 9.15 Role of evaporation of the rising liquid (*n*-nonane) on the capillary progression in a glass fabrics, expressed as the variation of the square of the capillary height h^2 versus time. (From Sénécot, J. M., *Etude De l'imprégnation Capillaire De Tissus De Verre*, PhD Thesis, Mulhouse (France), 2002.)

immediately appears in Figure 9.16 that linear relationships in semi-logarithmic scales are obtained in very good agreement with Equation 9.35, for a series of *n*-alkane rising into a glass fabrics (the same as this considered in Figure 9.15). The slope of these straight lines is related for each liquid used to the parameter κ, which allows the determination of the contact angle of this liquid onto the fibers constituting the fabrics.

Finally, it is worth mentioning that fabrics are highly nonisotropic materials compared to a column of packed powders due to their weaving

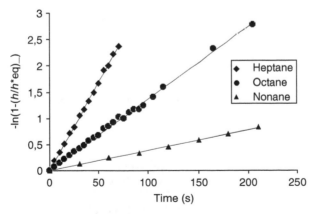

FIGURE 9.16 Capillary progression of three different *n*-alkanes into glass fabrics, analyzed according to Equation 9.35, and taking into account the evaporation of the rising liquid. (From Sénécot, J. M., *Etude De l'imprégnation Capillaire De Tissus De Verre*, PhD Thesis, Mulhouse (France), 2002.)

pattern and, particularly, to the presence of warp and weft structures. Moreover, to a first approximation two populations of pores have to be considered, since the mean size of pores located between yarns (macropores) is largely higher than this of pores located between monofilaments into the yarns (micropores). As a consequence, the capillary rise can be greatly affected by such a structure.[73] This can lead, for example, to the existence of two liquid fronts during the capillary rise, each of them exhibiting its own rate of progression.[83] Such a phenomenon has been attributed to the particular capillary transfer of the liquid from the warp to the weft yarns (or vice versa), which takes a given time leading to the observation of two different liquid fronts.

The horizontal capillary progression of liquids into fabrics has also been studied[85,86] in the literature. However, contrary to the vertical capillary rise, it is carried out generally by means of liquid drops directly deposited on the fabrics. These drops can be considered therefore as small and finite reservoirs which are totally drained at the end of the experiments. As suggested by Gillespie,[87] the horizontal capillary progression can proceed in two steps (Figure 9.17).

The first step can be related to a capillary impregnation from an infinite reservoir, as in the case of vertical capillary rise. For the second step, the reservoir becomes finite and tends to be totally drained into the fabrics.

To a first approximation and from the equations previously proposed by Gillespie,[87] Kissa[85] has established the following expression for the rate of liquid imbibition into the fabrics:

$$A(t) = K \left(\frac{\gamma}{\eta} \right)^u V^m t^n \qquad (9.37)$$

where A is the surface area of the impregnated zone, K is a coefficient, γ and η are respectively the surface tension and the viscosity and, finally, V is the volume of the liquid drop. Concerning step 1, the exponents involved in Equation 9.37 are equal to $u = 0.5$, $m = 0$ and $n = 0.5$. Therefore, in

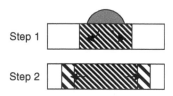

FIGURE 9.17 Horizontal capillary progression from a liquid drop deposited onto fabrics: schematic representation of the two steps of progression.

agreement with previous discussion, Equation 9.37 corresponds to Washburn's Equation (Equation 9.29). For the second step and for *n*-alkanes spreading on polyethylene terephthalate/cotton, polyethylene terephthalate, and cotton fabrics, Kissa has proposed the following experimental values of these exponents: $u = 0.3$, $m = 0.7$ and $n = 0.3$. Kawase et al.[86] have confirmed such a result in the case of other fabrics. Finally, the transition between the two steps 1 and 2 is easily determined by plotting A versus t in logarithmic scales (Equation 9.37), the slope n of the straight lines obtained varying from 0.5 for step 1 to about 0.3 for step 2.

9.3.5 OTHER WETTABILITY MEASUREMENTS ON DIVIDED SOLIDS

Beyond well-known Wilhelmy's and capillary rise approaches previously described, other methods allowing the assessment of surface energy of divided solids through wettability measurements have to be mentioned. This is the case in particular for immersional wetting method, which consists of dipping entirely a solid (powders, fibers or fabrics) into a liquid. The fact that the solid/vapor (or air) interface (SV) is replaced by a solid/liquid interface (SL) corresponds to a free energy change of immersion ΔG_{im}, which is equal to the difference of the interfacial energies, i.e., $\Delta G_{im} = \gamma_{SL} - \gamma_{SV}$. Therefore, this free energy change can be combined with both Young's Equation (Equation 9.1) and Gibbs equation to lead to the enthalpy of immersion,[8,88] also called heat of immersion ΔH_{im}, as follows:

$$\Delta H_{im} = \Delta G_{im} - T\left(\frac{\partial \Delta G_{im}}{\partial T}\right) = T\gamma_{L}\frac{\partial \cos \theta}{\partial T} - \left(\gamma_{L} - T\frac{\partial \gamma_{L}}{\partial T}\right)\cos \theta \quad (9.38)$$

where T is the absolute temperature. This thermodynamic approach can be applied to wettability measurements on powders and fibers by means of microcalorimetry methods and by determining the heat of immersion at different temperatures, provided that the variations of liquid surface tension with temperatures are known.

Another useful method is based on the adsorption of vapor on the solid surface and the analysis of the adsorption isotherms. As described above (see Section 9.2.1), the spreading pressure π of a vapor onto a solid surface is given by $\pi = \gamma_{S} - \gamma_{SV}$, where γ_{SV} represents the surface free energy of the substrate after equilibrium adsorption of vapor and γ_{S} is the surface free energy of the solid in vacuum. Therefore, when the vapor obeys the ideal gas law, the spreading pressure can be determined as follows:[8,88,89]

$$\pi = \gamma_S - \gamma_{SV} = RT \int_0^{p_{ve}} \Gamma \cdot \partial(\ln p_v) \qquad (9.39)$$

where R is the ideal gas constant, p_v is the vapor pressure, p_{ve} is the equilibrium vapor pressure, and Γ is the surface concentration of adsorbed vapor. The dispersive and nondispersive components of the surface energy of divided solids can therefore be estimated from Equation 9.39 by adsorption of different vapors: respectively nonpolar vapors, like n-alkanes, then polar vapors, such as water.

All of these wettability methods are briefly presented and discussed in a recent review by Lazghab et al.[90]

Finally, another approach is largely used nowadays to determine the thermodynamic surface properties of fibers and powders, i.e., the well-known inverse gas chromatography (IGC) technique. The principle of this method is based on the adsorption of gaseous probes onto divided solids placed into a chromatographic column. Even if wettability and IGC approaches are often compared, this latter is obviously based on different physical concepts. Moreover, IGC is fully described and analyzed in another chapter of the present book.[91]

9.4 ROLE OF SURFACE PROPERTIES OF FIBERS AND FILLERS, DETERMINED BY WETTABILITY MEASUREMENTS, ON THE MICROMECHANICAL BEHAVIOR OF THE INTERFACES

In numerous studies, it has been shown that the thermodynamic surface properties of solids (in other words, their surface energy) play a major role on the ability of such solids to interact, from a physicochemical point of view, with their environment. It is particularly true in the field of composite and reinforced materials, where a given level of adhesion has to be established at the interface between fibers (or filler) and the matrix, in order that the resulting multi-materials exhibit good mechanical or use properties. The aim of the present section is to give a few examples of relationships between surface energy of fibers or powders and mechanical (micromechanical) behavior of fiber/matrix or filler/matrix systems. The surface energy parameters of the divided solids are determined by the wettabilty techniques described in the previous sections.

The first example deals with different types of untreated and oxygen plasma treated carbon fibers. Paiva et al.[92] have determined the surface energy of pitch and PAN-based carbon fibers, and more particularly the nondispersive

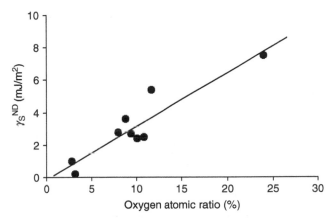

FIGURE 9.18 Relationship between the nondispersive component γ_S^{ND} of the surface energy of untreated and plasma treated carbon fibers and the ratio of oxygen atoms on the fiber surface as determined by XPS. (From Paiva, M. C., Bernardo, C. A., and Nardin, M., *Carbon*, 38,1323, 2000.)

component γ_S^{ND} of this surface energy, by means of tensiometric method in both one and two liquid phase methods. The liquid used were α-bromo-naphthalene and water/alkanes systems respectively and, therefore, γ_S^{ND} is related to the polar interactions with water and was calculated using Equation 9.7 in particular. Very low values of γ_S^{ND} are obtained for untreated carbon fibers, which therefore appear to be almost nonpolar materials. On the contrary, oxygen plasma treatments greatly increase the polar component of the surface energy of the fibers. In addition to wettability measurements, the authors have determined the ratio of oxygen atoms present on the surface for the different carbon fibers by means of x-ray photoelectron spectrometry (XPS). Figure 9.18 shows that a linear relationship can be established between γ_S^{ND} and this surface oxygen atomic content. Such a result clearly indicates that wettability measurements lead to a thermodynamic characterization of solid surfaces which is in very good agreement with other types of characterization, like the electronic spectroscopy used in the present case.

Moreover, the authors have analyzed the micromechanical properties of the interface between these untreated and plasma treated carbon fibers and a polymeric thermoplastic matrix, i.e., polycarbonate. More precisely, the fiber–matrix interfacial shear strength τ was determined by means of a fragmentation test, performed on single filament composites.[93] The most important point is that the interfacial shear strength can be considered, to a first approximation, as an estimation of the fiber–matrix adhesion. Figure 9.19 shows that the fiber–matrix interfacial shear strength increases, more or less linearly, with the

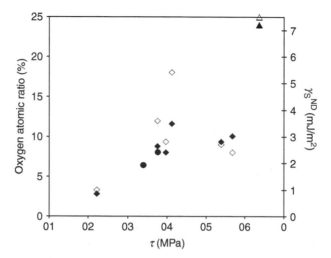

FIGURE 9.19 Relationships between the fiber–matrix interfacial shear strength τ and the nondispersive component γ_S^{ND} of the surface energy of carbon fibers (open symbols) or the oxygen atomic ratio on the fiber surface (filled symbols). (From Paiva, M. C., Bernardo, C. A., and Nardin, M., *Carbon*, 38,1323, 2000.)

nondispersive component of the surface energy of the carbon fibers γ_S^{ND}, or in other words, with the oxygen atomic ratio of the fiber surface in agreement with Figure 9.18.

Such a result clearly confirms that the adhesion established at the fiber–matrix interface is one of the most important parameters in controlling the interfacial behavior, and particularly, the interfacial shear strength τ, or the capacity of the interface to transfer the stress from the matrix to the fiber. We can therefore expect that such a fiber–matrix adhesion plays a major role on the mechanical performance and/or use properties of composite materials.[94]

From a general point of view, concerning the micromechanical analysis of fiber–matrix interfaces in model composites, the following linear relationship between the interfacial shear strength and the reversible work of adhesion W established at these interfaces, was proposed about ten years ago:[95]

$$\tau = \left(\frac{E_m}{E_f}\right)^{1/2} \frac{W}{\lambda} \qquad (9.40)$$

where E_m and E_f are elastic moduli of the matrix and the fiber respectively and λ is a distance. The interfacial shear strength was mostly measured by a fragmentation test on single fiber composites, whereas W was estimated from the surface properties of both constituents in contact, considering that only dispersive interactions and electron acceptor–donor interactions (see Section 9.2) are involved between both solids. Moreover, the components of the surface energy of fiber and matrix were determined either by wettability techniques or by inverse gas chromatography. The most important points concerning determination and calculation of τ and W are fully described elsewhere.[95]

The remarkable result is that in Equation 9.40, the average distance λ, equal to about 0.5 nm, corresponds to an equilibrium intermolecular centre-to-centre distance[8] involved in molecular interactions, such as van der Waals interactions, for example. Thus, apart from a mechanical term $(E_m/E_f)^{1/2}$, it appears that the interfacial shear strength (a mechanical quantity) is completely determined by the level of physicochemical interactions at the fiber–matrix interface, or in other words, by the reversible work of adhesion (a thermodynamic quantity) between fibers and matrix.

Experimentally, Equation 9.40 is valid, to a first approximation, for numerous types of fiber–matrix systems, such as carbon or glass fibers in thermosetting, thermoplastic, or elastomeric matrices (Figure 9.20). However, it does not apply directly to cases where properties of polymers are modified near the fiber; in other words, when interfacial layers exhibiting particular properties exist in the vicinity of the fiber surface.[96] Therefore, the mechanical properties, in particular the elastic modulus, of these interfacial

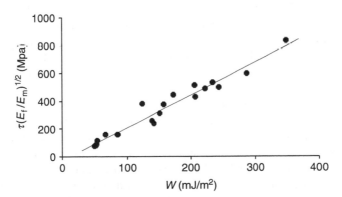

FIGURE 9.20 General relationship according to Equation 9.40 between the interfacial shear strength τ, normalized by the mechanical term $(E_f/E_m)^{1/2}$, and the reversible work of adhesion W for different fiber–matrix systems: carbon and glass fibers into thermosetting (epoxy and vinylester resins) and thermoplastic (polyethylene, poly (ether–ether–ketone)) polymeric matrices. (From Nardin, M. and Schultz, J., *Compos. Interfaces*, 1, 177, 1993.)

layers have to be taken into account in Equation 9.40 in place of this of the bulk matrix.

According to Israelachvili,[8] the work of adhesion W can be expressed as $W = P_A \lambda$, where P_A is the adhesive pressure at the interface. Therefore, Equation 9.40 becomes:

$$\tau = \mu \cdot P_A \qquad (9.41)$$

with $\mu = (E_m/E_f)^{1/2}$. It appears that the form of Equation 9.41 is equivalent to a law of friction (Amonton's law)[97] and, therefore, μ could be considered as a friction coefficient of static type essentially. The theoretical bases of Equation 9.40 and Equation 9.41 and the role played by the intermolecular interfacial distance λ as a function of the nature of the interactions involved at the interface were also examined.[98] However, such a relationship clearly indicates that, in the absence of external mechanical constraint, the interfacial shear strength is different from zero due to the existence of the adhesive pressure stemming from molecular interactions at the interface. This confirms that the thermodynamic surface properties of the materials in contact and determined by wettability measurements plays a major role, at least on the micromechanics, and certainly on the macromechanics of reinforced materials. Finally, it is worth mentioning that the previous analysis, based on Equation 9.40, has been very recently examined by Baley et al.[44] in the case of new types of composites based on vegetable fibers, such as flax fibers in particular.

Similar approaches can also be applied to filler–matrix systems. As for fiber reinforced composites, the level of adhesion at the interface between fillers and a matrix is of major importance in determining the mechanical performance of the resulting materials. For example, in a pioneer work concerning glass-bead filled polymers, Dekkers and Heikens[99–104] have clearly shown that the degree of interfacial adhesion determines to a large extent the tensile behavior of the composites. In polycarbonate matrix in particular, they observed that a poor level of particle–matrix adhesion can lead to dewetting cavitation at the interface, which contributes to inelastic deformations. On the contrary, in the case of good adhesion, shear deformation is found to be the only significant deformation mechanisms. These authors have also fully analyzed the craze formation in glass-beads reinforced glassy polymers, such as polystyrene, as a function of wetting or adhesive properties of filler surface (glass beads). They observed that, in the case of good adhesion, the crazes form near the pole of the glass bead (Figure 9.21a), whereas with poor interfacial adhesion, dewetting first occurs at the interface near the pole and propagates around the beads (Figure 9.21b), then crazes form in the matrix at an angle of about 60° from the poles relatively to the stress direction (Figure 9.21c).

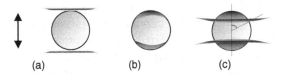

(a)   (b)   (c)

FIGURE 9.21 Schematic representation of craze formation process in polystyrene matrix around (a) an excellent adhering glass bead and (b,c) at two successive stages of a poor adhering glass bead, (b) dewetting during deformation, and (c) craze pattern. The arrow indicates the direction of the applied tension. (Redrawn from Dekkers, M. E. J. and Heikens, D., *J. Mater. Sci.*, 18, 3281,1983, Dekkers, M. E. J. and Heikens D., *J. Mater Sci. Lett.*, 3, 307, 1984.)

Several investigators have shown that the strength of the particle–polymer interfaces affects the yield stress of the reinforced materials and that poor adhesion tends to decrease this yield stress.[105] More recently, Kawagachi and Pearson,[106–109] have fully analyzed the effect of particle–matrix adhesion, and particularly the effect of moisture on the mechanical behavior of reinforced epoxies, such as yield behavior, cohesive strength, fracture toughness, and crack propagation. Whatsoever, it is now well accepted that the wettability properties of filler, which are related to their surface energy and therefore to their ability to establish adhesive interactions with their environment and polymeric matrix in particular, are of major importance in this field of reinforced materials, and more generally, multi-materials.

9.5 CONCLUSION

In the first part of the present chapter, it is shown that the thermodynamic surface characteristics of a solid play a major role in determining its adhesive properties, or in other words, its ability to establish physico-chemical interactions (van der Waals dispersive interactions, acid–base interactions, …) when it is in contact with another materials. It has been shown that these surface properties, i.e., surface and interfacial free energies and, consequently, the thermodynamic reversible work of adhesion, can be assessed by wettability measurements. From both experimental and theoretical point of views, it is more difficult to carry out and analyze wettability measurements on divided solids, like fibers and powders, than on well-defined and flat surfaces.

Therefore, in the second part of this chapter, we focused our attention on the description of the main techniques specially devoted to wettability measurements on such divided solids. These techniques are: the measurements of contact angle of liquid drops directly deposited on a fiber (drop-on-fiber system), the weight determination of the liquid meniscus on a fiber (Wilhelmy's method), and the capillary progression into a column of packed powders or into fiber fabrics. Other approaches are more briefly described, such as horizontal capillary imbibition, immersional wetting and microcalorimetry and spreading pressure determination trough vapor adsorption.

Finally, the role of surface properties of fibers and powders, determined by wettability measurements, on the mechanical behavior of fiber–matrix and filler–matrix interfaces in reinforced composites are briefly illustrated through two or three examples in the last part of this chapter.

List of Symbols

A surface area or cross section area (of a fiber, of a porous medium)

AN, DN Gutman's electron acceptor and electron donor numbers of liquids

c factor involved in weight gain measurements for capillary rise

C_a capillary number

C^A, E^A Drago's parameters characterizing the acidic character (superscript B for basic character) of a given system

E_m, E_f elastic modulus of matrix and fiber respectively

f factor converting enthalpy into free energy

F force

$F_{a/air}$ wetting force acting on a fiber at alkane/air interface

$F_{a/w}$ wetting force acting on a fiber at alkane/water interface

$F(\phi,k)$,

  $E(\phi,k)$ elliptic integrals of the first and second type, respectively

g gravitational acceleration

ΔG_{im} free energy change of immersion

h immersion length or height of capillary rise

h_{eq} maximum height of capillary rise (without evaporation)

h_{eq}^* maximum height of capillary rise (with evaporation)

$-\Delta H^{ab}$ variation of enthalpy corresponding to the establishment of an acid–base interaction

ΔH_{im} enthalpy or heat of immersion

K_A, K_D electron acceptor and electron donor numbers for a solid surface

ℓ length of a liquid drop (undoloïd) onto a fiber

n^{ab} number of acid–base bonds per unit interfacial area

p fiber perimeter

p_v vapor pressure

p_{ve} equilibrium vapor pressure

P_A adhesive pressure

ΔP	pressure difference
r	inner radius of a column
R	ideal gas constant
R_D	mean hydrodynamic radius of pores
R_S	mean static radius of pores
S	spreading coefficient
t	time
T	absolute temperature
v	rate of capillary penetration of a liquid into a porous medium
V	volume
w	weight
W, W_{12}	reversible work of adhesion between two media 1 and 2; subscripts 1 and 2 being replaced by subscripts S or L for solid and liquid respectively, by W for water, by A for alkane, ...
W^D	dispersive component of the reversible work of adhesion
W^{ND}	nondispersive component of the reversible work of adhesion
W^{ab}	acid–base (= nondispersive) component of the reversible work of adhesion
x_1, x_2	fiber radius and radius of a liquid drop (undoloïd) onto a fiber, with $n = x_2/x_1$
α	specific evaporation rate
γ	surface tension or surface energy
γ^D	dispersive component of the surface energy
γ^{ND}	nondispersive component of the surface energy
γ^{ab}	acid–base (nondispersive) component of the surface energy
γ^+, γ^-	acid and basic components respectively of the surface energy
Γ	surface concentration of vapor adsorbed on a solid
ε	porosity
η	viscosity
θ	contact angle
θ_{eq}	contact angle at equilibrium
θ_a	advancing contact angle
θ_r	receding contact angle
κ	evaporation coefficient given by: $\kappa = \varepsilon A/\alpha p$
λ	intermolecular distance
μ	friction coefficient
ξ	term considered as the tension of adhesion and given by: $\xi = \frac{F_{a/w}}{F_{a/air}}\gamma_A$
π	spreading pressure of a vapor onto a solid surface
ρ	density
τ	interfacial shear strength
χ	tortuosity of pores

REFERENCES

1. Kinloch, A. J., *Adhesion and Adhesives. Science and Technology*, Chapman & Hall, London, 1987.
2. Lee, L.-H., Ed., *Fundamentals of Adhesion*, Plenum Press, New York and London, p. 1, 1991.
3. Schultz, J. and Nardin, M., In *Adhesion Promotion Techniques*, Pizzi, A. and Mittal, K. L., Eds., Marcel Dekker Inc., New York, p. 1, 1999.
4. Sharpe, L. H. and Schonhorn, H., *Chem. Eng. News*, 15, 67, 1963.
5. Young, T., *Philos. Trans. Roy. Soc.*, 95, 65, 1805.
6. Dupré, A., *Théorie Mécanique De La Chaleur*, Gauthier-Villars, Paris, 1869.
7. Fowkes, F. M., *J. Phys. Chem.*, 67, 2538, 1963.
8. Israelachvili, J. N., *Intermolecular and Surface Forces*, 2nd ed., Academic Press, New York, 1992.
9. Fowkes, F. M., *Ind. Eng. Chem.*, 56, 40, 1964.
10. Owens, D. K. and Wendt, R. C., *J. Appl. Polym. Sci.*, 13, 1740, 1969.
11. Kaelble, D. H. and Uy, K. C., *J. Adhes.*, 2, 50, 1970.
12. Shanahan, M. E. R., Carré, A., Moll, S., and Schultz, J., *J. Chem. Phys.*, 83, 351, 1986.
13. Andrade, J. D., Gregonis, D. E., and Smith, L., In *Surface and Interfacial Aspects of Biomedical Polymers, vol. 1*, Andrade, J. D., Ed, Plenum Press, New York, p. 15, 1985.
14. Wu, S., *J. Polym. Sci.*, 34, 19, 1971.
15. Peper, H. and Berch, J., *J. Phys. Chem.*, 68, 1586, 1964.
16. Schultz, J., Tsutsumi, K., and Donnet, J. B., *J. Colloid Interface Sci.*, 59, 272, 1977.
17. Schultz, J., Tsutsumi, K., and Donnet, J. B., *J. Colloid Interface Sci.*, 59, 277, 1977.
18. Schultz, J. and Nardin, M., In *Modern Approaches to Wettability: Theory and Applications*, Schrader, M. E. and Loeb, G., Eds., Plenum Press, New York, p. 73, 1992.
19. Shanahan, M. E. R., Cazeneuve, C., Carré, A., and Schultz, J., *J. Chem. Phys.*, 79, 241, 1982.
20. Jensen, W. B., *The Acid–Base Concepts: An Overview*, Wiley-Interscience, New York, 1980.
21. Cain, S. R., *J. Adhes. Sci. Technol.*, 4, 333, 1990.
22. Lee, L. H., *J. Adhes. Sci. Technol.*, 7, 583, 1993.
23. Fowkes, F. M. and Mostafa, M. A., *Ind. Eng. Chem. Prod. Res. Dev.*, 17, 3, 1978.
24. Fowkes, F. M., *Rubber Chem. Technol.*, 57, 328, 1984.
25. Fowkes, F. M., *J. Adhes. Sci. Technol.*, 1, 7, 1987.
26. Fowkes, F. M., Kaczinski, M. B., and Dwight, D. W., *Langmuir*, 7, 2464, 1991.
27. Mittal, K. L. and Anderson, H. R. Jr., Eds., *Acid–Base Interactions: Relevance to Adhesion Science and Technology*, VSP, Utrecht, 1991.

28. Drago, R. S., Vogel, G. C., and Needham, T. E., *J. Am. Chem. Soc.*, 93, 6014, 1971.
29. Drago, R. S., Parr, L. B., and Chamberlain, C. S., *J. Am. Chem. Soc.*, 99, 3203, 1977.
30. Gutmann, V., *The Donor–Acceptor Approach to Molecular Interactions*, Plenum Press, New York, 1978.
31. Saint-Flour, C. and Papirer, E., *Ind. Eng. Chem. Prod. Res. Dev.*, 21, 337–666, 1982.
32. Papirer, E., Balard, H., and Vidal, A., *Eur. Polym. J.*, 24, 783, 1988.
33. Schultz, J., Lavielle, L., and Martin, C., *J. Adhes.*, 23, 45, 1987.
34. van Oss, C. J., Chaudhury, M. K., and Good, R. J., *Adv. Colloid Interface Sci.*, 28, 35, 1987.
35. van Oss, C. J., Chaudhury, M. K., and Good, R. J., *Chem. Rev.*, 88, 927, 1988.
36. Good, R. J., Chaudhury, M. K., and van Oss, C. J., In *Fundamentals of Adhesion*, Lee, L. H., Ed., Plenum Press, New York, p. 153, 1991.
37. Good, R. J. and van Oss, C. J., In *Modern Approaches to Wettability. Theory and Applications*, Schrader, M. E. and Loeb, G. I., Eds., Plenum Press, New York, p. 1, 1992.
38. Carroll, B. J., *J. Colloid Interface Sci.*, 57, 488, 1976.
39. Yamaki, J. I. and Katayama, Y., *J. Appl. Polym. Sci.*, 19, 2897, 1975.
40. Song, B., Bismarck, A., Tahhan, R., and Springer, J., *J. Colloid Interface Sci.*, 197, 68, 1998.
41. Carroll, B. J., *J. Colloid Interface Sci.*, 97, 195, 1984.
42. Nardin, M. and Ward, I. M., *Mater. Sci. Technol.*, 3, 814, 1987.
43. Walliser, A., *Caractérisation Des Interactions Liquid–Fibre élémentaire Par Mouillage*, PhD Thesis, Mulhouse (France), 1992.
44. Baley, C., Busnel, F., Grohens, Y., and Sire, O., *Composites A*, in press (available online).
45. Wei, M., Bowman, R. S., Wilson, J. L., and Morrow, N. R., *J. Colloid Interface Sci.*, 157, 154, 1993.
46. Bascom, W. D., In *Modern Approaches to Wettability: Theory and Applications*, Schrader, M. E. and Loeb, G., Eds., Plenum Press, New York, p. 359, 1992.
47. Ramé, E., *J. Colloid Interface Sci.*, 185, 245, 1997.
48. Sauer, B. B., *J. Adhes. Sci. Technol.*, 6, 955, 1992.
49. Schultz, J., Cazeneuve, C., Shanahan, M. E. R., and Donnet, J. B., *J. Adhes.*, 12, 221, 1981.
50. Bismarck, A., Wuertz, C., and Springer, J., *Carbon*, 37, 1019, 1999.
51. Park, S. J. and Kim, B. J., *Mater. Sci. Eng. A*, 408, 269, 2005.
52. Pluddeman, E. P., *Silane Coupling Agents*, Plenum Press, New York, 1982.
53. González-Benito, J., Baselga, J., and Aznar, A. J., *J. Mater. Process. Technol.*, 92–93, 129, 1999.
54. Shen, W., Brack, N., Ly, H., Parker, I. H., Pigram, P. J., and Liesegang, J., *Colloids Surf. A*, 176, 129, 2001.

55. Renner, M. and Bueno, M. A., In *Powders and Fibers: Interfacial Science and Applications*, Nardin, M. and Papirer, E., Eds., Taylor & Francis, New York, Chapter 8, p. 327, 2006.
56. Saïhi, D., El-Achari, A., Ghenaim, A., and Cazé, C., *Polym. Test.*, 21, 615, 2002.
57. Huang, F., Wei, Q., Wang, X., and Xu, W., *Polym. Test.*, 25, 22, 2006.
58. Lingström, R., Wågberg, L., and Larsson, P. T., *J. Colloid Interface Sci.*, 296, 396, 2006.
59. Pasquini, D., Belgacem, M. N., Gandini, A., and Curvelo, A. A. S., *J. Colloid Interface Sci.*, 295, 79, 2006.
60. Van de Velde, K. and Kiekens, P., *Angew. Makromol. Chem.*, 272, 87, 1999.
61. Cantero, G., Arbelaiz, A., Llano-Ponte, R., and Mondragon, I., *Composites Sci. Technol.*, 63, 1247, 2003.
62. Tavisto, M., Kuisma, R., Pasila, A., and Hautala, M., *Ind. Crops Prod.*, 18, 25, 2003.
63. Lucas, R., *Kolloid Zh.*, 23, 15, 1918.
64. Washburn, E. W., *Phys. Rev.*, 17, 273, 1921.
65. Rideal, E. K., *Philos. Mag.*, 44, 1152, 1922.
66. Poiseuille, J. L. M., *Ann. Chim. Phys.*, 21, 78, 1842.
67. Kilau, H. W. and Pahlman, J. E., *Colloids Surf.*, 26, 217, 1987.
68. Varadaraj, R., Bock, J., Brons, N., and Zushma, S., *J. Colloid Interface Sci.*, 167, 207, 1994.
69. Siebold, A., *Contribution à l'étude Des Interfaces Dans Les Matériaux à Matrice Ciment*, PhD Thesis, Mulhouse (France), 1998.
70. Siebold, A., Walliser, A., Nardin, M., Oppliger, M., and Schultz, J., *J. Colloid Interface Sci.*, 186, 60, 1997.
71. Chaudhury, M. K. and Whitesides, G. M., *Langmuir*, 7, 1013, 1991.
72. Hodgson, K. T. and Berg, J. C., *J. Colloid Interface Sci.*, 121, 22, 1988.
73. Pezron, I., Bourgain, G., and Quéré, D., *J. Colloid Interface Sci.*, 173, 319, 1995.
74. Hoffman, R. L., *J. Colloid Interface Sci.*, 50, 228, 1975.
75. Ngan, C. G. and Dussan, V. E. B., *J. Fluid Mech.*, 35, 27, 1982.
76. Siebold, A., Walliser, A., Nardin, M., Oppliger, M., and Schultz, J., *Colloids Surf. A*, 161, 81, 2000.
77. Chibowski, E. and Holysz, L., *Langmuir*, 8, 710, 1992.
78. Holysz, L. and Chibowski, E., *J. Colloid Sci.*, 164, 245, 1994.
79. Chibowski, E. and Holysz, L., *J. Adhes. Sci. Technol.*, 11, 1289, 1997.
80. Chibowski, E. and Perea-Carpio, R., *J. Colloid Interface Sci.*, 240, 473, 2001.
81. Karagüzel, C., Can, M. F., Sönmez, E., and Çelik, M. S., *J. Colloid Interface Sci.*, 285, 192, 2005.
82. Bico, J., *Imprégnation De Membranes Poreuses*, DEA Report, ESPCI, Paris (France), 1996.
83. Sénécot, J. M., *Etude De l'imprégnation Capillaire De Tissus De Verre*, PhD Thesis, Mulhouse (France), 2002.

84. Sénécot, J. M., Nardin, M., Reiter, G., Schultz, J., Christou, P., and Henrat, P., *Le Vide: Sci., Tech. et Appl.*, 296, 440, 2000.
85. Kissa, E., *J. Colloid Interface Sci.*, 83, 265, 1981.
86. Kawase, T., Sekoguchi, S., Fujii, T., and Minagawa, M., *Textile Res. J.*, 409, 1986.
87. Gillespie, T., *J. Colloid Interface Sci.*, 13, 32, 1958.
88. Roquerol, F., Roquerol, J., and Sing, K., *Adsorption by Powders and Porous Solids*, Academic Press, New York, 1999.
89. Bangham, D. H. and Razouk, R. I., *Trans. Faraday Soc.*, 33, 1459, 1937.
90. Lazghab, M., Saleh, K., Pezron, I., Guigon, P., and Komunjer, L., *Powder Technol.*, 157, 79, 2005.
91. Brendlé, E. and Papirer, E., In *Powders and Fibers: Interfacial Science and Applications*, Nardin, M. and Papirer, E., Eds., Taylor & Francis, New York, Chapter 2, p. 47, 2006.
92. Paiva, M. C., Bernardo, C. A., and Nardin, M., *Carbon*, 38, 1323, 2000.
93. Herrera-Franco, P. J. and Drzal, L. T., *Composites*, 23, 2, 1992.
94. Drzal, L. T. and Madhukar, M., *J. Mater. Sci.*, 28, 569, 1993.
95. Nardin, M. and Schultz, J., *Compos. Interfaces*, 1, 177, 1993.
96. Schultz, J. and Nardin, M., *J. Adhes.*, 45, 59, 1994.
97. Tabor, D., In *Microscopic Aspects of Adhesion and Lubrication*, Georges, J. M., Ed., Elsevier, New York, p. 651, 1982.
98. Nardin, M. and Schultz, J., *Langmuir*, 12, 4238, 1996.
99. Dekkers, M. E. J. and Heikens, D., *J. Mater. Sci.*, 18, 3281, 1983.
100. Dekkers, M. E. J. and Heikens, D., *J. Mater. Sci. Lett.*, 3, 307, 1984.
101. Dekkers, M. E. J. and Heikens, D., *J. Mater. Sci.*, 19, 3271, 1984.
102. Dekkers, M. E. J. and Heikens, D., *J. Mater. Sci.*, 20, 3865, 1985.
103. Dekkers, M. E. J. and Heikens, D., *J. Mater. Sci.*, 20, 3873, 1985.
104. Dekkers, M. E. J. and Heikens, D., *J. Appl. Polym. Sci.*, 30, 2389, 1985.
105. Amdouni, N., Sautereau, H., and Gerard, J. F., *J. Appl. Polym. Sci.*, 46, 1723, 1992.
106. Kawaguchi, T. and Pearson, R. A., *Polymer*, 44, 4229, 2003.
107. Kawaguchi, T. and Pearson, R. A., *Polymer*, 44, 4239, 2003.
108. Kawaguchi, T. and Pearson, R. A., *Compos. Sci. Technol.*, 64, 1981, 2004.
109. Kawaguchi, T. and Pearson, R. A., *Compos. Sci. Technol.*, 64, 1991, 2004.

10 Flow Micro-Calorimetry (FMC) Applied to the Study of Polymer Adsorption on Divided Matters

R. Al-Akoum, C. Vaulot, M. Owczarek,
Alain Vidal, and Bassel Haidar

CONTENTS

10.1 INTRODUCTION

Fine particle or powder science concerns research domains that are intermediate between colloidal and classical physics science. Sharp boundaries cannot be drawn between the three domains, but in general, fine particles have dimensions of the order of 0.1–2,000 μm. At such a scale, despite of a few direct observation methods such as sieving and microcopies, most characterization procedures are based on indirect approaches among which "adsorption" earns a prominent position. *Adsorption*, in general, denotes the enhancement of one component in an interfacial layer between two phases. The forces involved are of two types: physical (for *physisorption*), that depends on weak intermolecular attractive forces and chemical (for *chemisorption*), governed by strong chemical bond formation between the adsorbed species (*probe*) and the solid surface. Here again, no absolute sharp distinction can be made between the two cases, i.e., a strong hydrogen bond may overwhelm a weak charge transfer process [1].

Polymer adsorption on particles is of high scientific and technological concern in a tremendous number of applications and fields where surfaces and interfaces are critical criterion. Material science and reinforcement, colloid stability and flocculation, pigments, inks, pharmaceutical, and cosmetic preparations are few of many concerned purposes. The segment density profile of the adsorbed chain and especially the extension of its loops far from the surface are determining factors in all of these applications. The conformational conduct of a macromolecule in the adsorbed state is a direct result of the chemical structure, concentration, molar masse, and distribution of the polymer and of the surface chemistry, surface energy, and morphology of the solid.

From a theoretical point of view, the study of the polymer adsorption is generally treated from solution; there is almost no theoretical treatment of the adsorption from the melt [2]. Many theoretical treatments have been made to describe the conformation and the concentration profile of adsorbed polymers from its solution at the solid surface. Theories based on conformational statistics such as conformational change of a statistical coil [3], Monte Carlo simulation [4], mean field lattice model [5], and train-loop-tail model [6] have been used in these treatments. In addition, treatments based on density profile inspired by Flory Huggins theory [7] as well as scaling theory of de Gennes have also been used [8].

Different approaches to study a neutral polymer adsorption on solid surface are described in the literature. Experimental methods are based on several principals: isotherm determination, thickness and volume profile, and polymer segmental bounding. All stipulate the removal of the excess polymer from the surface and to separate reactants by extraction, centrifugation, sedimentation, etc. It has to be ensured that no desorption take place in this procedure.

Adsorption isotherms can be determined in a direct way which consist of a straight measuring of the amount of the adsorbed polymer: on a flat surface by attenuated total reflection (ATR) [9] or by ellipsometry [10], by surface plasma resonance [11], Raman scattering [12], or quartz crystal microbalance [13]. Indirect methods are based essentially on the polymer concentration measurement of the polymer solution before and after the adsorption process. The concentration can be determined by titration [14], by fluorescence, IR [15], FTIR, NMR, refractometry [16], radiometry [17], and by turbidity [18].

X-ray and neutron scattering of deuterated polymers [19] as well as ellipsometry and reflectometry [20] are currently used to determine the volume and thickness profile of the adsorbed layer.

The strength of polymer segmental bounding as well as the structure of the adsorbed layer can be deduced from the determination of the bound segment's fraction. In fact, properties of adsorbed segment differ from those of a free segment; such difference can be detected by infrared spectrometry, the intensity and the frequency of the adsorption bonds of the polymer segments, as well as by the surface group's change to extents depending on strength of interaction and the amount of bound segment, respectively, [21]. The changes of NMR resonance frequency, its chemical shift or lifetime of the nuclear states, spin–spin and spin–lattice relaxation can also be used to differentiate between rigid adsorbed segment and mobile loop segments [22]. Similarly, the broadening of the ESR lines of spin labeled polymers reflects a mobility reduction of adsorbed segment [23]. Electrochemical charge in the immediate contact with a surface, said to be located in the Stern layer, is changed by train segments adsorption and thus train density can be detected by titration curve of oxide surfaces [24]. However, microcalorimetry is a unique method to get direct insight into the energetics of polymer adsorption—it is particularly appropriated for the measurement of the number of adherent segments and force of adhesion, since each segment–surface binding is accompanied by a measurable enthalpic contribution. Enthalpies of wetting and adsorption, as well as replacing solvent by segments, are used for this purpose [25]. Still, the remarkable potential of this approach is not fully explored.

We will attempt in this contribution to put emphasis on the way calorimetric measurements can monitor not only adsorption of macromolecules on solid particles but also the conformation of the adsorbed chain. A special attention will be drawn on the influence of particle shape on the adsorption procedure. Three case studies will be discussed: *fine particles*, powder represented by a series of silicas of different specific surface areas; *laminates*, clay platelets more or less modified organically; and *fiber-like* solids, which constitute multi-wall nanotubes of carbon.

10.2 THEORY

10.2.1 THEORETICAL BASE FOR FMC

From an experimental standpoint, the determination of solution adsorption isotherms is less demanding than is the measurement of gas adsorption data, which requires the handling of vacuum, gauges, and other gas equipments. On the other hand, the interpretation of solution adsorption measurements is more problematic: indeed at least two components, the probe and the solvent, need to be considered. An adsorption competition often intervenes in complex physical phenomena.

Flow adsorption microcalorimetry (FMC) is a powerful tool for the combined evaluation of both the amount and heat of adsorption under dynamic flow conditions. The calorimeter (Figure 10.1) consists of a thermal insulator made of heavy block surrounding a chemically inert tube (PTFE) in which the calorimetric cell is attached. An injected liquid percolates continuously through a plug of powder placed inside the tube. A small quantity of the solute in solution is then injected into the fluid stream and the resultant change in temperature, due to the heat evolved, is detected by thermistors connected to a Wheatstone bridge and controlling unites. The calibration is made by introducing a small heating coil in the measurement cell. Thereafter, a controlled current is applied to the coil during a fixed time. The detection limit for such a setting is approximately equal to 2 mJ. The adsorption of solute generates a peak as a function of time with an area that is proportional to the total heat released or to the absorbed amount (Q) to be analyzed by adequate software.

Generally, the adsorption reaction is termed by convention *exothermic* when the arrangement of molecules in the *products* after adsorption is more stable, and thus possesses less energy than the arrangement of the molecules in the *reactants* before adsorption. The difference between the enthalpy of

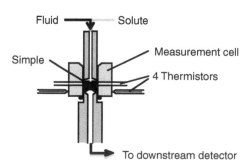

FIGURE 10.1 Scheme of FCM apparatus.

reaction products (H_P) and enthalpy of reaction reactants (H_R) is the enthalpy change (ΔH):

$$H_P - H_R = \Delta H$$

Therefore, an exothermal reaction generates a negative value of ΔH. When the converse conditions hold, the reaction is said to be endothermic and ΔH is positive. Usually, for low molecular weigh probes, an exothermal adsorption peak is followed by a pulse in the opposite direction as the solute is desorbed from the powder surface. Nonadsorbed and/or desorbed probes are detected by a downstream appropriate (refractometer or UV) detector (Figure 10.2) [26].

Hence, FMC measures the resulting net heat change of all the different events and effects that simultaneously take place in the system. In some cases, the studied effects may mask complex underlying factors that are not observed [27]. At least three different thermal events may occur all together during the probe adsorption step.

1. Since the displacement of solvent molecules from the adsorbent surface by the solute molecules contributes also to the measured enthalpy, FMC does not allow to determine the actual enthalpy of adsorption [28,29]. Rather, the enthalpy of replacement of pre-adsorbed solvent by the solute is measured. The adsorption enthalpy, ΔH^{ads}, can be extracted if the enthalpy of displacement, ΔH^{dis}, and provided that the number of solvent molecules displaced by one solute molecule are known; but this is usually not the case (at least with enough high accuracy) [30]. Therefore, FMC measures enthalpy of replacement, rather than the true adsorption enthalpy, ΔH^{ads}.

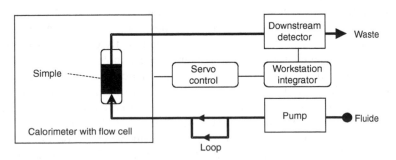

FIGURE 10.2 Schematic presentation of the experimental set.

2. The experimental calorimetric peak, Q, should be corrected by extracting the enthalpy associated with solution dilution, ΔH^{dil}, that is usually determined by a blank experiment conducted on an inert powder (polytetrafluoroethylene or Teflon), instead of the studied one, with the same solution. However, considering Teflon particles as reference particles for the studied powder, one omits the true effect of the actual particle size, distribution and arrangement on the dilution process.

3. The heat built up during the polymer adsorption may have—in addition to that associated with the formation of polymer-surface contact (trains, tr), ΔH^{tr}—a second contribution associated with the accumulation of segments in the adsorbed layer with respect to the bulk solution, ΔH^{ac}. We may assume, in agreement with Killman and Winter, that ΔH^{acc} is negligible [31].

Thus, the total heat released or (Q) may have different additive contributions that may be summarized as follows:

$$Q = \Delta H^{tr} + \Delta H^{dis} + \Delta H^{dil} + \Delta H^{acc}$$

We consider

$$\Delta H^{ads} = Q$$

the adsorption enthalpy change, strictly speaking, the enthalpy change involved with replacement, as long as ΔH^{acc} is negligible and ΔH^{dil} (despite its correction using an inert adsorbent) is of the same order of magnitude for a given set of comparable solute/adsorbents systems.

10.2.2 THEORETICAL BASE FOR POLYMER ADSORPTION

Flexible polymers have a high degree of liberty; therefore, their adsorption is characterized by a large number of conformational possibilities. This turns out to be an essential aspect of polymers adsorption behavior. As a matter of fact, the fraction of polymer segments, θ, interacting directly with a surface (*train*) reflects the actual conformation of a given adsorbed molecular weight, M_W. A large fraction of train is indicative of a highly flattened molecule, whereas a low θ value indicates that the adsorbed molecule forms large loops in a rather coil-like conformation. Evidently, θ is too partial to offer alone an integrally comprehensive model of the conformation since it does not discriminate refined conformations such as one long train from several short ones. But it remains a helpful parameter since it is easily accessible to experimental analyses.

The polymer/surface interaction is not the only driving force behind the conformation of an adsorbed polymer chain. Other factors may play a decisive role in such a process. For instance, the concentration (c) of the polymer solution determines the degree of chain–chain overlapping while approaching the solid surface. In a dilute solution, the coils are separate whereas in a semi-dilute solution coils do overlap. The border between the two regimes is defined by the overlap concentration threshold, $c*$. As $c = c*$, which follows a scaling low proportional to M_W^{α}, α is equal to 1/2 in a theta solvent. $c*$ decreases in better solvents since α goes to 3/5 [32]. Upon adsorption, another critical concentration, c_t, is predicted by theory. This theory is based on a self-similar polymer concentration profile on the surface of colloidal grains showing fractal surface [33]. Accordingly, the adsorbance, which is the total number of monomers per unit surface belonging to adsorbed chains, Γ, obeys a two terms equation:

$$\Gamma \sim 1/a^2[1 + N^{1/2}ca^3)^{7/8}]$$

"a" being the monomer length, N the number of monomer units per polymer chain and M the monomer molecular weight. This leads clearly to two regimes:

- Under a dilute solution regime, the first term is predominant. Thus, $\Gamma \sim 1/a^2$ is independent of both molecular weight and c, designing a flattened adsorbed molecule.
- In a more concentrated solution, the second term is dominant. Thus, $\Gamma \sim N^{1/2}ca^3)^{7/8}$ increases with both concentration and molecular weight, reflecting adsorbed molecule to be more randomly coiled.

The crossover between these two regimes occurs at a concentration, c_t, predicted to depend on M_W as: $c_t \sim a^{-3}. M_W^{-4/7}$. That is, after all, quite close to $c* \sim M_W^{-3/5}$, predicted for the overlapping polymer concentration in free solution.

It is clear from the above that the conformation (at constant c) may change with increasing molecular weight from flattened to coiled at a given $M_W = M_{W*}f(c)$. It is worth noticing from such a vision that the surface area controls mainly the total amount of adsorbed polymer and there is, apparently, no effect of the surface area on the polymer conformation. Furthermore, Scheutjens and Fleer's theory [34] tends to calculate the average conformation of adsorbed polymer chain. It predicts that for not too short chains, the "surface occupancy," or the total amount of "trains" at a saturated monolayer (ΔH^{ads} at surface saturation) depends only weakly on M_W of the polymer. This prediction has been contradicted by experimental results, since the measured adsorption enthalpies (micro calorimeter of the batch, conduction with a sample and a reference two cell type calorimeter) showed a strong decrease with increasing molecular weight [25]. In order to explain this contradiction, a miss estimation of the ΔH^{acc} contribution was evoked, but this explanation does not sustain the comparison of measured enthalpies in different solvents. Instead, they linked this inconsistency to a possible effect of the rate at which polymer surface contacts are formed. If the rate of adsorption of high M_W becomes too slow, the heat effect is smeared out over considerably longer times and part of the heat effect escapes the calorimeter detection. In such case, the use of FCM for polymer adsorption studies should be limited to moderate molecular weights, unless one can get the rate of adsorption to a reasonable value.

In practice, the FCM operation techniques used for adsorption studies are illustrated by the way a solid surface is brought into contact with the solute. The major procedures introduce the adsorbate solution in:

1. *A pulse of* "infinite-dilution," i.e., an infinitesimal amount of injected solute, compared to the total surface adsorption capacity (surface-excess conditions)
2. *A cumulative pulses of increasing volume* (or concentration), leading to a saturation curve similar to a Langmuir's isotherm
3. *An excess of solution* containing a given concentration of adsorbate, leading to a saturation of the surface (probe-excess conditions)

In the present instance, the pulse technique (*A*), in which adsorption takes place under surface excess conditions, will be discussed. This approach presents the advantage of focusing on a simple case, i.e., the adsorption behavior of free chains: it minimizes the contribution of ΔH^{acc}, allows a reasonable monitoring of the adsorption rate and a precise quantification of the amount of permanent adsorption, and permits an indirect measurement of the dilution of polymer solution. It suffers, however, from approaching the lower limit detection of the FMC.

Thus, adsorption (replacement) enthalpy measured by FMC, ΔH^{ads} (joule) normalized by the masses (gram) of the solid, m_s, and the actual adsorbed polymer, m_p, leads to the adsorption enthalpy per molecule, ΔH_{mol}:

$$\Delta H_{mol} \equiv \Delta H^{ads} M_W / (m_s m_p)$$

ΔH_{mol} should be proportional to the sum of train adsorbed segments of the chain tr_{mol}. In order to be able to compare different molecular weights, ΔH_{mol} may be normalized by a dimensional factor related to the gyration radius R_g:

$$\Delta H_{R_g} = \Delta H_{mol} M_W^{-\alpha}$$

where α is a scaling factor in the 0.5–1 range.

10.3 EXPERIMENTAL

In this work, all adsorption isotherms and displacement enthalpies were determined using liquid flow microcalorimetry (FMC, Microscal with a Perkin–Elmer Totalchrom Workstation integrator) [29,35]. The constant flow rates were insured thanks to a Gilson 307 pump. The adsorption isotherms expressed as the reduced surface excess were measured by the chromatographic method, which has thoroughly been described in the literature [36,37]. In this experiment, the adsorbent is placed in the flow cell and brought in contact with n-heptane, which was used as a solvent, constantly maintained under inert gas (Argon). 10 g/L solute in solutions were prepared using the same solvent and introduced to FMC through a 20 μL volume loop. Different siloxanes were purchased from GELEST–ABCR—polydimethyl-siloxane (PDMS) having variable molecular weights (M_W) will be referred-to in the text as P–X, where X is a number representing the M_W in kg/mol (P-4, P-29, P-420… for polymers with M_W equal to 4,000, 29,000 and 420,000… (g/mol, respectively) as well as three PDMS-oligomers: hexamethyldisiloxane $(CH_3)_3Si–O–Si(CH_3)_3$ (HMDS), octamethyltrisiloxane $(CH_3)_3SiO–(CH_3)_2$-$SiO–Si(CH_3)_3$ (OMTS), and decamethyltetrasiloxane $(CH_3)_3SiO–[(CH_3)_2$ $SiO]_2–Si(CH_3)_3$ (DMTS).

The polymer content (weight fraction) at the calorimeter output was monitored using a downstream differential refractometer detector (LDC Analytical refractometer CAL) and recorded using the integrator (Perkin–Elmer, Totalchrom Workstation integrator).

Three types of solid particulates, selected essentially for their dissimilar geometrical forms, will be presented. *Particles*, six grades of fumed silica samples (from Wacker Chemie) with different surface areas, were selected for this study. The structure as revealed by electron microscope is shown in Figure 10.3 for one of the silicas.

Silanized samples, surface treated by dimethyldichlorosilane, with increasing degrees of silanization were also considered. The surface area (S) values of the untreated silicas, in the 50–400 m^2 range, are gathered in Table 10.1, the size of corresponding equivalent elementary sphere is in the 100–10 nm range, respectively. [38]. The issue in this case is to determine how a polymer chain, with respect to its molecular weight, binds the surface, with how many contacts and whether the adsorption occurs by inter- or intra-particles links.

10.3.1 LAMINATES

Clays were selected for their platelet form and nanometric scale of their dimension. The used clays are lab-made aluminosilicate montmorionites

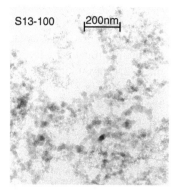

FIGURE 10.3 TEM cliché of S30 silica.

(Mmt). This category of clay is characterized by a moderate negative surface charge expressed as the cation exchange capacity, CEC, and referred to in meq/100 g as schematically presented in Figure 10.4 [39].

The aluminosilicate was rendered organophilic by treatment with increasing amounts of alkyl ammonium, octadecyltrimethylammonium bromide (ODTMABr) corresponding to the exact stochiometric amount 1 CEC up to 8-fold excess of the exchangeable Na+ ions, ($\times$) in g in Table 10.2.

The effect of such treatment is also increasing enlargement of the interlayer distance with the ODTMABr organic content. The matter here is to find if a polymer chain is able to breaks into the gallery and establish its links in such confined spaces.

The third type of investigated solid has a *tubular* form—it belongs to carbon nanotubes (CNTs) family, multiwalls carbon nanotubes (MWNTs).

TABLE 10.1
Surface Areas of Silica Samples

Reference	$S\ (m^2/g)$
S05	59
S13	129
S15	154
S20	195
S30	307
S40	389

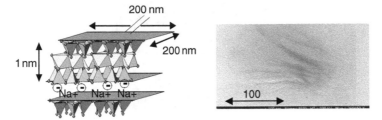

FIGURE 10.4 Schematic presentation of montmorillonite structure and TEM portrait.

In this case, two of the solid dimensions are in a nanometric scale. The used MWNTs are commercial "thin," 15 nm, and "very thin," 10 nm, materials, purchased from Nanocyl. A typical view of the material is shown in Figure 10.5.

The questions in this figure are: Can a polymer chain be adsorbed on such a nano-object? Does it unfold or wind in order to be adsorbed?

10.4 RESULTS AND DISCUSSION

10.4.1 TEMPORAL AND PERMANENT ADSORPTION

Generally, when pulse FMC procedure is used, low-molecular-weight probes [such as (HMDS) and (OMTS)] exhibit an exothermal adsorption peak followed by an endothermal peak in the opposite direction as the solute is desorbed from the silica surface (only HMDS results are presented in Figure 10.6). The amount of solute injected in the FCM, by pulse through the loop, is almost entirely detected at the output by the downstream detector (the dashed line in Figure 10.6, as compared to the peak of reference obtained

TABLE 10.2
Degree of Organic Treatment of the Clay in g/g of Clay (See Text in Section 10.3.1) and in the Corresponding Equivalence CEC

In Equivalence CEC	(x) In g of ODTMABr
1 CEC	0.294
2 CEC	0.588
3 CEC	0.882
4 CEC	1.176
8 CEC	2.355

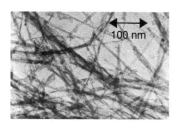

FIGURE 10.5 TEM cliché of the "very thin" MWNTs.

when the same pulse is injected directly into the downstream refractometer detector, the full line in the same figure). The adsorption is reversible.

However, a polymer such as polydimethylsiloxane (PDMS in Figure 10.6), used as a solute, shows a peculiar behavior. Although it exhibits as much enthalpy as do HMDS oligomers, the fraction of the retained polymer is very close to the injected amount. The output refractometer detects a very small amount of polymer (dashed line) compared to the injected reference (full line). In that case, the adsorption is said to be mostly permanent.

In fact, oligomers adsorption enthalpies, per gram of solute ΔH_p, do not differ fundamentally (~ 35 J/g) from those of macromolecules (~ 70–10 J/g depending on the molecular weight) as shown in Table 10.3. Nevertheless, when adsorption enthalpy is reported to one mole of solute, ΔH_{mol}, a difference of several orders of magnitude is observed between the two types of solutes—few Joules for oligomers compared to hundreds even thousands of kJ for macromolecules, as shown in Table 10.3. This result explains the cause of the permanent adsorption of a macromolecule on silica surface in opposition to the reversible adsorption of oligomers of comparable chemical structure, even though each individual contact of the two kinds of solute

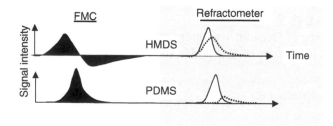

FIGURE 10.6 HMDS reversible and PDMS permanent adsorption on silica. Hachured area represents FCM thermograms, full line for references injected directly in the refractometer and dashed line for desorbed solutes.

TABLE 10.3
Enthalpy of Adsorption Reported to One Gram,
ΔH_p, **and One Mole,** ΔH_{mol}, **of Adsorbed Solute**

Solute	ΔH_p, J/g	ΔH_{mol}, kJ/mol
HMDS	39	0.006
OMTS	34	0.008
P4000	68	307
P29000	35	1.006
P420000	14	6.089

exhibits comparable amount of heats. Permanent adsorption of a macro-molecule on silica occurs as a result of a large number of links on the solid surface; each one of these multiple contacts is associated with a rather moderate polymer-segment/surface energy exchange.

10.4.2 ADSORPTION ISOTHERMS AND SATURATION

In pulsed FMC technique, surface saturation by a solute is hardly attainable before perpetrating a large number of injections. This is clearly demonstrated in Figure 10.6, which shows the evolution of cumulative enthalpy of adsorption as a function of cumulative amount of adsorbed polymer.

We notice, as it has already been shown for ΔH_p in Table 10.3, that $\Sigma \Delta H_p$ decreases with increasing molecular weights. This (if we assume ΔH^t of each individual contact is independent of M_W, only the number of contacts is able to fluctuate) reflects the enhancement of loops dimension in the adsorbed macromolecule with M_W. In addition, adsorption does not seem to attain a true threshold value, even after an extensive injection of polymer solution in continuous flow (3 h at 3 mL/h flow rate of 10 g/L solution, followed by rinsing with a flow of pure solvent until equilibrium; full symbols on Figure 10.7). Instead, an asymptotic approach to equilibrium is observed. In order to explain additional adsorptions, once the supposed surface saturation treatment was completed, one should admit that at least a partial recovery of uncontaminated surface does occur. Three mechanisms can be evoked to explain the recovery of such a clean surface after saturation: (i) desorption of adsorbed chains during solvent extraction step [40], (ii) long-term reorganization (by folding) of adsorbed chain and/or (iii) by slow and deliberate replacement of adsorbed molecule by a new one from more dilute polymer solution [32] and/or of higher M_W [41].

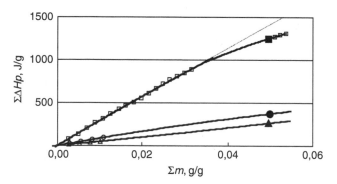

FIGURE 10.7 Cumulative enthalpy of adsorption, $\Sigma\Delta H_p$, as a function of cumulative amount of adsorbed polymer, Σm, for P4, squares; P29, circles; and P420, triangles; on silica S05.

10.4.3 Effects of Form Factors

As it has been mentioned before, three case studies associated with three different form factors will be discussed: *fine particles*, *laminates*, and *tubular* forms.

10.4.3.1 Fine Particles Case

Fine particles have a strong tendency to agglomerate, therefore, in most of the fine particles cases, we acutely do not deal with individual particles but with complex aggregate structures. Furthermore, in the particular experimental configuration settlement of the FMC, aggregates are positioned in a "close packing" arrangement and a tortuous channel pass is created, through which solvent and solute do infiltrate.

Figure 10.7 shows as an example of powders behavior, the ΔH_{mol} values obtained with a series of silica as a function of its surface area and M_Ws of the PDMS used in solution as a solute. Molar heat of adsorption increases slightly, with increasing M_W in its low range, and considerably when M_W exceeds certain value. It appears that when dealing with adsorption of polymer of variable molecular weights, neither ΔH_p, the enthalpy per gram, nor ΔH_{mol}, the enthalpy per mole, of adsorbed polymer are accurately representatives of the way a macromolecule is actually adsorbed (Figure 10.8). The former reflects the number of contacts per unit of mass (or length), thus, the length of loops formed by adsorbed polymer, the latter, the total contacts number involved in the adsorption of a given M_W.

The outcomes of these results are definitely different when expressed in term of the reduced molar heat of adsorption ΔH_{R_g}. In fact, Figure 10.9

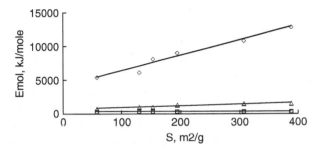

FIGURE 10.8 Molar enthalpy of adsorption, ΔH_{mol}, as a function of surface area for P4, square, P29 triangle and P420, lozenge.

represents ΔH_{R_g} normalized by its value in the case of the lowest surface area (considering the 50 m^2/g surface as a flat surface completely accessible independently of the surface area). Two behaviors are clearly distinguished: at low M_W, the reduced molar heat of adsorption is independent of the surface area, which means that macromolecules of this size perceive the different surfaces as completely smooth, two-dimensional and flat no matter what the actual surface is; at higher M_W, the reduced molar heat of adsorption increases with the surface area, and the increase is independent of M_W. This implies first that heat of adsorption per mol increases with the surface area. To explain this behavior, one should remember that particles in FMC cell are in a configuration of "densely packed" powders. The stream of the solvent and polymer solution permeates through three-dimensional tortuous channels, therefore, at M_W of size above certain limits (approaching the channels size) macromolecules start to interact with a complex, three-dimensional surface; the higher the surface area, the more complex and tortuous is the channel's

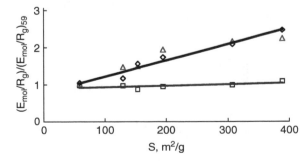

FIGURE 10.9 Reduced molar enthalpy in reference to S05 as a function of S for different molecular weights: P4, square; P29 triangle and P420, lozenge.

architecture and more frequents are the polymer–surface contacts and thus the heat of adsorption.

The second aspect is evidenced, in Figure 10.10, by the slopes of the line presented in Figure 10.9 (for 6 different M_W), which can be defined as a polymer/particles matching factor.

Sharp transition between two behaviors is identified in a relatively precise molecular weight for this family of silica. The increase of the polymer-surface contact number with increasing surface area is proportional to gyration ratio— at M_W above this transition, all polymers follow the same high matching coefficient, i.e., adsorption occurs on three-dimensional complex surface, at M_W below the transition the matching coefficient is low (near zero), i.e., adsorption is considered to occur on flat surfaces.

Thus, in this close backing arrangement of fine particles, the adsorption behavior of macromolecules differs the most for high M_W, Figure 10.8. It is worth noticing that this is not the case for other types of experimental settlements. For instance, when polymer adsorption takes place on discon-nected aggregates by mechanical melt compounding at infinite silica dilution in PDMS matrices, the amount of bound polymer (per masse unite of silica, after extraction with a good solvent referred to in Figure 10.11 as Bp), increases with M_W, as expected and similarly to the heat adsorption presented in Figure 10.8. However the adsorption converging point is positioned at the high M_W side in Figure 10.11 where a large macromolecule perceive the same external surface whatever the actual surface area is. In this case, contrary to Figure 10.8, the adsorption behavior of macromolecules differs the most at low M_W where a macromolecule is capable to access and distinguish different surface areas.

Thus, adsorption under FMC conditions proceeds in the inter-aggregates distances, whereas melt adsorptions in dilute aggregates systems act within the aggregate in the intra-aggregate spaces. Sheering forces, size of the swollen coil, salvations, and kinetics are some of many factors that may explain the discrimination between the two systems.

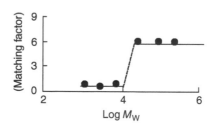

FIGURE 10.10 Matching factor as a function of M_W.

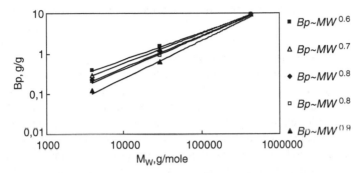

FIGURE 10.11 Bound polymer at infinite silica dilution, Bp, as a function of M_W for different silicas: S05, $\blacktriangle$; S13, $\square$; S15, $\blacklozenge$; S20, $\triangle$; S30, $\blacksquare$.

10.4.3.1.1 Effect of Silica Surface Modification on Enthalpy of Adsorption

Silica surface is markedly hydrophilic but can routinely turn organophilic by reaction with an appropriate silane. The consequence of the silanization reaction is to reduce the enthalpy of adsorption of PDMS on the silica surface. FMC allows not only measuring such heat reduction but also any consequence on the amount of adsorbed molecules. Figure 10.12a presents relative enthalpy of adsorption, $\Delta H/\Delta H_0$ (heat of adsorption on modified silica normalized by the unmodified one), as a function of the number of grafts per unit surface area, n_g, for three different polymers with M_Ws: 4,000, 29,000, and 420,000 g/mol (three polymers are mostly indistinguishable on the figure) and silicas. Upon surface modification, ΔH decreases with all surfaces and all M_Ws, however, the relative reduction of enthalpy, $\Delta H/\Delta H_0$, remains proportional to the graft surface density, n_g. Results of the adsorption of all M_Ws on all surface areas fell on a single, fairly good, master curve.

Figure 10.12b, shows the corresponding relative amount of adsorbed polymer, Q/Q_0. In opposition to the enthalpy of adsorption, the relative

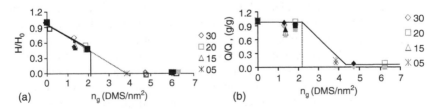

FIGURE 10.12 (a) Relative enthalpy of adsorption on modified silica. (b) Relative amount of adsorbed polymers as a function of grafts density.

amount of adsorbed polymer remains unaffected by the surface modification until a critical value of approximately $n_g \sim 2$ is reached. Above this limit, a sharp decrease of Q towards zero adsorption is observed. It can be assumed, at this point, that the decreases observed for both ΔH and Q beyond $n_g \sim 2$ occur most likely by sharp transitions towards zero enthalpy and adsorption, respectively. However, it is unambiguously evident that despite an important decrease of ΔH with increasing surface modification and n_g, an adsorbed macromolecule remains linked to the silica surface as long as the adsorption enthalpy is greater than a critical value, ΔH_c. Underneath this value adsorption enthalpy seems to be no longer sufficient to link polymer chains to the surface. This provokes a complete detachment of the macromolecule from the solid surface. Thus, while enthalpy of adsorption decreases monotonically with surface hydrophobization, the amount of adsorbed molecules remains unchanged up to the point where the thermal energy (kT) overcomes the enthalpy of adsorption. Only then the adsorption goes to zero by a sharp transition.

10.4.3.2 Laminates Case

Clays are basically hydrophilic materials, therefore, they cannot be blended properly with hydrophobic matrices such as polymer unless the clay is converted into an organophilic material by cation exchange treatment, for instance with alkyl ammonium, and/or the polymer is grafted by specific species having natural affinity for the clay, for instance by using hydroxy-terminated polymers. The way organic modifiers intercalate the galleries and enhance their size is discussed thoroughly in the literatures [42]; but the way a macromolecule moves toward and links the clay surface is less known. FMC can help to get some more insight knowledge of this process. One particularity of this system is its ability to adsorb an amount of organic modifier exceeding its natural CEC level. The adsorption occurs not only by cation exchange procedure but also by physical adsorption through specific or nonspecific interactions. Figure 10.13a presents (in a similar way to Figure 10.12) the relative enthalpy of adsorption, $\Delta H/\Delta H_0$, as a function of the modification ratio, expressed in equivalent CEC, for polymers with different M_ws. Upon surface modification and in a flagrant opposition to the silica behavior shown in Figure 10.12, ΔH remains unaffected by the surface modification until a critical value of approximately 3 CEC is reached. Above this limit a sharp decrease of $\Delta H/\Delta H_0$ towards zero adsorption is observed and thus for all M_ws. This amply that with increasing modification, fewer chains are adsorbed but each adsorbed one should progressively stretch itself out, searching a way to preserve the same number of attachments for all adsorbed chains at all modification ratios.

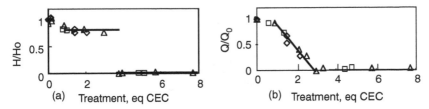

FIGURE 10.13 (a) Relative enthalpy of adsorption on modified clay. (b) Relative amount of adsorbed polymers as a function of modification degree.

The stretching process may be possible if the adsorption occurs in a confined space, such as the one schematically presented in Figure 10.14 [43]. Of two surfaces separated by a distance equal to D, surface modification provokes a reduction of the empty space in which a chain is able to infiltrate, as if the distance D is reduced and so is the size of the "blobs" forming the chain [44].

The relative amount of adsorbed polymer, Q/Q_0, decreases with surfaces modification for all M_Ws in a proportional way to the surface modification. Thus, it is evident in this case that ΔH does not decreases with increasing surface modification—an adsorbed macromolecule remains linked to the silica surface with the same total forces. However, the number of chains that have the chance to find their way to the surface and ensure such a total linkage decreases monotonically and goes eventually to zero when the coverage is complete.

10.4.3.3 Tubulars Case

High aspect ratio nanoscopic filler particles, such as carbon nanotubes, NTCs, have the curious characteristic of being nanometric in size with the exception of one of its three dimensions. When introduced into polymer media, NTCs potentially provide targets about which the chains can wind, as depicted schematically in Figure 10.15. It is not clear to what extent winding about filler particles occurs in this system or if supposed adsorption could result from such winding. The creation of stable polymer-nanotube links depends on good wetting interaction between the two, which is essentially polymer-specific, and depends in particular on chain conformation [45].

FIGURE 10.14 Drawing of a linear chain occluded in a confined space of width D.

FIGURE 10.15 Polymer chain winding around rigid tube.

It has been established that interaction between a conjugated polymer and NTCs [46] depends on defects in the nanotubes structure that nucleate crystal growth and affect the conformation of the polymer segments forming the first layer, and the next layer's stability is provided by the crystallization. Similarly, nanohybrid shish kebab structure has been reported [47], as shown in Figure 10.16.

For linear amorphous polymers adsorption, PDMS, FMC measurements provide two types of unexpected results. First, the fraction of adsorbed polymer of each injection is rather low compared to other solids that are almost total, as shown in Table 10.4, even for form- or nature-related solids such as carbon fibers or carbon blacks.

It is noteworthy that compacting ratio, or apparent density of solid, inside the FMC measurements cell depends inversely on the form-factor of the used solid. In this respect, clay particles behave as formless particles rather than as platelet ones. Therefore, when the apparent density decreases, like in the case

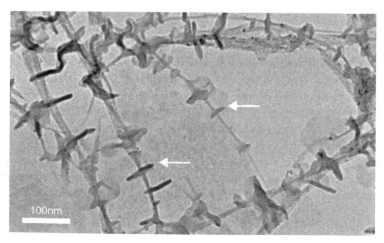

FIGURE 10.16 TEM cliché of NTCs–PA6,6 hybrid shish-kabob structure and its schematic representation. (From Li, L., Kodjie, L. S., and Li, C. Y., *PMSE*, 93, 842, 2005.)

TABLE 10.4
Q Values and Bulk and Apparent Densities for Different Divided Solids

Filler	Fraction of Adsorbed Polymer	Bulk Density	Apparent Density in FMC Cell
NTCs	~0.20	1.12	0.15
Unmodified Clay	> 0.90	1.64	0.25
Silica	> 0.97	2.19	0.35
Carbon Fibers[a]	> 0.80	1.78	0.18
Carbon black[a]	> 0.97	1.82	0.32

[a] Unpublished results.

of NTCs, the number of junction points where more than one tube meet (see Figure 10.5) decreases as well, and so does the adsorption of macromolecules on such center. Thus, under FMC experimental conditions, polymer adsorption on nanotubes seems to occur predominantly on individual tubes.

The second type of FMC results on NTCs is related precisely to the opportunity for a macromolecule to be adsorbed on a single tube. Figure 10.17 displays the enthalpy of adsorption for a several PDMS with different M_W. It is striking that adsorption takes place in a relatively small window of M_W within which adsorptions are permanents and associated with a substantial quantity of heat. Outside these limits, adsorptions are reversible and heat exchange is low.

It is recognized, as mentioned before, that molecules with low molecular weights are not adsorbed permanently on solid because of its low heat exchange with the surface, which seems to be the case with NTCs. However, high-M_W molecules are not adsorbed either presumably because of its failure to wind around the nanotube and incapacity to insure a number of adsorption contacts high enough to insure permanent adsorption. Only

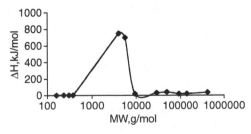

FIGURE 10.17 Enthalpies of adsorption of PDMS on NTC.

intermediate M_Ws are short enough to be stretched on the surface without too many penalties from the entropy point of view, but with offering enough enthalpy exchange to link the small chain permanently on the nanotube.

REFERENCES

1. Sing, K. S. W., In *Characterization of Powder Surfaces*, Parfitt, G. D. and Sing, K. S. W., Eds., Academic Press, New York, 1976.
2. Vuillaume, K., Haidar, B., and Vidal, A., *e-Polymers*, 032, 2003.
3. Di Matzio, E. A., *J. Chem. Phys.*, 42, 2101, 1965.
4. Carmesin, I. and Kremer, K., *Macromolecules*, 21, 2819, 1988.
5. Fleer, G. J., Cohen Stuart, M. A., Scheutjens, J. M. H. M., Cosgrove, T., and Vincent, B., *Polymer at Interface*, Chapman and Hall, London, UK, 1993.
6. Hong, K. M. and Noolandi, J., *Macromolecules*, 14, 1229, 1981.
7. Weber, T. A. and Helfand, E., *Macromolecules*, 9, 311, 1976.
8. de Gennes, P.-G. and Pincus, P., *J. Phys.*, 44, L241, 1983.
9. Dijt, J. C., Cohn Stuart, M. A., Hofman, J. E., and Fleer, G. H., *Colloids Surf.*, 54, 141, 1990.
10. Takahashi, A., Kawaguchi, M., Hirota, H., and Kato, T., *Macromolecules*, 13, 884, 1980.
11. Tassin, J. F., Seimens, R. L., Tang, W. T., Hadziiouannou, G., Swalen, J. D., and Smith, B. A., *J. Phys. Chem.*, 93, 2106, 1989.
12. Creigton, J. A., Blatchford, C. G., and Albrecht, M. G., *J. Chem Sci. Faraday Trans, II*, 75, 790, 1979.
13. Fu, T. Z. and Durning, C. J., Polymer preprints, ACS, *Div. Poly. Chem.*, 31, 519, 1990.
14. Inou, D., Kurosu, H., Chen, Q., and Ando, I., *Acta Polymer*, 46, 420, 1995.
15. Ottewill, T. H., Rochester, C. H., and Smith, A. L., Eds., *Adsorption from Solution*, Academic press, London, UK, 1983.
16. Yeh, S. L. and Frisch, H. J. L., *J. Polymer Sci.*, 27, 149, 1958.
17. Mills, A. K. and Hockey, J. A., *J. Chem. Soc. Faraday I*, 71, 2384, 1975.
18. Kilmann, E., Wild, T., Gütting, N., and Maier, H., *Colloids Surf.*, 18, 241, 1986.
19. Fleer, G. J., Cohn Stuart, M. A., Bijsterbosch, B. H., Cosgrove, T., and Vincent, B., *Polymer at Interfaces*, Chapman and Hall, London, UK, 1993.
20. Azzam, R. M. A. and Bashara, N. M., *Elipsometry and Polarized Light*, Amzsterdam, North Holland, 1989. Churaev, N. V. and Nikologorskaja, E. A., *Colloids Surface Sci.*, 5971.
21. Inou, D., Kurosu, H., Chen, Q., and Ando, I., *Acta Polymer* 46, 420, 1995. Ottewill, T. H., Rochester, C. H., and Smith, A. L., Eds., Academic Press, London, UK, 1995.
22. Blum, F., *Colloids Surf.*, 45, 361, 1990.
23. Sakai, H., Fujimuti, T., and Imamura, Y., *Bull. Chem. Soc. Jpn.*, 53, 3457, 1980.
24. Koopal, L. K. and Lyklema, J., *J. Electroanal. Chem.*, 100, 895, 1979.

25. Cohn Stuart, M. A., Fleer, G. J., and Bijsterbosch, B. H., *J. Colloid Interface Sci.* 90, 310, 1982 see also page 321; Ottewill, T. H., Rochester, C. H., and Smith, A. L., Eds., Academic Press, London, UK, 1982.
26. Allen, T., *Particle Size Measurement*, Chapman and Hall, London, UK, 1968.
27. Pettersson, A. and Rosenholm, J. B., *Langmuir*, 18 (22), 8447, 2002.
28. Rouquerol, J., *Pure Appl. Chem.*, 57, 69, 1985.
29. Van Os, N. M. and Haandrikman, G., *Langmuir*, 3, 1051, 1987.
30. Thibaut, A., Misselyn-Baudin, A. M., Broze, G., and Jérôme, R., *Langmuir*, 16, 9841, 2000.
31. Killmann, E. and Winter, K., *Angew. Makromol. Chem.*, 43, 53, 1975.
32. De Genne, P. J., In *Scaling Concept in Polymer Physics*, Cornell University Press, Ithaca, NY, 1979.
33. Marques, M. and Joanny, J. F., *J. Phys. France*, 49, 1103, 1988.
34. Scheutjens, J. M. H. M. and Fleer, G. J., *J. Phys. Chem.*, 83, 1619, 1979.
35. Groszek, A. J., *Thermochim. Acta*, 312, 133, 1998.
36. Noll, L. A. and Gall, B. L., *Colloids Surf.*, 54, 41, 1991.
37. Noll, L. A., *Colloids Surf.*, 26, 43, 1987.
38. Al Akoum, R., Haidar, B., and Vidal, A., *Macromol. Symp.*, 211, 271, 2005.
39. Da Silva, C., Haidar, B., Vidal, A., Miehe-Brendle, J., Le Dred, R., and Vidal, L., *J. Mater. Sci.*, 40, 1813, 2005.
40. Vuillaume, K., Haidar, B., and Vidal, A., *e-Polymers*, 026, 2003.
41. Vuillaume, K., Haidar, B., and Vidal, A., *e-Polymers*, 068, 2005.
42. Le Baron, P. and Pinnavaia, T. J., *Chem. Mat.*, 13, 3760, 2001; Lagaly, G., *Clay Minerals*, 16, 1, 1981.
43. Sakaue, T. and Raphaël, E., *Macromolecules*, ASAP Web Release Date. 03-Mar-2006.
44. Daoud, M. and de Gennes, P.-G.J., *Phys. (Paris)*, 38, 85, 1977.
45. Richardson, D. G. and Abrams, C. F., *Macromolecules*, 39 (6), 2330, 2006.
46. McCarthy, B., Coleman, J. N., Czerw, R., Dalton, A. B., in het Panhuis, M., Maiti, A., Drury, A., Bernier, P., Nagy, J. B., Lahr, B., Byrne, H. J., Carroll, D. L., and Blau, W. J., *J. Phys. Chem. B*, 106 (9), 2210, 2002.
47. Li, L., Kodjie, L. S., and Li, C. Y., *PMSA Preprints*, 93, 842, 2005; Li, L, Li, C. Y., and Ni, C., *J. Am. chem. soc.*, 128(5), 1692, 2006.

11 Scanning Probe Microscopies: Principles and Applications to Micro- and Nanofibers

Bernard Nysten

CONTENTS

11.1 INTRODUCTION

Scanning probe microscopies (SPMs) appeared in the beginning of the eighties with the development of the scanning tunneling microscope (STM) in 1981 by G. Binnig and H. Rorher [1]. For this important discovery, they received in 1986 the Nobel Prize for physics. Since then, many other microscopies based on the same concepts were developed: the atomic force microscopy (AFM) in 1986 [2] and the other techniques that derive from it, lateral force microscopy (LFM), magnetic force microscopy (MFM), ... and scanning near-field optical microscopy (SNOM) [3].

SPMs encountered a remarkable breakthrough over the two past decades and became essential techniques in the field of material and surface sciences [4,5]. This extraordinary development is certainly linked to their very high resolution capability that enables them to visualize the morphology and the microstructure of material surfaces from the micrometer scale down to the nanoscale (molecular or atomic scale) [6–14]. But, it is also largely due to the ability of these techniques to measure and map at the nanoscale physical, physicochemical, and chemical properties that cannot be measured at this scale by other classical surface analysis techniques or spectroscopies.

For instance, scanning tunneling spectroscopy (STS) enables one to locally measure the surface electronic properties, such as the surface density of states (surface DOS), of materials [15–19]. Atomic force microscopies also permit the measurement and the mapping of physical or chemical properties such as the mechanical properties (elastic or viscoelastic modulus) with force-curve measurements, force modulation microscopy (FMM), intermittent-contact atomic force microscopy (IC-AFM) with phase detection microscopy (PDM) [5,20–24], the frictional and adhesive properties with LFM [25–27], the chemical composition with chemical force microscopy (CFM) [28–32], the surface potential and charges with electrostatic force microscopy (EFM) or Kelvin probe microscopy (KPM) [33,34], or the magnetic domains or domain walls with MFM [35], ...

Our aim here is to illustrate how SPMs, mainly STM and AFM, can be used to characterize the surface structure and morphology and the mechanical properties of micro- and nano-sized fibers. In the first part, the working principles of the STM and AFM techniques will be briefly introduced. In the second part, their application to the characterization of the surface morphology and structure of carbon and polymer fibers will be presented. This second part will also present some of the works that were realized with the AFM to measure the mechanical properties of nano-sized fibers: carbon nanotubes, polymer nanotubes, and metallic nanowires.

11.2 PRINCIPLES OF SCANNING PROBE MICROSCOPIES

11.2.1 GENERAL PRINCIPLES

The general principle of SPMs is apparently very simple and is presented in Figure 11.1. It consists of raster scanning an ultrafine probe close to the surface that has to be analyzed and measuring point by point the local interactions between the surface and the probe atoms as a function of the probe displacement.

Ideally, the probe apex should be a single atom. In fact, the tip apex radius of curvature varies between 1 and 2 nm for the sharper ones up to several tens of nanometers. This will indeed affect the ultimate resolution of the technique. The typical distance, d, between the probe apex and the surface depends on the considered technique. In the case of contact mode atomic force microscopies, the probe is in physical contact with the surface and d is equal to a few Angstrom, the typical inter-atomic distance in solids. In the case of non-contact mode techniques, the probe-surface distance can vary from about 1 nm to a few tens of nanometers.

Since many different kinds of interactions can be measured between the probe and the sample, a large variety of techniques were developed. In Table 11.1, a nonexhaustive list of interactions and derived techniques are given. In the following, some of these techniques will be described in more details. Only the technique used for the application discussed in the second section will be presented.

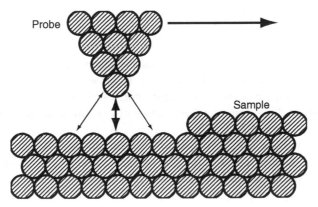

FIGURE 11.1 Schematic representation of the basic principles of scanning probe microscopies.

TABLE 11.1
Examples of Probe-Sample Interactions That Can Be Measured and of Related SPM Techniques

Interactions	Techniques and Abbreviations
Tunneling current	Scanning tunneling microscopy (STM)
	Scanning Tunneling Spectroscopy (STS)
Inter-atomic repulsive forces	Contact-mode atomic force microscopy (C-AFM)
	Lateral force microscopy (LFM)
	Force modulation microscopy (FMM)
Attractive van der Waals forces	Intermittent-contact atomic force microscopy (IC-AFM)
	Noncontact atomic force microcopy (NC-AFM)
Adhesion forces	Adhesion force microscopy
Chemical interactions	Chemical force microscopy
	Single molecule force spectroscopy
Electrostatic forces	Electrostatic force microscopy (EFM)
	Capacitive force microscopy (CFM)
	Kelvin probe microscopy (KPM)
Magnetic forces	Magnetic force microscopy (MFM)
Optical evanescent field	Near-field Scanning Optical Microscopy (SNOM)

11.2.2 SCANNING TUNNELING MICROSCOPY

11.2.2.1 Working Principles

The three main parts of an STM are the probe, a conductive tip, a piezoelectric scanner, and a control unit (Figure 11.2). Two configurations are possible—either the tip is placed on the scanner and the sample is fixed, as depicted in Figure 11.2, or the tip is fixed and the sample is placed on the scanner and is raster scanned under the tip. The piezoelectric scanner thus enable the raster scan of either the tip or the sample in the horizontal plane (x,y) by applied V_x and V_y voltage. It also permits their vertical displacement in the z direction by applying a voltage V_z.

A bias voltage, V_{bias}, is applied between the tip and the electrically conductive sample. The tip is then brought close to the surface until a tunneling current, I_t, can be measured. The typical tip-surface distance is then in the Angstrom range. Generally, the bias voltage varies between a few tens of mV for metals or semimetals up to 1 or 2 V for semiconductors. The tunneling current ranges from ~ 10 pA to ~ 10 nA.

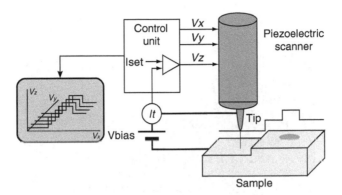

FIGURE 11.2 Scheme of the working principles of a scanning tunneling microscope.

While the tip is raster scanned above the sample surface, the tunneling current is measured and compared to a value fixed by the experimentalist. A feedback loop continuously adjusts the vertical position by varying V_z to maintain the tunneling current constant and equal to the set point. The lateral and vertical movements of the tip are recorded and treated to generate a tridimensional image of the surface topography. This working mode is called the "constant current" mode. It is generally used to analyze rough and large surface areas. Its drawback is that it necessitates a relatively slow scanning speed to enable the feedback loop to follow the surface topography. In another working mode, the "constant height" mode, the gains of the feedback loop are reduced to their minimum values. The tip is thus scanned at an almost constant vertical position, and the variations of the tunneling current that reflect the surface structure variations are measured as a function of the tip displacement. This working mode is restricted to small areas of surfaces flat at the atomic level. Its advantage is that it allows fast scanning thus reducing the effects of low frequency mechanical noise and of thermal drift. In practice, it is used to generate images with molecular or atomic resolution (Figure 11.3) [36].

11.2.2.2 Tunneling Current and Resolution

In a crude approximation, the tunneling current, I_t, measured between the tip and the surface, can be expressed with the following expression [37,38]:

$$I_t \propto V_{bias}\rho_S(0, E_F)\exp(-2\kappa\delta) \tag{11.1}$$

where $\rho_S(0,E_F)$ is the surface electronic density of states at the Fermi level, E_F, below the tip apex, κ is the tunneling decay length which depends on the work

FIGURE 11.3 STM image Si(111) 7×7 surface showing the ad-atoms of this reconstructed structure as well as a mono-atomic step. Point defects such as vacancies can be also clearly observed (From http://www.uark.edu/misc/mbestm/images/stm/si7x7.JPG.)

function of the sample, and δ is the tip-surface distance. This exponential dependence of the tunneling current on the tip-surface distance explains the very high resolution that can be achieved with STM. Indeed, an increase of δ by about 1 Å results in a decrease of I_t by a factor of ten. Therefore, the lateral resolution in STM is around 0.1 Å and the vertical resolution even smaller, ~ 0.05 Å. This high resolution does not only enable the imaging of the surface structure down to the atomic scale, but it makes STM a real local probe with the capability to visualize small scale defects at the surface such as point defects like vacancies or interstitial atoms (Figure 11.3) [8,36,39], dislocations [40], or adsorbed atoms or molecules [41–43].

It is important to mention here that the tunneling current does not depend on the surface topography, but on the surface electronic density of state. This means that STM images do not really show surface atoms but more precisely the variation of the surface electronic DOS. This was demonstrated by the visualization of the electronic wave function in the "quantum corral" [44].

Since STM requires the measurement of a current, it is limited to the analysis of electrically conductive materials like metals, semimetals, and semiconductors. The study of organic molecules is also possible provided they are deposited on a conducting substrate like gold or highly oriented pyrolytic graphite (HOPG) [43,45].

11.2.2.3 STM Tips

STM tips are generally fabricated with tungsten (W) or alloys of platinum with iridium (Pt/Ir). Tungsten tips are generally devoted for analyzes in ultra-high vacuum since they can easily be oxidized leading to unstable currents. Pt/Ir tip can be used either under vacuum or under ambient conditions.

Two main methods are used to prepare STM tips. The simplest one consists of cutting a metallic wire under tension. This method generally leads to tips with multiple apexes (Figure 11.4). The second method consists of the chemical or electrochemical erosion of metallic wires. This second methods gives tips with more regular and more reproducible shapes (Figure 11.4) and

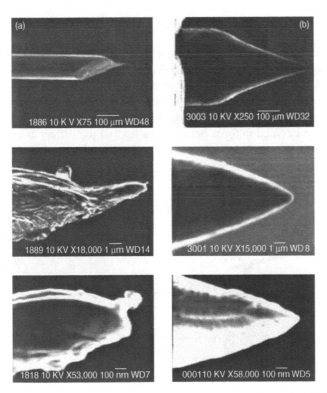

FIGURE 11.4 STM tips: (a) mechanically cut Pt/Ir tip; (b) electrochemically etched W tip.

can be optimized to obtain very sharp tips [46]. Nevertheless, STM tip are often far from being ideal, i.e., with one single atom forming the apex. Sharper tips will indeed provide better images on rough surfaces. However, it is the exponential dependence of the tunneling current on the inter-atomic distance which is actually responsible for the molecular and atomic resolution capabilities of STM.

11.2.3 Atomic Force Microscopies (AFMs)

Since the use of STM was limited to electrically conductive materials, researchers quickly tried to develop a local probe technique that would enable to analyze with the same resolution the surface of insulating materials such as ceramics, polymers, and biological samples. The first AFM was thus developed in 1986 [2]. Thereafter, a series of derived techniques were progressively developed: LFM [47], NC-AFM [48], and then in Inter-mittent-Contact or Tapping® mode, IC-AFM [49], MFM [50], FMM [51], ...

11.2.3.1 Principles of Contact-AFM

An atomic force microscope is very similar to a STM (Figure 11.5). Generally, the sample is placed on the piezoelectric scanner. A measurement head contains a support on which a soft microcantilever with an ultra-fine tip at its end is mounted and an optical detection system enables the measurement of the deflection of the cantilever. The tip is brought in contact with the sample surface. The interaction force between the tip and the surface is evaluated by measuring the vertical deflection of the microcantilever, d. The interaction

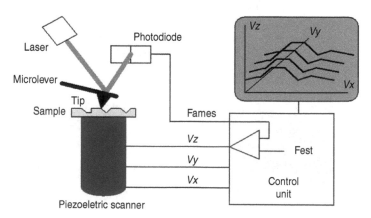

FIGURE 11.5 Scheme of the working principles of contact mode atomic force microscopy.

force, F, is then given by the classical Hooke's law,

$$F = k_N \delta \qquad (11.2)$$

where k_N is the vertical deflection spring constant of the microcantilever. To measure the vertical deflection, the beam of a laser diode is focused on the extremity of the cantilever and is reflected to a position sensitive photodetector (PSPD). This PSPD is made of two photodiodes. When the cantilever is deflected, the reflection of the laser beam moves on the PSPD and induces a variation of the photo-current (or voltage drop) measured between the two photodiodes.

As in the case of STM, two working modes may be used: the "constant force" mode and the "constant height" mode.

In the "constant force" mode, the cantilever deflection signal is measured while the sample is raster scanned under the tip. This signal is sent to the control unit where it is compared to a set-point signal. A feedback loop modifies the vertical position of the sample to maintain the deflection signal, i.e., the contact force, constant and equal to the set-point value. The lateral and vertical displacements of the sample are treated by the computer to generate a quantitative tridimensional image of the sample surface. This mode is generally used to generate images at the large scale (micrometer or submicrometer scale). Relatively slow scanning speeds are required to permit the feedback loop to follow the surface topography, especially on rough surfaces.

In the "constant height" mode, the gains of the feedback loop are minimized or even set to zero. In this case, it is the vertical deflection of the cantilever which is recorded as a function of the lateral displacements of the sample to generate an image of the surface microstructure. This working mode is, as in the case of STM, limited to the imaging of atomitically flat surface at small scale, i.e., nanometer scale. In this mode, it is possible to obtain molecularly or atomitically resolved images on crystalline samples such as mica or HOPG [52] or polymer crystals (Figure 11.6) [12,53].

At this point, it is worth noting that C-AFM is not really a local technique as STM though atomitically resolved images can be obtained. Indeed, most of the images published up to now never showed small scale defects such as vacancies, interstitials, dislocations,... This effect may be mostly explained by the fact that in C-AFM, the contact area is large compared to atom size, i.e., several square nanometer. The tip thus interacts with a large number of atoms at once. The fact that images reflecting the surface atomic structure can be obtained is due to the convolution of the interaction of the tip with a periodic surface. This has often been explained by a stick–slip mechanism [52]. This mechanism erases point defects. "True" atomic resolution was achieved using noncontact atomic force microscopy (NC-AFM), now more accurately called frequency-modulated atomic force microscopy (FM-AFM) [54–56].

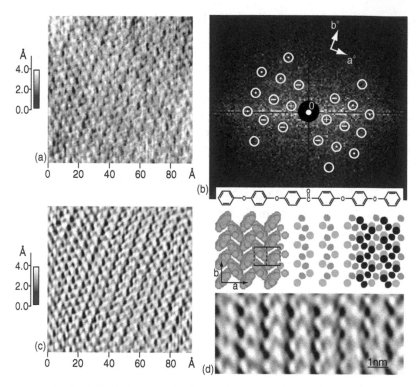

FIGURE 11.6 C-AFM images obtain on the (111) plane of ether-ether-ketone oligomer crystal. (a) Rough AFM image. (b) 2D power spectrum density of (a) showing several spots corresponding to the structure of the (111) surface. (c) Reconstructed image obtained from (b) after selection of the spots marked with white circles. (d) Chemical structure of the oligomer and detail of (c) compared to a molecular model obtain from x-ray diffraction data. (Adapted from Dupont, O., Jonas, A. M., Nysten, B., Legras, R., Adriaensens, P., and Gelan, J. *Macromolecules*, 33, 562–568, 2000.)

11.2.3.2 Other AFM Techniques

As already mentioned above, several other techniques based on AFM were soon developed.

In the contact mode techniques, LFM was developed to measure and map the friction force acting between the tip and the surface [47]. This friction force induces a torque on the tip that produces torsion of the cantilever. To measure the torsion, the two quadrants PSPD is replaced by a four quadrants PSPD. This technique enables the mapping of the variation of the chemical composition [14,26], the physicochemical properties (hydrophobicity or hydrophilicity) [25,27], or mechanical properties of surfaces [20]. Also in contact mode, FMM was developed to map the mechanical properties (stiffness and damping) of material surfaces [20,51]. To achieve this goal, the vertical position of the sample or of the cantilever support is vertically modulated with amplitude in the nanometer range and at a frequency between 1 and 10 kHz. The cantilever deflection at this frequency is measured with a lock-in amplifier. The mapping of the amplitude gives an image of the surface stiffness: when the tip-surface contact is stiff, the amplitude is high; it is lower when the contact is more compliant. The mapping of the phase shift between the modulation signal and the cantilever deflection oscillation gives a map of the viscoelastic properties of the surface.

It was soon noticed that C-AFM was not suitable to analyze the surface of soft materials such as elastomers or biological samples in air due to the high pressure in the contact area (up to several GPa). This high pressure indeed induced modifications and wear of the surfaces. To avoid this problem, NC-AFM [48] and IC-AFM [49] were developed. These techniques are now more accurately known as frequency-modulated AFM (FM-AFM) and amplitude-modulated AFM (AM-AFM) [57]. They are based on the fact that the resonance frequency of a harmonic oscillator is modified when it is submitted to a force field. It decreases when the force gradient is positive, i.e., for attractive forces, and it increases when the gradient is negative, i.e., for repulsive forces. In FM-AFM, generally used in vacuum, the cantilever is forced to vibrate at its resonance frequency. The frequency shift due to tip-surface interactions is measured and the feedback loop adjusts the sample vertical position to maintain this shift constant and to generate a topographic image of the surface. The energy necessary to maintain the vibration amplitude constant is simultaneously recorded to map the energy dissipation at the tip-surface contact. In AM-AFM, generally used in air or in liquid, the cantilever is forced to vibrate close to its resonance frequency. The vibration amplitude drop due to the shift of the resonance peak is measured and the feedback loop adjusts the sample vertical position to maintain the amplitude constant and to generate a topographic image of the surface. The phase shift between the excitation signal and the cantilever vibration is simultaneously recorded to map the energy dissipation at the tip-surface contact. Both techniques are able to map simultaneously surface topography and elastic, viscoelastic or adhesive properties [57–59].

11.2.3.3 AFM Tips and Cantilevers

Most of the probes used in AFM consist of silicon or silicon nitride cantilevers with integrated tips (Figure 11.7). Triangular-shaped cantilevers or cantilever beams are used. The cantilever length varies between ~ 50 and ~ 300 μm, the

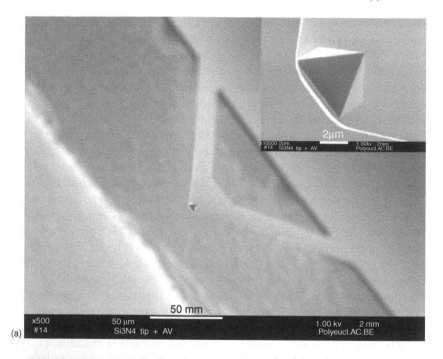

(a)

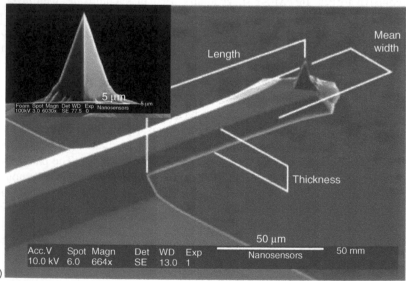

(b)

FIGURE 11.7 SEM images of typical AFM cantilevers and tips: (a) triangular silicon nitride cantilever with pyramidal integrated tip for C-AFM; (b) silicon cantilever beam with integrated pyramidal tip for AM- or FM-AFM.

beam or arm width between 10 and 50 μm and the thickness between ~0.5 and ~5 μm.

For C-AFM, soft cantilever are used with a spring constant varying between ~0.01 and 1 N m^{-1}. For FMM, EFM, MFM, cantilevers with a spring constant of a few N m^{-1} are used and for AM-AFM or FM-AFM, stiffer cantilevers with a spring constant of a few tens of Newton per meter are used.

Typically, the tip apex radius of curvature varies between ~50 nm for Si$_3$N$_4$ tips (Figure 11.7a) and ~10 nm for silicon tips (Figure 11.7b). The tip shape and size is critical for the ultimate resolution that can be achieved in AFM. The tip half-opening angle of these tips typically varies between 35 and 10°. For special applications, sharper or smaller tips are developed. "High aspect ratio" tips with tip half-opening angles smaller than 3° are developed to image sharp surface features, especially for micro-electronics applications [60]. "High-resolution" or "super-sharp" tips with apex radius of curvature as low as 1 nm are also developed to image very small surface features [61]. To enhance the size and the aspect ratio, techniques to stick or to grow carbon nanotubes at the apex of the tip were also developed [62].

11.2.3.4 Force Spectroscopy

It was very soon noticed that the high lateral resolution and high force resolution (less than 1 pN) will enable AFM to measure very precisely at the local scale surface properties like the elastic modulus or the adhesion by measuring the tip-surface interaction force [63]. This is achieved by measuring the cantilever deflection as a function of the sample vertical displacement, a procedure called measurement of force curves or approach-retraction curves. (Figure 11.8a). A saw-tooth voltage is applied to the z electrodes of the scanner to move the sample up and down and the cantilever vertical deflection is measured simultaneously. When the tip is far from the surface, the deflection remains constant (Figure 11.8 ①), then, when the attractive force gradient becomes larger than the cantilever spring constant, the tip jumps into contact with the surface (snap-in point, Figure 11.8 ②). While the sample continues to move up, the deflection increases (Figure 11.8 ③), then, when the sample moves down, it decreases (Figure 11.8 ④). Due to adhesion forces, chemical interaction, or capillary forces, the cantilever deflection can become largely negative when the sample is retracted (Figure 11.8 ⑤). Finally, when the attractive force gradient becomes smaller than the cantilever spring constant or the force applied by the cantilever becomes larger than the chemical bond, the cantilever jumps back to its equilibrium position (snap-out point, Figure 11.8 ⑥). This gives rise to deflection versus sample displacement curves as the one schematized

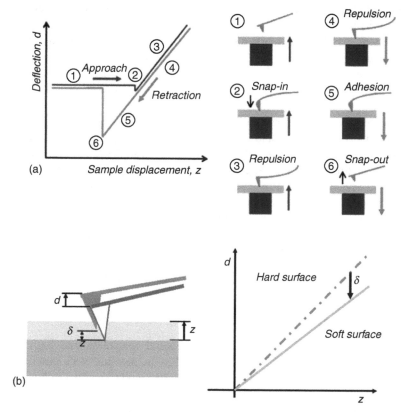

FIGURE 11.8 Schematic presentation of (a) the principles of approach–retraction or force curve measurements and (b) of nano-indentation with an AFM.

on Figure 11.8. From these curves, information on the tip-surface contact stiffness can be obtained using parts ③ and ④ [22,63,64], while data on the adhesion properties of the chemical composition can be deduced from the value of the force at the snap-out point (⑥) [26,27,30–32].

To deduce the stiffness of the tip-surface contact and hence the elastic modulus of the surface, the approach-retraction curves relating the cantilever deflection, d, to the sample vertical displacement, z, have to be transformed into curves relating the load applied by the cantilever, F, to the indentation depth of the tip into the surface, δ. If the cantilever normal spring constant, k_N, was calibrated, the load can be directly deduced using the classical Hooke's law (relation 2). As illustrated in Figure 11.8b, the indentation depth can be estimated by calculating the difference between the deflection expected on an infinitely hard surface, i.e., the sample vertical displacement, and the deflection measured on the compliant surface. It is thus given by

$$\delta = z - d \qquad (11.3)$$

The obtained load versus indentation depth can then be fitted with models developed for the contact mechanics to deduce the surface elastic modulus [65–69]. This approach was used by several authors to measure the surface elastic modulus of soft materials such as polymers [22,64,70–74] or biological samples [75–77].

A drawback of this technique is that it requires the knowledge of the shape and size of the AFM tip and the careful calibration of the sensitivity of the PSPD and of the cantilevers spring constant. Another one is that, due to the cantilever and the AFM geometries, the load is never applied perfectly perpendicular to the surface. Moreover, the tip can slide horizontally on the surface. Therefore, shear deformation, friction, and parasitic deflections of the cantilever (buckling, torsion) can interfere in the force curves [78]. However, in particular cases, accurate force spectroscopy can be performed to quantitatively measure the elastic properties of nanomaterials such as nanowires or nanotubes.

11.3 STUDIES OF MICRO- AND NANOFIBERS BY SPMS

In this section, we will briefly present and discuss a few applications of SPMs (STM and AFM) for the study of the surface morphology and structure of microfibers, i.e., carbon fibers and polymer fibers, and of the elastic modulus of polymer and metallic nanowires or nanotubes.

11.3.1 CARBON AND POLYMER FIBERS

11.3.1.1 STM on Graphite

The most common crystalline form of carbon in nature is hexagonal graphite. Graphite, or HOPG, is one of the layered materials that has been the most widely studied by STM [79]. Its crystalline structure is presented in Figure 11.9a. It is built up by layers with a honeycomb arrangement of sp^2 hybridized carbon atoms strongly linked together by covalent bonds, graphene layers. The inter-atomic distance is equal to 1.42 Å while the in-plane lattice constant, a_0, is equal to 2.46 Å. The graphene layers are stacked in such a way that neighboring layers are shifted relative to each other by an inter-atomic distance leading to an ABABAB⋯ stacking sequence. They are linked together by weak van der Waals forces. The inter-layer distance is equal to 3.354 Å and the ABABAB⋯ stacking leads to a c-axis lattice constant,

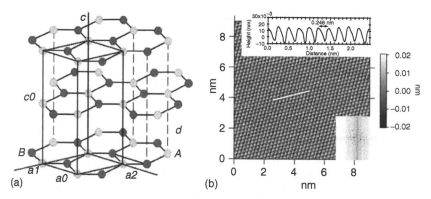

FIGURE 11.9 (a) Crystallographic structure of graphite showing the ABA stacking of the graphene layers; the unit cell is also shown with the following parameters: $a_0 = 0.246$ nm and $c_0 = 0.6708$ nm; the interlayer distance, d, is equal to 0.3354 nm; light grey circles correspond to A atoms with neighbor atoms in the upper and lower layers and dark grey circle to B atoms without neighbor in the upper and lower layers. (b) Example of STM image of the (0001) surface of HOPG obtained in constant height mode with a bias voltage of 20 mV and a tunneling current of 5 nA; the inset shows the power spectrum of the image revealing the spots corresponding to the hexagonal symmetry; the line section reveals the 0.246 nm inter-distance between the spots observed by STM.

c_0, equal to 6.708 Å perpendicular to the layers. The stacking sequence gives rise to two nonequivalent atomic sites within the graphene layers: carbon atoms (A-sites) have a neighboring atom directly below in the next layer (light gray atoms in Figure 11.9a), whereas the other atoms (B-sites) are located above the center of the sixfold carbon ring in the next layer (dark gray atoms in Figure 11.9a).

A typical STM image obtained on the surface of HOPG is presented in Figure 11.9b. On this figure, the power spectrum of the image reveals the hexagonal symmetry of the graphene layer. The section along one row of bright spots shows that the inter-maxima distance is equal to 2.46 Å, i.e., to the in-plane lattice constant a_0. This means that only one atom on two appears as a protrusion on the STM image. STM is thus able to distinguish between the two nonequivalent A and B sites. In fact, the B atoms give rise to a higher tunneling current and thus to bright spots on the STM images [79]. This site asymmetry is nearly independent on the bias polarity and decreases with increasing amplitude of the bias voltage. It confirms that STM images generally do not reflect the surface atomic structure but are strongly influenced by the local surface density of states.

11.3.1.2 STM on Carbon Fibers

According to the pristine material used for their fabrication (polyacrylonitrile, rayon, pitch, decomposed gaseous hydrocarbons, …) and to the thermal treatment (graphitization process) applied after the carbonization process, carbon fibers may exhibit a large range of structures from an almost amorphous carbon to a well-graphitized crystalline structure [80]. Various surface treatments are also used to increase the compatibility and the interface strength in carbon fiber-based composite materials [81].

STM and AFM were thus used with other surface analysis techniques to study the surface of carbon fibers. The main objectives were the characterization of their surface structure and morphology as a function of the pristine material used to fabricate the fibers, as a function of their thermal history (graphitization treatment) or as a function of the applied surface treatment [82].

a. Effect of graphitization degree. The effects on the surface morphology and structure of the pristine material and of the graphitization treatment were studied by several authors using STM [82–92]. As a general trend, it was found that polyacrylonitrile (PAN)-based carbon fibers had a rougher surface than pitch-based carbon fibers [82,85]. They exhibited rock-like stacks on their surfaces [82,87]. On the contrary, pitch-based carbon fibers had a ribbon-like morphology on their surfaces [82,87].

Most of the studies that were conducted also confirmed that a better graphitic structure can be obtained with pitch compared to PAN and that higher graphitization temperatures lead to larger and more perfect graphitic regions on the fiber surface [82]. For instance, Donnet et al. showed that the mesophase content of the pitch influenced the surface microstructure [92]—a larger mesophase content gave rise to a more graphitized surface structure.

The various graphite-like structures that were observed on the surface of the various types of carbon fibers are well illustrated by those that were observed on polyimide films carbonized and graphitized at various temperatures (Figure 11.10) [93].

On PAN-based carbon fibers and on pitch-fibers treated at lower temperatures, small regions with a graphitic organization were observed as the one presented in Figure 11.10a. These graphite-like regions did not exceed a few square-nanometers in size [83,86–88]. These regions were rather defective, as illustrated in Figure 11.10b, where the honeycomb graphene layer also presents heptagons and triangles. On more graphitized carbon fibers such as pitch-based fibers or vapor-grown fibers treated at high temperatures (above 2000°C), larger well-graphitized zones were observed on the fiber surface [85–87,91], similar to the one presented in Figure 11.10f.

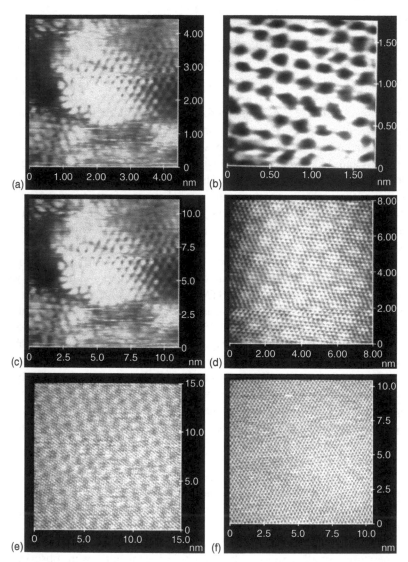

FIGURE 11.10 Evolution of the graphitic structure of polyimide based films as a function of the graphitization temperature. (a) Appearance of a graphitic structure on a film treated at 1800 C. (b) Graphitic zone on a film treated at 2000 C with defects such as pentagons and heptagons. (c) $\sqrt{3} \times \sqrt{3}R30°$ super-structure observed on a graphene layer close to a defective region (HTT = 2200 C). (d) and (e) Moiré patterns observed on a film treated at 2400 C. (f) Well-graphitized zone observed on a film treated at 2600 C. (Adapted from Nysten, B., Roux, J.C., Flandrois, S., Daulan, C., and Saadaoui, H., *Phy. Rev. B*, 48, 12527–12538, 1993.)

However, the graphitic regions were often defective with steps [87], point defects [87,90,92], or stacking defects [91,92]. The presence of these defects leads to the observation of different kinds of superstructures in the STM images of the graphitic zones.

First, the presence of point defects or steps gave rise to the observation of $(\sqrt{3} \times \sqrt{3})R30°$ modulation of the surface density of state similar to the one observed in Figure 11.10c [90,92]. This modulation did not correspond to a surface reconstruction with displacement of atoms at the surface but to a modulation of the surface density of state due to the reflection of the electron wave functions by the defects. This leads to the appearance of surface standing waves corresponding to interferences between the incident and the reflected wave functions [94].

Second, the presence of stacking defects, i.e., turbostratic graphite, lead to the observation of Moiré patterns like the ones shown in Figure 11.10d and Figure 11.10e [91,92]. These Moiré patterns are characterized by a large scale hexagonal modulation in the atomic STM images. They were attributed to rotational misorientation of the two surface graphene layers corresponding to a turbostratic structure [93,95]. The periodicity of the Moiré pattern, R, enabled to estimate the rotation angle, θ, between the two graphene layers with the following relation

$$\sin\frac{\theta}{2} = \frac{a_0}{2R} \tag{11.4}$$

On vapor-grown carbon fibers, rotational angles between 3 and 20° were observed [91].

b. Effects of surface treatments. The effects of surface treatments on the morphology and the microstructure of carbon fiber surfaces were also extensively studied by STM [81,96–103]. Various physical and chemical surface treatments were studied—oxidation in air, plasma treatments with inert gases (argon, …) or reactive gases (oxygen, nitrogen, …), electrochemical surface treatments, chemical oxidation with nitric acid, ….

Plasma treatments were found to roughen the fiber surface with the appearance of pitting, holes, and trenches [96,97,102]. On the contrary, electrochemical treatments (anodic oxidation) did not modify extensively the fiber surface morphology [98,99]. At the atomic scale, they yielded a modest oxidation of the surface with a preferential attack at the edges of the graphitic zones. The microstructure of the oxidized fibers was very similar to that of weakly graphitized carbon fibers. Oxidation in nitric acid also strongly modified the surface morphology of carbon fibers [96,103]. The surface roughness was increased with the appearance of either a rock-like surface morphology or of pitting.

11.3.1.3 AFM on Polymer Fibers

To study the surface morphology and microstructure of polymer fibers, it was necessary to use AFM since these materials were not electrically conductive. Several studies were carried out to characterize the surface morphology and/or the polymer chain orientation in polymer fibers or in oriented polymer films.

Polyimide fibers were studied by Patil et al. [104]. They showed that the surface microstructure at the molecular level was formed by straight backbone fragments arranged in the (1–10) planes. The molecular backbone was found to be oriented at a definite angle with respect to the fiber axis.

AFM was also used to study poly(p-phenyleneterephthalamide) or Kevlar® fibers [105–107]. These fibers exhibited microfibrillar morphology. At the molecular scale, the polymer chains were organized in a crystalline microstructure. The positions of the phenyl rings with respect to the hydrogen-bonded sheets were visualized and the observed structures were consistent with those determined by x-ray diffraction.

Oriented polyethylene fibers and films were studied by AFM [108–111]. On oriented polymer films, it was shown that a fibrillar morphology is developed during the drawing process [108,109]. This oriented morphology is illustrated in Figure 11.11, where AFM images taken on polymer films with draw ratios of 8 and 72, respectively, are presented. At the molecular scale, polymer chains oriented in the drawing direction could be observed (Figure 11.12). Similar morphologies and microstructures were observed on polyethylene fibers [110,111]. On melt drawn polyethylene films, the "shish kebab" crystals were visualized by AFM [112].

Starting from a classical spherulitic morphology, a similar fibrillar morphology as the one observed on polyethylene films and fibers (Figure 11.11) was observed on melt-spun polypropylene filaments after stretching [113].

11.3.2 ELASTIC MODULUS OF NANOTUBES AND NANOWIRES

Over the last thirty years, there has been an increasing interest in the study of materials with reduced dimensions and dimensionality (2-D, 1-D, and 0-D materials), such as thin films, nanowires, and metallic clusters. The large interest for these materials is due to the exceptional properties they can exhibit compared to those of bulk materials (3-D materials). Therefore, many studies concerning the effect of reduced dimensionality on the physical properties of nanomaterials are in progress. Among these properties, the mechanical ones are particularly important because they are essential for material functionality and application domains. In the case of thin films, the most frequently

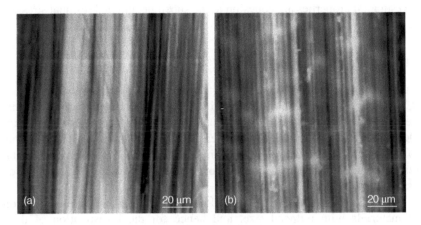

FIGURE 11.11 Morphology of oriented polyethylene films. (a) Melt pressed film with a draw ratio of 8; (b) solution cast film with a draw ratio of 72. The drawing direction is vertical for both films.

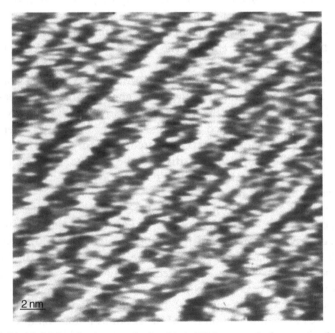

FIGURE 11.12 Molecular image obtained by AFM on a solution cast polyethylene film with a draw ratio of 72.

investigated properties are adhesion to the substrate, internal residual stresses, and tensile properties [114].

11.3.2.1 Carbon Nanotubes

Up to now, the most widely studied nanomaterials were carbon nanotubes (CNTs) because they exhibit extraordinary electrical and mechanical properties and they are supposed to have many potential applications in nano-electronics and nano-mechanical systems. Although technical difficulties due to their small size (such as their manipulation) exist, several experimental results concerning the measurement of their elastic modulus were obtained using AFM.

The invention of the AFM offered new possibilities for the manipulation of materials with reduced size, and hence for the investigation of their properties [115–118]. With AFM tips, individual carbon nanotubes can be manipulated and placed at predefined position by applying appropriate lateral forces. As concerns the mechanical properties, the AFM tip allows the direct application of a force to the probed material. Based on this principle, tensile-load experiments were performed on multi-wall nanotubes (MWNTs) [119] and single-wall nanotube (SWNT) ropes [120] using AFM tips inside a SEM. A MWNT was attached at both ends to two AFM tips by electron beam deposition of carboneous material. The experiments were recorded on video by operating in situ inside a SEM. The MWNT was then loaded with tensile constraints by moving the top AFM tip upwards. The stress-strain response was measured and the Young's modulus was calculated from the slope of this relation. The authors assumed that the load is applied only to the outer shells of the MWNTs, and therefore the actual Young's modulus of the investigated MWNTs cannot be determined by this method. The method was then applied to SWNTs ropes. The authors considered that only the SWNTs at the perimeter of the rope were sensitive to the applied load. In this case, the measured average values of the Young's modulus ranged from 320 to 1470 GPa for 8 probed ropes.

The lack of mechanical measurement techniques applicable to materials with nanometer size was overcome using AFM. Indeed, the AFM tip cannot only accurately apply forces to the studied material but also measure the resulting deformation. First, MWNTs were dispersed on an ultrafiltration membrane and the nanotubes suspended over pores were selected [121]. The selected nanotubes were then imaged in contact mode with different applied loads. The geometrical dimensions of the nanotubes were determined from the AFM images as well as the profile of the beam as a function of the applied load. The elastic modulus was deduced using the classical formula of beam deflection. As the nanotube-membrane adhesion seemed to be sufficiently high, clamped-beam conditions were assumed for the nanotubes. No correlation between the elastic modulus and the nanotube diameter was found for the eleven probed MWNTs. The obtained average value for the elastic modulus was 810 GPa. This value could be underestimated because the nanotubes were not considered as hollow beams.

Another alternative approach was proposed by the same authors for SWNT ropes [23]. This approach consisted of selecting nanotubes suspended over pores and carrying out a three-point bending test at the midpoint along the suspended length with the AFM tip. In this case, the nanotube deflection resulting from the applied force was measured directly by AFM. With the geometrical dimensions of the suspended nanotube obtained from the AFM images and assuming the boundary conditions of clamped beams, the elastic modulus was determined from the classical formula of beam deflection. For the ten probed SWNT ropes, the resulting elastic modulus increased strongly when the rope diameter decreased. The highest elastic modulus was 1.3 TPa for a rope diameter of 3 nm whereas a value of 67 GPa was measured for a rope diameter of 20 nm.

In the same way, the mechanical properties were determined by measuring the lateral bending of MWNTs [122]. Nanotubes were randomly dispersed over a flat surface of MoS_2, which has a low friction coefficient, allowing one to neglect the nanotube-substrate interaction in order to more easily deduce their elastic modulus. One end of the probed nanotubes was pinned to the substrate by depositing pads of a rigid oxide (SiO_2). The nanotubes were then laterally deflected across the substrate surface at different positions along the unpinned lengths while measuring the lateral force versus the lateral displacement. Before tip-nanotube contact, the lateral force (friction) was measured. After contact with the nanotubes, the measured lateral force increased linearly as long as the nanotube was elastically deformed. From the acquired lateral force versus lateral displacement data, the elastic modulus of the probed nanotubes was deduced using the classical formula of beam deflection of a pinned-free beam. The resulting average value for six different MWNTs with diameters ranging from 26 to 76 nm was 1.28 ± 0.59 TPa.

The results for the Young's modulus in the various experiments were spread over a large range of values. This was due mainly to differences in the inter-wall or inter-tube cohesion. Indeed, this cohesion strongly influenced the stiffness of MWNTs or SWNT bundles and hence the elastic modulus values. A lack of cohesion between adjacent nanotubes induced a low transfer of load to the inner part of the structure and resulted in a dramatic loss of stiffness. Therefore, the inter-wall cohesion and the stress intensity applied to the structure played a major role in the determined value of the elastic modulus [123]. In the case of MWNTs submitted to small forces [121], the cohesion was sufficiently high to allow a good load transfer. When the applied force increased, decohesion could occur between the inner and outer shells and resulted in a lower measured elastic modulus.

11.3.2.2 Polymer Nanotubes

Among polymer materials, nanotubes made of electrically conductive polymers such as polypyrrole (PPy) exhibit peculiar properties [124–127]. It was observed that when the nanotube outer diameter decreases, the electrical conductivity increases dramatically by more than one order of magnitude. Therefore, the mechanical properties of these nanotubes were measured using AFM to check whether effects similar to those observed for electrical conductivity occur. PPy nanotubes were thus mechanically tested in nanoscopic three-point bending and the elastic modulus was measured as a function of the nanotube outer diameter and the synthesis temperature [24].

The polymer nanotubes were synthesized using a recently developed template-based method that uses the pores of polycarbonate track-etched membranes as "nanoreactors" [127]. The template membranes had a pore density of $10^9 \, \text{cm}^{-2}$. In order to obtain nanomaterials with different outer diameters, membranes with pore size ranging between 30 and 250 nm were used. A metallic layer serving as the working electrode was evaporated onto one side of the membrane. A 10–20 nm adhesion layer of Cr was first evaporated, followed by a Au film with a thickness from 500 nm to 1 µm. The electro-polymerization was performed at room temperature as well as at 5 and $-10°C$, in a conventional one-compartment cell with a Pt counter electrode and an Ag/AgCl reference electrode. A 0.1 M pyrrole/0.1 M LiClO$_4$ solution was used and polypyrrole was generated by polarizing the working electrode at 0.8 V vs. Ag/AgCl.

After synthesis, the membrane was dissolved by immersion in a dichloromethane solution containing dodecyl sulfate as surfactant [128] and the suspension was placed in an ultrasonic bath during one hour to separate the nanostructures from the gold film previously evaporated on the backside of the membrane. The suspensions were then filtered through poly(ethylene terephthalate) (PET) membranes with pore diameters ranging between 0.8 and 3 µm. In order to remove any contaminant from the nanomaterial surface, the samples were thoroughly rinsed with dichloromethane. To minimize shear deformations in the experiments, the ratio between the suspended length of the tube, L, and its outer diameter, D_{out}, should be higher than 16 [129]. To achieve this, each series of nanowires or nanotubes synthesized in a template membrane with a specific pore diameter was dispersed on a corresponding PET membrane with a pore diameter satisfying this criterion.

Large-scale images (typically up to $80 \times 80 \ \mu m^2$) were first acquired in order to select nanotubes suspended over pores that could be used to measure their mechanical properties (Figure 11.13a). Once a suspended nanotube was located, an image of it at lower scale (down to $1 \times 1 \ \mu m^2$) was then realized to precisely determine its dimensions, i.e., its suspended length, L, and its outer diameter, D_{out} (Figure 11.13b). The outer diameter is determined by the measurement of its height with respect to the supporting membrane to avoid tip artifacts. The inner diameter, D_{in}, of the PPy nanotubes was estimated using a previously established calibration curve relating the outer and inner diameters obtained by characterizing a large amount of nanotubes by SEM.

After selection of a nanotube crossing a hole, the AFM tip was accurately located at the midpoint along the suspended length. The nanotube was then mechanically tested in nanoscopic three points, bending by performing force-curve measurement using the close-loop detectors of the scanner to measure the actual sample displacement and working in the linear dynamic range of the photo-detector as described in Section 11.2.3.4. This procedure yielded curves relating the force exerted by the cantilever to the imposed sample vertical displacement, z (Figure 11.14a). It is important to note that, to obtain reliable quantitative results, the spring constant of each used cantilever had to be calibrated. This was done by deflecting it against a reference cantilever of known spring constant [130]. The force versus sample displacement curves previously obtained were then converted into force versus nanotube deflection, δ, curves using the relation 3. The force curve measured over a portion of a nanotube lying on the membrane (dotted line) shows that nanotube flattening out and tip indentations in the nanotube were negligible in the range of applied forces. Indeed, in this case, the cantilever deflection was equal to the sample

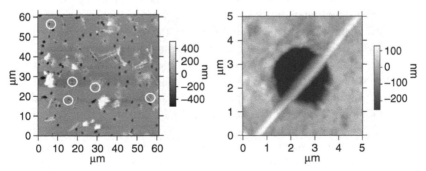

FIGURE 11.13 (a) Large-scale AFM image of polypyrrole nanotubes dispersed on a PET membrane; nanotubes crossing pores are marked with white circles. (b) Small-scale AFM image of a nanotube crossing a pore.

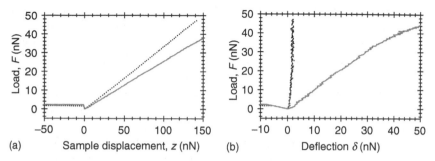

FIGURE 11.14 Load versus sample vertical displacement (a) and load versus deflection and (b) curves measured respectively on the membrane (black dotted line) and on a suspended nanotube (grey solid line).

displacement as revealed by the infinite slope of the force–deflection curve in the contact zone (Figure 11.14b).

In order to deduce the elastic modulus, the slope of the force-deflection curve was first determined by linear regression, giving access to the nanotube stiffness, $k_t = \partial F / \partial \delta$. The latter was measured over the linear portion of the curves at low force and deflection values so as to only take into account pure bending of the tube and to avoid possible interference of buckling or tip indentation. In three-point bending tests, the nanotube deflection involves both tensile/compressive deformations and shear deformations. Therefore the superposition principle implies that the total deflection corresponds to the sum of the deflections due to tensile/compressive and shear deformations. To obtain the tensile modulus, E, from the measured force and deflection, it was necessary to minimize shear deformations. This was achieved by dispersing the nanotubes on PET membranes with pore diameter satisfying this criterion. For the determination of the elastic modulus, it was also necessary to know the clamping conditions of the nanotubes on the pore edges. This indeed determined the choice of model used to deduce the elastic modulus from the tube stiffness: simply supported- or clamped-beam. Many images were recorded with various angles between the nanotube axis and the fast scan direction as well as between the nanotube axis and the cantilever long axis due to the random dispersion of the tubes over the PET membranes. No nanotube was ever displaced during these experiments. The nanotube adhesion to the membrane thus seemed to be sufficiently high to prevent any lift-off during the bending test. This justified the use of the clamped-beam model to calculate the elastic modulus.

According to this model, the reduced elastic modulus, E_r, was given by the following relationship [129] when the force was applied to the midpoint of the suspension length:

$$E_r = k_t \frac{L^3}{192I} \qquad (11.5)$$

with I the cross-section momentum of inertia:

$$I = \frac{\pi}{64} \left(D_{out}^4 - D_{in}^4 \right) \qquad (11.6)$$

The elastic modulus measured for a large number of nanotubes was reported as a function of the outer diameter and the synthesis temperature (Figure 11.15). The elastic modulus obtained for nanotubes with $D_{out} >$ 100 nm varied between 1.2 and 3.2 GPa. These values were comparable to those obtained for polypyrrole films [131,132]. When the nanotube diameter decreased ($100 > D_{out} > 70$ nm), the elastic modulus slowly increased up to 5 GPa. When it further decreased ($D_{out} < 70$ nm), the elastic modulus strongly increased and finally reached values up to more than 100 GPa. The elastic modulus measured on PPy nanotubes differed significantly from that of films. These results indicated that the elastic modulus depends strongly on the nanotube thickness or diameter. The behavior observed for the elastic modulus was very similar to that previously observed for the electrical conductivity [125,127]. It was indeed previously shown that the electrical conductivity of PPy nanotubes synthesized in the same conditions increased by more than one

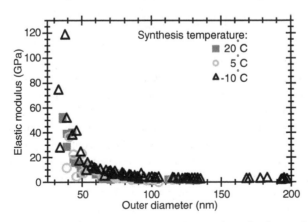

FIGURE 11.15 Elastic modulus measured on three series of polypyrrole nanotubes synthesized at various temperatures reported as a function of their outer diameter.

order of magnitude when the outer diameter decreased from 120 nm down to 35 nm [127].

Generally, physical properties such as the elastic modulus or the electrical conductivity are directly related to the material structural perfection. The observed variation of the modulus suggested that the degree of order of the nanotube structure increases with decreasing thickness. The behavior observed could be explained via assumptions based on the ratio of ordered chains in the nanotube. The polymer chains could be better aligned along the nanotube axis at the outer surface and could also contain fewer defects than the polypyrrole chains in the inner part of the nanotube. The hypothesis concerning the reduction of defects in the polymer chains with decreasing thickness was supported by analyses using Raman spectroscopy that provided information on the relative conjugation length [124,125,127]. Raman analyses were conducted on PPy nanotubes with various diameters. The ratio between the intensity of a band sensitive to the PPy oxidation state (band at 1595 cm^{-1}) and the intensity of the skeletal band (band at 1500 cm^{-1}) provided a relative measurement of the conjugation length. An increase of the π-conjugation length, normally correlated to a decrease of chain defects, was observed with a decrease of the nanotube diameter. The better alignment of the PPy chains at the surface could be due to the fact that the pore surface of the PC membrane is negatively charged. Since PPy polymerization was oxidative, electrostatic interactions could favor the growth of the polymer chains along the tube axis. This orientation effect was more important for nanotubes with smaller thickness because the surface to volume ratio increases when the thickness decreases.

Martin et al. have also shown that the electrical conductivity of PPy nanotubes increases when the temperature synthesis decreases [124,133]. Using FTIR and UV-visible-NIR spectroscopies, these authors demonstrated that the polymer chains are better oriented for the smallest diameter and that the conjugation length in PPy increases with decreasing synthesis temperature [124,133]. Therefore, in order to check whether this effect was also observed for the elastic modulus, measurements were conducted for nanotubes synthesized at different temperatures (-10, 5, and 20°C). The three curves obtained for the elastic modulus as a function of the outer diameter exhibited the same behavior, similar to that previously presented (Figure 11.15). The effect of the synthesis temperature on the measured elastic modulus did not clearly appear without significant differences between the elastic modulus values obtained for the various temperatures. This suggested that the apparent increase of the measured elastic modulus should be due to another effect. This behavior was later explained by introducing surface tension effects in the deformation mechanism of the nanotube [134].

11.3.2.3 Metallic Nanowires

Though metallic and semiconducting nanowires are of major importance in the fields of nanoelectronics, MEMS, and NEMS, their mechanical properties were less studied with AFM compared to CNTs. However, the study of the mechanical properties of such nanomaterials is also interesting on the fundamental point of view since several theoretical works have predicted that the increase of the surface to volume ratio at this scale should modify these properties, mainly due to effects of surface tension [135–142].

Combining AFM and nanoindentation, several authors have studied the hardness and the elastic modulus of metallic and semiconducting nanowires. Li et al. [143] characterized silver nanowires and found that they had comparable hardness and elastic modulus values to those of bulk silver. The properties of ZnS nanobelts with thickness varying between 50 and 100 nm were also studied [144]. It was found that their hardness is about twice that of bulk ZnS while their elastic modulus decreased by about 50%. Feng et al. measured the hardness and the elastic modulus of ZnO and GaN nanowires deposited on a silicon substrate with the same techniques [145]. They found that the hardness of GaN nanowires is larger than that of bulk GaN while that of ZnO nanowires is smaller than that of bulk ZnO. For both materials, the elastic modulus was estimated to be lower for the nanowires than for the bulk materials.

Lateral force microscopy was also used to measure the mechanical properties of nanowires. Wu et al. measured the Young's modulus and the strength of silver nanowires (16–36 nm) [146] and gold nanowires (40–250 nm) [147] held in a double-clamped beam configuration. The nanowires were positioned above trenches and were laterally bended with the AFM tip. The elastic modulus measured for Ag nanowires with a diameter ranging between 10 and 20 nm was found to be equal to that of bulk silver. For the gold nanowires, the measured elastic modulus was also equal to that of the bulk material but the yield strength was much larger. LFM was also used to measure the elastic properties of vertically aligned ZnO nanowires or nanorods [148]. By simultaneously acquiring the topography and the lateral force images in contact mode, the elastic modulus of the individual nanowires was derived. The measurement was based on quantifying the lateral force required to induce maximal deflection of the nanowires. An elastic modulus equal to 29 ± 8 GPa was found for nanowires with diameters around 45 nm.

Similar nanoscopic bending testing like the test used for the polypyrrole nanotubes was used to measure the elastic modulus on thin chromium cantilevers [149], on amorphous SiO_2 nanowires [150], and on silicon nanobridges [151]. On the Cr cantilevers, it was found that the measured Young's modulus decreased when the cantilevers get thinner. The SiO_2 nanowires were dispersed on a substrate with 1 µm wide trenches and three-point bending tests were performed. A Young's modulus of 77 ± 7 GPa for nanowires with a diameter ranging between 50 and 100 nm was measured, i.e., a value similar to that of bulk silica.

Cuenot et al. have developed an alternative method to measure the elastic modulus of suspended nanotubes or nanowires, referred as resonant contact AFM [152]. In this method, an oscillating external electric field was applied between the sample holder and the microscope head containing the cantilever holder and induced the cantilever vibration due to the polarization forces acting on the tip. By varying the intensity and the frequency of the electric field, it was possible to completely characterize the resonance spectrum of cantilevers while the tip contacts the sample surface or not. The electrostatic resonant contact method was based on the measurement of the resonance frequency of a cantilever in contact with the sample surface. When the tip contacts the sample, the resonance frequencies of the cantilever-sample system shifted to higher values compared to the resonance frequencies of the free cantilever. They showed that the resonance frequency of a cantilever in contact with a surface depends strongly on the contact stiffness [153]. Thus, by measuring the resonance frequency of cantilevers in contact with silver and lead nanowires, they were able to determine their stiffness and, hence, their elastic modulus. As previously observed for the PPy nanotubes and as recently measured with another resonant technique [154], the measured elastic modulus increased dramatically when the diameter decreased. This apparent increase of the elastic modulus was explained by introducing the effect of surface tension that contributed more and more to the mechanical response of the nanowires or nanotubes deformed in three-point bending when the diameter decreased [134,153]. For nanowires, the apparent measured elastic modulus, E_{app}, was found to be given by the following relation when surface deformation was taken into account.

$$E_{app} = E + \frac{8}{5}\gamma(1-\nu)\frac{L^2}{D^3} \qquad (11.7)$$

In this relation, E is the bulk Young's modulus, γ is the surface tension, ν is the Poisson's ratio, L is the suspended length of the nanowires, and D is its diameter. Using this equation to analyze the variation of the measured elastic modulus as a function of the ratio L^2/D^3 obtained for Ag and Pb nanowires and for PPy nanotubes, a Young's modulus corresponding to that measured on bulky samples was found. Also, this method enabled the determination of the surface tension of solid materials, which should not be confused with the surface energy [136] and for which very few techniques exist.

11.4 CONCLUSIONS

In this chapter, the main principles of STM and AFM were presented. The most generally used techniques were especially described. Then, some applications on micro- and nanofibers were presented.

We have shown that STM and AFM are very powerful tools to characterize the surface morphology and microstructure of microfibers like carbon fibers and polymer fibers. The main advantages of these techniques are that they enable the precise characterization of the surface topography from the micrometer scale down to the nanometer scale and the atomic or molecular scale, they do not necessitate heavy sample preparation, and they can be used under ambient conditions. Moreover, they do not only enable the imaging of surface morphology and structure but they also permit the mapping and the measurement of physical or chemical properties. This capacity was illustrated by a short review of the measurement of the elastic modulus of carbon and polymer nanotubes and of metallic nanowires using the AFM.

REFERENCES

1. Binnig, G., Rohrer, H., Gerber, C., and Weibel, E., *Phys. Rev. Lett.*, 49, 57–61, 1982.
2. Binnig, G., Quate, C. F., and Gerber, C., *Phys. Rev. Lett.*, 56, 930–933, 1986.
3. Pohl D.W., In *Scanning Tunneling Microscopy II*, Wiesendanger R. and Güntherodt H. J., Eds., Springer, Berlin, pp. 233–271, 1992.
4. Magonov, S. N. and Whangbo, M. H., *Surface Analysis with STM and AFM*, VCH, Weinheim, 1996.
5. Magonov, S. N., In *Encyclopedia of Analytical Chemistry*, Meyers, R. A., Ed., Wiley, Chichester, pp. 7432–7491, 2000.
6. Wiesendanger, R., Güntherodt, H. J., Güntherodt, G., Gambino, R. J., and Ruof, R., *Phys. Rev. Lett.*, 65, 247–250, 1990.
7. McGonigal, G. C., Bernhardt, R. H., and Thomson, D. J., *Appl. Phys. Lett.*, 57, 28–30, 1990.
8. Wiesendanger, R., *J. Vac. Sci. Technol. B*, 12, 515–529, 1994.
9. Nysten, B., Roux, J. C., Flandrois, S., Daulan, C., and Saadaoui, H., *Phys. Rev. B*, 48, 12527–12538, 1993.
10. Stone, V. W., Jonas, A. M., Nysten, B., and Legras, R., *Phys. Rev. B*, 60, 5883–5894, 1999.
11. Ivanov, D. A., Nysten, B., and Jonas, A. M., *Polymer*, 40, 5899–5905, 1999.
12. Dupont, O., Jonas, A. M., Nysten, B., Legras, R., Adriaensens, P., and Gelan, J., *Macromolecules*, 33, 562–568, 2000.
13. Dreezen, G., Ivanov, D. A., Nysten, B., and Groeninckx, G., *Polymer*, 41, 1395–1407, 2000.
14. Pallandre, A., Glinel, K., Jonas, A. M., and Nysten, B., *Nano Lett.*, 4, 365–371, 2004.

15. Feenstra, R. M., Stroscio, J. A., and Fein, A. P., *Surf. Sci.*, 181, 295–306, 1987.
16. Feenstra, R. M. and Stroscio, J. A., *J. Vac. Sci. Technol. B*, 5, 923–929, 1987.
17. Zhang, Z. and Lieber, C. M., *Appl. Phys. Lett.*, 62, 2792–2794, 1993.
18. Olk, C. H. and Heremans, J., *J. Mater. Res.*, 9, 259–262, 1994.
19. Chen C. J., In *Scanning Tunneling Microscopy III*, Güntherodt H. J. and Wiesendanger R., Eds., Springer, Berlin, pp. 141–178. 1992.
20. Nysten, B., Legras, R., and Costa, J. L., *J. Appl. Phys.*, 78, 5953–5958, 1995.
21. Burnham, N. A., Behrend, O. P., Ouveley, F., Gremaud, G., Gallo, P. J., Gourdon, D., Dupas, E., Kulik, A. J., Pollock, H. M., and Briggs, G. A. D., *Nanotechnology*, 8, 67–75, 1997.
22. Tomasetti, E., Legras, R., and Nysten, B., *Nanotechnology*, 9, 305–315, 1998.
23. Salvetat, J. P., Briggs, G. A. D., Bonard, J. M., Bacsa, R. R., Kulik, A. J., Stöckli, T., Burnham, N. A., and Forró, L., *Phys. Rev. Lett.*, 82, 944–947, 1999.
24. Cuenot, S., Demoustier-Champagne, S., and Nysten, B., *Phys. Rev. Lett.*, 85, 1690–1693, 2000.
25. Nysten, B., Verfaillie, G., Ferain, E., Legras, R., Lhoest, J. B., Poleunis, C., and Bertrand, P., *Microsci. Microanal. M.*, 5, 373–380, 1994.
26. Frisbie, C. D., Rozsnyai, L. F., Noy, A., Wrighton, M. S., and Lieber, C. M., *Science*, 265, 2071–2074, 1994.
27. Noy, A., Frisbie, C. D., Rozsnyai, L. F., Wrighton, M. S., and Lieber, C. M., *J. Am. Chem. Soc.*, 117, 7943–7951, 1995.
28. Sinniah, S. K., Steel, A. B., Mille, C. J., and Ruett-Robey, J. E., *J. Am. Chem. Soc.*, 118, 8925–8931, 1996.
29. Noy, A., Vezenov, D. V., and Lieber, C. M., *Ann. Rev. Mater. Sci.*, 27, 381–421, 1997.
30. Duwez, A. S., Poleunis, C., Bertrand, P., and Nysten, B., *Langmuir*, 17, 6351–6357, 2001.
31. Duwez, A. S. and Nysten, B., *Langmuir*, 17, 8287–8292, 2001.
32. Duwez, A. S. and Nysten, B., Novel methods to study interfacial layers, In *Studies in Interface Science*, Möbius, D. and Miller, R., Eds., Elsevier Science, Amsterdam, pp. 137–150, 2001.
33. Martin, Y., Abraham, D. W., and Wickramasinghe, H. K., *Appl. Phys. Lett.*, 52, 1103–1105, 1988.
34. Terris, B. D., Stern, J. E., Rugar, D., and Mamin, H. J., *J. Vac. Sci. Technol. A*, 8, 374–377, 1990.
35. Grütter, P., Mamin, H. J., and Rugar, D., In *Scanning Tunneling Microscopy II*, Güntherodt, H. J. and Wiesendanger, R., Eds., Springer, Berlin, pp. 151–208, 1992.
36. http://www.uark.edu/misc/mbestm/images/stm/si7x7.JPG.
37. Tersoff, J. and Hamann, D. R., *Phys. Rev. B*, 31, 805–813, 1985.
38. Chen, C. J., *Introduction to Scanning Tunneling Microscopy*, Oxford University Press, New York, 1993.
39. Hamers, R. J., In *Scanning Tunneling Microscopy I*, Wiesendanger, R. and Güntherodt, H. J., Eds., Springer, Berlin, pp. 83–129, 1992.

40. Daulan, C., Derre, A., Flandrois, S., Roux, J. C., and Saadaoui, H., *J. Phys. I*, 5, 1111–1117, 1995.
41. Eigler, D. M. and Schweizer, E. K., *Nature*, 344, 524–526, 1990.
42. Wintterlin, J. and Behm, R. J., In *Scanning Tunneling Microscopy I*, Wiesendanger, R. and Güntherodt, H. J., Eds., Springer, Berlin, pp. 40–82, 1992.
43. De Feyter, S. and De Schryver, F., *J. Phys. Chem. B*, 109, 4290–4302, 2005.
44. Crommie, M. F., Lutz, C. P., and Eigler, D. M., *Science*, 262, 218–220, 1993.
45. Chiang, S., In *Scanning Tunneling Microscopy I*, Wiesendanger, R. and Güntherodt, H. J., Eds., Springer, Berlin, pp. 181–205, 1992.
46. Libioulle, L., Houbion, Y., and Gilles, J. M., *Rev. Sci. Instrum.*, 66, 97–100, 1995.
47. Mate, C. M., McClelland, G. M., Erlandsson, R., and Chiang, S., *Phys. Rev. Lett.*, 59, 1942–1945, 1987.
48. Martin, Y., Williams, C. C., and Wickramasinghe, H. K., *J. Appl. Phys.*, 61, 4723–4729, 1987.
49. Zhong, Q., Innis, S., Kjoller, K., and Elings, V. B., *Surf. Sci. Lett.*, 290, 688–692, 1993.
50. Martin, Y. and Wickramasinghe, H. K., *Appl. Phys. Lett.*, 50, 1455–1457, 1987.
51. Maivald, P., Butt, H. J., Gould, S. A. C., Prater, C. B., Drake, B., Gurley, J. A., Elings, V. B., and Hansma, P. K., *Nanotechnology*, 2, 103–106, 1991.
52. Meyer, E. and Heinzelmann, H., In *Scanning Tunneling Microscopy II*, Wiesendanger, R. and Güntherodt, H. J., Eds., Springer, Berlin, pp. 99–149, 1992.
53. Stocker, W., Schumacher, M., Graff, S., Lang, J., Wittmann, J. C., Lovinger, A. J., and Lotz, B., *Macromolecules*, 27, 6948–6955, 1994.
54. Giessibl, F. J., *Science*, 267, 68–71, 1995.
55. Loppacher, C., Bammerlin, M., Guggisber, M., Schär, S., Bennewitz, R., Baratoff, A., Meyer, E., and Güntherodt, H. J., *Phys. Rev. B*, 62, 16944–16949, 2000.
56. Giessibl, F. J., *Rev. Mod. Phys.*, 75, 943–949, 2003.
57. Garcia, R. and Pérez, R., *Surf. Sci. Rep.*, 47, 197–301, 2002.
58. Leclère, P., Lazzaronni, R., Brédas, J. L., Yu, J. M., Dubois, P., and Jérôme, R., *Langmuir*, 12, 4317–4320, 1996.
59. Martin, P., Marsaudon, S., Aimé, J. P., and Bennetau, B., *Nanotechnology*, 16, 901–907, 2005.
60. See for instance http://www.nanosensors.com/products_overview.html#HAR, http://www.olympus.co.jp/en/insg/probe/en/specsiaspectE.html, http://www.nanoandmore.com/afm_probes.php?catID = 10.
61. See for instance http://www.nanoandmore.com/afm_probes.php?catID = 9, http://www.nanosensors.com/products_overview.html#SSS, http://www.spmtips.com/products/cantilevers/datasheets/hi-res/.
62. Hafner, J. H., Cheung, C. L., and Lieber, C. M., *J. Am. Chem. Soc.*, 121, 9750–9751, 1999.
63. Burnham, N. A. and Colton, R. J., *J. Vac. Sci. Technol. A*, 7, 2906–2913, 1989.

64. Nysten, B., Meerman, C., and Tomasetti, E., Microstructure and tribology of polymer surfaces, In *ACS Symposium Series No 741*, Tsukruk, V. and Wahl, K. J., Eds., American Chemical Society, Washington, D. C., pp. 304–316, 2000.

65. Hertz, H., *J. Reine. Angew Math.*, 92, 156–171, 1882.

66. Sneddon, I., *Int. J. Eng. Sci.*, 3, 47–57, 1965.

67. Johnson, K. L., Kendall, K., and Roberts, A. D., *Proc. Roy. Soc. London A*, 324, 301–313, 1971.

68. Derjaguin, B. V., Muller, V. M., and Toporov, Y. P., *J. Colloid Inter. Sci.*, 53, 314–326, 1975.

69. Oliver, W. C. and Pharr, G. M., *J. Mater. Res.*, 7, 1564–1583, 1992.

70. Heuberger, M., Dietler, G., and Schlapbach, M., *Nanotechnology*, 5, 12–23, 1994.

71. Vallat, M. F., Giami, S., and Coupard, A., *Rubber Chem. Technol.*, 72, 701–711, 1999.

72. Bischel, M. S., Vanlandingham, M. R., Eduljee, R. F., Gillepsie, J. W., and Schultz, J. M., *J. Mater. Sci.*, 35, 221–228, 2000.

73. Du, B., Zhang, J., Zhang, Q., Yang, D., He, T., and Tsui, O. K. C., *Macromolecules*, 33, 7521–7528, 2000.

74. Viville, P., Daoust, D., Jonas, A. M., Nysten, B., Legras, R., Dupire, M., Michel, J., and Debras, G., *Polymer*, 42, 1953–1967, 2001.

75. Radmacher, M., Fritz, M., Cleveland, J. P., Walters, D. A., and Hansma, P. K., *Langmuir*, 10, 3809–3814, 1994.

76. Radmacher, M., Fritz, M., Kacher, C. M., Cleveland, J. P., and Hansma, P. K., *Biophys. J.*, 70, 556–567, 1996.

77. Touhami, A., Nysten, B., and Dufrêne, Y. F., *Langmuir*, 19, 4539–4543, 2003.

78. Sasaki, M., Hane, K., Okuma, S., and Bessho, Y., *Rev. Sci. Instrum.*, 65, 1930–1934, 1994.

79. Wiesendanger, R. and Anselmetti, D., In *Scanning Tunneling Microscopy I*, Wiesendanger, R. and Güntherodt, H. J., Eds., Springer, Berlin, pp. 131–179, 1992.

80. Oberlin, A., Bonnamy, S., and Lafdi, K., In *Carbon Fibers*, Donnet, J. B., Wang, T. K., Peng, J. C. M., and Rebouillat, S., Eds., Marcel Dekker, New York, pp. 85–159, 1998.

81. Peng, J. C. M., Donnet, J. B., Wang, T. K., and Rebouillat, S., In *Carbon Fibers*, Donnet, J. B., Wang, T. K., Peng, J. C. M., and Rebouillat, S., Eds., Marcel Dekker, New York, pp. 161–229, 1998.

82. Wang, T. K., Donnet, J. B., Peng, J. C. M., and Rebouillat, S., In *Carbon Fibers*, Donnet, J. B., Wang, T. K., Peng, J. C. M., and Rebouillat, S., Eds., Marcel Dekker, New York, pp. 231–309, 1998.

83. Brown, N. M. D. and You, H. X., *Surf. Sci.*, 237, 273–279, 1990.

84. Magonov, S. N., Cantow, H. J., and Donnet, J. B., *Polym. Bull.*, 23, 555–562, 1990.

85. Hoffman, W. P., Hurley, W. C., Liu, P. M., and Owens, T. W., *J. Mater. Res.*, 6, 1685–1694, 1991.

86. Hoffman, W. P., *Carbon*, 30, 315–331, 1992.
87. Donnet, J. B. and Qin, R. Y., *Carbon*, 30, 787–796, 1992.
88. Donnet, J. B. and Qin, R. Y., *Carbon*, 31, 7–12, 1993.
89. Quin, R. Y. and Donnet, J. B., *Carbon*, 32, 323–328, 1994.
90. de Bont, P. W., Scholte, P. M. L. O., Hottenhuis, M. H. J., van Kempen, G. M. P., Kerssemakers, J. W., and Tuinstra, F., *Appl. Surf. Sci.*, 74, 73–80, 1994.
91. Endo, M., Oshida, K., Kobori, K., Takeuchi, K., Takahashi, K., and Dresselhaus, M. S., *J. Mater. Res.*, 10, 1461–1468, 1995.
92. Donnet, J. B., Wang, T. K., and Shen, Z. M., *Carbon*, 34, 1413–1418, 1996.
93. Nysten, B., Roux, J. C., Flandrois, S., Daulan, C., and Saadaoui, H., *Phy. Rev. B*, 48, 12527–12538, 1993.
94. Kondo, S., Lutwyche, M., and Wada, Y., *Jpn J. Appl. Phys.*, 33, L1342–L1344, 1994.
95. Saadaoui, H., Roux, J. C., Flandrois, S., and Nysten, B., *Carbon*, 31, 481–486, 1993.
96. Hoffman, W. P., Hurley, W. C., Owens, T. W., and Phan, H. T., *J. Mater. Sci.*, 26, 4545–4553, 1991.
97. Smiley, R. J. and Delgass, W. N., *J. Mater. Sci.*, 28, 3601–3611, 1993.
98. Donnet, J. B., Park, S. J., and Wang, W. D., *Polym. Adv. Technol.*, 5, 395–399, 1994.
99. Qin, R. Y. and Donnet, J. B., *Carbon*, 32, 323–328, 1994.
100. Zielke, U., Hüttinger, K. J., and Hoffman, W. P., *Carbon*, 34, 983–998, 1996.
101. Daley, M. A., Tandon, D., Economy, J., and Hippo, E. J., *Carbon*, 34, 1191–2000, 1996.
102. Oyama, H. T. and Wightman, J. P., *Surf. Inter. Anal.*, 26, 39–55, 1998.
103. Figueiredo, J. L., Serp, P., Nysten, B., and Issi, J. P., *Carbon*, 37, 1809–1816, 1999.
104. Patil, R., Tsukruk, V. V., and Reneker, D. H., *Polym. Bull.*, 29, 557–563, 1992.
105. Snétivy, D., Vancso, G. J., and Rutledge, G. C., *Macromolecules*, 25, 1037–7042, 1992.
106. Li, S. F. Y., McGhie, A. J., and Tang, S. L., *Polymer*, 34, 4573–4575, 1993.
107. Li, S. F. Y., McGhie, A. J., and Tang, S. L., *J. Vac. Sci. Technol. A*, 12, 1891–1893, 1994.
108. Magonov, S. N., Sheiko, S. S., Deblieck, R. A. C., and Möller, M., *Macromolecules*, 26, 1380–1386, 1993.
109. Shönherr, H., Vancso, G. J., and Argon, A. S., *Polymer*, 36, 2115–2121, 1995.
110. Wawkuschewski, A., Cantow, H. J., and Magonov, S. N., *Polym. Bull.*, 32, 235–240, 1994.
111. Wawkuschewski, A., Cantow, H. J., Magonov, S. N., Hewes, J. D., and Kocur, M. A., *Acta Polymer*, 46, 168–177, 1995.
112. Jandt, K. D., Buhk, M., Miles, M. J., and Petermann, J., *Polymer*, 35, 2458–2462, 1994.
113. Hauttojärvi, J. and Leijala, A., *J. Appl. Polym. Sci.*, 74, 1242–1249, 1999.
114. Vinci, R. P. and Vlassak, J. J., *Ann. Rev. Mater. Sci.*, 26, 431–462, 1996.

115. Falvo, M. R., Clary, G. J., Taylor, R. M. II, Chi, V., Brooks, F. P. Jr., Washburn, S., and Superfine, R., *Nature*, 389, 582–584, 1997.

116. Avouris, P., Hertel, T., Martel, R., Schmidt, T., Shea, H. R., and Walkup, R. E., *Appl. Surf. Sci.*, 141, 201–209, 1999.

117. Falvo, M. R., Clary, G. J., Helser, A., Paulson, S., Taylor, R. M. II, Chi, V., Brooks, F. P. Jr., Washburn, S., and Superfine, R., *Microsci. Microanal.*, 4, 504–512, 1999.

118. Yu, M. F., Yakobson, B. I., and Ruoff, R. S., *J. Phys. Chem. B*, 104, 8764–8767, 2000.

119. Yu, M. F., Lourie, O., Dyer, M. J., Moloni, K., Kelly, T. F., and Ruoff, R. S., *Science*, 287, 637–640, 2000.

120. Yu, M. F., Files, B. S., Arepalli, S., and Ruoff, R. S., *Phys. Rev. Lett.*, 84, 5552–5555, 2000.

121. Salvetat, J. P., Kulik, A. J., Bonard, J. M., Briggs, G. A. D., Stöckli, T., Metenier, K., Bonnamy, S., Beguin, F., Burnham, N. A., and Forró, L., *Adv. Mater.*, 11, 161–165, 1999.

122. Wong, E. W., Sheehan, P. E., and Lieber, C. M., *Science*, 277, 1971–1975, 1997.

123. Salvetat-Delmotte, J. P. and Rubio, A., *Carbon*, 40, 1729–1734, 2002.

124. Menon, V. P., Lei, J., and Martin, C. R., *Chem. Mater.*, 8, 2382–2390, 1996.

125. Duchet, J., Legras, R., and Demoustier-Champagne, S., *Synth. Met.*, 98, 113–122, 1998.

126. Demoustier-Champagne, S., Duchet, J., and Legras, R., *Synth. Met.*, 101, 20–21, 1999.

127. Demoustier-Champagne, S. and Stavaux, P. Y., *Chem. Mater.*, 11, 829–834, 1999.

128. De Vito, S. and Martin, C. R., *Chem. Mater.*, 10, 1738, 1998.

129. Timoshenko, S. P. and Gere, J. M., *Mechanics of Materials*, Van Nostrand, New York, 1972.

130. Tortonese, M. and Kirk, M., *Micromachining and Imaging SPIE Proceedings Series*, 3009, 53, 1997.

131. Gandhi, M., Spinks, G. M., Burford, R. P., and Wallace, G. G., *Polymer*, 36, 4761–4765, 1995.

132. Sun, B., Jones, J. J., Burford, R. P., and Skyllas-Kazacos, M., *J. Mater. Sci.*, 24, 4024–4029, 1989.

133. Cai, Z., Lei, J., Liang, W., Menon, V. P., and Martin, C. R., *Chem. Mater.*, 3, 960–967, 1991.

134. Cuenot, S., Fretigny, C., Demoustier-Champagne, S., and Nysten, B., *Phys. Rev. B*, 69, 165410, 2004.

135. Gurtin, M. E. and Murdoch, A., *Arch. Ration. Mech. Anal.*, 57, 291, 1975.

136. Cammarata, R. C., *Prog. Surf. Sci.*, 46, 1–38, 1994.

137. Cammarata, R. C., *Mater. Sci. Eng. A*, 237, 180–184, 1997.

138. Miller, R. E. and Shenoy, V. B., *Nanotechnology*, 11, 139–147, 2000.

139. Shenoy, V. B., *Int. J. Sol. Struct.*, 39, 4039–4052, 2002.

140. Sharma, P., Ganti, S., and Bhate, N., *Appl. Phys. Lett.*, 82, 535–537, 2003.

141. Villain, P., Beauchamp, P., Badawi, K. F., Goudeau, P., and Renault, P. O., *Scripta. Mater.*, 50, 1247–1251, 2004.
142. Dingreville, R., Qu, J., and Cherkaoui, M., *J. Mech. Phys. Sol.*, 53, 1827–1854, 2005.
143. Li, X., Gao, H., Murphy, C. J., and Caswell, K. K., *Nano Lett.*, 3, 1495–1498, 2003.
144. Li, X., Wang, X., Xiong, Q., and Eklund, P. C., *Nano Lett.*, 5, 1982–1986, 2005.
145. Feng, G., Nix, W. D., Yoon, Y., and Lee, C. J., *J. Appl. Phys.*, 99, 074304, 2006.
146. Wu, B., Heigdelberg, A., Boland, J. J., Sader, J. E., Sun, X. M., and Li, Y. D., *Nano Lett.*, 6, 468–472, 2006.
147. Wu, B., Heidelberg, A., and Boland, J. J., *Nature Mater.*, 4, 525–529, 2005.
148. Song, J., Wang, X., Rieda, E., and Wang, Z. L., *Nano Lett.*, 5, 1954–1958, 2005.
149. Nilsson, S. G., Borrisé, X., and Montelius, L., *Appl. Phys. Lett.*, 85, 3555–3557, 2004.
150. Ni, H., Li, X., and Gao, H., *Appl. Phys. Lett.*, 88, 043108, 2006.
151. Tabib-Azar, M., Nassirou, M., Wang, R., Sharma, S., Kamins, T. I., Islam, M. S., and Williams, R. S., *Appl. Phys. Lett.*, 87, 113102, 2005.
152. Cucnot, S., Frétigny, C., Demoustier-Champagne, S., and Nysten, B., *J. Appl. Phys.*, 93, 5650–5655, 2003.
153. Nysten, B., Frétigny, C., and Cuenot, S., Testing, reliability, and application of micro-and nano-material systems III, In *Proc. SPIE Conf*, Geer, R. E., Meyendorf, N., Baaklini, G. Y., and Michel, B., Eds., SPIE, Bellingham, pp. 78–88, 2005.
154. Chen, C. Q., Shi, Y., Zhang, Y. S., Zhu, J., and Yan, Y. J., *Phys. Rev. Lett.*, 96, 075505, 2006.

Section IV

Interfaces and Interphases in
Biological Processes

12 Biocompatibility of Powdered Materials: The Influence of Surface Characteristics

Patrick Frayssinet and Patrice Laquerriere

CONTENTS

12.1 INTRODUCTION

The natural or synthetic materials existing in the environment interact differently with living organisms depending on their form. Compared to bulk material, powders and fibers do not generally show the same toxicity or biocompatibility. In nature, bulk material generally interacts with the skin while powdered material also interacts with the lungs, the eyes, and the digestive tract, due to its ability to remain in suspension in air or water. The characteristics of the material influence the route of contamination.

Particulate material has been known to be involved in pathological processes and diseases since the antiquity. Today, it is a major cause of health concern. Lung fibrosis has been identified as a typical disease of mine workers. It was demonstrated a long time ago that it is due to the inhalation of silica particles while the same compound is not toxic when under bulk material form.

More recently, asbestos fibers have been shown to be the cause of asbestosis, and they are a major concern for public health in several developed countries.

Other pathological states are associated with the presence of powders or fibers in tissues. Urate crystals or pyrophosphate stones in the kidney are responsible for lithiasis. Calcium pyrophosphate dihydrate crystals formed in the joints cause chondrocalcinosis. Hydroxyapatite crystals trigger an acute or chronic arthritis when they are found in leukocytes and synovial fluid. Urate crystals trigger gout and provoke the appearance of tophi.

The use of biomaterials for replacement of body parts such as hip or knee joints has created biological troubles related to the release of wear particles by the materials. They are composed of metals, ceramics, or polymers and trigger a foreign body reaction responsible for aseptic loosening of the prosthesis due the lysis of the surrounding bone. This consequence of particle release is a very frequent cause of surgical intervention in orthopedic surgery.

Thus the influence of particles, powders, and fibers in the tissue is of primary interest in several fields of public health.

12.2 THE BODY RESPONSE AGAINST EXOGENOUS MATERIALS

In order to preserve its integrity, the mammal organism is equipped to recognize and eliminate any virus, bacteria, or foreign body having penetrated it. The body's response to a foreign body may be divided into two categories—specific and nonspecific. Macrophages play an important role in both responses. The nonspecific response is based on the local destruction of the foreign body by macrophages and there is no intensity increase upon re-exposure. Exogenous materials can cause local acute toxicity, inducing cell injury or death, and are at the origin of a coordinated inflammatory and foreign body reaction, which involves several cell lineages in a restricted area. The specific immune response requires the stimulation of a special lineage of white blood cells—the lymphocytes—which keep the memory of a contact with a particular material and can react very quickly when further contacts are made. The immune response is not limited to the region of implantation of the foreign body.

12.2.1 CELL INJURY

The presence of particulate materials in contact with the cells of the organism frequently induces cell injuries which can heal, trigger an inflammatory reaction, and/or kill the cell.

The mechanisms by which the cell metabolism is affected by powders are numerous. A lesion of the cell membrane by channel blocks, lipid per-oxidation, protein crosslinking, phospholipase activation, or even by mechanical alteration is often involved. The interaction between the particles and enzymes, and in particular the complexation of mineral elements necessary for enzyme function, can also be responsible. Damages can result from the powder itself or from toxins leached from the particles.

12.2.2 INFLAMMATION AND FOREIGN BODY REACTION

Inflammation is a reaction of the microcirculation characterized by movement of fluid and circulating cells from the blood into the extravascular sector. These cells attempt to eliminate the pathogenic insult, as such as a foreign body, and the injured tissue components, if there are any.

The presence of a foreign body leads to the production of inflammatory mediators, which can be divided into vasoactive mediators and chemotactic factors. The vasoactive mediators increase the vascular permeability. The chemotactic factors recruit and stimulate the inflammatory cells in the inflammation site. These different factors are proteins or eiconasoids which are either secreted by the local cells or circulated in an inactive state in the plasma and activated by a change of conformation. These molecules are often organized in molecule cascades. When a molecule turns active, it activates another molecule in the cascade. The plasma contains three major enzyme cascades composed of a series of sequentially activated proteases which are interrelated: the coagulation cascade, kinin generation, and complement.[1]

The complement system consists of a group of 20 plasma proteins. In addition to being a source of vasoactive mediators, components of the complement system are an integral part of the immune system and play an important role in host defence against bacterial infection. Activation of the alternative pathway of the complement system can be caused by a foreign body. The material characteristics responsible for this activation are not known (Figure 12.1).

Hageman factor (clotting factor XII), of the coagulation cascade, is an additional source of vasoactive mediator which is activated when exposed to negatively charged surfaces (Figure 12.1).

The key cell of the body response against a material which does not show an acute toxicity is the macrophage. The macrophage originates from the monocyte lineage which is produced in the bone marrow of the adult. The macrophages can be divided into two cell populations. One population resides in different tissues after its migration from the blood stream. The second is able to migrate through the tissues from the blood when activated by chemotactic factors.

The macrophage is a cell able to internalize the particles by phagocytosis in special internal compartments in order to destroy them by physico-chemical

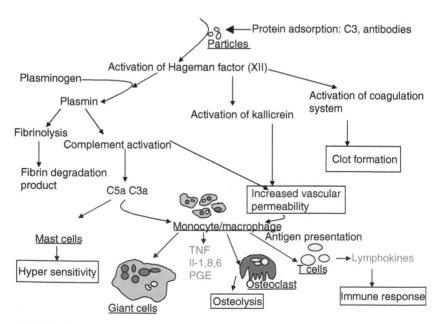

FIGURE 12.1 Activation of the different cascades involved in the inflammatory reaction and leading to a foreign body reaction, either accompanied by an immune response or not.

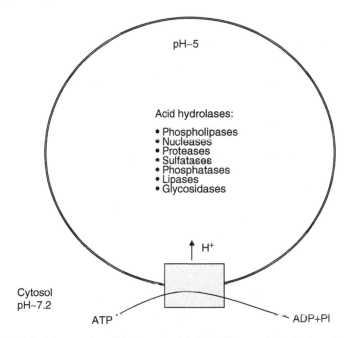

FIGURE 12.2 Scheme of a cell lysosome with the different acid hydrolases it contains. The low pH is maintained by a proton pump.

mechanisms (Figure 12.2 and Figure 12.3). Phagocytosis is the ingestion of large particles, minerals, microorganisms, or dead cells in large vesicles called phagosomes. These cells can also fuse to form multinucleated giant cells. Interleukin-4 (IL-4) is able to induce foreign body giant cells from human monocytes and macrophages.[2] This effect is optimized with granulocyte macrophage–colony stimulating factor (GM–CSF) or IL-3 (Interleukin-3), dependent on the concentration of IL-4.

In mammals, three classes of white blood cells act as professional phagocytes—macrophages, neutrophils, and dendritic cells. These cells all develop from hemopoietic stem cells. They defend us against infection and clear the organism from dead or senescent cells.

The phagosomes have diameters determined by the size of the particle ingested and can be almost as large as the phagocytic cell itself.

Once formed, the phagosomes fuse with another kind of cytoplasmic vesicle and the lysosomes and the ingested material is degraded. The particles end up in the lysosomes which are membrane-enclosed cell compartments filled with hydrolytic enzymes used for the intracellular digestion of macromolecules (Figure 12.2 and Figure 12.3). They contain about 40 types of

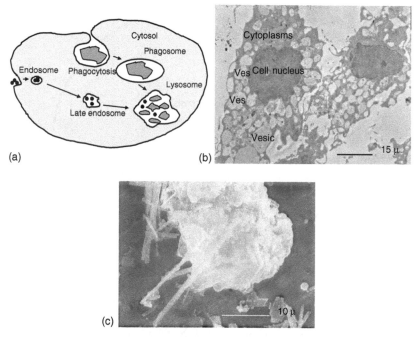

FIGURE 12.3 (a) Pathways to degradation in lysosomes. By membrane invagination, both lead to the phagocytosis of particles of different sizes in the lysosome. (b) TEM sections of bone marrow stromal cells having phagocytosed calcium phosphate particles dissolved in EDTA. The particles were in the numerous vesicles (ves) contained in the cytoplasm. (c) SEM of a macrophage grown in vitro in contact with calcium phosphate fibers. The fibers are located at the membrane surface and fragments are phagocytosed.

hydrolytic enzymes, all of which are acid hydrolases requiring an acid environment to be active. The pH in the lysosome is maintained low by an H+ ATPase.

Under a critical size (around 100 nm), the particles are internalized in pinocytic vesicles. Virtually all eukaryotic cells continually ingest bits of their plasma membrane in the form of small vesicles which may contain nanoparticles. These small particles also end up in the lysosomes through the endosomes.

When the characteristics of the material do not allow its complete degradation in the cell vesicles, the remnants of the materials can be released in the extracellular fluids and/or the cell can be involved in a process of programmed death called apoptosis. Macrophage death is accompanied by the release of hydrolytic enzymes and inflammatory factors in the extracellular medium causing damage to the surrounding tissues and increasing the inflammatory reaction.

The macrophages do not solely ingest particles. Phagocytosis is responsible for the activation of the cell which synthesizes and secretes various proteins and glycoproteins, such as growth factors and cytokines, which in turn activate other cells with specific activities, such as immune cells or osteoclasts, which digest the bone matrix (Figure 12.4). This synthesis activity is responsible for the amplification of the inflammatory reaction. Macrophages also secrete metalloproteases, a family of proteolytic enzymes that are capable of degrading all major components of the extracellular matrix.

To be phagocytosed, a particle must bind to the macrophage surface and trigger the phagocytosis process though receptors on the cell membrane. Antibodies specifically link to molecules that they recognize at the particle surfaces. A particular domain of the antibody (Fc region) is exposed to the exterior and recognized by some receptors of the macrophage membrane. Thus they can bridge the particles to membrane receptors and trigger their internalization (opsonization).

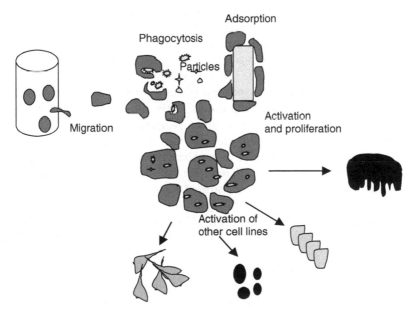

FIGURE 12.4 The macrophages originate from the monocytes migrating from the blood vessels. They are activated by the phagocytosis of particles and synthesize proteins acting locally on several cell lineages like osteoclasts, fibroblasts, or immune cells.

Components of the complement can also bridge the particle to the cells. They can be adsorbed at the material surface due to its chemical properties. Exposure to blood during implantation permits extensive opsonization with the labile fragment C3b and the rapid conversion to its hemolytically inactive, but more stable, form C3bi.[3] C3b binds to the monocyte receptor CD35. C3bi is also recognized by CD11b/CD18 and CD11C/CD18, which are receptors found at the surface of macrophages.

Regarding the surface chemistry of the particles, numerous proteins can adsorb at their surface and probably find a receptor on the macrophage membrane. It is not clear if a direct adhesion of the macrophage membrane at the surface of the material can occur.

The tophi appearing around the joints of patients suffering from gout is a good example of the response of the organism against crystals. Long, needle-shaped urate crystals may be found intracellularly in the leucocytes of the synovial fluid. Extracellular soft tissue deposits of these crystals are surrounded by foreign-body giant cells, and they are in association with inflammatory responses of mononuclear cells. Sodium urate crystals absorb fibronectin, complement, and a number of other proteins on their surfaces. On phagocytosing these protein-coated crystals, neutrophils release activated oxygen species and lysosomal enzymes which mediate tissue injury. Furthermore, neutrophils that have phagocytosed urate crystals also secrete leukotriene B4, kinins, collagenase, kallikrein, prostaglandins, and Il-1. All of these factors are promoters of the inflammatory response.

The role of macrophages in silicosis, asbestosis, or focal dust emphysema has been known for a long time. Pathologists have hypothesized that silicon hydroxide groups on the surface of the particles form hydrogen bonds with cell membrane phospholipids and proteins. This interaction would lead to macrophage death, releasing free silica particles and fibrogenic factors. Asbestos fibers are long (~ 100 μm) and thin (0.5–1 μm). They deposit in the distal airways and many of the larger fibers penetrate the interstitial space.[4] Some cases of diffuse interstitial lung fibrosis were attributed to rare earth particles showing that the types of particles that can lead to lung fibrosis could be more numerous than initially thought.[5]

At the molecular level, Soloviev et al.[6] demonstrated the rapid activation of the NFκB transcription factor by the phagocytosis of titanium alloy particles.

It must be noted that the particles do not stay in the location in which they are implanted nor do they enter the organism. Once they have been phagocytosed, they migrate with the cells in which they are trapped. They migrate the same way as usual, and the particles follow lymph vessels to the lymph nodes. Systemic dissemination of large quantities of metallic wear and corrosion products occurs in patients having malfunctioning joint replacement prostheses. These products can be in the form of wear particles, colloidal organometallic complexes, inorganic metal salts, oxides, or other species. Dissemination via the lymphatic and blood circulatory systems results in significant elevations in serum and urine metal concentrations and deposition of wear particles in organs of the reticuloendothelial system (liver, spleen, lymph nodes), which contain a lot of macrophages and act as filters. Metallic deposits of chromium were even detected in the renal tubules of patients who had been equipped with cobalt–chromium–molybdenum joint prostheses for some time. They did not have the characteristics of wear particles suggesting a precipitation process.[7]

A common complication of the presence of particles of an exogenous material in the body is the aseptic loosening of hip or knee prosthesis. These implants are made of a metal or ceramic surface sliding against another surface made of ceramic or polyethylene. These joint surfaces are held by a metal stem inserted into the bone marrow cavity which is filled with polymethacrylate to grip the stem. It is easy to understand that the stress acting at the interface of these different materials produces numerous particles which are released in the surrounding tissues.

Within a few months or years, a very thick (several centimeters) membrane-like rubber can be formed at the interface between the prosthesis and the bone. The formation of this membrane is accompanied by the destruction of the bone, which can almost disappear. When these membranes are sectioned and observed under a microscope, they are made exclusively of macrophages organized in several histological layers with a small proportion of conjunctive tissue in which blood vessels are found (Figure 12.5). The histological layer at the interface of the prosthesis is made of macrophages that look like synovial cells—this is the reason why these membranes are named pseudosynovial membranes. The bone destruction is due to the activation of the osteoclasts which are visible all along the bone trabeculae surface. All the macrophages contain debris. Some of them can have phagocytosed several kinds of debris, polyethylene, polymethacrylate, metal, and even alumina. When the size of the debris reaches several dozens of μm, the macrophages fuse and form giant cells. The phagocytosis of particles triggers the activation of the macrophages which synthesize different molecules: PGE_2, collagenase, TNF, IL-1, IL-6, PDGF, metalloproteinase, cyclo-oxygenase[8,9] (Figure 12.4). These molecules act both on the osteoblasts and osteoclasts and will lead to the proliferation and activation of osteoclasts, leading to bone destruction (osteolysis). Recently, Xu et al. demonstrated that macrophages found in the proximity to hip prostheses and containing polyethylene, methacrylate, or metal debris (titanium or cobalt–chromium alloys) also synthesized TGF-α and EGF.[10]

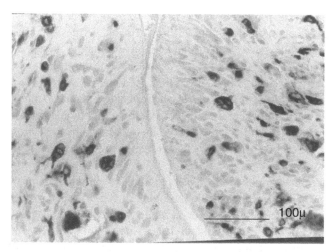

FIGURE 12.5 Section of a pseudosynovial membrane made of macrophages and giant cells (labeled in black with CD68 antibodies) surrounding failed hip prostheses.

The foreign body reaction taking place at the contact of the orthopaedic material is another good example of the effects of particulate materials in the organism. All of these materials are known to be well tolerated in their bulk form while they trigger a dramatic foreign body reaction when in a powdered form. The molecular events occurring after the phagocytosis of the particles and leading to the bone destruction have been described, however the influence of the material characteristics were not. Horowitz et al.[11] have shown that the phagocytosis of PMMA or titanium particles by macrophages did not trigger the synthesis of the same molecules.

The macrophages and giant cells may exhibit different phenotypes in contact with foreign particles. Frayssinet et al.[12] demonstrated that the location of acid phosphatase activity in the macrophage was different in granulomas formed around hip stems. Four different phenotypes were evidenced regarding their phosphatase activity: nucleus + cytoplasm +, cytoplasm + nucleus −; nucleus − cytoplasm +, nucleus − cytoplasm −. No relationship was established between the particle characteristics and a particular phenotype.

The role of the macrophages and giant cells is not limited to the phagocytosis of particles. The first cells in contact with bulk materials once the blood clot is eliminated are the cells of the monocyte lineage. These cells precede the loose connective tissue of the healing reaction in the pores of calcium phosphate ceramics and are involved in the ossification occurring at their surface.[13]

12.2.3 THE IMMUNE PART OF THE FOREIGN BODY REACTION

The macrophage is not the only cell type which can be found in contact with exogenous material. Lymphocytes, plasmocytes, and mastocytes can also be shown. The lymphocytes and plasmocytes are visible if a component of the material presented by the macrophages to the T cells is recognized by the T cells and has already been in contact with them. Macrophages and dendritic cells which are part of the monocyte/macrophage lineage are able to stimulate the immune response against specific antigens. After initial adhesion, antigen-specific recognition, and phagocytosis, these cells process the antigen then incorporate peptides derived from the antigens in the MHC class I and II molecules located in the cell membrane. These cells come in contact with T cells which are activated against the antigen borne at the surface of the cell membrane (antigen presentation). Uptake of micro-particle-adsorbed proteins triggers prolonged, efficient antigen presentation. This property is used to increase the immune power of vaccines by adsorption of the vaccine antigens on nano or microparticles. It must be noted that components of exogenous materials (organic or minerals) can be associated with proteins to constitute a hapten and be recognized by immune cells.

Mastocytes are often evidenced in the proximity of biomaterial or its debris. These cells are filled with histamine granules that can be released in the extracellular matrix under special stimuli, such as fixation of particles or molecules on specific antibodies (IgE), which are found at the surface of membrane. These molecules specific for a given material provoke the release of the vesicle contents in the extracellular medium which leads to various effects on the tissues causing the symptomatology of hypersensitivity. These cells are at the origin of hypersensitivity reactions and could be responsible for hypersensitivity to materials. The organism must have already been in contact with the molecule or the material for the mastocytes to possess the IgE specific for the molecules leading to cell degranulation.

Hypersensitivity to inorganic materials has already been described.[14] Hypersensitivity to various metals including cobalt, chromium, and nickel does occur and can generally be determined by appropriate skin testing. Recent studies suggest that more than half of the patients presenting for removal or resealing of prosthetic components are sensitive to in vitro cell migration inhibition tests.

Neutrophils may interact with foreign bodies. Several polymers are known to increase the production of reactive oxygen intermediates such as the superoxide anion (O_2^-).[15,16] This effect is however marginal as there are very few or no neutrophils in contact with most of the foreign bodies when there is no infection.

12.2.4 CARCINOGENESIS

Although the results have not been universally accepted, many animal experiments have shown a direct correlation between the initiation of sarcomas and the injection of particulate metal debris. This appears to be related to the concentration as well as the physical nature of the metal implanted.[17] The mechanism of the carcinogenicity is not known, but metal ions, particularly cobalt, chromium, and nickel, are known to induce infidelity of DNA synthesis by causing the pairing of noncomplimentary nucleotides, thereby creating a misinterpretation of the genetic code.[18]

12.2.5 ACTIVATION OF COAGULATION

Inflammation and coagulation cannot be considered separate processes as several connecting points exist and make them part of the host response. Vascular endothelium damaged during inflammation is a surface where proteins involved in both coagulation and the development of inflammation are expressed. At the site of injury, platelets are activated and release mediators modify the tissue integrity. Factor XII, present in the plasma, is activated at the contact of negative surface of either biological structure or exogenous material and activates the coagulation cascade, which results in thrombin formation known to be mitogenic and chemotactic for the cells of the inflammatory reaction.

12.3 CHANGES IN CELL BEHAVIOR INDUCED BY MATERIALS

12.3.1 CELL ADHESION

Adhesion to exogenous surfaces has profound effects on the structure and behavior of cells, especially anchorage-dependent cells. Most of the cell lines grown in vitro attach to the surface of the culture dishes or carriers placed in the dish and are thus anchorage-dependent. Adhesion to a substrate is a complex multistage process including cell attachment and spreading, focal adhesion formation, extracellular matrix deposition, and rearrangement.

Tissues are made of cells and extracellular proteins which constitute the extracellular matrix, which consists of an intricate network of macro-molecules. This matrix is composed of proteins and glycoproteins secreted and assembled by the cells to which they are bound.

The linkage of the extracellular matrix to the cell requires transmembrane cell adhesion proteins that act as matrix receptors and link the matrix to the cell's cytoskeleton. These molecules are the integrins which are transmembrane heterodimers composed of two noncovalently associated transmembrane glycoprotein subunits called α and β.

Once synthesized, the extracellular matrix can be adsorbed at the surface of the powders or fibers and thus bridge the cell to the material. They can also constitute a meshwork, engulfing the particles, and mechanically linking the cells to the particles.

Meyer et al.[19] demonstrated significant differences in the rate of cell attachment to biomaterials during the first 7 h of cell culture. They noticed that the more hydrophobic a material was, the slower the cells attached to the surface. However, this was challenged by other authors.[20] It is probable that the role played by the hydrophobicity of the material depends on the material's characteristics and on those of the cell line used.

Altering the chemical composition of the material surface modifies the adhesion of human bone-derived cells. Titanium ion implantation at the surface of Co–Cr alloy increased cell adhesion, the implant topography and surface energy being in the same range.[21] The mechanism of differential attachment to the substrata with different chemistries was in part attributed by Howlett and colleagues to the folding of the attachment proteins, as well as to the altered metabolism in the cells.[22]

More generally, at the molecular level, it is believed that the state and species of adsorbed proteins significantly influence the extent of material-mediated inflammatory responses. Tang et al. have demonstrated that adsorbed fibrinogen (Fg) is primarily responsible for the accumulation of phagocytes on implant surfaces[23] and the pro-inflammatory activity of adsorbed Fg resides in the D domain. More recent results indicated that a particular peptide sequence, P1 (γ90–202), within Fg D30, is responsible for the phagocyte accumulation.[24] It was more recently reported that a second peptide, P2 (γ377–395) is also involved in this material response.[25] Tests on a variety of commonly used biomaterials suggest a strong correlation between the degree of material-induced P1/P2 exposure and the extent of biomaterial-mediated inflammatory response.[26] It indicates that materials with a high surface area, potentially able to adsorb large quantities of P1/P2, are more potent in triggering an inflammatory response.

It is, however, not known why adsorbed and not solution-phase Fg is inflammatory. It is probable that the adsorption of the Fg on the material changes its conformation and makes the P1/P2 available for the phagocytes.

12.3.2 CELL DEATH

The death of cells can be the result of cell injuries, leading to the collapse of the cell functions, or can be triggered by a process of programmed cell death (apoptosis), which is under the dependence of the cell itself and which leads to true cell suicide.

Cell death resulting from the phagocytosis of particles has been well documented for the debris from hip prostheses. Several types of debris are reported to be associated with cell death, such as metals, polyethylene, or PMMA.[27,28] No such necrosis was found with alumina debris. However, the characteristics of the materials were not further specified.

One of the first steps of cell injury is the variation of ionic concentrations in the cell cytoplasm—in particular, the Na/K ratio which is maintained thanks to membrane sodium pumps. Injurious agents can interfere with this ratio by increasing the permeability of the membrane itself, damaging the Na/K pumps, or interfering with the generation of the ATP necessary to drive the pump. Laquarriére et al.[29,30] studied the Na/K of macrophages when they phagocytosed calcium phosphate particles of various characteristics, but with the same composition. They showed that the ionic variation was dependent on the surface area and the shape of the particles.

12.3.3 CELL PROLIFERATION

The phagocytosis of particles can trigger a modification of cell proliferation without showing any signs of toxicity. This property has been demonstrated with calcium phosphate particles.[31] It is probable that the influence of these particles on the proliferation is due to the adsorption of proteins such as growth factors at the particle surface. Maloney et al.[32] showed that titanium, titanium–aluminum, and chromium stimulated cell proliferation (^{3}H-thymidine uptake; particles 0.1–10 μm in size at low concentration). Higher concentrations and cobalt particles were toxic as evidenced by the absence of thymidine uptake.

12.3.4 CELL DIFFERENTIATION

Differentiation, which is the ability of a cell to synthesize specific molecules for a particular cell function in a tissue or organ, is also altered, at least in vitro, by contact with particles. Vermes et al.,[33] for example, demonstrated that phagocytosable particles of cp titanium, titanium alloy, and cobalt–chromium (only the particle size was given) significantly reduced procollagen α1 gene expression, whereas other osteoblast-specific genes (osteonectin, osteoclacin, and alkaline phosphatase) did not show significant changes.

Morais et al.[34] showed that the corrosion products generated by stainless steel decreased the consumption of Ca and P of rat osteoblasts and induced deposition of Ni and Cr in the cell matrix. They concluded that the metal ions associated with the corrosion products affected both the proliferation and differentiation of the cells (Figure 12.1 and Figure 12.4).

12.3.5 Cell Transformation

The reports concerning the cell transformation induced by materials are confusing, as there is no specific in vitro test for the evaluation of cell transformation by solid materials. It is necessary to refer to the tumorogenicity tests done in rats (see Hypersensitivity and Carcinogenicity chapters).

12.3.6 Cell Migration

The contact or the phagocytosis of particles with macrophages or white blood cells are known to induce the synthesis of molecules attracting (to the site of inflammation) other cells of different or identical lineage.

12.4 MATERIAL CHARACTERISTICS INVOLVED IN THE MODIFICATION OF CELL AND TISSUE BIOLOGY

The different effects described above, induced by particles in the organism, vary with the particles' characteristics.

Compared to bulk materials, powders have the essential properties needed to migrate in the organism and have a high surface area. In the international literature, the best studied characteristic is the chemical composition, particularly for foreign body reaction around implants, probably because it is the easiest to study. It must be kept in mind that even if the particles are well characterized, it is difficult to vary only one or two characteristics at a time and the effect of the particles is almost always concentration dependent.

12.4.1 Composition

The composition of the particles is not stable during the period of residence in the tissue and cells. Processes of dissolution-precipitation occurring in the periphery or in sites remote from the particles may affect the cell and tissue response to the material. It is well known, for example, that bioactive materials used in orthopedic surgery, once implanted, are coated by a layer of carbonated apatite appearing by epitaxial growth.[35] Corrosion products are generated from the metal surfaces and precipitates have been evidenced even in sites remote from the particles.[36]

The composition of coal dust involved in coal worker pneumoconiosis has been studied. Sorensen et al.[37] stated that the increase of concentration of Fe,

Pb, Cu, and Ni in the coal had an influence on its cell toxicity. Christian et al.[38] demonstrated that the concentration of Ni and Zn was very important to the pathogenicity of coal dust.

Studies on the effect of particle composition on macrophage synthesis and secretion are numerous and show that it is a key factor for the amount of cytokine secretion. At high doses, TiAlV particles do not affect the cell viability while stainless steel and CoCrMo are toxic.[39] In peripheral blood monocytes, metallic particles (CoCr and Ti) were found to be more stimulatory for PGE2, IL-1, TNF-α, and IL-6 than for polyethylene particles.[40,41] The comparison of the induction of macrophage apoptotic cell death by alumina, zirconia, and polyethylene particulates showed the response to be concentration and size dependent. In this case, particle composition had no effects.[42]

As stated above, ion modification of commonly used orthopedic material affects the attachment of human bone-derived cells. This effect occurs through passivated oxide layers, suggesting that it is due, at least in part, to ion release. The release of toxic compounds by particles can indeed affect their tolerance. The effect of the composition is enhanced by the surface area and thus the exchange rate between the organism and the material.

A good example of the effect of ion release is given by alumina coatings deposited at the surface of some hip prostheses by plasma spraying. The spraying conditions induce the formation of a gamma alumina phase, which is less stable than the alpha phase used for hip balls. Corrosion particles are formed which release aluminum and provoke a lesion in the bone comparable to osteomalacia in the proximity of the coating. There is aluminum deposition in the bone accompanied by an anomaly of the mineralization of the bone matrix.[43]

Several metal ions released by particles that can be found in implanted materials were shown to interact with cell biology. Nickel, cobalt, and chromium interfere with the metabolism of white blood cells and influence their migration.[44] These ions may exist in the biological fluids in different valency states. Chromium released from implant debris can be incorporated into organometallic complexes as CrIII or VI, Cr in the higher oxidation state being more biologically active than when in the lower one. CrIII is found almost exclusively in serum while CrVI is found bound to the blood cells.[45]

12.4.2 PROTEIN ADSORPTION

The surface modifications occurring at the surface of calcium phosphate ceramics or in a particular class of glasses have been particularly well studied because they are thought to be the reason why these materials are bioactive—i.e., elicit a modification of the biology of the tissue in contact with their surface—and do not trigger a foreign body reaction. They include dissolution, precipitation, and ion exchange accompanied by absorption and incorporation of biological molecules.[46] The bioactivity of apatite ceramics is due to the formation of a carbonated apatite layer by epitaxial growth at the material surface paralleled by protein adsorption. Ducheyne et al.[46] have shown that serum proteins adsorb in tandem with the occurrence of solution-mediated reactions, leading to the formation of a silica gel at the surface of bioactive glasses.

The higher local concentration of adhesion proteins in the adsorbed layer is not the only factor causing an increase in efficacy of cell adhesion. The adsorption can cause the exposure of novel binding sites that are normally hidden in the soluble proteins.[47] Several studies have shown the formation of hidden sites detected with monoclonal antibodies after adsorption of fibrinogen[48] and fibronectin.[49]

12.4.3 SIZE AND SURFACE AREA

The main characteristics of the powders and fibers which influence their biocompatibility are the size of the particles and their surface area, which are interdependent.

At first, the size of the particle rules their ability to be internalized by the cells involved in the foreign body reaction. There is a limit in the size of the particles that can be phagocytosed by the mononuclear cells, macrophages, or monocytes. It is considered that above the size of 30–40 μm, depending on the shape, the particles are not phagocytosed by these cells. Over this size, the particles can be phagocytosed by multinuclear cells which result from the fusion of the monocytes or macrophages. However, the size must not exceed 50–60 μm.

Under an approximate size of 60–70 nm, it is considered that virtually all the cells can internalize the particles, even the cells which are not specialized in this function.

Nevertheless, particles of sizes compatible with their cell internalization can be found in the extracellular matrix of the tissue outside the cells. It is not clear why some of these particles are phagocytosed and others are not.

On the other hand, surface phenomena such as dissolution or precipitation are proportional to the surface area, as is the rate of exchanges between the material and the cells it is in contact with.

Furthermore, when precipitation occurs at the surface of the material, it is demonstrated that the composition of the extracellular fluids can be modified and modify the cell behavior proportionally to the surface area.[50]

Concerning lung diseases, particle size and concentration in the air were proven to be key factors in the induction of lung fibrosis. Razzaboni et al.[51] reported that thermal and chemical treatment of respirable size silica dust samples induced marked changes in their toxicity. The cytotoxicity of crystalline α-quartz and fumed silica particles is decreased by calcinations and can be related to the dehydroxylation of the surface, detectable using photo-acoustic infrared spectroscopy and zeta-potential measurements. They also found that cytotoxicity was strongly dependent on particle size, with large surface area (small size) particles being more toxic.

The sintering temperature of particles and fibers of calcium phosphate ceramics has been shown to influence its toxicity on cultured monocytes grown in their contact and measured by Na/K ratio.[52]

Regarding the carcinogenicity of metals, Memoli et al.[53] implanted 22 groups of Sprague–Dawley rats with a variety of orthopedically relevant alloys in solid rod, powdered, and sintered aggregate forms and the animals were observed until their death (30 months). A slight increase of sarcomas was noted in Co, Cr, and Ni containing implants, compared with sham operated animals or controls. Tumors were more common in rats that received metal powder compared with those that received rods or sintered implants.

12.4.4 WETTABILITY, SURFACE CHARGE, AND ZETA POTENTIAL

These properties influence cell and protein adsorption on the surface. Complement, fibrinogen, serum proteins, and antibody adsorption are at the origin of a chemotactic attraction of macrophages and/or lymphocytes in contact with the material.

It has been hypothesized by Ducheyne et al.[54] that the zeta potential of calcium phosphate ceramics is directly related to the surface reactivity governing osteoconductivity. High zeta potentials (20 mV) could lead to the deposition of a mineral layer at the material surface, enhancing biocompatibility.

The behavior of cells and tissues in contact with particles of bioactive materials has been studied because these materials are particularly well tolerated, although they are degradable and release numerous particles with various characteristics. An interesting study was done by Möller et al.[55] They showed that zeta potential and interfacial tension (measured as hydrophobicity) had a distinct influence on cell reactions and that the zeta potential of poly-L-lactide/calcium phosphate composites was higher than interfacial tension for the differentiation of a culture of osteoblasts.

Surface charge is known to activate the coagulation cascade. Initiation of the intrinsic pathway occurs when prekallikrein, high-molecular-weight kininogen, factor XI, and XII are exposed to a negatively charged surface.

The surface charge of a material is known to be a determinant factor of tolerance. However, the published results are very confusing. There is no real study taking into account the texture, the charge of the surface and its density in its influence on cells and tissue biology. The cells have a negative surface charge due to the polarization of their membrane. Thus, the effects of the surface charges are directly propagated to the cells as well as indirectly via ionic and macromolecular substances adsorbed onto the material. Recently, Kobayashi et al.[56] demonstrated that the negative surface charge of polarized hydroxyapatite ceramics enhance the formation of bone on contact.

Tabata and Ikada[57] have studied the effect of the size and surface charge of polymer microspheres on their phagocytosis by macrophages (mouse peritoneal macrophages). It was shown that the maximal phagocytosis of the microspheres took place when their size was in the range 1.0–2.0 μm. Microspheres with hydrophobic surfaces were more readily phagocytosed than those with hydrophilic surfaces.

12.4.5 Organization of the Crystal Surface

Hanein et al.[58] have demonstrated that two structurally distinct faces of the same crystal, calcium (R,R)-tartrate tetrahydrate, differ greatly in their capacity to serve as adhesive substrate. These prismatic crystals are delimited by two different face types, denoted {011} and {101}, which are chemically equivalent but differ in their structural organization. This binding is presumably independent of exogenous proteins, as it is not affected by Arg-Gly-Asp (RGD) peptides or the presence of serum in the medium. In contrast, cell adhesion to the {101} faces is relatively slow (>24 h), is promoted by serum proteins, and can be inhibited by RGD peptides. The rate of adhesion to the {011} faces is faster than that observed on "conventional" tissue culture substrate.

The same authors examined enantiomorphous crystals of calcium (R,R)- and (S,S)-tartrate tetrahydrate. They did not detect any effects on cells grown in a solution saturated with respect to either the enantiomers. However, examination of the cells at the crystal surface using light and SEM showed that after 10 mn of incubation, the {011} faces of the (R,R) crystal form were densely covered by adherent A6 cells, whereas almost no cells were observed on the equivalent faces of the (S,S) crystals. Both the extent and kinetics of the "slow" cell adhesion to the {101} faces of the two enantiomeric crystals, which was apparent ~24 h after plating, were identical. The authors checked several cell lines for their stereospecificity (A6, MDCK, MCF7, RAT1, CEF). The cells revealed different specificities regarding the faces for which they showed affinity.

12.4.6 Shape and Topography

Laquarerre et al.[59] have studied the role of the characteristics of hydro-xyapatite particles in the synthesis and secretion of different molecules (Il-6, Il-1 and TNFα) by macrophages after their phagocytosis in vitro. The secretion of theses molecules indicated the inflammatory power of the particles. The particles had very different characteristics—size, surface area, shape, solubility—but they had the same composition. The cells were grown in the presence of the powder for 8 days. The Na/K ratio, which is tightly regulated during the cell life and ensures the membrane voltage in the cell cytoplasm, was measured during the experiment and indicated the cytotoxicity of the powder.

Several groups of hydroxyapatite particles were individualized:

- Cytotoxic and inflammatory. This is the fiber group.
- Poorly inflammatory and poorly cytotoxic. This is the group of the undefined shape and spherical powders.
- Not inflammatory and not cytotoxic. These are the larger particles which cannot be phagocytosed.

The large particles which cannot be phagocytosed were very well tolerated in vivo. This is consistent with all the histological reports about the biocompatibility of hydroxyapatite ceramics.

The most inflammatory and cytotoxic particles were fibers, even though they might have a surface area comparable to particles that were much better tolerated. There was thus an influence of the shape on the powder biocompatibility.

The influence of phagocytosis on the secretion of metalloprotease by monocytes in vitro is important because they play a role in the development of osteolysis and implant loosening. Chou et al.[60] observed variations in the production of MMP-2 on a smooth titanium surface after a few hours (1.5 and 3) compared to rough surfaces, suggesting that surface area and shape are key factors in the production of metalloproteases by the fibroblast cell line they used. In Laquarrière's study, all the different kinds of HA-particles induced a transitory increase in MMP-2 and -9 secretion while only fibers increased a MMP-9 secretion for a longer time, indicating that simply the shape was able to modify secretion.

Regarding the topography, Keller et al.[61] noticed a significantly larger level of cell attachment on rough surfaces. Microgrooved materials elicited a modification of cell orientation. It is not known what the influence of the topography at the molecular level is, although it could change integrin adsorption, in particular on macrophage behavior.

12.4.7 MAGNETISM

Magnetic particles can be used in cancer therapy,[62] blood purification,[63] lymph node[64] imaging, or hyperthermia.[65] Apart from the other characteristics of the particles, their magnetism can be assumed to influence their biocompatibility. The effect of magnetic fields on molecules and cells has been extensively studied in recent years.[66] Although transcription alterations can occur as a result of cell exposure to electromagnetic fields, no apparent disruption in routine physiological processes such as growth and division is apparent.

A study of the interaction of magnetic microspheres has been carried out using the in vitro MTT assay and intrathecal injection in rats.[67] The magnetic particles were embedded within polymers. Both magnetic and nonmagnetic microspheres led to similar cell growth inhibition. It must be pointed out that the studies regarding the effects of magnetic fields on cell biology are confusing, as the characteristics of the magnetic fields are very diverse as are the cell lines tested.

12.5 CONCLUSIONS

This study is not exhaustive—the diversity of the material characteristics evaluated by the different authors and the number of protocols which have been used does not allow for a short report.

Although particulate materials are a major concern in public health, little is known about the influence of their surface chemistry on how they are tolerated by cells and tissues. It is known that particulate materials have an influence on the behavior of isolated cells or tissues and even organisms. However, papers studying the relationship between the response of cells and tissues and a distinct surface characteristic are very few. Furthermore, there is no general rule which can be attributed to surface characteristics in the tolerance of a powder or fiber. The various types of influence that the characteristics have are interdependent and it is difficult to assess the role of one particular characteristic on tolerance.

It is clear, however, that each of the characteristics does durably affect macrophage biology and can provoke their apoptotic death. It seems that surface area and shape are very often cited as factors involved in the intensity of the foreign body reaction.

It is unfortunate that in most of the papers dealing with particles in contact with cells and tissues, the particle characteristics are not fully reported. This impairs any comparison between the studies reported.

Abbreviations

PGE2   Prostaglandin E2
TNF   Tumor necrosis factor
PDGF   Platelet derived growth factor
TGF-α   Tumor growth factor-α
EGF   Epithelial growth factor
PMMA   Polymethyl methacrylate
IgE   Immunoglobulin E
DNA   Deoxy ribonucleic acid
HA   Hydroxyapatite
MMP2   Matrix metalloproteinase 2
MTT   3-(4,5-dimethylthiazole-2-yl)-2,5-diphenyl tetrazolium bromide

REFERENCES

1. Fantone, J. C. and Ward, P. A., In *Pathology*, Rubin, E. and Farber, J. L., Eds., 2nd ed., JB Lippincott, Philadelphia, PA, pp. 33–66, 1994.
2. McNally, A. K. and Anderson, J. M., *Am. J. Pathol.*, 147, 1487–1499, 1995.
3. McNally, A. K. and Anderson, J. M., *Proc. Natl Acad. Sci. U.S.A*, 91, 10119–10123, 1994.
4. Rubbin, E. and Farber, J. L., In *Pathology*, Rubin, E. and Rubin, J. L., Eds., JB Lippincott, Philadelphia, PA, pp. 557–617, 1994.
5. Maier, E. A., Dietemann-Molard, A., Rastegar, F., Heimburger, R., Ruch, C., Maier, A., Roegel, E., and Leroy, M. J., *Clin. Chem.*, 32(4), 664–668, 1986.
6. Soloviev, A., Schwarz, E. M., Puzas, J., Rosier, R. N., and O'Keefe, R. J., *Transaction of 47th Annual Meeting of Orthopaedic Research Society*, 0003, 2001.
7. Urban, R. M., Jacob, J. J., Tomlinson, M. J., and Galante, J. O., *Transaction of 47th Annual Meeting of Orthopaedic Research Society*, 0165, 2001.
8. Shanbhag, A. S., Jacobs, J. J., Black, J., Galante, J. O., and Glant, T. T., *J. Arthroplasty*, 10(4), 498–506, 1995.
9. Giant, T. T., Jacobs, J. J., Molnàr, G., Shanbhag, A. S., Valyon, M., and Galante, J. O., *J. Bone Miner. Res.*, 8(9), 1071–1079, 1993.
10. Xu, J-W., Ma, J., Li, T-F., Waris, E., Alberty, A., Santavirta, S., and Konttinen, Y. T., *Ann. Rheum. Dis.*, 59, 822–827, 2000.
11. Horowitz, S. M. and Gonzales, J. B., *Calcif. Tissue Int.*, 59, 392–396, 1996.
12. Frayssinet, P., Gineste, L., Primout, I., and Guilhem, A., *Bull. de l'Assoc. des Anatomistes*, 82(256), 9–11, 1998.
13. Heinemann, D. E. H., Siggelkow, H., Ponce, L. M., Viereck, V., Wiese, K. G., and Peters, J. H., *Immunobiology*, 202, 68–81, 2000.
14. Evans, E. M., Freeman, M. A. R., Muller, A. J., and Vernon-Roberts, B., *J. Bone Joint Surg. [Br.]*, 56, 626–642, 1974.

15. Kaplan, S. S., Basford, R. E., Jeong, M. H., and Simmons, R. L., *J. Biomed. Mater. Res.*, 26, 1039–1051, 1992.
16. Nathan, C. F., *J. Clin. Invest.*, 80, 1550–1560, 1987.
17. Swanson, S. A. V., Freeman, M. A. R., and Heath, J. C., *J. Bone Joint Surg. [Br.]*, 55, 759–773, 1973.
18. Sirover, M. A. and Loeb, L. A., *Science*, 194, 1434, 1976.
19. Meyer, U., Szulczewski, D. H., Möller, K., Heide, H., and Jones, D., *Cells Mater.*, 3, 129–140, 1993.
20. van Kooten, T. G., Schakenraad, J. M., van der Mei, H. C., and Busscher, H. J., *Cells Mater.*, 1, 307–316, 1991.
21. Howlett, C. R., Zreiquat, H., Wu, Y., McFall, D. W., and McKenzie, D. R., *J. Biomed. Mater. Res.*, 45, 345–354, 1999.
22. Howlett, C. R., Zreiqat, H., O'Dell, R., Noorman, J., Evans, P., Dalton, B. A., McFarland, C., and Steel, J. G., *J. Mater. Sci. Mater. Med.*, 5, 715–722, 1994.
23. Tang, L. and Eaton, J. W., *J. Exp. Med.*, 178, 2147–2156, 1993.
24. Tang, L., Urugova, T. P., Plow, E. F., and Eaton, J. W., *J. Clin. Invest.*, 97, 1329–1334, 1996.
25. Ugarova, T. P., Solovjov, D. A., Zhang, L., Yee, V. C., Medved, L. V., and Plow, E. F., *J. Biol. Chem.*, 273, 22519–22527, 1998.
26. Hu, W.-J., Eaton, J. W., and Tang, L., *Sixth World Biomaterials Congress Transactions*, 403, 2000.
27. Bos, I. and Willmann, G., *Acta Orthop. Scand.*, 72(4), 335–342, 2001.
28. Boss, J. H., Shajrawi, I., Luria, I., and Mendes, D. G., *VCOT*, 8, 107–113, 1995.
29. Laquerriere, P., Kilian, L., Bouchot, A., Jallot, E., Grandjean, A., Guenounou, M., Balossier, G., Frayssinet, P., and Bonhomme, P., *J. Biomed. Mater. Res.*, 58, 238–246, 2001.
30. Laquerriere, P., Grandjean-Laquerriere, A., Jallot, E., BaPossier, G., Frayssinet, P., and Guenounou, M., *Biomaterials*, 24, 2739, 2003.
31. Cheung, H. S. and Haak, M. H., *Biomaterials*, 10, 63–67, 1989.
32. Maloney, W. J., Smith, R. L., Castro, F., and Schurman, D. J., *J. Bone Joint Surg.*, 75-A(6), 835–844, 1993.
33. Vermes, C., Chandrasekaran, R., Jacobs, J. J., Galante, J. G., Roebuck, K. A., and Glant, T. T., *J. Bone Joint Surg. 83-A*, 2, 201–211, 2001.
34. Morais, S., Sousa, J. P., Frenandes, M. H., Carvalho, G. S., de Bruijn, J. D., and van Blitterswijk, C. A., *J. Biomed. Mater. Res.*, 42, 199–212, 1998.
35. Heughebaert, M., LeGeros, R. Z., Gineste, M., and Guilhem, A., *J. Biomed. Mater. Res.*, 22, 257–268, 1988.
36. Urban, R. M., Jacobs, J. J., and Gilbert, J. L., *J. Bone Joint Surg.*, 76-A, 1345–1359, 1994.
37. Sorensen, J. J., Kober, T. E., and Petering, H. G., *Am. Ind. Hyg. Assoc. J.*, 35, 93–98, 1990.
38. Christian, R. T., Nelson, J. B., Cody, T. E., Larson, E., and Bingham, E., *Environ. Res.*, 20, 358–365, 1979.
39. Haynes, D. R., Boyle, S. J., and Rogers, S. D., *Clin. Orthop.*, 352, 223–230, 1998.

40. Blaine, T. A., Pollice, P. F., and Rosier, R. N., *J. Bone Joint Surg.*, 76B, 53–59, 1994.

41. Shanbhag, A. S., Jacobs, J. J., and Black, J., *J. Orthop. Res.*, 13, 792–801, 1995.

42. Catelas, I., Huk, O. L., and Petit, A., *Biomaterials*, 20, 625–630, 1999.

43. Frayssinet, P., Tourenne, F., Rouquet, N., Conte, P., and Bonel, G., *J. Mater. Sci.: Mater. Med.*, 5, 491–494, 1994.

44. Rennes, A. and Williams, D. F., *Biomaterials*, 13, 731–743, 1992.

45. Merrit, K., Brown, S. A., and Sharkey, N. A., *J. Biomed. Mater. Res.*, 18, 1005–1015, 1984.

46. Ducheyne, P. and Qiu, Q., *Biomaterials*, 20, 2287–2303, 1999.

47. Ginsberg, M. H., Xiaoping, D., O'Toole, T. E., Loftus, J. C., and Plow, E. F., *Thromb. Haemost.*, 70, 87–93, 1993.

48. Soria, C., Mirshahi, M., Boucheix, C., Aurengo, A., Perrot, J. Y., Bernadou, A., Samama, M., and Rosenfeld, C., *J. Colloid Interface Sci.*, 107, 204–208, 1985.

49. Ugarova, T. P., Zamarron, C., Veklich, Y., Bowditch, R. D., Ginsberg, M. H., Weisel, J. W., and Plow, E. F., *Biochemistry*, 34, 4457–4466, 1995.

50. Frayssinet, P., Rouquet, N., Fages, J., Durand, M., Vidalain, P. O., and Bonel, G., *J. Biomed. Mater. Res.*, 35, 337–347, 1997.

51. Razzaboni, B. L. and Bolsaitis, P., *Environ. Health Perspect.*, 87, 337–341, 1990.

52. Laquerriere, P., Kilian, L., Bouchot, A., Jallot, E., Grandjean, A., Guenounou, M., Balossier, G., Frayssinet, P., and Bonhomme, P., *J. Biomed. Mater. Res.*, 58(3), 238–246, 2001.

53. Memoli, V. A., Urban, R., Alroy, J., and Galante, J. O., *J. Orthop. Res.*, 4, 346–355, 1986.

54. Ducheyne, P., Kim, C. S., and Pollack, S. R., *J. Biomed. Mater Res.*, 26, 147–168, 1992.

55. Möller, K., Meyer, U., Szulczewski, D. H., Heide, H., Priessnitz, B., and Jones, D. B., *Cells Mater.*, 4, 263–274, 1994.

56. Kobayashi, T., Nakamura, S., and Yamashita, K., *J. Biomed. Mater. Res.*, 57, 477–484, 2001.

57. Tabata, Y. and Ikada, Y., *Biomaterials*, 356–362, 1988.

58. Hanein, D., Geiger, B., and Addadi, L., *Sciences*, 263, 1413–1416, 1994.

59. Laquarrière, P. PhD thesis, Université de Reims, 2000.

60. Chou, L., Firth, J. D., Uitto, V. J., and Brunette, D. M., *J. Biomed. Mater. Res.*, 39, 437–445, 1998.

61. Keller, J. C., Collins, J. G., Niederauer, G. G., and McGee, T. D., *Dent. Mater.*, 13, 62–68, 1997.

62. Häfeli, U. O., Pauer, G. J., and Roberts, W. K., In *Scientific and Clinical Applications of Magnetic Carriers*, Häfeli, U., Schutt, W., Teller, J. et al., Eds., Plenum, New York, p. 501, 1997.

63. von Appen, K., Weber, C., and Losert, U., *Artif. Organs*, 20, 420, 1996.

64. Kresse, M., Wagner, S., and Taupitz, M., In *Scientific and Clinical Applications of Magnetic Carriers*, Häfeli, U., Schutt, W., Teller, J. et al., Eds., Plenum, New York, p. 545, 1997.
65. Jordan, A., Scholz, R., and Wust, P., *Int. J. Hyperthermia*, 13, 587, 1997.
66. Goodman, E. M., Greenebaum, B., and Marron, M. T., *Int. Rev. Cytol.*, 158, 279–338, 1995.
67. Häfeli, U. O. and Pauer, G. J., *J. Magn. Magn. Mater.*, 194, 76–82, 1999.

Section V

The Solid-Liquid Interface

13 The Solid–Electrolyte Interfacial Region

Fabien Thomas, Bénédicte Prelot, and Jérôme F. L. Duval

CONTENTS

13.1 INTRODUCTION

By definition, powders and fibers are made of small particles, the size of which often lies in the nanometric (or colloidal) domain. The small size confers such particles a high specific surface area, and thus a predominance of the surface properties over the bulk properties. Among the surface properties, the electric charge is a key quantity for understanding the interfacial characteristics of numerous colloidal systems of environmental and technological significance (e.g., froth flotation, paints, inks, food industry, cosmetics, or water treatment). The density of the charge, its distribution around the particle, and its way of compensation in the adjacent neighborhood all determine the selectivity of ionic solutes adsorbed at the interface, and control the stability and flow behavior of colloidal dispersions via long-range interparticular interactions. Although these phenomena occur at the nanometric or micrometric scale, their consequences are reflected at the macroscopic scale in environmental and industrial processes. Therefore, measuring and modeling the electric charge at the solid–liquid interface is a mandatory prerequisite for catching the basic interfacial and colloidal features pertaining to powders and fibers, and for describing multiphasic systems on a mechanistic level.

In the numerous industrial domains involving powders and fibers, the surface charge has a crucial importance in regard to many aspects. Reinforcement of polymers by mineral or carbon-based materials requires strong adhesion between the matrix and the dispersed particles. To improve the fiber–matrix adhesion, it is necessary to adequately adjust the polarity of the surface to create a favorable environment for hydrogen bonding or covalent linking [1,2]. This is generally achieved by modifying the nonpolar carbon fibers according to wet-chemical, electrochemical, or plasma methodologies [3]. In the dyeing of textile fibers, penetration of the dyestuffs into the fibers is strongly promoted by the presence of reactive groups entering the composition of the fibers [4]. In papermaking, the removal of water from aqueous cellulose slurry strongly depends on the ionization degree of the acidic groups that influence the formation of fiber networks and aggregates [5]. Natural fibers are increasingly used in the building of materials and structural parts in automotive application, essentially because of their low cost, light weight, and environmental friendliness. The wide panel of reactive groups they offer is a considerable advantage for insertion into polar matrices. To that purpose, the cellulose fibers must be purified from pectins and waxy substances [6]. On the other hand, polar groups cause swelling processes in aqueous media, which must be prevented [7]. In all of the above cases, detection, identification, and quantification of the charged groups at the surface of the unit particles is a prerequisite to understanding and controlling the behavior of powder and fiber materials.

The current chapter describes the general concepts and quantitative methods most commonly applied to define and probe the electrical and chemical properties of charged surfaces. The subject is not new and the principles are now well understood—at least for 'ideal' particles which are uniformly charged, isotropic in geometry, and hard. In practice, 'real' particles do generally not comply with these constraints. They mostly present chemical or topographic surface heterogeneities due to their molecular or crystalline structure and their morphology. Moreover, the interface itself is rarely sharp, especially in the case of organic or biological particles or when organic molecules are adsorbed. Therefore, the description given here of the conventional electric double layer and surface complexation models is followed by a survey of the electrokinetic concepts methods that can be used to describe nonideal systems, especially with regard to their charge heterogeneity and surface softness.

13.2 ORIGIN OF THE SURFACE CHARGE

In solid matter, the covalent bonds between the constituting atoms maintain the cohesion of crystalline or amorphous networks but strongly restrict the mobility of the atoms. For this reason, fluctuations in order or composition cannot be compensated by local rearrangements, as in the liquid or gas phase. Such fluctuations may originate from:

- **Stoichiometric imbalance between the bulk and the surface of the solid**

  This situation is encountered in systems such as Ag_2S water or AgI water, or in cases where the outermost atoms are partially soluble.

- **Valency or stoichiometry defects in the crystal lattice**

  These defects are recurrent in such systems as zeolites or phyllosilicates (clays) where lattice substitutions by cations of lower valency ultimately result in a negative overall charge. The opposite situation is encountered in layered double hydroxides, which bear a negative charge. The lattice charge determines a cationic (or anionic) exchange capacity, which is permanent and independent of the external conditions.

- **Vacant coordination bonds on the outermost atoms**

  From a structural point of view, the coordination number of the atoms exposed at the surface of a crystal is necessarily lower than that of the atoms in the solid bulk. In water, or under atmospheric moisture conditions, completion of the coordination shell is achieved by chemisorbtion of dissociated water molecules, which in turn leads to the formation of hydroxyl surface groups (Figure 13.1). The latter, noted $\equiv$SOH, interact with hydronium or hydroxyl ions, similarly to soluble hydroxo complexes:

$$\equiv S\text{–}O^- + H_3O^+ \rightleftarrows \equiv S\text{–}OH + H_2O \rightleftarrows \equiv S\text{–}OH_2^+ + HO^- \qquad (13.1)$$

Chemical equilibria as those depicted in Equation 13.1 are basically responsible for the presence of a surface charge on metal oxides and salts. For organic materials, the surface charge results from the protonation/ deprotonation of functional entities such as amine, sulphonate, carboxyl, or hydroxyl groups.

Equation 13.1 illustrates the amphoteric behavior of hydroxyl surface groups, the dissociation degree of which is primarily governed by the pH of the electrolytic solution. To put it in a nutshell, the resulting surface charge is basically positive at acidic pH values, negative at alkaline pH values, and zero at the so-called point of zero charge (PZC).

The gradient of chemical potential, as met between the charged surface and the bulk solution, tends to be cancelled by ionic migration towards the bulk solution in the form of dissolution of the solid, accumulation of

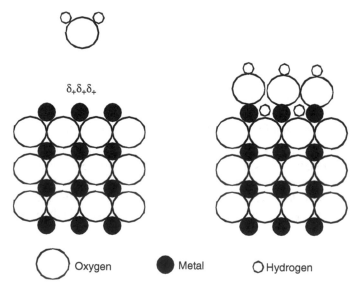

FIGURE 13.1 Schematic view of the surface hydroxylation of metal oxides.

oppositely charged ions (counterions), adsorption of dissolved molecules, and interparticular interaction.

- **Accumulation of charged species at hydrophobic surfaces**

    Unlike the hydrophilic surfaces, to which a great deal of work has been devoted in the past decades (see aforementioned list of items commonly considered as being exhaustive), the hydrophobic surfaces have been and are still the subject of various analyses. Organic materials of natural or synthetic origin, including nonionizable polymers, systematically display a negative charge at neutral pH [8–11]. That negative surface charge stems from the specific adsorption of potential-determining anions (OH^- or Cl^-), with the following decreasing series in affinity: $OH^- > H_3O^+ \gg Cl^- = K^+$. Hydroxyl ions form relatively unstable hydration structures [9], which considerably favors their adsorption onto hydrophobic surfaces. On the other hand, the mobility of water molecules is higher in the vicinity of hydrophobic surfaces than for their hydrophilic counterparts [12–14]. This, in turn, leads to a positioning of the shear plane directly adjacent to the hydrophobic surface whereas for hydrophilic surfaces strong structuration of two to three layers of water molecules [15] repels the shear plane to about 1 nm from the surface.

13.3 ELECTRIC PROPERTIES OF CHARGED SURFACES

13.3.1 THE DOUBLE LAYER MODEL

The earliest descriptions proposed for the charged surface–solution interface were given in terms of point charges and spatial distribution for the electric potential.

The Helmholtz model [16] depicts the charged interface as a molecular capacitor of capacitance C and charge $\pm\sigma$. The plates associated to this capacitor correspond to the planes where the dissociated surface groups and the layer of counterions accumulated at the interface are thought to be located. These are separated by a distance d. The plate carrying the charge $+\sigma$ (>0) is located at $x = 0$, with x being the distance from this reference plate. Since the space in between the two planes is free of charges, the therein spatial distribution is linear and varies according to

$$\psi(x) = \psi(x = 0) - \frac{\sigma}{\varepsilon}x \qquad (13.2)$$

The potential drop across the capacitor $\Delta\psi$ is given by $\Delta\psi = (\sigma d)/\varepsilon$.

As an alternative to the above simplistic picture, Chapman [17] and Gouy [18] considered that the counterions form a cloud (diffuse layer) around the charged particle, and describe the corresponding spatial distribution of the potential on the basis of the Poisson–Boltzmann equation. For the flat geometry, the latter is written:

$$\frac{d^2\psi}{dx^2} = \kappa^2\psi \quad \text{with} \quad \kappa^2 = \frac{e^2 \sum_i \left(Z_i^2 n_{io}\right)}{\varepsilon_o \varepsilon_r k_B T} \tag{13.3}$$

with

ψ	local electrostatic potential
x	distance from the surface
n_{io}	bulk concentration of ion i expressed in m^{-3}
e	elementary charge of the electron
Z_i	valency of the ion i
ε_r	relative dielectric constant of the medium
ε_o	permittivity of the vacuum
κ	reciprocal screening Debye length
k_B	Boltzmann constant
T	absolute temperature

The parameter κ^{-1} has the dimension of a length and basically characterizes the spatial extension of the electric double layer from the charged solid surface. The potential distribution is exponential with a characteristic decay length κ^{-1}. At $x = \kappa^{-1}$, the potential is reduced by a factor $e \approx 2.7$.

Incongruities inherent to the aforementioned models were pointed out by Stern [19]. In the capacitive model, the accumulation of the counterions in a rigid layer is not realistic because of their thermal motion and the electrostatic repulsion between them. In the diffuse layer model, the concentration of the counterions increases exponentially when moving from the bulk solution to the interface and reach unrealistic high values incompatible with experimental data. Therefore, Stern [19], and later, Grahame [20], proposed a combination of the two preceding models to better describe the ionic distribution close to and away from the surface.

The resulting formalism, known as the double layer model (DLM), still has authority today. It reproduces satisfactorily the balance between the thermodynamic limiting cases of the Helmholtz model (minimum energy) and the diffuse layer model (maximum entropy).

Figure 13.2 is a schematic picture that reconciles the electrical description of the double layer on the basis of the Poisson–Boltzmann equation (diffuse

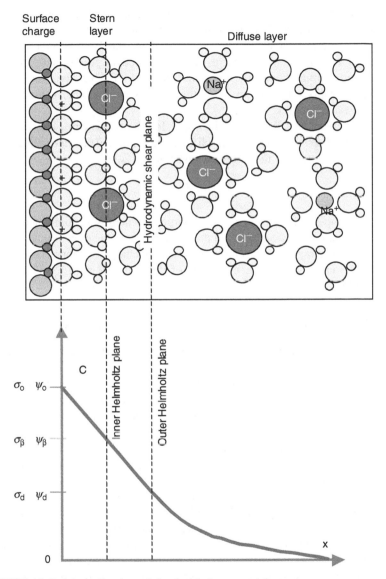

FIGURE 13.2 Schematic view of the double layer model.

layer model valid for *x* larger than about 1 nm) and of the Helmholtz molecular condensator (valid in the close vicinity of the charged surface). The scheme depicted in Figure 13.2 further accounts for the molecular dimension of the species present in solution, especially that of the water molecules.

The counterions complexed with the surface sites (which are responsible for the overall surface charge σ_0) build up the Stern layer. The characteristic plane crossing the centers of the counterions is called the inner Helmholtz plane and the corresponding charge density is noted σ_β. Additional planes can be defined within the Stern layer, according to the relative size of the complexed cations and anions or to the inner-sphere or outer-sphere nature of the surface complexes [21].

The Stern layer is limited by the outer Helmholtz plane beyond which the charge (of density noted σ_d) is diffuse in nature. The electroneutrality condition is written for the interface taken as a whole:

$$\sigma_o + \sigma_\beta + \sigma_d = 0$$

The compensation of the surface charge is generally uncompleted due to the electrostatic repulsion between the ions, except in the case of inner-sphere complexes or at high ionic strength. The residual charge is compensated by the diffuse ion layer. There, the evolution of the potential with the distance is described by Equation 13.3.

13.3.2 Methods for Measuring the Electrokinetic (ζ) Potential

When a pressure gradient is applied along a charged and fixed substrate or when a charged particle is subjected to an external electric field, the interfacial electric double layer is caused to move parallel to the surface. It is now commonly accepted that in the vicinity of the surface there is a plane of shear where the mobile diffuse layer moves independently from the rigid Stern layer (Figure 13.3).

The position of the shear plane can be inferred by careful analysis of the structuration of water molecules at a given hydroxylated surface. Immersion calorimetry and ^{1}H NMR measurements [15] have shown that water molecules are structured by hydrogen bonding with the surface hydroxyls ordered by the crystal lattice. This ordering influences the orientation of the water molecules over 1–3 molecular layers, which approximately corresponds to 3–10 Å. Beyond this distance, the looseness of the hydrogen bonds allows the water molecules to possess the mobility of bulk water. Such structuration of water molecules by the surface, enhanced by the hydration shell of the counter-ions in the Stern layer, introduces a physical discontinuity at a nanometric distance from the surface. As a consequence, the position of the hydrodynamic shear plane may be at a quite fixed nanometric distance from the surface, and the value of the electrokinetic potential noted ζ assimilated for practical reasons to that of the potential at the onset of the diffuse layer region, noted ψ_d.

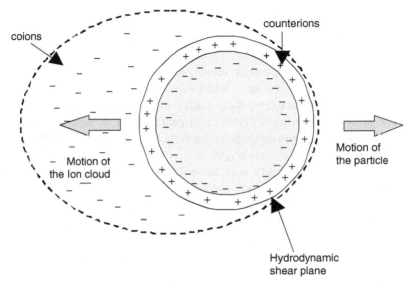

FIGURE 13.3 Schematic principle of electrophoresis.

Depending on the size of the particles and the way the motion is created, different types of electrokinetic phenomena can be distinguished [22,23] (Table 13.1).

The principles and central equations of these techniques are shortly described below.

TABLE 13.1
Main Electrokinetic Phenomena and Measurement Methods

Technique	Gradient	What Moves	Measurement	Type of Particles
Electrophoresis	Electric	Particles	Velocity	Colloidal
Electro-osmosis	Electric	Liquid	Velocity	Colloids, salts
Streaming potential	Mechanical	Liquid	Potential	Coarse particles, fibers, films
Sedimentation potential	Gravity	Particles	Potential	Colloids
Acoustophoresis	Pulsed electric /acoustic	Particles	Phase shift	All sizes

13.3.2.1 Electrophoresis

Electrophoresis is the motion of charged colloidal particles or polyelectrolytes in an electrolyte solution under the action of an external electric field E (Figure 13.4). The electrophoretic velocity of the particles, v_e (m s^{-1}), may be measured according to various methods such as photon correlation spectroscopy, Doppler velocity, or image analysis. To compare data obtained for diverse systems following different experimental methodologies, one classically normalizes v_e with respect to the field to obtain the electrophoretic mobility μ_e (m^2 V^{-1} s^{-1}): $\mu_e = v_e/E$.

The sign of the mobility is positive if the particles move towards the cathode, and negative in the opposite case.

The calculation of the ζ potential from the electrophoretic mobility was initiated by Von Smoluchowski [24] and successively completed by the work of Hückel [25], Henry [26], O'Brien, and White [27]. The different formulations are collected in Table 13.2, and their domains of application are further illustrated in Figure 13.5.

The choice of the appropriate formula is determined by the dimensionless parameter κa, where a is the radius of the particle.

The formulation by O'Brien and White [27] allows the numerical conversion of the electrophoretic mobility into the electrokinetic potential without any restriction on the magnitude of that potential, on the value of κa, or on the valence of the electrolyte considered. Also, various adaptations were made in order to account for particular conditions such as extreme salt concentrations [28,29] or suspensions with high solid fraction of particles [30,31]. The impact of surface conductivity and double-layer polarization on the electrophoretic mobility were also taken into account [32,33]. In the past

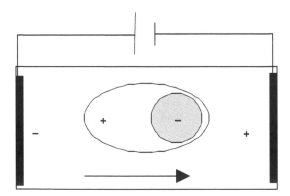

FIGURE 13.4 Principle of colloidal electrophoresis.

TABLE 13.2
Expressions for the Conversion of the Electrophoretic Mobility into ζ Potential

Author	Formula	Conditions of Application
Von Smoluchowski	$\mu = (4\pi\varepsilon_0)D\zeta/4\pi\eta = \varepsilon\zeta/\eta$	$\kappa a \gg 1$
Hückel	$\mu = (4\pi\varepsilon_0)D\zeta/6\pi\eta = 2\varepsilon\zeta/3\eta$	$\kappa a \ll 1$
Henry	$\mu = (4\pi\varepsilon_0)D\zeta/6\pi\eta f_1(\kappa a)$ $= 2\varepsilon\zeta/3\eta f_1(\kappa a)$	For $\kappa a \gg 1$
		$f_1(\kappa a) = 3/2 - 9/2\kappa a + 75/2\kappa^2 a^2 - 330/\kappa^3 a^3$
		For $\kappa a \ll 1$ $f_1(\kappa a) = 1 + (\kappa a)^2/16 - 5(\kappa a)^3/48 - (\kappa a)^4/96 + (\kappa a)^5/96 - [(\kappa a)^4/8 - (\kappa a)^6/96]e^{\kappa a}\int_{\kappa a}^{\infty}(e^{-t}/t)dt$
O'Brien	Plot of $E = (6\pi\eta e/\varepsilon kT)\mu$ versus $y = e\zeta/kT$	No restriction in κa (numerical approach)

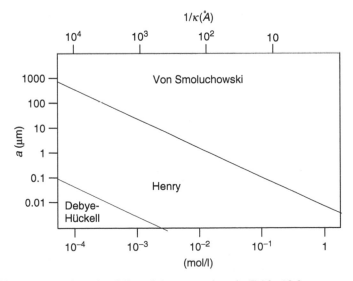

FIGURE 13.5 Domains of validity of the expressions in Table 13.2.

decade, much attention has been devoted to the electrophoresis of soft particles—that is, particles covered with a permeable polymeric layer [34,35]. Due to the importance of such soft colloids in various fields of science as biology or environmental physical chemistry, we shall present in Section 13.3.3 the basic features related to the modeling of soft interfaces and accompanied electrokinetic properties.

Electrophoresis is by far the most widely used technique to probe surface properties of colloidal particles of mineral, organic, or biologic nature.

13.3.2.2 Electro-Osmosis

In electro-osmosis, the solid phase (in the form of a plug or a capillary) is stationary and the motion of the liquid phase results from the motion of the ions in an externally applied electric field (Figure 13.6). The relevant physical quantities are the electro-osmotic velocity and the electro-osmotic flow rate.

The latter can be evaluated either from the mass of liquid collected in an open system or from the velocity of marker particles in a closed system. The ζ potential is calculated according to the following equation:

$$\frac{Q_e}{E} = -\frac{\varepsilon_0 \varepsilon_r \zeta A}{\eta L} \tag{13.4}$$

where the minus sign indicates that the direction of the flow is that of the electric field for $\zeta < 0$.

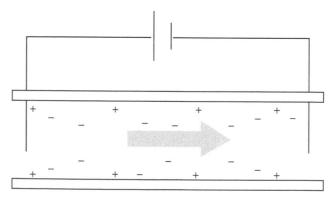

FIGURE 13.6 Principle of electro-osmosis.

with

Q_e flow rate
E electric field
A cross-section area of the capillary
L length of the capillary

which is strictly valid for sufficiently thin double layers ($\kappa a \gg 1$), with a the characteristic size of the cross-section where electro-osmosis takes pace.

Electro-osmotic measurements are performed on samples such as membranes [36].

13.3.2.3 Streaming Potential

Within the framework of that technique, a pressure gradient ΔP is applied on the liquid, in a porous medium located between two plates or in a capillary. This causes a displacement of charges in the EDL as the liquid flows tangentially to the solid, thus creating an upflow electric field (Figure 13.7).

The corresponding potential difference ΔV is called the streaming potential and, in a steady state, it is linearly related to the applied pressure according to the Helmholtz–Smoluchowski equation:

$$\frac{\Delta V}{\Delta P} = \frac{\varepsilon_0 \varepsilon_r \zeta}{\eta} \frac{R}{\lambda R_\infty} \qquad (13.5)$$

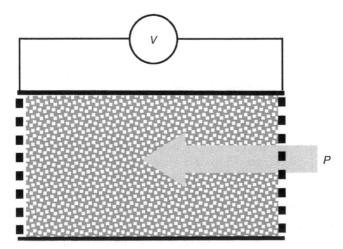

FIGURE 13.7 Principle of the streaming potential.

with

> λ conductivity of the solution
> R resistance of the plug at conductivity λ
> R_∞ resistance of the plug at high conductivity
> η viscosity

Alternatively, the streaming current can be measured, together with the liquid flow rate, and the ζ potential can be calculated from a relationship similar to that given in Equation 13.5.

The streaming potential technique is often used for the study of interfaces of various shapes such as plates [37], membranes [38], porous plugs [39,40], and capillaries [41]. Comparison between the results obtained with the streaming potential method and with the electrophoretic method generally leads to good correlations [42].

13.3.2.4 Sedimentation Potential

Upon sedimentation of charged colloidal particles in a fluid under gravity or in a centrifugation field, a potential difference is produced. This difference is called the sedimentation potential, basically because the ion cloud is displaced by the motion of the particle. Negatively charged particles generate a negative potential in the direction of their motion, and a steady state results from a back-flow of cations (Figure 13.8).

The electric field thus generated is converted into the ζ potential by

$$E = \frac{4\zeta\varepsilon_0\varepsilon_r(\Delta\rho)g}{3\eta\kappa} \tag{13.6}$$

with

> $\Delta\rho$ density difference between the particles and the fluid
> g gravitational constant

13.3.2.5 Electro-Acoustics

The coupling of an electric and an acoustic field generates electroacoustic phenomena [43] (Figure 13.9). These are of two closely related types. A sound wave passing through a colloidal suspension generates an alternating electromotive force due to the vibration of the charged particles relative to the fluid. A colloid vibration current/potential (CVC/CVP) can then be measured. Reciprocally, an alternating electric field applied on a colloidal suspension produces a sonic wave, called electrokinetic sonic amplitude (ESA) [44].

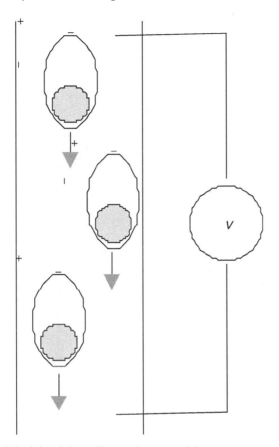

FIGURE 13.8 Principle of the sedimentation potential.

These phenomena can be considered as the A.C. analogues of sedimentation potential/current and electrophoresis, respectively.

The ESA is related to the dynamic mobility of the charged particles, μ_d:

$$\text{ESA} = A(\omega)\phi\frac{\Delta\rho}{\rho}B\mu_d \tag{13.7}$$

with

$A(\omega)$	calibration function
ϕ	volume fraction of particles
$\Delta\rho$	difference in density between the particles and the liquid of density ρ
B	determined by the acoustic impedance of the suspension

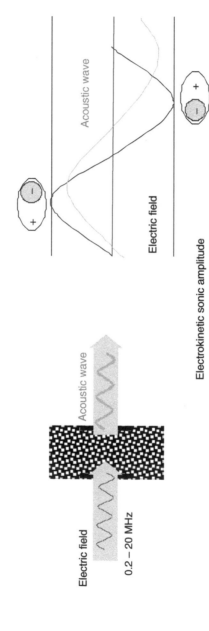

FIGURE 13.9 Principle of ESA.

The ζ potential is calculated from the dynamic mobility:

$$\mu_d = \frac{2\varepsilon\zeta}{3\eta} G(a, \omega)[1 + f] \tag{13.8}$$

with

    η viscosity of the suspension medium
    G depends on the particle size and the measuring frequency ω
    f complex function

For all electrokinetic phenomena in general, interpretation of the results requires, beforehand, careful examination of the distributions at the interface for the electrostatic potential and hydrodynamic velocity (electro-hydro-dynamic coupling). These analyses, which have lead to the now well-accepted electrokinetic theories for hard surfaces, are still in progress for the soft systems. We shall devote special attention to these systems in Section 13.3.3.

13.3.2.6 The Use of Electrokinetic Phenomena in Microfluidics

Microtechnology has decisively dominated technological progress since the middle of the last century. Many new disciplines were thus developed, such as microelectronics, micromechanics, and microfluidics. In this context, advantage was taken from electrokinetic phenomena for the purpose of moving selectively charged particles, ions, or fluids in an externally applied electric field (electrophoresis and electro-osmosis) without using mechanical devices [45]. Electrophoresis (see Section 13.3.2.1.) is widely applied in life science as a separation technique for colloids and macromolecules [46]. In information technology, electronic ink (e-ink) is gaining considerable importance due to liquid crystal displays (LCD) or light emitting diode displays (LED), thanks to very low power consumption, high contrast, and wide viewing angle. The principle is depicted in Figure 13.10. A thin film of microcapsules containing charged and colored particles (say negatively charged white particles and positively charges colored particles) constitutes a pixel pattern. It is integrated between a transparent electrode (viewing direction) and a substrate of addressable electrodes. When a voltage is applied between the top and the bottom electrode, the oppositely charged colored and white particles migrate in the corresponding directions [47].

In the so-called "lab-on-a-chip," increasingly applied in analytical biochemistry, molecules and reagents are moved through microchannels and microchambers where several operations are carried out, such as separation,

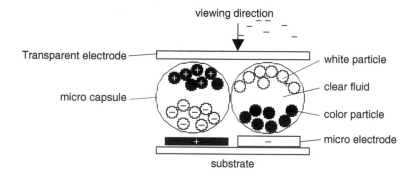

FIGURE 13.10 Principle of the electronic ink display (example of white and colored particles).

filtration, mixing, reaction, and detection. Electrokinetically driven transport and flow control in microchannels by electro-osmosis (see Section 13.3.2.2) is superior compared to pressure-driven microsystems because of higher speed and smaller sample consumption [45].

13.3.3 THE "SOFT" INTERFACES

In the past decades, major efforts have been invested into the development of theories for the electrokinetics of soft interfaces, e.g., interfaces pertaining to particles or more generally colloidal systems coated with *permeable, generally charged polymer layers* [48–62]. These studies are motivated by the increasing number of experimental analyses aimed at the electrical and hydrodynamic characterizations of various colloidal/biocolloidal systems, such as adsorbed polyelectrolytes [63–69], bacteria [70–72], or environmental colloids like humic substances [73]. Soft or "hairy" spherical particles usually consist of a hard core covered by an adsorbed polyelectrolyte layer characterized by a three-dimensional distribution of hydrodynamically stagnant, ionogenic groups. The resulting charged layer is sometimes called "charged surface layer," even if the qualification "surface" loses its well-defined meaning here. In the complete absence of a particle core, the soft particle becomes a spherical polyelectrolyte (porous sphere). Gel-like surfaces also fall in the category of "soft surfaces" in the sense that they consist of a defined network of polymer segments (of given charged groups density) which is penetrable for ions or small colloids. It has long been recognized that the electrophoretic behavior of soft particles and the electrokinetics (streaming potential, streaming current) of gel layers substantially deviate from that of hard (i.e., rigid) particles and hard flat surfaces, respectively. The reasons for this are essentially electrostatic and hydrodynamic in nature. In the interfacial region, the distribution of fixed charged groups in the ion-permeable layer takes place within distances comparable to, if not larger than, the Debye length, thus considerably modifying the electric potential distribution as derived from hard-surface models. Also, electro-osmotic flow and/or penetration of hydrodynamic flow inside the layer may lead to orders of magnitude discrepancy between the observed electrokinetic response and that expected on the basis of the classical Helmholtz–Smoluchowski approach.

13.3.3.1 Double Layer Modeling at a Soft Interface

Figure 13.11 depicts a schematic representation of a system covered with a soft polymeric layer of thickness δ. For the sake of simplicity, planar geometry is adopted in the developments below, which thus hold for planar soft systems or large soft particles of radius a with $\kappa a \gg 1$. At equilibrium, the electric potential across the soft interface satisfies the Poisson equation, written

$$\frac{d^2\psi(x)}{dx^2} = -\frac{\rho_{\text{fix}}(x) + \rho_{\text{el}}(x)}{\varepsilon_o \varepsilon_r} \tag{13.9}$$

where x stands for the dimension perpendicular to the interface of interest. The space charge density $\rho_{\text{el}}(x)$ originates from the mobile electrolyte ions distributed within and around the soft layer. Assuming that Boltzmann statistics applies, $\rho_{\text{el}}(x)$ is simply given for a symmetrical $z{:}z$ electrolyte by

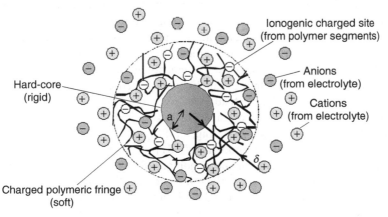

FIGURE 13.11 Schematic representation of a soft particle composed of a hard-core of radius a covered with a (negatively) charged polymeric layer of thickness d permeable to ions and water molecules.

$$\rho_{el}(x) = -2zen_o \sinh(ze\psi(x)/k_B T) \qquad (13.10)$$

where n_0 is the bulk volumic concentration of the electrolyte ions. The local space charge density, noted $\rho_{fix}(x)$, stems from the charges carried by the polymer segments constituting the soft layer. In Equation 13.9, it is tacitly assumed that the dielectric permittivity in the bulk soft layer equals that of water. This is appropriate for highly hydrated layers, as commonly reported in the literature. In the general case, the hard core of the particle carries a surface charge density noted σ_0 so that the electric field at $x = 0$ must satisfy

$$\left. \frac{d\psi}{dx} \right|_{x=0} = -\frac{\sigma_0}{\varepsilon_0 \varepsilon_r} \qquad (13.11)$$

The far-field condition (at $x \to \infty$) simply expresses the electroneutrality met in the bulk electrolytic solution, that is

$$\frac{d\psi}{dx}\Big|_{x\to\infty} = 0 \quad \text{and} \quad \psi(x \to \infty) = 0$$

$$(13.12)$$

(choice of the reference for the potentials)

Equation 13.11 and Equation 13.12 are the two boundary conditions associated with the Poisson–Boltzmann Equation 13.9 and Equation 13.10. To further proceed on a quantitative level, knowledge of the spatial distribution of the polymer segments (and hence of the space charge density $\rho_{fix}(x)$) is required. The latter information may be obtained provided that the type of interactions existing within the system {polymer segments–solvent molecules–electrolyte ions} and the eventual details concerning the preparation of the soft structure and subsequent grafting on the boundary surface are known. Implementation of these elements into molecular dynamic (MD) or mean-field self-consistent calculations may then allow for the determination of $\rho_{fix}(x)$ [74–77]. Instead, we give in Figure 13.12 a schematic representation of the potential distribution for a uniformly charged ("interfacial step-function modeling," [48]) and heterogeneously charged ("interfacial diffuse layer modeling," [51], [67–68]) polymer layer. For polymeric systems satisfying the condition $\kappa\delta \gg 1$, the potential, which is asymptotically reached in the bulk soft layer, is called the Donnan potential, commonly noted ψ^D. For uncharged hard core, $(\sigma_0 \equiv 0)$, ψ^D is derived from the local electroneutrality condition in the bulk soft layer, written

$$-2zen_0 \sinh\left(ze\psi^D/k_B T\right) + \rho_{fix}(x \to 0) = 0 \qquad (13.13)$$

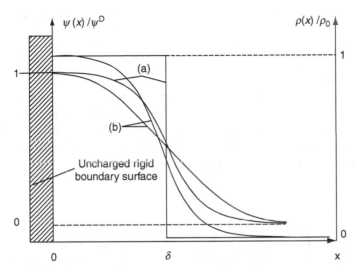

FIGURE 13.12 Schematic representation of the potential distribution (black lines) across a soft layer for two types of charge distribution (red lines): (a) "step-function modeling" (homogeneous layer) and (b) "diffuse interface modeling" (heterogeneous layer).

which reduces to $\psi^D = \rho_0 k_B T / 2 z^2 e^2 n_0$ for a soft layer of uniform charge density ρ_0 and sufficiently low Donnan potentials (Debye-Hückel approximation). It is emphasized here that the notion of Donnan potential is irrelevant for soft systems where δ is of the order of or lower than κ^{-1}.

13.3.3.2 On the Notion of ζ-Potential for Soft Systems?

For hard surfaces, the ζ-potential is defined as the potential at the (slip) plane where, under given electrokinetic conditions, the hydrodynamic velocity is zero. For soft surfaces, the gradual penetration of flow within the permeable soft structure renders impossible the location, *a priori*, of any slip plane. Rigorous quantitative analysis of the hydrodynamic velocity distribution, as obtained from resolution of the pertaining Navier–Stokes equation, confirms that there is no unambiguous way to define a position within the soft layer where the fluid velocity is zero [67–68]. This has been long recognized for spherical polyelectrolytes [48], but the conclusion actually holds for any soft structure. Interpretations of electrokinetic data for various soft materials in terms of ζ-potentials should necessarily integrate beforehand a careful analysis of the hydrodynamic profiles so as to clearly define the positioning chosen for the virtual slip plane. Given the preceding considerations and putting aside the necessarily incorrect concept of electrokinetic potential, the hydrodynamic modeling of soft systems is commonly done within the framework of the Debye-Bueche theory [78]. In the latter, polymer segments within the soft layer are viewed as immobile spherical resistance centers that exert frictional forces on the fluid flowing across and around the soft layer. A key quantity therein is the so-called soft parameter, commonly noted λ_0 [48]. $1/\lambda_0$ has a dimension of a length and basically corresponds to the typical degree of penetration for the fluid into the charged layer. In the limiting case where $\lambda_0 \to \infty$, the particle is hard; whereas for $\lambda_0 \to 0$, the particle is fully permeable. Intermediate values of λ_0 pertain to partially permeable particles (Figure 13.13).

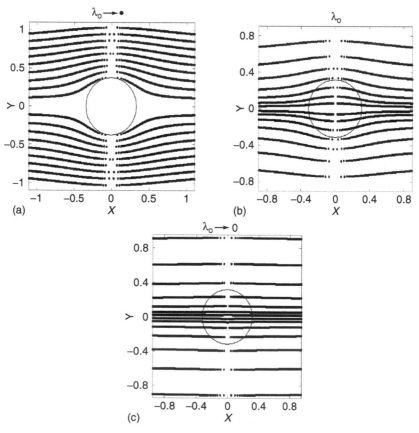

FIGURE 13.13 Typical flow streamlines (a) around and/or across a rigid particle, (b) a partially, and (c) completely permeable particle migrating under the action of an applied dc electric field.

Having in mind the basic features related to the electrostatics (Section 13.3.3.1) and hydrodynamics (above alinea) of soft structures, electrokinetics of these systems can be tackled. It is beyond the scope of the current chapter to present an exhaustive description of the plethora of electrokinetic models developed for soft systems. To that purpose, the reader is referred to the excellent work by Ohshima on the dc electrophoresis of soft particles [48], the recent modeling of ac/dc electrophoresis for soft particles by Hill et al. [51,52,61,62], and the streaming potential/current analysis for planar soft interfaces [67–69].

Described below is a methodology according to which surface properties of spherical soft systems may be evaluated from electrophoretic mobility (noted μ) data.

Commonly, μ is measured over a wide range of electrolyte concentrations, say from 0.1 mM to 1 M or higher. The specific signature of the softness of the object examined is found in the peculiar behavior of μ in the high concentration regime where screening of the electric double layer is large ($\kappa^{-1} \to 0$ and $\psi \to 0$). Whereas the mobility of a hard particle is zero under these conditions, that of a soft particle with homogenous charge distribution approaches a finite nonzero value given by

$$\mu \to \frac{\rho_0}{\eta(\lambda_0)^2} \tag{13.14}$$

where the hydrodynamic and electrostatic parameters, λ_0 and ρ_0, respectively, have been defined above. Equation 13.14 is valid for systems where the thickness of the soft, permeable material verifies $\delta >> \lambda_0$. In the electrolyte concentration regime, which corresponds to sufficiently low electrostatic potentials ($\psi < RT/F$), Equation 13.14 becomes [48]

$$\mu = \frac{\rho_0}{\eta\lambda_0^2} + \frac{\varepsilon_0\varepsilon_r}{\eta} \frac{\psi_0/\kappa_m + \psi^D/\lambda_0}{1/\kappa_m + 1/\lambda_0} \tag{13.15}$$

where the potential ψ_0 (which is the potential at the position corresponding to the location of the outer boundary of the surface layer) and the Debye layer κ_m within the polymeric fringe depend on the space charge density ρ_0, according to the following expressions [48]:

$$\psi_0 = \psi^D - \frac{k_B T}{ze} \tanh\left(\frac{e\psi^D}{2zk_B T}\right) \tag{13.16}$$

$$\kappa_m = \kappa \left\{\cosh\left(\frac{e\psi^D}{k_B T}\right)\right\}^{1/2} \tag{13.17}$$

Equation 13.14 through Equation 13.17 are further strictly applicable for systems where curvature effects are negligible, that is $\kappa\delta \gg 1$. Mobility data are traditionally interpreted by fitting the mobility-ionic strength curve, in terms of Equation 13.14 through Equation 13.17, after adjustment of the unknown variables λ_0 (which is a measure of the hydrodynamic permeability of the system) and ρ_0 according to a least-square method. Deviations of the theoretical calculations from experimental data are commonly reported for low ionic strengths. The reasons for these are numerous: (i) the formalism by Ohshima as aforementioned does not take into account polarization and relaxation of the electric double layer under the action of the applied electric field, these phenomena being of prime importance at low electrolyte concentrations where the electric potentials may largely exceed RT/F; (ii) Equation 13.14 through Equation 13.17 neglect the effect of sizes since they refer to systems where $\kappa\delta \gg 1$ and $\lambda_0\delta \gg 1$ only; (iii) the interfacial modeling underlying the derivation of Equation 13.14 through Equation 13.17 is step function-like (Figure 13.2), which is obviously a simplistic representation where heterogeneous distribution of the charges and polymer segments are omitted; and, last but not least, (iv) Equation 13.14 through Equation 13.17 do not take into account the necessary dependence of the space charge density ρ_0 on the local potential distribution across the soft surface layer or equivalently, on the electrolyte concentration n_0. This dependence has long been experimentally and theoretically demonstrated within the framework of interfacial double layer formalism for rigid particles, which may be regarded as particular soft particles with $\lambda_0 \to \infty$, and by potentiometric titrations of soft systems like bacteria. Whereas Equation 13.14 through Equation 13.17 constitute the basis for a sound physical analysis of mobility data collected for sufficiently large electrolyte concentrations, their shortcomings for lower ionic strength conditions necessarily require the consideration of more advanced models. These rely on the numerical solution of the governing electrostatic and transport equations without any restriction on the size, charge, double layer thickness, and polymer segment distribution [51,52,61,62,73,79]. Recent analyses have further pointed out the advantage of combining electrophoretic measurements with independently obtained data pertaining to the size of the soft systems under consideration (via diffusion coefficient measurements for example) and/or to their charge (potentiometric titration) [73,79].

13.4 SURFACE COMPLEXATION

Unlike the early electric descriptions given for the double layer at the solid–solution interface (Section 13.3.1), chemical models aiming at the representation of the surface chemistry of the solid appeared only in the 1960s. These chemical models were developed for the evaluation of the thermodynamic constants associated with the surface ion exchange equilibria as in protonation and complexation reactions. The models have been so far essentially developed for metal oxides.

13.4.1 MODELS OF SURFACE CHARGING

The protonation–deprotonation equilibria which generate the surface charge are expressed within the formalism of the mass-action law. Historically, two types of approaches were followed:

- A *global approach*, developed by Parks [80,81], Davis and coworkers [82–84], and Stumm and coworkers [85,86], which accounts for the amphoteric behavior of charged surfaces:

$$= S\text{–OH}_2^+ \rightleftarrows\ = S\text{–OH} + H_s^+ \quad K_{a1} \tag{13.18}$$

$$= S\text{–OH} \rightleftarrows\ = S\text{–O}^- + H_s^+ \quad K_{a2} \tag{13.19}$$

The dissociation constants are

$$K_{a1} = \frac{| = \text{SOH}|.|H_s^+|}{| = \text{SOH}_2^+|} \quad \text{and} \quad K_{a2} = \frac{| = \text{SO}^-|.|H_s^+|}{| = \text{SOH}|} \tag{13.20}$$

with

$\|$	activities of the species
$-$SO	a surface group composed of a metallic atom linked to the solid network and of an oxygen atom. Concerning the ions, the subscript s indicates that the species is located at the surface.

The above formulation, which assumes a two-step deprotonation of the sites with a unit charge, is called the 2-pK model. The point of zero charge (PZC) is $\text{pH}_{PZC} = 1/2(pK_{a1} + pK_{a2})$.

- A *local approach* is based on the principle of formal bond valence and was first introduced by Pauling [87–89], and later used for metal oxides by Parks [81], Brown [90], and Yoon [91]. It takes into account the coordination shell of the atoms for the calculation of the dissociation constants of surface groups in metal (hydr)oxides. The bond valence v of the metal ion is the ratio between its valency Z and its coordination number

CN: $\nu = Z/CN$. It basically represents the mean charge per bond. The charge of a surface oxygen ion is not completely compensated by the bonds with ions in the underlying layers, and a partial formal charge δ remains $\delta = n \cdot \nu - 2$, where n is the number of metal ions bound to the oxygen.

The protonation of the surface sites is then represented by the following equilibria:

$$= S_n - O^{n\nu-2} + H_s^+ \rightleftarrows = S_n - OH^{n\nu-1} \tag{13.21}$$

$$= S_n - OH^{n\nu-1} + H_s^+ \rightleftarrows = S_n - OH_2^{n\nu} \tag{13.22}$$

According to Bolt and van Riemsdijk [92] and van Riemsdijk et al. [93], these equations can be reduced to the protonation of a singly coordinated hydroxo group:

$$= S - OH^{-1/2} + H_s^+ \rightleftarrows = S_n - OH_2^{+1/2} \tag{13.23}$$

The model then reduces to a 1-pK representation, which has the merit to introduce a limited number of parameters. The protonation constant is then equivalent to the point of zero net charge (PZNC).

On this basis, the multi-site complexation (MUSIC) model was introduced by Hiemstra et al. [94–96] in order to predict the dissociation constants of metal (hydr)oxides from the local environment of a surface site (Figure 13.14).

The local dissociation constant can be derived from the interatomic distances with the following equations:

$$\log K = A - Bn\frac{\nu}{L},$$

where

$$A = \frac{Z_H Z_{O(H)} N e^2}{2.3RT 4\pi\varepsilon_1 r} - \frac{\Delta G_*^o}{2.3RT} \quad \text{and}$$

$$B = \frac{Z_H N e^2}{2.3RT 4\pi\varepsilon_2 L} \tag{13.24}$$

with

Z_H	valency of hydrogen $= 1$
$Z_{O(H)}$	valency of oxygen $(= -2)$, or hydroxyl $(= -1)$
N	Avogadro number
r	oxygen–proton distance
l	metal–proton distance
$\varepsilon_1, \varepsilon_2$	effective microscopic dielectric constants
ΔG_*^0	nonelectrostatic contributions

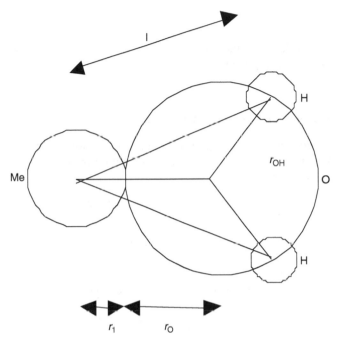

FIGURE 13.14 Geometric model of a surface site of a metal (hydr)oxide. After Heimstra, T.,Van Riemsdijk, W. H., and Bold, G. H., *J.Colloid Interface Sci.*, 133, 91, 1989.

For the sake of illustration, the dissociation constants for titanium oxide are [96]:

$$Ti_3\text{--}O^- + H_s^+ \rightleftarrows Ti_3\text{--}OH \qquad\qquad \log K_{1,2} = -7.5$$

$$Ti\text{--}OH^{-1/3} + H_s^+ \rightleftarrows Ti\text{--}OH_2^{+2/3} \qquad \log K_{1,2} = 6.3$$

$$Ti_2\text{--}O^{-2/3} + H_s^+ \rightleftarrows Ti_2\text{--}OH^{+1/3} \qquad \log K_{1,2} = 5.3$$

Later developments were added to the original MUSIC model [96]. In the refined version of the model, the proton affinity of an oxygen is related both to the undersaturation of the oxygen valency and to the hydrogen bonds.

All the above thermodynamic constants are apparent and are thus called conditional constants. Intrinsic constants are obtained by taking into account the coulombic term:

$$K_{xint} = K_{xapp} \exp\left(-\frac{ze\psi_0}{kT}\right)$$

with

z valency of the ion
ψ_0 surface potential.

13.4.2 MODELS OF SURFACE COMPLEXATION

The interaction between charged sites and counter-ions from the solution (anions A^- and cations C^+) result in the formation of surface complexes.

In the 2-pK model:

$$= S\text{--}OH_2^+ + A_s^- \rightleftharpoons = S\text{--}OH_2^+A^- \qquad (13.25)$$

$$= S\text{--}O^- + C_s^+ \rightleftharpoons = S\text{--}O^-C^+ \qquad (13.26)$$

In the 1-pK model:

$$= S\text{--}OH^{(1-n)+} + A_s^- \rightleftharpoons = S\text{--}OHA^{n-} \qquad (13.27)$$

$$= S\text{--}O^{n-} + C_s^+ \rightleftharpoons = S\text{--}O^-C^{(1-n)+} \qquad (13.28)$$

The surface complexation models (SCM) take into account the above chemical surface reactions, the nature of the surface complex (inner- or outer-sphere complex), and the description of the electric potential at charged interfaces (see Section 13.3.1). The four main SCM are represented in Figure 13.15. They differ mainly by the EDL model they are based on and by the different numbers of corresponding adjustable parameters, namely the capacitance in the Stern layer, the site density, the dissociation and the complexation constants.

The constant capacitance model (CCM) is based on the flat capacitor model of Helmholtz [16]. Since the early work of Stumm and co-workers [85,86], it is commonly used for the study of metal sorption in aquatic media, where the ionic strength is low. The basic assumptions are the constancy of the capacitance C_1, and the inner-sphere nature of the surface complex. Therefore, the studied system must involve strong complexation and controlled ionic strength, which is the case in the above studies. In principle, 3 (or 4, depending on whether the 1-pK or 2-pK model is adopted) parameters are required.

The diffuse layer model (DLM) is based on the work of Gouy [18] and Chapman [17], and was developed by Stumm and coworkers [86]. The potential is considered constant through the Stern layer, and it decreases exponentially in the diffuse layer. This model accounts for the effect of the

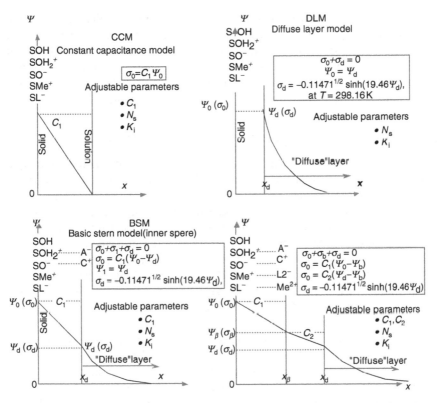

FIGURE 13.15 Schematic description of the surface complexation models (SCM). C_x, capacitances; N_s, site density; K_i, dissociation and complexation constants. (Adapted from Lützenkirchen, J., *Environ. Sci. Technol.*, 32, 3149, 1998.)

ionic strength on the surface dissociation and complexation equilibria. The DLM requires only 2–3 parameters—the site density and the complexation constants. The reaction constants can also be extrapolated at zero ionic strength to allow data comparison [83].

The Basic Stern Model (BSM) can be considered as a combination of the CCM and the DLM [19,20]. It was proposed to eliminate the problems of charge accumulation at the surface, due to the exponential form of the potential expression in the classical Gouy-Chapman theory (see Section 13.3.1). For high ionic strength, the BSM reduces to the CCM, and for low ionic strength it reduces to the DLM. Six or seven parameters are required in the BSM.

The Triple Layer Model (TLM) is based on the works of Yates et al. [98,99], and it is applied by Davis et al. [83]. Compared to the BSM, the TLM includes two capacitances in the Stern layer, which respects the electric double layer model. Therefore, the number of adjustable parameters grows to 7 or 8.

Comparing the results of the above 4 models, Lützenkirchen [97] showed that all of them give a satisfactory description of experimental results, provided that the SCM is chosen in accordance with the properties of the studied system (nature of surface complex, ionic strength).

13.4.3 Methods for Measuring the Surface Charge

The charge residing at the surface of solids is usually quantified by acid–base potentiometric titration, after the methodology developed in the early work by Bolt [92] and Parks and de Bruyn [100] for metal oxides. The amount and variation of the protonated or deprotonated surface hydroxyls when varying the pH and the ionic strength of the electrolyte are calculated for each titration point from the difference between the volume of titrant (acid or base solution) added to the suspension and a reference volume. The reference data are of three kinds: the concentration of H^+ and OH^- ions deduced from the equilibrium pH, the titration of the particle-free electrolyte solution, or the titration of the electrolyte solution previously equilibrated with the solid under investigation.

Two different titration procedures are reported in the literature: stepwise titration or batch titration. Stepwise titration consists of adding successive aliquots of titrant to a suspension, generally by means of an automated device (Figure 13.16). The equilibrium condition required before adding the next increment is then crucial. Short delays may result in nonequilibrium, whereas long delays possibly favor the dissolution of the solid and consequently increase the experimental error due to the drift of the pH electrode. An inherent advantage of the stepwise method is that the curves are obtained with a unique sample.

Batch titration consists of preparing different suspensions at various initial pHs and measuring the pH at equilibrium, most often after 24 h. Although the results are usually affected by higher experimental error as compared to those obtained from continuous titration (solid amount, volumes, carbonatation of the solutions), an advantage of batch titration is that different analytical methods can be used to quantify secondary processes involving proton/hydroxide consumption that occur during the equilibrium phase. For example, in the case of clays, the dissolution processes were estimated from the resulting speciation of elements in solution [101–104]. This calculation is based either on the resulting amounts of elements and their thermodynamic acid–base data, or from back-titration. The latter approach [101,105] consists of titrating the equilibrated supernatants back to the initial pH in order to take into

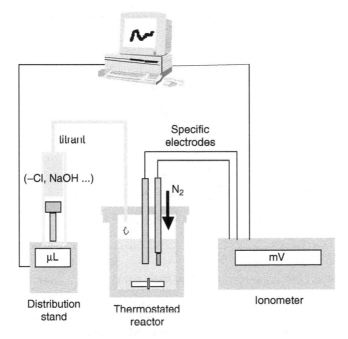

FIGURE 13.16 Typical setup for automated stepwise titration.

account the titrant consumption by the side reactions, quantified by difference between the initial amount of added titrant and the amount used for back titration.

The charge curve is calculated by the difference between the interpolated titration curves of the sample suspension and the blank electrolyte solution (Figure 13.17):

$$\sigma_0 = \frac{\Delta QNF}{MA_s} \tag{13.29}$$

with

σ_0 surface charge density expressed in Coulomb per unit surface area
ΔQ amount of adsorbed titrant
N Avogadro number
F Faraday constant
M Mass of titrated solid
A_s specific surface area of the solid

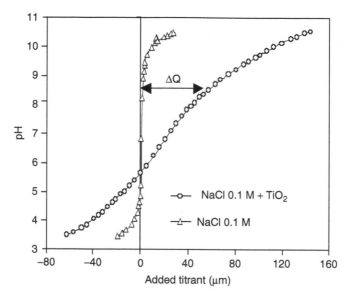

FIGURE 13.17 Stepwise titration curve of TiO_2 anatase in NaCl. (From Prélot, B., Mesure et modélisation de l'hétérogénéité énergétique à l'interface oxyde/électrolyte/métaux, PhD Thesis, University of Nancy-INPL, France, p. 234, 2001. With permission.)

For simple oxides in the presence of an indifferent (1:1) electrolyte, the curves corresponding to different ionic strengths should display characteristic intersection points that define the acid–base properties of the solid (Figure 13.18). The definitions generally used for these characteristic points are given in Table 13.3 [106–108].

There are other methods to infer the location of the PZC of a solid surface. In mass titration, the equilibrium pH of suspensions according to growing solid content is extrapolated to infinite concentration [109–111]. Another technique consists of extrapolation of the initial pH of a suspension to infinite ionic strength [112].

The total charge density can be evaluated using a linear regression analysis of the titration curve, the Gran plot [113]. The Gran function $F_G = V \times 10^{-pH}$ for $pH < 7$ and $F_G = V \times 10^{(pH-14)}$ for $pH > 7$ (with V: total volume) is plotted versus the added volume of titrant. The intercepts of the straight lines indicate the equivalence points (Figure 13.19). For the blank electrolyte, a unique equivalence point is observed, and this allows for accurate calculation of the concentration of the titrant. In the presence of dissociable species, the volume difference between the two equivalence points

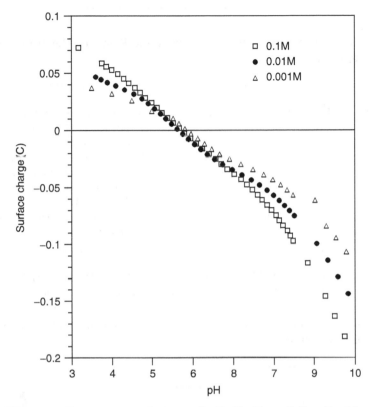

FIGURE 13.18 Charge curves of anatase in NaCl. (From Prélot, B., Mesure et modélisation de l'hétérogénéité énergétique à l'interface oxyde/électrolyte/métaux, PhD Thesis, University of Nancy-INPL, France, p. 234, 2001. With permission.)

corresponds to the surface consumption of titrant and can be converted into the surface charge using Equation 13.29.

The application of surface complexation models allows the estimation of the pK of the surface reactions. As an example, the 0.1 M titration curve of anatase was modeled using the CCM in the 2-pK mode (Figure 13.20). The model parameters are: apparent dissociation constants $pK_1 = 5.3$ and $pK_2 = 7.7$, and site density 0.4 sites/nm^2.

The PZC of the studied anatase (Figure 13.18) is located at pH 6.5, which fits within the "statistical" range of PZC for titanium oxide, obtained from more than 130 values published in the last four decades and collected by Kosmulski [114] (Figure 13.21). These values of PZC are strongly spread, although a pseudo-gaussian maximum appears around pH 6. The scatter is only partly attributed by the author to the polymorphs Anatase and rutile,

TABLE 13.3
Definitions of the Characteristic Intersection Points on the Titration and ζ Potential

Acronym	Name	Definition
PZNPC	Point of zero net proton charge	Intersection between raw titration curves of the blank solution and of the suspension
PZSE	Point of zero salt effect	Intersection between charge curves at different electrolyte concentrations
PZC	Point of zero charge	Common intersection point where both PZNPC and PZSE coincide
IEP	Isoelectric point	pH of zero ζ potential on electrokinetic curves

which show a statistical difference in PZC of about 0.5 pH units. Experimental aspects such as the temperature, the method (potentiometric titration, electrophoresis, or electroacoustics), the contamination of the sample by silica leached out of the glassware, or impurities in commercial samples are principally incriminated.

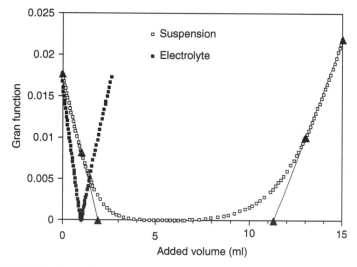

FIGURE 13.19 Gran plot of the titration data of anatase and electrolyte solution (NaCl). (Data from Prélot, B., Mesure et modélisation de l'hétérogénéité énergétique à l'interface oxyde/électrolyte/métaux, PhD Thesis, University of Nancy-INPL, France, p. 234, 2001. With permission.)

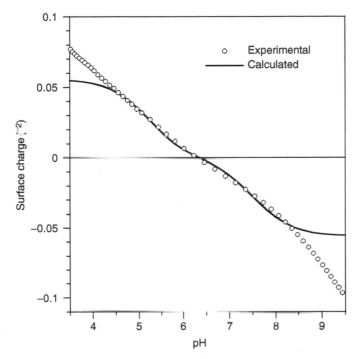

FIGURE 13.20 Example of modeling of the 0.1 M charge curve of anatase (in Figure 13.18) using the CCM in the 2-pK mode. $pK_1 = 5.3$; $pK_2 = 7.7$; $N_s = 0.4$ sites/nm^2.

However, another explanation can be proposed, namely, surface heterogeneity due to different local pK on the crystal faces of titanium oxide. Differences in morphology, by exposing various proportions of the crystal faces, could then contribute to the scattered PZC observed in Figure 13.18. Therefore, proper study of the surface charge of any powdered solid should include accurate knowledge of the shape and crystal-chemistry of the sample.

13.4.4 SURFACE HETEROGENEITY AND THE EDL

Due to the mass action law they are based on, the models of surface dissociation and complexation (see Section 13.4.2 and Section 13.4.3) implicitly postulate that the reactive surface sites are thermodynamically equivalent for a given solid. However, the question of surface heterogeneity at the solid-electrolyte interface arose together with the publication of the first

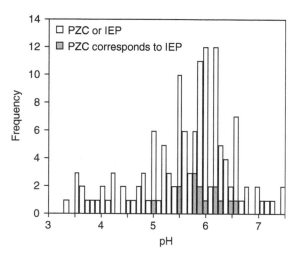

FIGURE 13.21 Distribution of the values of PZC and IEP of TiO_2, collected in the literature. (Adapted from Kosmulski, M., *Adv. Colloid Inteface Sci.*, 99, 255, 2002.)

surface complexation models [114–116]. Later, Davis and Leckie [117] and Benjamin and Leckie [118] proposed a more formal description of surface heterogeneity originating from differently coordinated surface groups. This assumption is strongly supported by crystallographic considerations, which clearly indicate that site density and local charge are characteristic of a given crystal plane, as shown by the application of the MUSIC model [94,95] (see Section 13.4.1). It should then be possible to predict the acid–base behavior of "perfect" crystalline solids.

On the other hand, surface heterogeneity also originates from disordered crystal lattices, defects in crystal planes, uneven distribution of dissociable functions, or chemical impurities. Given the fact that all of those imperfections are encountered for natural solids [119], it is impossible to derive the surface properties from any crystal-chemical characterization of the samples.

However, the experimental description of heterogeneous surfaces can be achieved if it is assumed that successive monoprotic dissociation occurs upon potentiometric titration. The formalism of affinity distributions can then be used. It was successfully applied to the solid–gas interface, for instance, by Villiéras et al. [120–122], and to the solid–electrolyte interface by Contescu et al. [123] and Prélot et al. [124]. In principle, a titration curve can be treated as an adsorption isotherm of protons or hydroxo ions (according to the titrant used), and the surface dissociation equilibria can be expressed in terms of Langmuir adsorption isotherm. The dissociation constant of a monoprotic reaction is $K = [=SOH]/[=SO^-][H^+]$. Introducing the coverage

$\theta = [\equiv SOH]/[Ns]$ and $1 - \theta = [\equiv SO^-][Ns]$ in this expression yields:

$$K = \frac{\theta}{1 - \theta[H^+]} \quad \text{and} \quad \theta = \frac{K[H^+]}{1 + K[H^+]} \tag{13.30}$$

A heterogeneous surface can be modeled with a linear combination of Langmuir equations, as proposed by Koopal [125]. Experimental determination of the complexation constants of heterogeneous surfaces is achieved from the analysis of the derivatives of the titration curves, by decomposition into local derivative isotherms. The derivative of the Langmuir-type Equation 13.30 is:

$$\frac{d\theta}{d \ln[H^+]} = \theta(1 - \theta) \tag{13.31}$$

When working with real solids, the assumption of true homogeneous domains must be tempered with a distribution term. A quasi-gaussian distribution can be used [126,127]. Equation 13.30 becomes then the so-called Langmuir–Freundlich equation. The term m introduced therein takes a value between 0 and 1 and represents the broadening of the affinity constant K around its mean value K':

$$\theta = \frac{K[H^+]^m}{1 + K[H^+]^m} \tag{13.32}$$

and its derivative is:

$$\frac{d\theta}{d \ln[H^+]} = m\theta(1 - \theta) \tag{13.33}$$

It has been shown that the gaussian term m may also express cooperative effect due to lateral interactions between adsorbed species. However, lateral interactions are better described by the Temkin isotherm. With $a = \omega/kT$ (ω the lateral interaction energy terms), this equation is:

$$\theta = \frac{K \exp(a\theta)[H^+]}{1 + K \exp(a\theta)[H^+]} \tag{13.34}$$

and its derivative is:

$$\frac{d\theta}{d \ln[H^+]} = \frac{\theta(1 - \theta)}{1 - a\theta(1 - \theta)} \tag{13.35}$$

The parameters of the local derivative isotherms are then used to generate linear combinations of Langmuir-type isotherms (Equation 13.30). The maximum of the derivative corresponds to the equilibrium constant of the local charging equilibrium, the area corresponds to the proportion of charged groups, and the broadening represents the energy dispersion (Equation 13.33) on the considered patch and/or the energy of the lateral interactions (Equation 13.35).

The high resolution charging curves on anatase at three background electrolyte concentrations ($NaClO_4$) are shown in Figure 13.18. They are characterized by the common intersection point which defines the PZC at pH = 5.7. The derivative curve (Figure 13.22) was fitted with three local isotherms featuring three energetic domains centerd at pK 5.2, 7.5, and 9.3, respectively. This distribution is in agreement with that determined in similar conditions by Contescu et al. [128]. However, it fits only one of the pK values predicted by the MUSIC model: -7.5, 5.3, and 6.3, which correspond respectively to the tridentate oxo- ($=Ti_3O$), bidentate oxo- ($=Ti_2O$), and monodentate hydroxo ($=TiOH$) groups on the *110* plane [96]. Better agreement is found with the scheme proposed by Ludwig and Schindler [129] from the metal adsorption behavior of anatase: pK 5.40, 7.75, and 9.46.

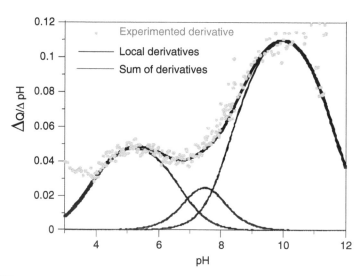

FIGURE 13.22 Decomposition of the derivative titration curve of anatase in 0.1 M NaCl. (From Prelot, B., Mesure et modelisation de l'hétérogénéité énergétique à l'interface oxyde/électrolyte/métaux, PhD Thesis University of Nancy-LNPL, France, p. 234, 2001.)

13.5 APPLICATIONS OF THE DOUBLE-LAYER CONCEPT TO POWDERS AND FIBERS

Aqueous dispersions of powders and fibers are encountered in a number of industrial domains. Whatever the process they are involved in, it is of great importance to control the reactivity at the solid–liquid interface and to control the stability of the colloidal suspensions in order to keep the process running. The double-layer characteristics of the solid–aqueous solution interface then constitute crucial data to predict and control the macroscopic behavior of those systems according to the environmental conditions (pH, ionic composition...). In that scope, the Point of Zero Charge (PZC), which is a specific signature of a given solid–solution system, represents a key parameter. It determines the sign of the surface charge according to the pH, and therefore the selectivity for ion adsorption and the stability of colloidal suspensions.

The two following examples illustrate a use that can be made of the isoelectric point as an indicator of the surface charge in order to control, respectively, the ion adsorption selectivity and the colloidal stability. The example in Figure 13.23 concerns the flotation of Beryl. Separation of metal ores is mainly performed by flotation, in which ionic surfactants are selectively adsorbed on the ore particles, which become hydrophobic enough to be collected by air bubbles [130]. The electrokinetic measurements done at two ionic strengths indicate that the IEP is close to pH 4.2. The use of two oppositely charged surfactants (dodecyl ammonium and dodecyl sulfonate) results in contrasted recovery rates. For pH < IEP, the positively charged surface is selective to the anionic surfactant, whereas for pH > IEP, the cationic surfactant is the better collector. It is then possible to select the best conditions for optimizing the industrial separation process—conditions that necessarily further depend on the properties of the other minerals associated to the ore of interest.

Figure 13.24 describes the interrelations between the electrokinetic potential and the flow behavior of a goethite suspension [131]; the suspension undergoes minimal electrostatic repulsion at pH around the IEP (pH 9.4), which results in aggregation and hence in the presence of a maximum for the viscosity of the system. Such behavior must be prevented in industrial processes where slurries are conveyed in order to avoid plugging of the pipes. Therefore, dispersants such as low molecular weight polyelectrolytes are generally added. Conversely, filtration or dewatering of dispersions is strongly enhanced in the domain where the suspension is destabilized, i.e., in the vicinity of the IEP. Coagulants such as partly hydrolyzed Al^{3+} or Fe^{3+} salts or macromolecular flocculants are used for this purpose.

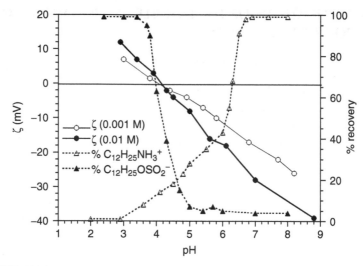

FIGURE 13.23 Froth flotation of Beryl. Dependence of zeta potential and recovery rate on pH and ionic strength (KCl). (After Cases, J. M., Les phénomènes physico-chimiques à l'interface, Application au procédé de la flottation, PhD Thesis, University of Nancy, France, p. 176, 1967.)

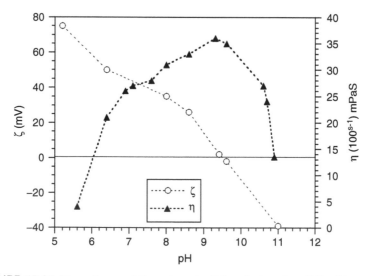

FIGURE 13.24 Dependence of the zeta potential and viscosity of Goethite on pH (NaCl 0.001 M). (Data from Blakey, B. C., and James, D. F., The viscous behavior and structure of aqueous suspensions of goethite, *Colloids and Surf. A, Physicochem. Eng. Aspects*, 231, 19, 2003.)

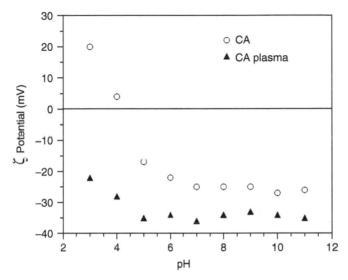

FIGURE 13.25 Effect of oxygen plasma treatment on the zeta potential of carbon fibers (in KCl 0.001 M). (Data from Bismarck, A., Kumru, M. E., and Springer, J., Influence of oxygen plasma treatment of PAN-based carbon fibers on their electrokinetic and wetting properties, *J. Colloid Interf. Sci.*, 210, 60, 1999.)

In composite materials, adhesion of reinforcement particles to polymers is a crucial parameter that influences the usage properties of the manufactured product. Adhesion is significantly strengthened in the presence of ionic reactive sites at the interface. Surface treatments of the reinforcement powders or fibers are often used in order to create such sites. Potentiometric titration gives access to the surface density of the created sites [133–135]. Furthermore, electrokinetic measurements are useful to indicate the evolution of the surface charges, as illustrated in Figure 13.25 [132]. An oxygen plasma treatment strongly increases the acidic properties of carbon fibers, as clearly shown by the increase in negative potential.

List of Symbols

=SOH surface hydroxyl group
A area of the capillary in streaming potential
a radius of the particle
$A(\omega)$ calibration function in ESA
A_s specific surface area of the solid

B	constant determined by the acoustic impedance of the suspension
C	capacitance
E	electric field
e	elementary charge of the electron
ESA	electric sonic amplitude
F	Faraday constant
G	factor depending on the particle size and the measuring frequency ω in ESA
g	gravitational constant
K_{ai}	dissociation constant
k_B	Boltzmann constant
L	length of the capillary in streaming potential
l	metal–proton distance
M	mass of titrated solid
N	Avogadro number
n_{io}	bulk volumic concentration of ion i expressed in m^{-3}
N_s	site density
Q_e	flow rate in a capillary in streaming potential experiments
r	oxygen–proton distance
R	resistance of the plug at conductivity ω
R_∞	resistance of the plug at high conductivity
T	absolute temperature
x	distance from surface
Z_H	valency of hydrogen
Z_i	valency of a given ion
$Z_{O(H)}$	valency of oxygen ($= -2$), or hydroxyl ($= -1$)
ΔP	pressure drop in streaming potential experiments
ΔV	voltage drop in streaming potential experiments
ΔG_*^0	nonelectrostatic contributions
ΔQ	amount of adsorbed titrant
$\Delta \rho$	difference in density between the particles and the liquid
$\Delta \psi$	potential drop
δ	thickness of polymeric layer
ε	dielectric constant
ε_r	relative dielectric constant of the medium
ε_0	permittivity of the vacuum
$\varepsilon_1, \varepsilon_2$	effective microscopic dielectric constants
ϕ	volume fraction of particles
η	dynamic viscosity of the suspending medium
κ_m	reciprocal Debye length within the polymeric fringe of soft particle
κ	reciprocal screening Debye length
λ	conductivity of the solution

λ_0 softness parameter ($1/\lambda_o$ has the dimension of a length)
μ_e electrophoretic mobility
ν_e electrophoretic velocity
θ surface coverage
ρ particle density
$\rho_{el}(x)$ local space charge density (that stems from mobile ions)
$\rho_{fix}(x)$ local space charge density (that stems from fixed charges)
ρ_0 uniform space charge density gravitational constant
σ charge density (in general)
σ_d charge density at the outer Helmholtz plane
σ_β charge density at the inner Helmholtz plane
σ_0 charge density at the surface of a hard particle
σ^0 charge density of a soft particle at the core
ψ_0 surface potential (case of hard particle)
ψ local electrostatic potential
ψ^D Donnan potential
ψ^0 potential at the position corresponding to the location of the outer
         boundary of the surface layer (case of soft particle)
ζ electrokinetic potential

REFERENCES

1. Bismarck, A., Emin Kumru, M., and Springer, J., Influence of oxygen plasma treatment of PAN-based carb on fibers on their electronic and wetting properties, *J. Colloid Interface Sci.*, 210, 60, 1999.
2. Campagne, C., Devaux, E., Perwueltz, A., and Cazé, C., Electrokinetic approach of adhesion between polyester fibers and latex matrices, *Polymer*, 43, 6669, 2002.
3. Bismarck, A. and Springer, J., Characterization of fluorinated PAN-based carbon fibers by zeta-potential measurements, *Colloids Surf. A: Physicochem. Eng. Aspects*, 159, 331, 1999.
4. Espinosa-Jimenez, M., Gimenez-Martin, E., and Ontiveros-Ortega, A., Effect of tannic acid on the ζ potential, sorption, and surface free energy in the process of dyeing of leacril with cationic dye, *J. Colloid Interface Sci.*, 207, 170, 1998.
5. Bhardwaj, N. K., Kumar, S., and Bajpai, P. K., Effect of zeta potential on retention and drainage of secondary fibers, *Colloids Surf. A: Physicochem. Eng. Aspects*, 260, 245, 2005.
6. Stana-Kleinschek, K. and Ribitsch, V., Electrokinetic properties of processed cellulose fibers, *Colloids Surf. A: Physicochem. Eng. Aspects*, 140, 127, 1998.
7. Bellmann, C., Caspari, A., Albrecht, V., Loan Doan T.T., Mäder, E., Luxbacher, T., and Kohl, R., Electrokinetic properties of natural fibers, *Colloids and Surf. A: Physicochem. Eng. Aspects.*, ASAP, 2005.

8. Van Wagenen, R. A., Coleman, D. L., King, R. N., Triolo, P., Brostrom, L., Smith, L. M., Gregonis, D. E., and Andrade, J. D., *J. Colloid Interface Sci.*, 84, 155, 1981.

9. Zimmermann, R., Dukhin, S., and Werner, C., Electrokinetic measurements reveal interfacial charge at polymer films caused by simple electrolyte ions, *J. Phys. Chem.*, 105, 8544, 2001.

10. Weidenhammer, P. and Jacobasch, H. J., Investigation of adhesion properties of polymer materials by atomic force microscopy and zeta potential measurements, *J. Colloid Interface Sci.*, 180, 232, 1996.

11. Hermitte, L., Thomas, F., Bougaran, R., and Martelet, C., Contribution of the comonomers to the bulk and surface properties of methacrylate copolymers, *J. Colloid Interface Sci.*, 272, 82, 2004.

12. Ulberg, D. E. and Gubbins, K. E., Water adsorption in microporous graphite carbons, *Mol. Phys.*, 84, 1139, 1995.

13. Hartning, C., Witschel, W., and Spohr, E., Molecular dynamics study of electrolyte-filled pores, *J. Phys. Chem. B*, 102, 1241, 1998.

14. Kitano, H., Ichikawa, K., Fukuda, M., Mochiuzuki, A., and Tanaka, M., The structure of water sorbed to polymethoxyethylacrylate film as examined by FT-IR spectroscopy, *J. Colloid Interface Sci.*, 242, 133, 2001.

15. Fripiat, J., Cases, J. M., François, M., and Letellier, M., Thermodynamic and microdynamic behavior of water in clay suspensions and gels, *J. Colloid Interface Sci.*, 89, 378, 1982.

16. Helmholtz, H., Studien über elektrische grenzenschichten, *Annales der Physik und Chemie*, 7, 337, 1879.

17. Chapman, D., A contribution to the theory of electrocapillarity, *Phil. Mag.*, 25, 475, 1913.

18. Gouy, M. G., Sur la fonction électrocapillaire, *Ann. Phys.*, 9, 129, 1917.

19. Stern, O., Zur theorie der elektrischen doppelschicht, *Z. Elektrochemie*, 30, 508, 1924.

20. Grahame, D. C., The electric double-layer and the theory of electrocapillarity, *Chem. Rev.*, 41, 441, 1947.

21. Charmas, R., Piasecki, W., and Rudzinski, W., Four layer complexation model for ion adsorption at electrolyte/oxide interface: theoretical foundations, *Langmuir*, 11, 3210, 1995.

22. Hunter, R. J., *Foundations of Colloid Science*, Vol. 1, Oxford University Press, New York, 1986.

23. Hunter, R. J., In *Zeta Potential in Colloid Science. Principles and Applications*, Ottewill, R. and Rowell, R., Eds., Academic Press, New York, 1981.

24. Von Smoluchowski, M., Towards a mathematical theory of the coagulation kinetics of colloidal solutions, *Z. Phys. Chem.*, 92, 129, 1918.

25. Hückel, E., Die kataphorese der kugel, *Phys. Z.*, 25, 204, 1924. Henry, D. C., The cataphoresis of suspended particles, *Proc. Roy. Soc. London A*, 133, 106, 1931.

26. Henry, D. C., The cataphoresis of suspended particles. Part I: The equation of cataphoresis, *Proc. Roy. Soc. London A*, 133, 106, 1931.

27. O'Brien, R. W. and White, L. R., Electrophoretic mobility of a spherical colloid particle, *J. Chem. Soc. Faraday Trans.*, 2(78), 1607, 1978.
28. Zukoski, C. F. and Saville, D. A., The interpretation of electrokinetic measurements using a dynamic model of the Stern layer. I. The dynamic model, *J. Colloid Interface Sci.*, 114, 32, 1986.
29. Zukoski, C. F. and Saville, D. A., The interpretation of electrokinetic measurements using a dynamic model of the Stern layer. II. Comparisons between theory and experiment, *J. Colloid Interface Sci.*, 114, 45, 1986.
30. Ohshima, H., Electrophoretic mobility of spherial colloidal particles in concentrated suspensions, *J. Colloid Interface Sci.*, 188, 480, 1997.
31. Ennis, J. and White, L. R., Electrophoretic mobility of a semi-dilute suspension of spherical particles with thick double-layers and low zeta potentials, *J. Colloid Interface Sci.*, 185, 157, 1997.
32. Lee, E., Chu, J. W., and Hsu, J. P., Electrophoretic mobility of a concentrated suspension of spherical particles, *J. Colloid Interface Sci.*, 209, 240, 1999.
33. Grosse, C. and Shilov, V. N., Electrophoretic mobility of colloidal particles in weak electrolyte solutions, *J. Colloid Interface Sci.*, 211, 160, 1999.
34. Ohshima, H., Electrophoretic mobility of soft particles, *Colloids and Surf. A: Physicochem. Eng. Aspects.*, 103, 249, 1995.
35. Ohshima, H., Electrophoretic mobility of soft particles in concentrated suspensions, *J. Colloid Interface Sci.*, 225, 233, 2000.
36. Labbez, C., Fievet, P., Thomas, F., Szymczyk, A., Vidonne, A., Foissy, A., and Pagetti, P., Evaluation of the "DSPM" model on a titania membrane: measurements of charged and uncharged solute retention, electrokinetic charge, pore size, and water permeability, *J. Colloid Interface Sci.*, 262, 200, 2003.
37. Scales, P. J., Grieser, F., Healy, T. W., White, L. R., and Chan, D. Y. C., Electrokinetics of the silica-solution interface: a flat plate streaming potential study, *Langmuir*, 8, 965, 1992.
38. Bowen, W. R. and Cao, X., Electrokinetic effects in membrane pores and the determination of zeta potential, *J. Membrane Sci.*, 141, 267, 1998.
39. Minor, M., Van der Linde, A. J., and Lyklema, J., Streaming potentials and conductivities of latex plugs in indifferent electrolytes, *J. Colloid Interface Sci.*, 203, 177, 1998.
40. Minor, M., Van der Linde, A. J., Van Leeuwen, H. P., and Lyklema, J., Streaming potentials and conductivities of porous silica plugs, *Colloids Surf. A*, 142, 165, 1998.
41. Guzev, I. and Horvath, C., Streaming potential in open and packed fused-silica capillaries, *J. Chromatogr. A*, 948, 203, 2002.
42. Johnson, P. R., A comparison of streaming and microelectrophoresis methods for obtaining the zeta potential of granular porous media surfaces, *J. Colloid Interface Sci.*, 209, 264, 1999.
43. Debye, P., A method for the determination of the mass of electrolyte ions, *J. Chem. Phys.*, 1, 13, 1933.

44. O'Brien, R. W., Cannon, D. W., and Rowlands, W. N., Electroacoustic determination of particle size and zeta potential, *J. Colloid Interface Sci.*, 173, 406, 1995.
45. Ehrfeld, W., Electrochemistry and microsystems, *Electrochem. Acta*, 18, 2857, 2003.
46. Andrews, A. T., *Electrophoresis: Theory, Techniques and Biochemical and Clinical Applications*, Clarendon Press, Oxford, 1986.
47. Kim, C. A. et al., Microcapsules as an electrokinetic ink to fabricate color electrophoretic displays, *Synth. Metals*, 151, 181, 2005.
48. Ohshima, H., Electrophoresis of soft particles, *Adv. Colloid Interface Sci.*, 62, 189, 1995.
49. Ohshima, H., On the general expression for the electrophoretic mobility of a soft particle, *J. Colloid Interface Sci.*, 228, 190, 2000.
50. Ohshima, H., Electrophoretic mobility of soft particles, *J. Colloid Interface Sci.*, 163, 474, 1994.
51. Hill, R. J., Saville, D. A., and Russel, W. B., Electrophoresis of spherical polymer-coated colloidal particles, *J. Colloid Interface Sci.*, 258, 56, 2003.
52. Hill, R. J., Saville, D. A., and Russel, W. B., Polarizability and complex conductivity of dilute suspensions of spherical colloidal particles with charged (polyelectrolyte) coatings, *J. Colloid Interface Sci.*, 263, 478, 2003.
53. Saville, D. A., Electrokinetic properties of fuzzy colloidal particles, *J. Colloid Interface Sci.*, 222, 137, 2000.
54. Lopez-Garcia, J. J., Grosse, C., and Horno, J., Numerical study of colloidal suspensions of soft spherical particles using the network method: 1. DC electrophoretic mobility, *J. Colloid Interface Sci.*, 265, 327, 2003.
55. Lopez-Garcia, J. J., Grosse, C., and Horno, J., Numerical study of colloidal suspensions of soft spherical particles using the network method: 2. AC electrokinetic and dielectric properties, *J. Colloid Interface Sci.*, 265, 341, 2003.
56. Dukhin, S. S., Zimmermann, R., and Werner, C., Intrinsic charge and donnan potentials of grafted polyelectrolyte layers determined by surface conductivity data, *J. Colloid Interface Sci.*, 274, 309, 2004.
57. Wunderlich, R. W., The effects of surface structure on the electrophoretic mobilities of large particles, *J. Colloid Interface Sci.*, 88, 385, 1982.
58. Levine, S., Levine, M., Sharp, K. A., and Brooks, D. E., Theory of the electrokinetic behavior of human erythrocytes, *Biophys. J.*, 42, 127, 1983.
59. Sharp, K. A. and Brooks, D. E., Calculation of the electrophoretic mobility of a particle bearing bound electrolyte using the non-linear Poisson–Boltzmann equation, *Biophys. J.*, 47, 563, 1985.
60. Ohshima, H., Approximate expression for the electrophoretic mobility of a spherical colloidal particle covered with an ion-penetrable uncharged polymer Layer, *Colloid Polym. Sci.*, 283, 819, 2005.
61. Hill, R. J. and Saville, D. A., "Exact" solutions of the full electrokinetic model for soft spherical colloids: electrophoretic mobility, *Colloids Surf. A*, 267, 31, 2005.

62. Hill, R. J., Hydrodynamics and electrokinetics of spherical liposomes with coatings of terminally anchored poly(ethylene glycol): numerically exact electrokinetics with self-consistent mean-field polymer, *Phys. Rev. E*, 70, 051406, 2004.

63. Starov, V. and Solomentsev, Y. E., Influence of gel layers on electrokinetic phenomena: 1. Streaming potential, *J. Colloid Interface Sci.*, 158, 159, 1993.

64. Starov, V. and Solomentsev, Y. E., Influence of gel layers on electrokinetic phenomena: 2. Effect of ions interaction with the gel layer, *J. Colloid Interface Sci.*, 158, 166, 1993.

65. Donath, E. and Voigt, A., Streaming current and streaming potential on structured surfaces, *J. Colloid Interface Sci.*, 109, 122, 1986.

66. Donath, E., Budde, A., Knippel, E., and Bäumler, H., "Hairy Surface Layer" concept of electrophoresis combined with local fixed surface charge density isotherms: application to human erythrocyte electrophoretic fingerprinting, *Langmuir*, 12, 4832, 1996.

67. Duval, J. F. L. and van Leeuwen, H. P., Electrokinetics of diffuse soft interfaces. I. Limit of low Donnan potentials, *Langmuir*, 20, 10324, 2004.

68. Duval, J. F. L., Electrokinetics of diffuse soft interfaces. II. Analysis based on the non-linearized Poisson–Boltzmann equation, *Langmuir*, 21, 3247, 2005.

69. Yezek, L., Duval, J. F. L., and van Leeuwen, H. P., Electrokinetics of diffuse soft interfaces. III Interpretation of data on the polyacrylamide/water interface, *Langmuir*, 21, 6220, 2005.

70. Van der Wal, A., Electrochemical characterization of the bacterial cell surface, Ph.D. Thesis, Wageningen Universiteit, The Netherlands, 1996.

71. Poortinga, A.T., Electric double layer interactions in bacterial adhesion and detachment, PhD Thesis, Rijksuniversiteit Groningen, The Netherlands, 2001.

72. Bos, R., van der Mei, H. C., and Busscher, H. J., "Soft-particle" analysis of the electrophoretic mobility of a fibrillated and non-fibrillated oral streptococcal strain: *Streptococcus salivarius*, *Biophys. Chem.*, 74, 251, 1998.

73. Duval, J. F. L., Wilkinson, K., van Leeuwen, H. P., and Buffle, J., Humic substances are not hard-spheres: permeability determination from electrophoretic mobility measurements, *Environ. Sci. Technol.*, 39, 6435, 2005.

74. Van Male, J., Self-consistent-field theory for chain molecules: extensions, computational aspects, and applications, PhD Thesis, Wageningen University, The Netherlands, 2003.

75. Scheutjens, J. M. H. M. and Fleer, G. J., Statistical theory of the adsorption of interacting chain molecules. 1. Partition function, segment density distribution, and adsorption isotherms, *J. Phys. Chem.*, 83, 1619, 1979.

76. Van der Gucht, J., Equilibrium polymers in solution and at interfaces, PhD Thesis, Wageningen University, The Netherlands, 2004.

77. Zhulina, E. B., Klein Wolterink, J., and Borisov, O. V., Screening effects in a polyelectrolyte brush: Self-consistent-field theory, *Macromolecules*, 33, 4945, 2000.

78. Debye, P. and Bueche, A., Intrinsic viscosity, diffusion and sedimentation rate of polymers in solution, *J. Chem. Phys.*, 16, 573, 1948.

79. Duval, J. F. L., Busscher, H. J., van de Belt-Gritter, B., van der Mei, Henny, C., and Norde, W., Analysis of the interfacial properties of fibrillated and non-fibrillated oral streptococcal strains from electrophoretic mobility and titration measurements: evidence for the shortcomings of the 'classical soft-particle approach', *Langmuir*, 21, 11268, 2005.

80. Parks, G. A., The isoelectric points of solid oxides, solid hydroxides, and aqueous hydroxo complex systems, *Chem. Rev.*, 65, 177, 1965.

81. Parks, G. A., Aqueous surface chemistry of oxides and complex minerals, *Adv. Chem. Series*, 97, 121, 1967.

82. Davis, J. A., James, R. O., and Leckie, J. O., Surface ionization and complexation at the solid oxide–water interface. I. Computation of electrical double-layer properties in simple electrolytes, *J. Colloid Interface Sci.*, 63, 480, 1978.

83. Davis, J. A. and Leckie, J. O., Surface ionization and complexation at the oxide/water interface. II. Surface properties of amorphous iron oxyhydroxide and adsorption of metal ions, *J. Colloid Interface Sci.*, 67, 90, 1978.

84. Davis, J. A. and Leckie, J. O., Surface ionization and complexation at the oxide/water interface. III. Adsorption of anions, *J. Colloid Interface Sci.*, 74, 32, 1980.

85. Stumm, W., Huang, C. P., and Jenkins, S. R., Specific chemical interactions affecting the stability of dispersed systems, *Croatica Chemica Acta*, 53, 291, 1970.

86. Stumm, W., Kummert, R., and Sigg, L., A ligand exchange model for the adsorption of inorganic ligands on hydrous oxide interfaces, *Croatica Chemica Acta*, 53, 291, 1980.

87. Pauling, L., *The Nature of the Chemical Bond*, 3rd ed., Cornell University Press, Ithaca, NY, 1960.

88. Pauling, L., Atomic radii and interatomic distances in metals, *J. Am. Chem. Soc.*, 51, 1010, 1929.

89. Pauling, L., *The Nature of Electrostatic Bonds*, 3rd ed., Cornell University Press, Ithaca, NY, 1967. (Chap. 13–16).

90. Brown, I. D., Bound valences, a simple structural model for inorganic chemistry, *Chem. Soc. Rev.*, 7, 359, 1978.

91. Yoon, R. H., Salman, T., and Donnay, G., Predicting points of zero charge of oxides and hydroxides, *J. Colloid Interface Sci.*, 70, 483, 1979.

92. Bolt, G. H. and van Riemsdijkd, W. H., Ion adsorption on inorganic variable charge constituents, In *Soil Chemistry B. Physicochemical Models*, Bolt, G. H., Ed., Elsevier, Amsterdam, 1982.

93. Van Riemsdijk, W. H., Bolt, G. H., Koopal, L. K., and Blaakmeer, J., Electrolyte adsorption on heterogeneous surfaces: adsorption models, *J. Colloid Interface Sci.*, 109, 219, 1986.

94. Hiemstra, T., Van Riemsdijk, W. H., and Bolt, G. H., Multisite proton adsorption modeling at solid/solution interface of (hydr)oxides: a new approach. I. Model description and evaluation of intrinsic reaction constants, *J. Colloid Interface Sci.*, 133, 91, 1989.

95. Hiemstra, T., de Wit, J. C. M., and Van Riemsdijk, W. H., Multisite proton adsorption modeling at solid/solution interface of (hydr)oxides: a new approach. II. Applications to various important (hydr)oxides, *J. Colloid Interface Sci.*, 133, 105, 1989.

96. Hiemstra, T. and Van Riemsdijk, W. H., A surface structural approach of ion adsorption: the charge distribution model, *J. Colloid Interface Sci.*, 179, 488, 1996.

97. Lützenkirchen, J., Comparison of 1-pK and 2pK versions of surface complexation theory by the goodness of fit in describing surface charge data of (hydr)oxides, *Environ. Sci. Technol.*, 32, 3149, 1998.

98. Yates, D. E., Levine, S., and Healy, T. W., Site binding model of the electrical model at the oxide/water interface, *J. Chem. Soc. Faraday Trans. I*, 70, 1807–1818, 1974.

99. Yates, D. E., James, R. O., and Healy, T. W., Titanium dioxyde-electrolyte interface. Part 1. Gas adsorption and tritium exchange studies, *J. Chem. Soc. Faraday Trans.*, 76, 1, 1980.

100. Parks, G. A. and De Bruyn, P. L., The zero point of charge of oxides, *J. Phys. Chem.*, 66, 967, 1962.

101. Baeyens, B. and Bradbury, M. H., A mechanistic description of Ni and Zn sorption on Na-montmorillonite Part I: Titration and sorption measurements, *J. Contaminant Hydro.*, 27, 199, 1997.

102. Tournassat, C., Greneche, J. M., Tisserand, D., and Charlet, L., The titration of clay minerals: I. Discontinuous back titration technique combined with CEC measurements, *J. Colloid Interface Sci.*, 273, 224, 2004.

103. Duc, M., Gaboriaud, F., and Thomas, F., Sensitivity of the acid–base properties of clays to the methods of preparation and measurement: 1. Literature review, *J. Colloid Interface Sci.*, 273, 224, 2004.

104. Duc, M., Gaboriaud, F., and Thomas, F., Sensitivity of the acid–base properties of clays to the methods of preparation and measurement: 1. Literature review, *J. Colloid Interface Sci.*, 289, 148, 2005.

105. Du, Q., Sun, Z., Forsling, W., and Tang, H., Acid–base properties of aqueous illite surfaces, *J. Colloid Interface Sci.*, 187, 221, 1997.

106. Prélot, B., Mesure et modélisation de l'hétérogénéité énergétique à l'interface oxyde/électrolyte/métaux, PhD Thesis, University of Nancy-INPL, France, p. 234, 2001.

107. Sposito, G., *The Surface Chemistry of Soils*, Oxford University Press, Oxford, 1984.

108. Sposito, G., On points of zero charge, *Environ. Sci. Technol.*, 32, 2815, 1998.

109. Noh, J. S. and Schwarz, J. A., Estimation of the point of zero charge of simple oxides by mass titration, *J. Colloid Interface Sci.*, 130, 157, 1989.

110. Zalac, S. and Kallay, N., Application of mass titration to the point of zero charge determination, *J. Colloid Interface Sci.*, 149, 233, 1992.

111. Kallay, N. and Zalac, S., Charged surfaces and interfacial ions, *J. Colloid Interface Sci.*, 230, 1, 2000.

112. Hayes, K. F. and Leckie, J. O., Modeling ionic strength effects on cation adsorption at hydrous oxide/solution interfaces, *J. Colloid Interface Sci.*, 115, 564, 1987.

113. Gran, G., Determination of the equivalence point in potentiometric titrations. Part II, *Analyst*, 77, 661, 1952.

114. Kosmulski, M., The significance of the difference in the point of zero charge between rutile and anatase, *Adv. Colloid Interface Sci.*, 99, 255, 2002.

115. Kruyt, H. R., *Colloid Science*, Elsevier Publishing Co., New York, 1952.

116. Brown, I. D., Bound valences, a simple structural model for inorganic chemistry, *Chem. Soc. Rev.*, 7, 359, 1978.

117. Davis, J. A. and Leckie, J. O., Surface ionization and complexation at the oxide/water interface. II. Surface properties of amorphous iron oxyhydroxide and adsorption of metal ions, *J. Colloid Interface Sci.*, 67, 90, 1978.

118. Benjamin, M. M. and Leckie, J. O., Multisite adsorption of Cd, Cu, Zn and Pb on amorphous iron oxyhydroxide, *J. Colloid Interface Sci.*, 79, 209, 1981.

119. Nagashima, K. and Blum, F. D., Proton adsorption onto alumina: extension of multisite complexation (MUSIC) theory, *J. Colloid Interface Sci.*, 217, 28, 1999.

120. Villiéras, F., Cases, J. M., François, M., Michot, L., and Thomas, F., Texture and surface energetic heterogeneity of solids from modeling of low pressure gas adsorption isotherms, *Langmuir*, 8, 1789, 1992.

121. Villiéras, F., Michot, L., Bardot, F., Cases, J. M., François, M., and Rudzinski, W., An improved derivative isotherm summation method to study surface heterogeneity of clay minerals, *Langmuir*, 13, 1104, 1997.

122. Villiéras, F., Michot, L., Cases, J. M., Bérend, I., Bardot, F., François, M., Gérard, G., and Yvon, J., Static and dynamic studies of the energetic surface heterogeneity of clay minerals, In *Equilibria and Dynamics of Gas Adsorption on Heterogeneous Solid Surfaces*, Rudzinski, W., Steele, W. A., and Zgrablich, G., Eds., Vol. 104, Elsevier, Amsterdam, pp. 573–623, 1997.

123. Contescu, C., Jagiello, J., and Schwartz, J. A., Heterogeneity of proton binding sites at oxide/solution interface, *Langmuir*, 9, 1754, 1993.

124. Prélot, B., Charmas, R., Zarzycki, P., Thomas, F., Villiéras, F., Piasecki, W., and Rudzinski, W., Application of the theoretical 1-pK approach to analyzing proton adsorption isotherm derivatives on heterogeneous oxide surfaces, *J. Phys. Chem. B*, 106, 13280, 2002.

125. Koopal, L. K., Mineral hydroxides: from homogeneous to heterogeneous modeling, *Electrochem. Acta*, 41, 2293, 1996.

126. Bersillon, J. L., Villiéras, F., Bardot, F., Görner, T., and Cases, J. M., Use of the gaussian distribution function as a tool to estimate continuous heterogeneity in adsorbing systems, *J. Colloid Interface Sci.*, 240, 400, 2001.

127. Koopal, L. K., Van Riemsdijk, W. H., De Wit, J. C. M., and Benedetti, M. F., Analytical isotherm equations for multicomponent adsorption on heterogeneous surfaces, *J. Colloid Interface Sci.*, 166, 51, 1994.

128. Contescu, C., Popa, V. T., and Schwartz, J. A., Heterogeneity of hydroxyl and deuteroxyl groups on the surface of TiO$_2$ polymorphs, *J. Colloid Interface Sci.*, 180, 149, 1996.
129. Ludwig, C. and Schindler, P. W., Surface complexation on TiO$_2$. I. Adsorption of H$^+$ and Cu^{2+} ions onto TiO$_2$, *J. Colloid Interface Sci.*, 169, 284, 1995.
130. Cases, J. M., Les phénomènes physico-chimiques à l'interface, Application au procédé de la flottation, PhD Thesis, University of Nancy, France, p. 176, 1967.
131. Blakey, B. C. and James, D. F., The viscous behavior and structure of aqueous suspensions of goethite, *Colloids and Surf. A, Physicochem. Eng. Aspects*, 231, 19, 2003.
132. Bismarck, A., Kumru, M. E., and Springer, J., Influence of oxygen plasma treatment of PAN-based carbon fibers on their electrokinetic and wetting properties, *J. Colloid Interface Sci.*, 210, 60, 1999.
133. Toebes, M. L., van Heeswijk, J. M. P., Bitter, J. H., van Dillen, A. J., and de Jong, K. P., The influence of oxidation on the texture and the number of oxygen-containing surface groups of carbon nanofibers, *Carbon*, 42, 307, 2004.
134. Fras, L., Johansson, L. S., Stenius, P., Laine, J., Stana-Kleinschek, K., and Ribitsch, V., Analysis of the oxidation of cellulose fibers by titration and XPS, *Colloids and Surf. A, Physicochem. Eng. Aspects*, 260, 101, 2005.
135. Matsumoto, H., Wakamatsu, Y., Minagawa, M., and Tanioka, A., Preparation of ion-exchange fiber fabrics by electrospray deposition, *J. Colloid Interface Sci.*, 293, 143, 2006.

14 Polymer Dynamics at Solid–Liquid Interfaces

Emile Pefferkorn

CONTENTS

14.1 INTRODUCTION

The image of the macromolecule adsorbed at the solid–liquid interface has progressively been clarified according to new techniques and methods which have been implemented [1–12]. Presently, one has to determine the fate of the macromolecule from the moment it leaves the bulk solution and contacts the adsorbent surface to the final moment where the macromolecule is embedded within the polymer layer. Then, the established layer may be observed under different situations. First of all, it may be interesting to determine the layer characteristics when the polymer interface is rapidly washed out with solvent. Secondly, it may be interesting to get the information for the layer "at rest" in the presence of the bulk solution. Finally, in some instances, it could be important to determine the modifications of the layer characteristics when the layer faces polymer solutions of varying concentration.

The aim of this chapter is to present some dynamic characteristics of the macromolecules during layer formation and after layer establishment. Therefore, it was important to determine the processes that control the macromolecule transfers from bulk to interface and from interface to bulk [8–10,13–17]. Due to their great flexibility, macromolecules sustain reconformation during adsorption and the established layer may relax as long as equilibrium is not established. Reconformation during adsorption was investigated for different systems and the relaxation times were determined for charged and uncharged systems [18–20]. For established layers, overshoot was determined to occur, but the phenomenon behind the process remained highly non-elucidated [20,21].

Finally, if establishment of equilibrium conformations at solid–liquid interfaces requires bulk to surface and inverse transfers, equilibrium within established polyelectrolyte layers may be differently achieved by internal migration and spatial redistribution of counter-ions of the adsorbed polyelectrolyte [22,23].

The present results were obtained by implementing adsorption experiments with radio-labeled and nonlabeled polymer molecules of equal molecular weight. First of all, radio-labeling usually allows the detection of few nanograms of polymer material, and is particularly suited to the determination of the adsorption amounts as well as temporal changes in the adsorption amount. Secondly, concomitantly employing labeled and nonlabeled molecules in the same experiment allowed the rates and extents of interfacial transfers of polymer to be determined, in spite of the fact that the total amount of adsorbed polymer remained constant at adsorption equilibrium.

In order to allow a good reading of the report, the background of the experimental methods was presented first, and then the different interfacial phenomena were reported and discussed as being centered upon the different systems of interest. This presentation may further serve to illustrate that the interfacial characteristics are highly system dependent. The major conclusions of the study are given in the last paragraph.

14.2 EXPERIMENTAL METHODS

14.2.1 THE REACTOR DEVICE

The reactor device was employed to determine the rate of polymer adsorption under various experimental conditions [24]. By changing the rate of polymer supply to the adsorbent confined in the reactor, the device serves to determine reconformation of adsorbed molecules as far as a slow polymer supply allows a longer reconformation, whereas a fast rate of polymer supply strongly limits the extent of the process. The device was similarly employed to evidence interfacial transfers of polymer by first establishing a radio-labeled polymer layer and then replacing the radioactive supernatant with an identical solution of nonlabeled polymer. Analysis of the effluent thus provides information leading to the determination of the rate of interfacial transfers.

14.2.1.1 Kinetics of Adsorption

The dispersion of x gram of adsorbent (powder or fibers) was first confined in the reactor of volume V, and the polymer solution of concentration C_0 was injected at the controlled rate J_v for a given period T at the inlet aperture. Simultaneously, the liquid phase was collected at the outlet using a sample collector and the successive aliquots of volume V were analyzed for radioactivity content. From the balance between the amount of polymer being injected and the concentration C_{eff} of polymer in the effluent, one derives the adsorption increments ΔN_s corresponding to the successive sampling periods and finally the total adsorption N_s [24].

The kinetic coefficient $K(N_s)$ of adsorption, which provides information on the difficulty to put new molecules on the surface during adsorption, is determined as a function of the amount N_s of polymer already adsorbed [25–28].

$$K(N_s) = \Delta N_s / C_{eq} \qquad (14.1)$$

with

$$C_{eq} = vC_{eff}/x \qquad (14.2)$$

C_{eq}, C_{eff}, v, and x are expressed in mg/g adsorbent, mg/mL solution, mL, and g, respectively. This leads to express $K(N_s)$ in min^{-1}. Usually, different values of the adsorption N_s obtained under various experimental conditions were compared for a given value of $K(N_s)$.

14.2.1.2 Kinetics of Interfacial Transfers

14.2.1.2.1 Exchange Experiments at Constant Bulk Solution Concentration

Adsorption of radioactive polymer was carried out as above and the equilibrium concentration of nonadsorbed polymer was fixed at a value C_{eff}^* [13]. The adsorbent was then allowed to settle and the supernatant was carefully collected and immediately replaced with a nonlabeled solution at the same concentration C_{eff}. The interstitial solution was weighted to precisely calculate the remaining bulk radioactivity. After filling of the reactor with the same nonlabeled solution, this solution was injected at a slow rate into the reactor. The effluent was collected at the outlet aperture for successive periods Δt and the radioactivity of each sample was determined. Equation 14.3 provides the radioactivity balance for the polymer transfer ΔN_s during the time increment Δt, which was determined as a function of elution time t:

$$\Delta N_s^* \propto V\Delta C_{eff}^* \qquad (14.3)$$

The existence of an equilibrium situation was confirmed in an inverse experiment where the transfer characteristics were determined for the exchange of non-labeled adsorbed polymer with labeled bulk solution polymer [13]. In this situation, Equation 14.4 holds

$$\Delta N_s^* \propto -V\Delta\left(C_0^* - C_{eff}^*\right) \qquad (14.4)$$

where C_0^* is the concentration of the injection solution.

14.2.1.2.2 Exchange Experiment at Increasing Bulk Solution Concentration

In this experiment, labeled polymer was adsorbed, and at adsorption equilibrium, the supernatant polymer solution in the reactor was removed and replaced with solvent. The polymer concentration in the cell was then slowly increased by immediate injection of nonlabeled polymer at concentration C_0 and desorption was measured by collecting and analyzing the effluent as a function of time [24,29]. The desorption rate is given by Equation 14.3.

14.2.2 Surface Area Exclusion Chromatography

Chromatographic separation experiments were carried out using the stationary phase Whatman glass microfiber filters GF/B, whose retention is limited to particles above 1 μm [27]. Each filter had a diameter of 5 mm and a thickness of 1.4 mm, the mean diameter of the glass microfiber being estimated close to 6 μm. The area available for polymer adsorption was estimated to be equal to 1 dm^2. The column was constituted of 56 filters stacked within a glass tube fitted with inlet and outlet apertures. Two porous nonadsorbing Teflon filters were fitted on the column inlet in order to ensure the polymer solution would be uniformly distributed at the stationary phase entrance. Radio-labeled polymer was used for the determination of the amount of polymer adsorbed on the filters. We notice that in surface area exclusion chromatography (SAEC), interest is directed to the adsorption histogram which expresses the amount of polymer adsorbed on the successive filters i. The amount of polymer being injected did not saturate the column, so no polymer is usually detected in the effluent. The usual histogram shows the adsorption to be maximal at the inlet of the stationary phase, to progressively decrease with the distance from the entrance, and to sharply drop at a given filter.

14.3 THE POLYACRYLAMIDE/ALUMINOSILICATE/ WATER SYSTEM

14.3.1 The Adsorption Isotherm

The adsorption of neutral polyacrylamide was attributed to the formation of hydrogen and polypeptide bonds between the isolated silanol groups and the amide groups of the polymer [16,30,31]. Polypeptide bonds were shown to disappear above 40°C. Hydrogen bonding with surface groups produces maximal adsorption at pH 4.5 as shown in Figure 14.1 (lower dashed line).

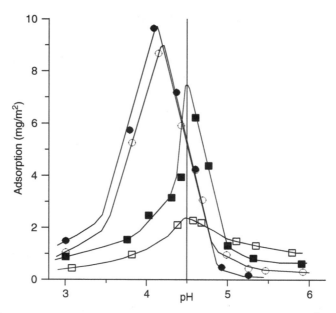

FIGURE 14.1 Representation as a function of pH of the amount (mg/m^2) of neutral and hydrolyzed polyacrylamide adsorbed after a period of 12 h on alumino-silicate beads suspended in water at 25°C. The degree of hydrolysis was 0 (— – —); 0.054 (□); 0.15 (■); 0.22 (○), and 0.24 (●).

14.3.2 THE ADSORPTION PROCESSES

The amount of polyacrylamide adsorbed is shown by a symmetrical curve peaked at pH 4.5. The rate of establishment of the polyacrylamide layer at the aluminosilicate/water interface, determined at the maximal coverage, shows two kinetic regimes which were previously interpreted on the basis of the Langmuir model, assuming two types of kinetics to be operative [16]. The recent interpretation of the experimental result based on the kinetic model of de Gennes shows the apparent kinetics coefficient $K(N_s)$ (mL/mol×min), which represents the instantaneous rate of adsorption to be a function of $1 - \theta$, θ being the relative surface coverage (Figure 14.2) [24,32]. Up to a degree of coverage θ close to 2/3, the rate of adsorption decreases $(1 - \theta)$ according to the Langmuir model or the random sequential adsorption (RSA) models. The rate of further adsorption accords to the theoretical model of de Gennes:

$$K(N_s) \propto (1 - \theta)^{1.5} \tag{14.5}$$

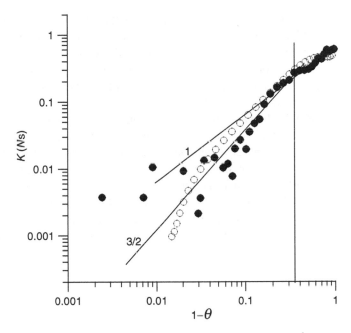

FIGURE 14.2 Representation of the kinetic coefficient $K(N_s)$ (min^{-1}) as a function of the fraction of free surface area (Equation 14.5). The dashed line of slope 1 corresponds to the random sequential adsorption model and is valid for $\theta < 2/3$ (dotted line) while the solid line of slope 1.5 determines the domain of validity of the de Genes theory. The two sets of experimental points were calculated using polynomial ($\bigcirc$) and eye fitting ($\bullet$) for interpolation of the temporal variation of the specific radioactivity C_{eff} of the reactor.

Equation 14.5 implies that the very low concentration of polymer in the bulk strongly limits the kinetics of adsorption (in agreement with the present experimental conditions), and neglects the backward flux of macromolecules (from the adsorbent to the bulk). In the model, the rate of adsorption is controlled by the migration of the molecule chains through the dense network of chain segments belonging to macromolecules that are already adsorbed. Macromolecules belonging to the two populations of adsorbed molecules possibly display different dynamic characteristics, as shown below.

14.3.3 EXCHANGE EXPERIMENTS AT CONSTANT BULK SOLUTION CONCENTRATION

Desorption of radioactive polymer did not occur in the presence of polymer-free solvent within the limits of experimental precision [13]. Conversely, at finite solution concentrations of polymer, a net flux of labeled polymer from surface to solution set in leading to apparition of radioactive polymer in the dispersion medium. Extrapolation of the radioactivity to time zero revealed a rapid, increasingly important rise in bulk radioactivity C_{eff}^* with increasing bulk solution concentration, as shown in Figure 14.3. Figure 14.3 also shows the variation with time of C_{eff}^* for solution concentrations of 2, 5, 10, and 17 mg/L at pH 4.0. The variation of C_{eff}^* as a function of time leads to a linear dependence of $-\mathrm{d}\ln N_s^*/\mathrm{d}t$ on the bulk concentration C_0, as shown in Figure 14.4. Initial increase and slow transfer of radioactive material demonstrate the existence of two populations. Finally, experiments using nonlabeled adsorbed polymer and radioactive bulk solution polymer showed the forward and backward fluxes to be equal [13]. One is led to conclude that the interfacial mobility of adsorbed molecules, including adsorbed chain segments, increases with the bulk concentration C_0 of the solution in contact with the interface.

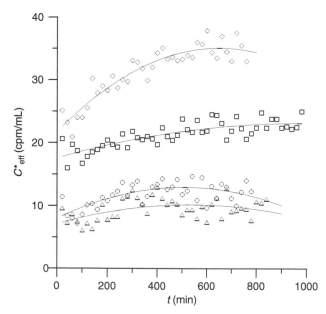

FIGURE 14.3 Representation as a function of time of the specific radioactivity C_{eff}^* of the effluent originating from the interfacial exchange with nonlabeled polyacrylamide of radio-labeled polyacrylamide adsorbed on alumino-silicate beads suspended in water at pH 4.0 and 25°C. In the different experiments, the constant total concentration of polyacrylamide was fixed at 2 (△), 5 (○), 10 (□), and 17 mg/L (◇).

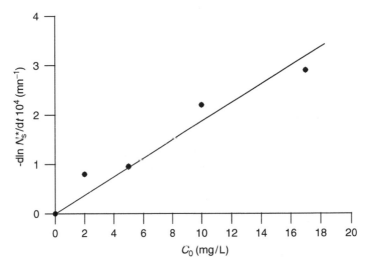

FIGURE 14.4 Representation of the specific exchange rate of radio-labeled poly-acrylamide as a function of the total polymer concentration in the reactor. The beads were dispersed in polymer solutions at 25°C and pH 4.0.

14.3.4 EXCHANGE EXPERIMENTS AT INCREASING BULK SOLUTION CONCENTRATION

Since the amount of radioactive polymer desorbed during the initial fast exchange increased with the solution concentration, the interfacial desorption of radioactive polymer was determined while the concentration C_{eff} was progressively increased by injection of a solution of concentration C_0, equal to 28 mg/L into the reactor containing the adsorbent initially dispersed in a polymer free system at a constant Jv of 0.05 mL/min. This experiment was repeated under different pH conditions since this parameter controls the amount of polymer adsorbed at equilibrium [24,29]. Figure 14.5 shows the variation with time of C_{eff}^* in experiments carried out at different pH. No desorption was observed at time zero at the given pH. As the polymer concentration C_{eff}^* progressively increases, the variation of $-(C_{eff}/C_{eff}^*) \times (dN_s^*/dt)$ is derived as a function of time. Assuming the existence of a fast interfacial exchange between labeled and non-labeled polymer, Equation 14.6 describes the expected variation of $-(C_{eff}/C_{eff}^*) \times (dN_s^*/dt)$ to be given by:

$$-\left(\frac{C_{eff}}{C_{eff}^*}\right)\left(\frac{dN_s^*}{dt}\right) = C_0 J_v + V C_{eff} \frac{d\ln}{dt}\left[\frac{C_{eff}^*}{C_{eff}}\right] \approx C_0 J_v \qquad (14.6)$$

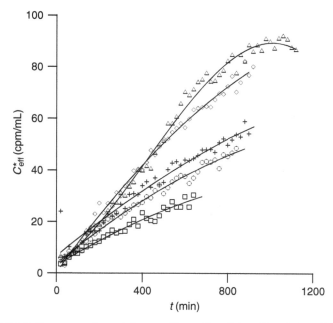

FIGURE 14.5 Representation of the specific radioactivity C^*_{eff} of the effluent as a function of time originating from interfacial exchange with non-labeled polyacrylamide of radio-labeled polyacrylamide adsorbed on alumino-silicate beads initially dispersed in water at 25°C and pH 3.0 ($\square$), 4.5 (+), 4.7 ($\diamond$), 5.5 ($\triangle$), and 7 ($\bigcirc$). The rate of supply of non-labeled polyacrylamide was 1.4×10^{-3} mg/min. The lines are polynomial fits which were employed to derive the rate of variation of C^*_{eff}.

In fact, the experimental variations of C^*_{eff} as a function of time shown in Figure 14.6 provides $-(C_{\text{eff}}/C^*_{\text{eff}}) \times (dN^*_s/dt)$ to be quite independent of time and equal to 1.4 ± 0.2 mg/min, in agreement with Equation 14.6.

14.3.5 CONCLUSION

A constant finding is that an absence of macromolecules in the supernatant precludes the desorption of adsorbed macromolecules. However, the system becomes unstable and the adsorbed polymer molecules appear to engage in a relatively fast exchange with bulk solution molecules with increasing polymer concentration in the supernatant.

Adsorption kinetics, interfacial exchanges between adsorbed and bulk solution polymer, have thus revealed the existence of two populations of adsorbed molecules when the situation is frozen at a constant concentration of the polymer in the bulk. Nevertheless, adsorbed molecules belonging to one population may exchange with molecules of the second population when the

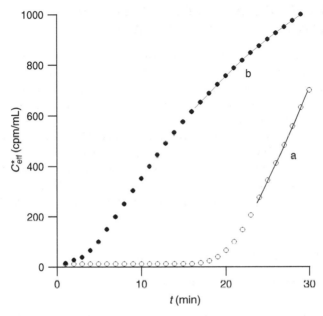

FIGURE 14.6 Representation of the specific radioactivity C_{eff}^* of the effluent as a function of time. Curve (a) corresponds to adsorption of hydrolyzed polyacrylamide following the injection of a solution of $C_0^* = 1894$ cpm/mL at $J_v = 0.429$ mL/min: the polymer adsorbed amounts to 6.16 mg/m². Curve (b) corresponds to adsorption of hydrolyzed polyacrylamide following the injection of a solution of $C_0^* = 1380$ cpm/mL at $J_v = 0.68$ mL/min: the polymer adsorbed amounts to 0.8 mg/m². The solid curves determine the period corresponding to the pure dilution regime (no adsorption).

situation thaws with increasing polymer solution concentration. One may conjecture that in the absence of bulk molecules, there is always an exchange between chains segments belonging to trains, loops, and tails of the same or a different adsorbed macromolecule, but no desorption occurs. When bulk solution molecules intervene in this exchange process, trains are preferentially replaced in the interface by chains belonging to solution molecules. When the system is frozen, a fast desorption affects a part of molecules being proportional to the bulk concentration and the rate of slow desorption is controlled by the polymer concentration in the supernatant. When the interfacial layer thaws under progressive increase in polymer concentration, the rate of desorption for all adsorbed molecules is fast, and a permanent renewal of adsorbed molecules sets in. This was visualized in the fast establishment of the isotopic equilibrium between labeled and nonlabeled macromolecules.

14.4 THE HYDROLYZED POLYACRYLAMIDE/ ALUMINOSILICATE/WATER SYSTEM

Figure 14.1 shows the adsorption of hydrolyzed polyacrylamide to increase with the degree of hydrolysis while the maximal adsorption is shifted towards pH values smaller than 4.5 [30]. The increased adsorption at lower pH values was attributed to the presence of negatively charged carboxylic acid groups which strongly interact with the positive groups of the aluminosilicate adsorbent. The depressed adsorption at higher pH values was attributed to the repulsive action of the negatively charged surface groups. Obviously, charge–charge attraction and repulsion strongly contribute to adsorption and correct the major interaction resulting from hydrogen bonding. These effects influence the adsorption mechanism and interfacial characteristics in a complex manner.

14.4.1 The Two-Step Adsorption Process

Three situations may be distinguished:

- At low pH, the net charge of the adsorbent is positive, and in addition to the usual $= $Al–OH-Acrylamide hydrogen bond responsible for the adsorption of neutral polyacrylamide [31,33], electrical forces between $=$Al$-$OH$_2^+$ and dissociated acrylic acid and acrylate groups contribute to increase the amount of polymer adsorbed. These interactions promote the adsorption of hydrolyzed polyacrylamide of a higher degree of hydrolysis and the adsorption equilibrium is rapidly established under such experimental conditions. As shown in Figure 14.6 (curve a), the polymer concentration C_{eff} in the bulk solution remains near zero for a relatively long time and adsorption stops rapidly. The shape of the variation of C_{eff} is characteristic of the mobile adsorption model [26,27].
- At high pH, the net charge of the adsorbent is negative and electrical forces between $= $Al–O$^-$ and acrylate groups tend to slow down the polymer adsorption. This leads to less adsorption of hydrolyzed polyacrylamide of higher degree of hydrolysis and the shape of curve b in Figure 14.6 is characteristic of the localized adsorption model.
- Near the zero point of charge, the more or less equal densities of positive and negative groups generate conflicting interactions which strongly inhibit the establishment of thermodynamic equilibrium, and the adsorption process successively resembles localized and mobile adsorption processes, as shown in Figure 14.7, where slow and fast adsorption regimes were determined to occur successively. This two-step adsorption process was attributed to interfacial ion-exchange reactions

involving poly-ions and small counter-ions of the adsorbent. In order to evidence this mechanism, an experiment was carried out using the radio-labeled polyion (β emitter) and radio-isotope $^{125}I^-$ (γ emitter) as tracer counter-ions of the adsorbent [19]. At time zero, the suspension at pH 5.0 contained a mixture of Cl^- and $^{125}I^-$, with the following stoichiometry:

$$[Cl^-] + [^{125}I^-] \cong mX = \left\lfloor = Al - OH_?^+ \right\rfloor \qquad (14.7)$$

Quantities in brackets are ionic concentrations in eq/mL of solution, m is the concentration in g/mL, and X is the number of Al groups per gram of adsorbent. The experiment was performed under controlled injection of radio-labeled polymer in the reactor. The concentrations of β and γ emitters were determined in the effluent as a function of time (Figure 14.7). It can be seen that the solution concentration of $^{125}I^-$ decreases strongly during the initial

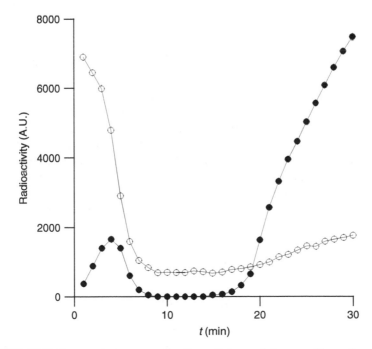

FIGURE 14.7 Representation as a function of time of the specific radioactivity (arbitrary unit) of the effluent for the hydrolyzed polyacrylamide (●) and the iodine ion (○) during adsorption on alumino-silicate beads initially dispersed in water at 25°C and pH 5.05.

slow adsorption of polyelectrolyte, but remains constant during the fast adsorption process and slowly increases as the adsorption slow down. This result may be interpreted as follows.

In the initial adsorption step, the negatively charged adsorbing polymer interacts with already existing $=Al-OH_2^+$ groups, and progressively the "wall" of the excess negatively charged carboxyl groups induces enhanced protonation of the neutral $= Al–OH$ surface groups. The resulting surface excess of positive charge is neutralized by rapid transfer of Cl^- and $^{125}I^-$ from the solution to the surface, as reflected by the decrease in the $^{125}I^-$ concentration in the bulk solution. Due to the preferential adsorption of polyelectrolyte, a supplementary transfer of polyions from solution to surface later leads to inversing the transfer of the small ions.

14.4.2 INTERFACIAL RECONFORMATION OF ADSORBED POLYELECTROLYTE

When isolated adsorbed polymer molecules sustain the strong attraction of the adsorbent for a relatively long time, conformational relaxation (reconformation and flattening) is expected to occur. At the impact time $t = 0$, the area $S(0)$ occupied by the macromolecule may be considered to be equal to the cross-sectional area of the solute macromolecule. The surface area $S(t)$ continuously increases with adsorption time t due to the interfacial reconformation. In the reactor method, macromolecules are continuously injected into the bulk—a given number of molecules at time intervals of Δt—and adsorbed without delay since no polymer was detected in the effluent at low pH values (see Figure 14.6, curve a). In this situation, the number of polymers which can be deposited on the unit surface area was determined assuming that full surface coverage is achieved when the N_sth polymer has been adsorbed.

Comparison of the specific adsorptions in model and experiments gives the following equations for the initial (0) and equilibrium (eq) situations. (eq) means that the relaxation of one adsorbed polymer approaches complete flattening in the absence of nearest neighbors.

$$(N_s - 1)\Delta t = N_s/J_v C_0 \qquad (14.8)$$

$$S(0) = [N_s(0)]^{-1} \qquad (14.9)$$

$$S(eq) = [N_s(eq)]^{-1} \qquad (14.10)$$

Since reconformation is expected to stop after a given adsorption period due to the limited size of the polymer chain, the variation with time of the surface area $S(t)$ is given by [20]:

$$\frac{S(t) - S(\text{eq})}{S(0) - S(\text{eq})} = \exp\left(-\frac{t}{\tau_R}\right) \tag{14.11}$$

where $S(\text{eq})$ corresponds to the area of one adsorbed polymer when no neighbor prevented full flattening. The final surface coverage N_s is related to the experimental parameters J_v and C_0 and the polymer characteristics $N_s(0)$, $N_s(\text{eq})$, and τ_R (min) are obtained through

$$\frac{\left(N_s/N_s(0)\right) - \left(N_s(\text{eq})/N_s(0)\right)}{1 - \left(N_s(\text{eq})/N_s(0)\right)}$$

$$= \frac{J_v C_0 \tau_R}{N_s(0)} \left\{ 1 - \exp\left(-\frac{N_s/N_s(0)}{J_v C_0 \tau_R/N_s(0)}\right) \right\} \tag{14.12}$$

Relative values of $N_s/N_s(0)$ are expressed as a function of $J_v C_0 \tau_R/N_s(0)$ in Equation 14.12 in order to illustrate the interdependence of experimental parameters and polymer characteristics. A typical experiment is analyzed using Equation 14.12, and the results are reported in Figure 14.8. Comparison of the experimental points with the different representations of $N_s/N_s(0)$ as a function of $J_v C_0 \tau_R/N_s(0)$ gives $S(\text{eq})/S(0) \approx 50$ and τ_R is determined to be close to 700 min [24].

Relaxation phenomena were observed during the slow adsorption regime near the point of zero charge of the adsorbent (Figure 14.7). In this situation, Equation 14.8 is no longer valid since the injected molecules exert several attempts before adsorption succeeds. Therefore, the growing disc model was developed to take into account that adsorbed molecules may sustain flattening for longer times since the establishment of the interfacial layer requires much more time than it did in the previous situation [34,35]. In order to evaluate the correcting time and coverage factors which have to be applied when a certain proportion of the injected macromolecules are not immediately adsorbed and further attempts have to be made, the rate of adsorption was assumed to be limited by the surface occupation as derived from the random sequential adsorption model. The limitation in surface coverage then results from polymer reconformation, as defined by the growing disc model.

Flattening of a macromolecule begins at the time of its contact with the solid surface. In simulation, the flattening of the ith adsorbed polymer (disc) is

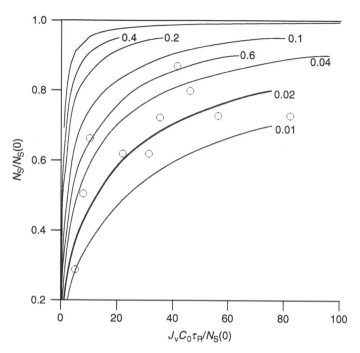

FIGURE 14.8 Representation of the reconformation $N_s/N_s(0)$ of the adsorbed hydrolyzed polyacrylamide as a function of $J_vC_0\tau_R/N_s(0)$. The different curves are generated using $N_s(eq)/N_s(0) \to \infty$ (upper dashed curve), and the values of $N_s(eq)/N_s(0)$ are indicated on the solid curves according to Equation 14.12. Open circles correspond to hydrolyzed polyacrylamide adsorbed on alumino-silicate beads initially dispersed in water at 25°C and pH 4.4. The parameters of the lower dashed curve were used to derive the reconformation characteristics of the polymer.

expressed in terms of its radius $r(i, t')$ by [34]:

$$r(i,t') = r\left\{\frac{S_{eq}}{S_0} + \left[1 - \frac{S_{eq}}{S_0}\right]\exp\left(-\frac{t'}{n\tau_R}\right)\right\}^{1/2} \qquad (14.13)$$

where t' is zero at the moment the disc comes into contact with the surface. In the algorithm, t' is incremented by one unit when n successively injected discs have attempted to adsorb on the surface, n representing the number of macromolecules supplied to the surface per unit time t'. This scheme is similar to the adsorption of one macromolecule with relaxation time $n\tau_R$.

The general conclusions are as follows:

- N_s increases with $J_vC_0\tau_R$. In a given experiment (constant rate of polymer supply), Ns increases with τ_R. Hence, if the reconformation is slow, the area available for further adsorption decreases slowly. Similarly, for a given polymer, Ns increases with the rate of polymer supply and maximal coverage is obtained with concentrated solutions.

- N_s increases as $S(\text{eq})/S(0)$ decreases. A reconformation of small amplitude leads to relatively slow surface saturation, whereas wide spreading of the macromolecules results in rapid surface blocking.

14.4.3 CONCLUSIONS

Adsorption characteristics of polyelectrolytes on oppositely charged surfaces have been thoroughly investigated [36–39]. However, the present paragraph addressed the mechanism of formation and the evolution of such systems with time. The present conclusions are relevant to systems for which the main adsorption results from a non-electrical specific interaction (such as hydrogen bonding in the present system, of hydrophobic interactions for proteins), which is corrected by effects of attractive or repulsive charge–charge interactions.

- When the surface charge density is small, attraction is exerted between punctual surface sites and polymer groups. The rate of establishment of the surface layer accords with the localized adsorption model, for which the adsorption only succeeds after many attempts—the number of unsuccessful attempts increasing with the progress of the surface coverage.
- When the surface charge density of the adsorbent is great, attractions are exerted between the "wall" of the surface sites and the polyions. This leads to a much stronger attraction and the rate of establishment of the layer corresponds to that derived from the mobile adsorption model. Up to coverage of one third, molecules systematically adsorb at the first attempt.
- When the net surface charge is close to zero, adsorption develops firstly with characteristics of localized adsorption and secondly with that of the mobile adsorption process. The transition from the first to the second step corresponds to the generation of new surface charges by charge induction effects.

For the hydrolyzed polyacrylamide/aluminosilicate/water system, the rate of polymer reconformation is slow while the extent of the process is extremely large. The experimental variation of the amount of polymer adsorbed as a function of the rate of polymer supply leads to an estimate of the amount of polymer adsorbed to be a fifty-fold decrease at infinitely slow polymer supply. This result calls for the following comment: actually, at instantaneous polymer supply (when the polymer solution is poured into the adsorbent dispersion and the system is rapidly homogenized), the adsorbed polymer amounts to

12 mg/m^2, which decreases to 0.24 mg/m^2 at flattening completion; the latter value corresponding to the adsorption value of polyelectrolytes at adsorption equilibrium [40,41].

14.5 THE HYDROLYZED POLYVINYLAMINE/CELLULOSE FIBER/WATER SYSTEM

Cellulose fibers present surface and internal carboxylic and some other negatively charged groups [42,43]. Polyvinylamine is composed of positively charged chain segments and strong electrical interactions were expected to occur on mixing the two materials. Due to the high molecular weight of the polymer, only polymer surface interactions were possible [43–45].

14.5.1 RECONFORMATION DURING ESTABLISHMENT OF POLYMER LAYER

The amount of polyvinylamine adsorbed on cellulose fiber was determined with the aid of the reactor device by altering various parameters [46]. The surface area available to polymer adsorption was first modified while holding constant the other experimental variables. In a second set of experiments, the mass of dispersed fibers was held constant while the rate of polymer supply was controlled by varying either the polymer concentration C_0 or the injection rate J_v. In all of these situations, the various amounts of adsorption N_s were determined for $K(N_s) = 0.01$ min^{-1} (near complete surface coverage) using the representation of $K(N_s)$ as a function of N_s. Under these conditions, it was equally difficult to put new molecules on the polymer coated fiber surface containing N_s molecules.

14.5.1.1 Influence of the Surface Area Available to Polymer Adsorption

A special situation is shown in Figure 14.9 which represents the kinetic coefficient $K(N_s)$ as a function of the degree of surface coverage N_s when adsorption experiments were carried out using 0.017, 0.05, or 0.09 g cellulose fibers as adsorbent while taking constant the rate of polymer supply. For $K(N_s) = 0.01$, the specific adsorption N_s unusually is a function of the surface area available. Moreover, transforming N_s in the amount of polymer adsorbed on the mass x of fiber dispersed in the reactor leads to a single total adsorption of 0.055 mg of polymer. This result also suggests that adsorbed macromolecules relax at the solid–liquid interface, but reconformation should be much faster than for the previous system since polymer injection only lasted 30 min.

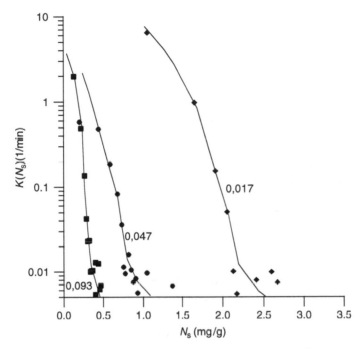

FIGURE 14.9 Representation of the kinetic coefficient $K(N_s)$ (min^{-1}) as a function of the amount of polyvinylamine adsorbed N_s (mg/g cellulose). The mass x (g) of cellulose dispersed in the reactor is indicated on the curve.

In the present situation, only a small portion of the injected molecules were determined to adsorb rapidly and progressively; the liquid phase became enriched in polymer. Since a quite similar adsorption level was observed independently of the dispersed fiber mass, thermodynamic and kinetic factors were expected to control the adsorption. Actually, polymer adsorbed on a large surface is able to reconform for a long time before neighboring adsorbed molecules impede the process, and thus the calculated specific adsorption N_s is small. Conversely, since coverage of a small surface area is accelerated, the reconformation is less extended and becomes rapidly blocked and the calculated specific adsorption is high.

14.5.1.2 Influence of the Rate of Polymer Supply

14.5.1.2.1 On the Amount of Polymer Adsorbed at the End of the Period of Injection

Adsorption values N_s corresponding to $K(N_s)$ equal to 0.01 min^{-1} were derived by varying the rate of polymer supply while maintaining the fiber

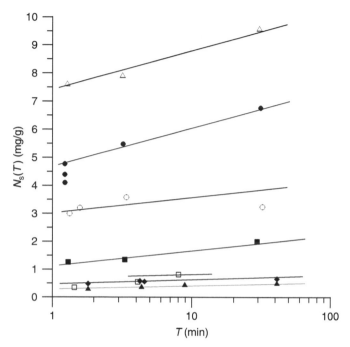

FIGURE 14.10 Representation of the amount of polyvinylamine adsorbed $N_s(T)$ as a function of the adsorption period T. The different curves correspond to the different concentrations C_0 (mg/mL) of 0.02 ($\blacktriangle$), 0.04 ($\blacklozenge$), 0.08 ($\square$), 0.16 ($\blacksquare$), 0.31 ($\bigcirc$), 0.6 ($\bullet$), and 1.2 ($\triangle$).

mass constant. For a given concentration of the polymer solution being injected, the injected volume was constant and the different injection periods T only resulted from changes in the injection rate J_v. Figure 14.10 shows the specific adsorption N_s as a function of injection time T for different polymer concentrations C_0.

Adsorption figures varied in a very large range, and in taking into account that with all of these figures the adsorption probability fell back to a constant low value, one has to assume that different injection conditions lead to different interfacial conformations of the adsorbed polymer. This only can result from fast and extensive reconformation during layer establishment. Furthermore, the rate of injection was found to not be a parameter of paramount importance. As shown in Figure 14.11, the amount of polymer adsorbed mainly depended on the amount of polymer that was supplied to the fibers during the period T.

$$N_s(T) = 0.4[J_v C_0 T]^{0.85} \qquad (14.14)$$

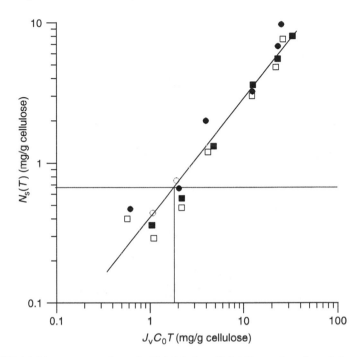

FIGURE 14.11 Representation of $N_s(T)$ (mg/g cellulose) as a function of the amount of polyvinylamine injected $J_v C_0 T$ (mg/g cellulose). The different symbols correspond to the different concentrations C_0 (mg/mL) of 0.02 ($\blacktriangle$), 0.04 ($\blacklozenge$), 0.08 ($\square$), 0.16 ($\blacksquare$), 0.31 ($\bigcirc$), 0.6 ($\bullet$), and 1.2 ($\triangle$). The dashed line gives the amount adsorbed at adsorption equilibrium.

14.5.1.2.2 On the Kinetics of Adsorption

The variation of N_s as a function of time has been thoroughly investigated under various rates of polymer supply which were implemented w separately setting the rate of injection J_v and the concentration C_0 of the injected polymer solution. Due to the progressive increase of the polymer concentration C_{eff} in the bulk solution within the reactor during injection, it was not possible to ascertain whether a given analytical function holds for the kinetics of adsorption. The best fit of the experimental variation of N_s vs. t was found to be $N_s(t) \propto \log t$, as evidenced in Figure 14.12 for injection at the constant rate of 0.06 mL/min of polymer solutions of different concentrations C_0. Finally, the specific influences of J_v and C_0 were separately extracted from the experimental results and Equation 14.15 holds for the kinetics of adsorption.

$$N_s(J_v, C_0, t) \propto \frac{C_0^{0.78}}{J_v^{0.12}} \log t \qquad (14.15)$$

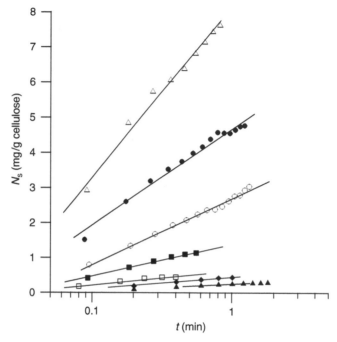

FIGURE 14.12 Representation of the adsorption N_s (mg/g cellulose) as a function of time t (log-normal scale) for injection of solutions of different polymer concentrations C_0 (mg/mL): The different curves correspond to the different concentrations C_0 (mg/mL) of 0.02 (▲), 0.04 (◆), 0.08 (□), 0.16 (■), 0.31 (○), 0.6 (●), and 1.2 (△).The solid lines accord to Equation 14.15.

Fast adsorption requires injection of solutions of high concentration C_0 at relatively slow rates J_v, although the latter parameter only exerts a minor role as estimated from the results presented in Figure 14.11.

For the present system, fast reconformation during adsorption strongly perturbs the kinetics of layer establishment since usually the rate of injection J_v and the polymer concentration C_0 similarly intervene in the rate of polymer supply $J_v C_0$ [26].

14.5.2 Relaxation Phenomena within the Interfacial Polymer Layer

Adsorption isotherms showed that the adsorption at equilibrium (the plateau values were determined for concentrations belonging to the range of C_{eff} values realized in reactor experiments after the various injection periods T) was close to 0.66 mg/g cellulose fibers (see the dashed horizontal line in Figure 14.11). This information led to determining the mechanism of the layer relaxation, which strongly reduces the adsorption levels to lower equilibrium values.

Therefore, SAEC experiments were implemented, but for technical reasons—cellulose fibers were replaced with glass fibers which are available in the form of filters. Interactions between the latter material and cellulose may be assumed of similar nature.

Typical experiments were carried out by setting the amount of polymer which was injected in the column to 0.16 ± 0.02 mg. The parameters which were varied are the polymer concentration C_{el} and the rate of injection J_v, and Figure 14.13 shows three typical chromatograms. The actual shapes do not agree with those previously obtained from the numerical simulation study or experiments with nonreconforming polymers (dashed lines) [40,47].

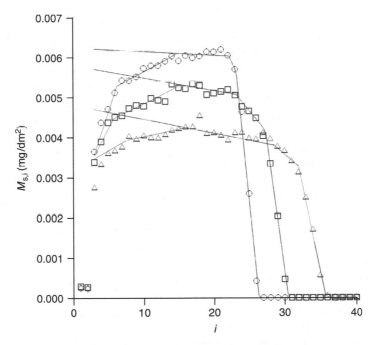

FIGURE 14.13 Representation of the histogram (adsorption of polyvinylamine per filter as a function of the filter number (i) corresponding to injections of a constant amount of polymer close to 0.157 mg during a constant period of 120 min under the following conditions for the concentration of the elutant C_{el} (mg/100 mL) and the injection rate (mL/min), respectively: 2 and 0.0578 (○), 1 and 0.144 (□), and 0.5 and 0.281 (△). The $M_{s,i}$ values extrapolated from the dashed lines to $i = 3$ (first adsorbing filter) were employed to derive $\sigma(0)$.

As indicated above, the different histograms, although being of similar shape, present characteristics that are functions of the concentration C_{el} in the elutant and the rate of injection J_v. However, the adsorption values determined for $i = 3$ (the first adsorbing filter) tend to a single smaller value of 0.0035 mg/dm^2 independently of the duration of the chromatography experiment varying from 22.5 (results not shown) to 117 min. Usually the adsorption is expected to be maximal on the first adsorbing filter (as schematized by the dashed line), to monotonously decrease with i and to drop sharply (as observed in Figure 14.13 as well). Maximal adsorption at $i = 3$ (dashed line) thus indicates jamming coverage to be established.

The experimental histogram results from the fact that the adsorbed layer sustains relaxation and that individual macromolecules reconform through expulsion from the layer of exceedingly adsorbed molecules (overshoot). Although changes in the amount of macromolecules adsorbed per filter i towards a single value close to 0.0035 mg/dm^2 are attributed to a great variation in the period of elution, the process cannot be simply interpreted on the basis of a time dependent relaxation of individual macromolecules. Characteristics of the liquid phase equilibrating the polymer interface should exert a major role in the layer relaxation. Actually, as indicated in Section 14.3.2 and Section 14.3.3, the stability of polymer interfaces (towards the surface to bulk exchange) is a function of the layer environment and becomes modified when the environment is progressively changed [24]. Accordingly, the interfacial exchange between adsorbed (radio-labeled) and solute molecules (nonlabeled) was determined to be a function of the concentration of the polymer solution facing the interface [13]. Therefore, flattening of entangled macromolecules is supposed to result from desorption of some macromolecules. This process is facilitated by the presence of solute macromolecules, possibly because the interfacial mobility of chain segments belonging to trains increases with the concentration of the solute molecules. Interfacial mobility supposes a permanent exchange between chain segments in trains and loops (or tails), which is possibly mediated by ion-exchange between discrete charged chain segments of the poly-ion and small counterions. Accordingly, the extent of reconformation should be a function of (i) the period for which the polymer interface effectively faces non-adsorbed polymer molecules and (ii) the concentration of these molecules. In order to take into account these two parameters, extent of reconformation is expressed as a function of the mass $J_v C_{el} t$ of the solute molecules, which effectively faces the adsorbed molecules at the different filters for a given period t. The period t is calculated taking into account the total period of injection and the average time required for filling each of the successive filters. $t = 0$ characterizes the filter i, for which the radioactivity drops to zero—molecules adsorbed on this last filter were never in contact with bulk macromolecules.

Macromolecular surface area relaxation is expressed by

$$\frac{\sigma}{\sigma_0} = \frac{M_{si0}}{M_{si}} = F(J_v C_{el} t) \qquad (14.16)$$

σ represents the area occupied by one macromolecule at a given filter when the polymer adsorbed on that filter has been in contact with an amount of $J_v C_{el} t$ solute macromolecules. For $t = 0$, the interface has not been in contact with solute macromolecules since the injected molecules only served to cover the filter area, and σ_0 thus represents the area attributed to one macromolecule before reconformation. Adsorption M_{si0} corresponding to values of the dashed lines shows that the amount of polymer adsorbed prior to relaxation slowly decreases with filter number i.

The temporal variation of σ/σ_0 as a function of $J_v C_{el} t$ quantitatively describes the flattening of the adsorbed macromolecule in the presence of neighbor molecules. Results corresponding to SAEC experiments for different injection times are summarized and a single curve of σ/σ_0 as a function of $J_v C_{el} t$ is shown. This observation implies that the extent and rate of reconformation are determined by the amount of polymer being supplied to the system for a given time (Figure 14.14).

14.5.3 CONCLUSION

The present system manifests characteristics that are certainly common to various charged adsorbent/polyelectrolytes systems. The originality of the present system is that reconformation processes are relatively fast without being instantaneous. Very different interfacial figures may be established by controlling the polymer supply, and interestingly, these figures may be saved when the interfaces are rapidly isolated from the bulk molecules. This assumption is based on SAEC experiments showing that interfaces do not relax in the absence of solute macromolecules. Actually, surface to bulk solution transfers were determined to be blocked too in the absence of solute molecules and the two phenomena should be related.

The major information gained is that reconformation affecting isolated macromolecules during establishment of the layer is a characteristic of the polymer molecules and that the extent of reconformation only depends on the area available to flattening. In addition, reconformation stops when the surface area is coated. Relaxation affecting established interfacial layer proceeds differently and requires the presence of solute molecules for a given time. Surface to bulk transfers were observed only in the presence of molecules in the liquid phase, and these two phenomena should be related as well. Finally, layer relaxation cannot be dissociated from concomitant surface to bulk and inverse transfers and polymer desorption.

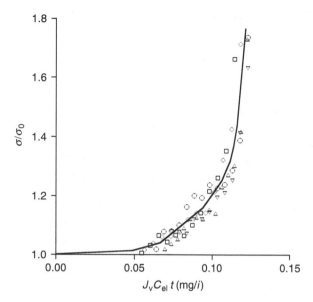

FIGURE 14.14 Representation of σ/σ_0 as a function of $J_v C_0 t$ (mg), which represents the amount of polymer supplied to the various filter i (mg/filter i) for experiments lasting 120 min ($\diamond$) (∇) ($\triangle$), 22.5 min ($\square$), and 15.1 min ($\bigcirc$). On the abscissa, the time $t =$ zero applies to the filter characterized by the drop in adsorption whereas the maximal contact period t applies to filter $i = 3$.

14.6 CONCLUDING REMARKS

Radio-labeling of macromolecules which slightly modifies one millionth of polymer chain segments and is usually without specific effects on the adsorption characteristics obviously constitutes the most efficient way to investigate the interfacial adsorption phenomena, although other methods have been implemented to the same objectives. The present results always concerned the relaxation characteristics of homogeneous layers, but the method was recently applied to heterogeneous layers composed of macromolecules of different molecular weight, hydrolysis grade, and/or conformational characteristics [48].

Results presented in this chapter do not concern the polyvinylpyridine/charged polystyrene latex/water system, which was also investigated in our group [20,34]. However, slow reconformation of polyvinylpyridine was determined to strongly influence the development of aggregation of lattices in the presence of polymer [49–51]. Fast reconformation was previously determined to limit the efficiency of orthokinetic aggregation when mixing of the constituent was delayed in time [52]. Additionally, slow internal redistribution of chain segments was determined to strongly modify the colloidal stability of hydrolyzed polyacrylamide/aluminum oxide/water systems for weeks [53,54].

One of the major advantages of employing reconforming polymers concerns the possibility to generate interfacial polymer systems characterized by diverse coatings and layer conformations. As diverse means, coating and conformation may sustain designed modifications with time under a controlled environment, or may be frozen at any step by washing out the nonadsorbed molecules. This may be particularly interesting in the manufacturing of polymer multilayers [55].

A second point concerns the synthesis of new organic/inorganic materials, which offers new opportunities to create original materials [56–58]. Nothing is known about systems for which the solid substrate is extremely thin and able to sustain some local curving on adsorption of a single macromolecule. Opportunity to control crystal growth by adding reconforming polymer has not been investigated at present.

Finally, it is worth mentioning that polymer adsorption studies have scarcely concerned the long term behavior, in spite of the fact that unexpected phenomena have been described in the literature [8,59–64].

ACKNOWLEDGMENTS

This research was supported by the IFP (Rueil-Malmaison, France), the BASF Aktiengesellschaft (Ludwigshafen, Germany), and the CNRS. The author is particularly indebted to A. Carroy, A.-C. Jean Chronberg, A. Shulga and J. Widmaier. G. Chauveteau (IFP), S. Champ, and H. Auwet (BASF Aktiengesellschaft) are acknowledged for stimulating discussions.

REFERENCES

1. Cohen Stuart, M. A., Cosgrove, T., and Vincent, B., Experimental aspects of polymer adsorption at solid/solution interfaces, *Adv. Colloid Interface Sci.*, 24, 143–239, 1986 (and references cited).
2. Leger, L., Hervet, H., and Rondelez, F., Reptation in entangled polymer solutions by forced Rayleigh light scattering, *Macromolecules*, 14, 1732–1738, 1981.
3. Dejardin, P., Variation with temperature of the thickness of an adsorbed polymer layer in the collapsed state, *J. Phys.*, 44, 537–542, 1983.

4. Kawaguchi, M., Hayakawa, K., and Takahashi, A., Adsorption of polystyrene onto a metal surface in good solvent conditions, *Macromolecules*, 16, 631–635, 1983.

5. Blum, F. D., Magnetic resonance of polymers at surfaces, *Colloids Surf.*, 45, 361–376, 1990.

6. Caucheteux, I., Hervet, H., Jerome, R., and Rondelez, F., Investigation by optical evanescent waves of the structure of adsorbed polymer layer in good solvents, *J. Chem. Soc. Faraday Trans.*, 86, 1369–1375, 1990.

7. Cosgrove, T., Prestidge, C. A., and Vincent, B., Chemisorption of linear and cyclic polymethylsiloxanes on alumina studied by Fourier-transform infrared spectroscopy, *J. Chem. Soc. Faraday Trans.*, 86, 1377–1382, 1990.

8. Frantz, P. and Granick, S., Kinetics of polymer adsorption and desorption, *Phys. Rev. Lett.*, 66, 899–902, 1991.

9. Dijt, J. C., Cohen Stuart, M. A., and Fleer, G. J., Kinetics of polymer adsorption and desorption in capillary flow, *Macromolecules*, 25, 5416–5423, 1992.

10. Enriquez, E. P. and Granick, S., Colloids surfaces A: Chain flattening and infrared dichroism of adsorbed poly(ethylene oxide), *Colloids Surf. A*, 113, 11–17, 1996.

11. Senden, T. J., Di Meglio, J. M., and Silberzan, I., The conformation of adsorbed polyacrylamide and derived polymers, *C.R. Acad. Sci. Paris*, t. 1 (Série IV), 1143–1152, 2000.

12. Afif, A., Chikhi, M., Hommell, H., and Legrand, A. P., Transport of a poly(ethylene oxide) solution through porous coated silica studied by spin labeling, *Polymer*, 42, 2711–2715, 2001.

13. Pefferkorn, E., Carroy, A., and Varoqui, R., Dynamic behavior of flexible polymers at a solid:liquid interface, *J. Polym. Sci.: Polym. Phys. Ed.*, 23, 1997–2008, 1985.

14. Johnson, H. E., Douglas, J. F., and Granick, S., Topological influences on polymer adsorption and desorption dynamics, *Phys. Rev. Lett.*, 70, 3267–3270, 1993.

15. Douglas, J. F., Frantz, P., Johnson, H. E., and Schneider, H. M., Regimes of polymer adsorption–desorption kinetics, *Colloids Surf. A*, 86, 251–254, 1994.

16. Pefferkorn, E., Carroy, A., and Varoqui, R., Adsorption of polyacrylamide on solid surfaces. Kinetics of the establishment of adsorption equilibrium, *Macromolecules*, 18, 2252–2258, 1985.

17. van de Ven, T. G. M., Kinetics of polymer and polyelectrolyte adsorption on surfaces, *Adv. Colloid Inter. Sci.*, 48, 121–140, 1994.

18. Pefferkorn, E., Jean-Chronberg, A-C. , and Varoqui, R., Structural relaxation modes of polyelectrolytes adsorbed at a solid–liquid interface, *C.R. Acad. Sci. Paris*, 308 (Série II), 1203–1208, 1989.

19. Pefferkorn, E., Jean-Chronberg, A-C. , and Varoqui, R., Conformational relaxation of polyelectrolytes at a solid–liquid interface, *Macromolecules*, 23, 1735–1741, 1990.

20. Pefferkorn, E. and Elaissari, A., Adsorption–desorption in charged polymer/colloid systems. Structural relaxation of adsorbed macromolecules, *J. Colloid Inter. Sci.*, 138, 187–194, 1990.

21. Johnson, H. E., Clarson, S. J., and Granick, S., Overshoots as polymers adsorb, *Polymer*, 34, 1960–1962, 1993.

22. Pefferkorn, E., Fragmentation of colloidal aggregates by polyelectrolyte adsorption, In *Physical Chemistry of Polyelectrolytes*, Radeva, T., Ed., Surfactant Science Series 99, Marcel Dekker, New York, pp. 509–566, 2001.

23. Pefferkorn, E., Ringenbach, E., and Elfarissi, F., Aluminum ions at polyelectrolyte interfaces. III. Role in polyacrylic acid/aluminum oxide and humic acid/kaolinite aggregate cohesion, *Colloid Polym. Sci.*, 279, 498–505, 2001.

24. Pefferkorn, E., Polyacrylamide at solid–liquid interfaces, *J. Colloid Interface Sci.*, 216, 197–220, 1999.

25. Pefferkorn, E., Haouam, A., and Varoqui, R., Thermodynamic and kinetic factors in adsorption of polymer on a plane lattice, *Macromolecules*, 21, 2111–2116, 1988.

26. Elaissari, A., Haouam, A., Huguenard, C., and Pefferkorn, E., Kinetic factors in polymer adsorption at solid–liquid interfaces. Methods of study of the adsorption mechanism, *J. Colloid Interface Sci.*, 149, 68–83, 1992.

27. Pefferkorn, E., Elaissari, A., and Huguenard, C., Adsorption processes in surface area exclusion chromatography, In *Interfacial Phenomena in Chromatography*, Pefferkorn, E., Ed., Surfactant Science Series 80, Marcel Dekker, New York, pp. 329–386, 1999.

28. Geffroy, C., Labeau, M. P., Wong, K., Cabane, B., and Cohen Stuart, M. A., Kinetics of adsorption of polyvinylamine onto cellulose, *Colloids Surf. A*, 172, 47–56, 2000.

29. Carroy, A., Thesis, Conformation et dynamique des chaînes de polyacrylamide à l'interface solide–liquide, Université Louis Pasteur, Strasbourg, 1986.

30. Pefferkorn, E., Jean-Chronberg, A-C., Chauveteau, G., and Varoqui, R., Adsorption of hydrolyzed polyacrylamide onto amphoteric surfaces, *J. Colloid Interface Sci.*, 137, 66–74, 1990.

31. Pefferkorn, E., Nabzar, L., and Carroy, A., Adsorption of polyacrylamide to Na kaolinite: Correlation between clay structure and surface properties, *J. Colloid Interface Sci.*, 106, 94–103, 1985.

32. de Gennes, P. G., Some dynamical features of adsorbed polymers, In *Molecular Conformation and Dynamics of Macromolecules in Condensed Systems*, Nagasawa, M., Ed., Studies in Polymer Systems, Vol. 2, Elsevier, Amsterdam, pp. 315–331, 1988.

33. Nabzar, L. and Pefferkorn, E., An experimental study of kaolinite crystal edge—polyacrylamide interactions in dilute suspensions, *J. Colloid Interface Sci.*, 108, 243–248, 1985.

34. Elaissari, A. and Pefferkorn, E., Polyelectrolyte adsorption at solid/liquid interfaces: A simple model for the structural relaxation and excluded area effects, *J. Colloid Interface Sci.*, 143, 85–91, 1991.

35. van Eijk, M. C. P., Cohen Stuart, M. A., Rovillard, S., and DE Coninck, J., Adsorption and spreading of polymers at plane interfaces; theory and molecular dynamics simulations, *Eur. Phys. J. B*, 1, 233–244, 1998.

36. Papenhuijzen, J., van der Schee, H. A., and Fleer, G. J., Polyelectrolyte adsorption. I. A new lattice theory, *J. Colloid Interface Sci.*, 104, 540–552, 1985.

37. Böhmer, M. R., Evers, O. A., and Scheutjens, J. M. H. M., Weak polyelectrolytes between two surfaces: Adsorption and stabilization, *Macromolecules*, 23, 2288–2301, 1990.

38. Blaakmeer, J., Böhmer, M. R., Cohen Stuart, M. A., and Fleer, G. J., Adsorption of weak polyelectrolytes on highly charged surfaces. Poly (acrylic acid) on polystyrene latex with strong cationic groups, *Macromolecules*, 23, 2301–2309, 1990.

39. van de Steeg, H. G. M., Cohen Stuart, M. A., de Keizer, A., and Bijsterbosch, B. H., Polyelectrolyte adsorption: A subtle balance of forces, *Langmuir*, 8, 2538–2546, 1992.

40. Huguenard, C., Widmaier, J., Elaissari, A., and Pefferkorn, E., Adsorption of polyelectrolytes on chromatographic columns. Simulated and experimental concentration profiles, *Macromolecules*, 30, 1434–1441, 1997.

41. Alince, B., Vanerek, A., and van de Ven, T. G. M., Effects of surface topography, pH and salt on the adsorption of polydisperse polyethyleneimine onto pulp fibers, *Ber. Bunsenges Phys. Chem.*, 100, 954–962, 1996.

42. van de Ven, T. G. M., A model for the adsorption of polyelectrolytes on pulp fibers: Relation between fiber structure and polyelectrolyte properties, *Nordic Pulp. Paper Res. J.*, 15, 494–501, 2000.

43. Gellersttedt, F. and Gatenholm, P., Surface properties of lignocellulosic fibers bearing carboxylic groups, *Cellulose*, 6, 103–121, 1999.

44. Wagberg, L., Polyelectrolyte adsorption onto cellulose fibres—A review, *Nordic Pulp. Paper Res. J.*, 15, 586–597, 2000.

45. Wagberg, L. and Ödberg, L., Polymer adsorption on cellulosic fibers, *Nordic Pulp. Paper Res. J.*, 2, 135–140, 1989.

46. Shulga, A., Widmaier, J., Pefferkorn, E., Champ, S., and Auweter, H., Kinetics of adsorption of polyvinylamine on cellulose fibers. I. Adsorption from salt free solutions., *J. Colloid Interface Sci.*, 258, 219–227, 2003.

47. Elaissari, A., Chauveteau, G., Huguenard, C., and Pefferkorn, E., Surface area exclusion chromatography. Influence of localized and mobile adsorption processes, *J. Colloid Interface Sci.*, 173, 221–230, 1995.

48. Oulanti, O., Pefferkorn, E., Champ, S., and Auweter, H., Relaxation phenomena of hydrolyzed polyvinylamine molecules adsorbed at the silica/water interface. II. Saturated heterogeneous polymer layers, *J. Colloid Interface Sci.*, 291, 105–111, 2005.

49. Elaissari, A. and Pefferkorn, E., Polyelectrolyte induced aggregation of latex particles: Influence of the structural relaxation of adsorbed macromolecules on the colloid aggregation mode, *J. Colloid Interface Sci.*, 141, 522–533, 1991.

50. Cohen Stuart, M. A., Adsorbed polymers in colloidal systems, *Polymer J.*, 23, 669–682, 1991.
51. Pelssers, E. G. M., Cohen Stuart, M. A., and Fleer, G. J., Kinetics of bridging flocculation. Role of relaxations in the polymer layer, *J. Chem. Soc. Faraday Trans.*, 86, 1355–1361, 1990.
52. Wagberg, L., Ödberg, L., and Glad-Nordmark, G., Conformation of adsorbed polymers and flocculation of microcrystalline cellulose and pulp suspensions, In *Fundamental of Papermaking*, Baker, C. F. and Punton, V. W., Eds., Mechanical Engineering Publications Limited, London, pp. 413–435, 1990.
53. Ringenbach, E., Chauveteau, G., and Pefferkorn, E., Aggregation/fragmentation of colloidal alumina. I Role of the adsorbed polyelectrolyte, *J. Colloid Interface Sci.*, 172, 203–207, 1995.
54. Ringenbach, E., Chauveteau, G., and Pefferkorn, E., Aggregation/fragmentation of colloidal alumina. II. Scaling laws of fragmentation, *J. Colloid Interface Sci.*, 172, 208–213, 1995.
55. Decher, G., Hong, J-D. , and Schmitt, J., Buildup of ultrathin multilayer films by a self-assembly process: III. Consecutively alternating adsorption of anionic and cationic polyelectrolytes on charged surfaces, Thin Solid Films, 210/211, 831–835, 1992.
56. Jada, A. and Pefferkorn, E., Smooth and rough spherical calcium carbonate particles, *J. Mat. Sci. Lett.*, 19, 2077–2079, 2000.
57. Rotstein, H. G. and Tannenbaum, R., Cluster coagulation and growth limited by surface interactions with polymers, *J. Phys. Chem. B*, 106, 146–151, 2002.
58. Amjad, Z., Ed., *Advances in Crystal Growth Inhibition Technologies*, Kluwer Academic/Plenum Publishers, Dordrecht, 2002.
59. Johnson, H. E. and Granick, S., New mechanism of nonequilibrium polymer adsorption, *Science*, 255, 966–968, 1992.
60. Frantz, P. and Granick, S., Exchange kinetics of adsorbed polymer and the achievement of conformational equilibrium, *Macromolecules*, 27, 2553–2558, 1994.
61. Voronov, A., Pefferkorn, E., and Minko, S., Adsorption of protonated polyvinylpyridine on silica. Correlation between interfacial conformation and colloidal stability, *Macromolecules*, 31, 6387–6389, 1998.
62. Voronov, A., Pefferkorn, E., and Minko, S., Oscillatory phenomena at polyvinylpyridine/silica interfaces, *Macromol. Rapid. Commun.*, 20, 85–87, 1999.
63. Minko, S., Voronov, A., and Pefferkorn, E., Oscillation phenomena at polymer adsorption, *Langmuir*, 16, 7876–7878, 2000.
64. Voronov, A., Minko, S., Shulga, A., and Pefferkorn, E., Non-equilibrium adsorption at solid/liquid interfaces from polyelectrolyte solutions, *Colloid Polym. Sci.*, 282, 1000–1007, 2004.

15 Pyrogenic Silica— Rheological Properties in Reactive Resin Systems

Herbert Barthel, Torsten Gottschalk-Gaudig, and Michael Dreyer

CONTENTS

15.1 INTRODUCTION

Finely divided silicas may be classified according to their manufacturing process—there are natural products, byproducts, and synthetic products. The product origin implies distinct differences of the properties of these silicas. Natural products as quartz powders or diatomaceous earth, and by-products that include fused silica, silica fume, or fly ashes from metallurgy and power plants are mostly crystalline silica products of micron size particles with surface areas up to $1 \text{ m}^2/\text{g}$. In contrast, synthetic silicas from wet processes or thermal pyrogenic reactions are typically amorphous silica products with a high surface area of $> 100 \text{ m}^2/\text{g}$. Wet processes that are based on the reaction of soluble silicates with aqueous acids lead to silica gels or precipitated silica.[1] By far, the most important thermal pyrogenic path is the flame hydrolysis of silanes, which gives access to pyrogenic silica.

Pyrogenic silica is a finely divided amorphous silicon dioxide produced by high temperature hydrolysis of silicon tetrachloride in an oxygen–hydrogen flame (Figure 15.1). In a first step, proto particles directly formed during the chemical reaction collide and coalesce to give primary particles at high flame temperatures.[2] At lower temperatures, collision, sticking, and partial fusion of primary particles result in the formation of stable aggregates (Figure 15.1a). While leaving the flame and cooling, the silica aggregates continue to collide, but their surfaces are now solid and agglomerates (Figure 15.1b) of aggregates formed by physico–chemical surface interactions.

Pyrogenic silica is widely used in industry as an efficient thickening agent, providing shear thinning and thixotropy to liquid media like sealants, adhesives, composite resins, coatings, and inks. Various parameters control the rheological performance of pyrogenic silica.

- The smoothness of the primary particle surfaces, which provides a maximum contact area for various types of interactions like H-bonding and van der Waals interactions of dipolar and dispersive character.

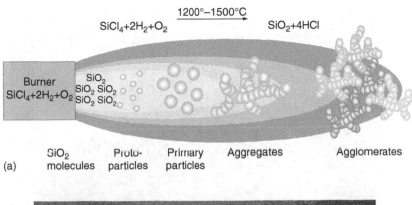

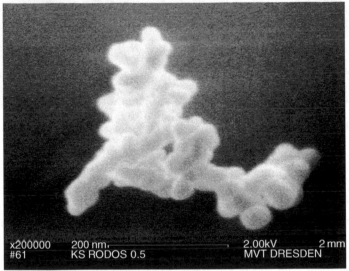

(b)

FIGURE 15.1 (a) Particle formation in a flame; (b) TEM micrograph of a pyrogenic silica aggregate. (*continued*)

- The space-filling structure of the aggregates with a low mass fractal dimension D_m, leading to fluffy structured agglomerates, typically with a bulk density $d_{bulk} = 50\text{--}100$ g/L, and agglomerate sizes > 1 μm.
- The high physico–chemical interaction potential of the pyrogenic silica surface is based on its reactive surface silanol groups (surface density 1.8 SiOH/nm^2), but is also due to its polar Si–O bonds containing bulk. By surface modification, most commonly surface silylation, these interactions can be controlled precisely.

(c)

FIGURE 15.1 (*continued*) (c) TEM micrograph of a pyrogenic silica agglomerate.

Particle interactions are the driving force for agglomerate and network formation, enabling pyrogenic silica to form percolating networks in liquid media.[3–5] Basically, two kinds of networks are possible: firstly, a network of pyrogenic silica particles or aggregates originating from direct particle–particle-contacts, and secondly, a network based on polymer bridging where aggregates are interacted by polymers at least on two particles. Real systems may consist of both and mixed types. At rest or very low shear rates, these networks are able to immobilize large volume fractions of liquids even at low pyrogenic silica loading (<5 wt%), resulting in very high viscosities or a yield point, respectively. Upon applying shear forces, the network structure is reversibly destroyed and the apparent viscosity of the mixture decreases with increasing shear rate. When the shearing stops, the system is able to recover the network structure. Figure 15.2 depicts this process schematically.

Control over the rheological properties of liquid systems is of enormous industrial importance. Pyrogenic silica has the advantage of being an effective rheological additive which will not undergo swelling and also exhibits chemical inertness. Additionally, thickening by pyrogenic silica results in a non-Newtonian system commonly accompanied by a yield point, reversible

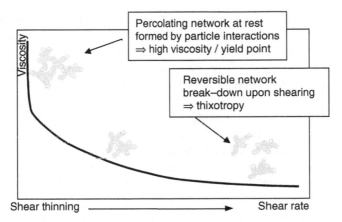

FIGURE 15.2 Gel evolution during shear processes.

shear thinning, and thixotropy. An understanding of the parameters which control thickening is of great technical interest.

Particle–particle interactions and, consequently, the formation of agglomerates or larger particle networks are related to a nonlinear increase in the apparent particle size and effective volume fraction. This is because particle clusters are usually less dense in mass than the primary particles themselves.[6] Therefore, the rheology of a dispersed solid phase in a liquid medium will depend both on its ability to undergo particle–particle interactions and on its specific particle structure.

15.2 PYROGENIC SILICA AS A RHEOLOGY CONTROL AGENT FOR REACTIVE RESIN SYSTEM

Many industrial and manufacturing processes face the requirement to apply a liquid solution, resin, or polymer onto vertical substrates as coatings, paints, lacquers, etc. During application, the liquid should be thin and easy to paint or spray, but consecutively the applied liquid film should stay on the substrate the time needed to accomplish crosslinking and hardening. Sag resistance is a most challenging request in modern coatings and adhesive systems.

15.2.1 SAG RESISTANCE AND RHEOLOGY STUDIES

In a coatings laboratory, sag resistance is monitored by applying a coating of defined wet film thickness onto a black and white panel, using a gage, increasing stepwise the wet-film thickness of the coating system under test.

Typically, the sag resistance is expressed in micro-meters of wet film thickness, where sag just does not occur. In general, correlations of sag resistance in application with rheology parameters from quality control laboratories are used. Most recently, it has been demonstrated that plotting the sag resistance and the critical shear stress (which is the shear stress at tan $\delta = 1$, which is the threshold where storage modulus G' is equal to loss modulus G'') of silica containing paint formulations versus the loading by the silica reveals the different regimes of the silica network—from a stage of isolated silica clusters to a percolated silica network. These results validate that sag in plant applications can be monitored and even predicted from advanced rheological laboratory data. (Figure 15.3a) and (Figure 15.3b) demonstrate the reversibility of shear thinning of a pyrogenic silica loaded coating formulation: after applying a high shear, the viscosity recovers rapidly to its initial high level. Obviously, the thickening effect and also

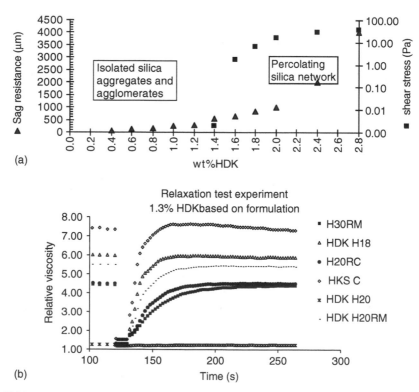

FIGURE 15.3 (a) Influence of silica weight fraction on sag resistance and critical shear stress; and (b) relaxation tests using different silica grades.

its time dependency strongly depend on the chemistry of the silica used: HDK® H20, a partially dimethylsiloxy groups-silylated silica of Brunauer–Emmett–Teller (BET) 200 m²/g, HDK® H30RM, a fully trimethylsiloxy groups-modified silica of BET 300 m²/g, HDK® H18, a fully polydimethyl-siloxane-grafted silica of BET 200 m²/g, HDK® H20RC, a fully *n*-octylsilane groups-modified and HDK® H20RH, a fully *n*-hexadecylsilane groups-modified silica of BET 200 m²/g.

Figure 15.4 reveals a strong relationship of sag resistance and silica loading, whereas Figure 15.5 indeed demonstrates the power of such rheology measurements, predicting sag resistance from controlled shear rate rheology measurements.

15.2.2 RHEOLOGICAL STUDIES

The rheological behavior of colloidal dispersions is mainly driven by the phase volume of the dispersed phase and attractive or repulsive particle–particle interaction forces, respectively.[7]

In the case of pyrogenic silica, two general types of silica are commonly used to govern the rheological properties of liquid media such as paints or adhesives. These are hydrophilic and surface treated hydrophobic grades, respectively. Hydrophilic silica is usually used in less polar resins dissolved in larger quantities of an organic solvent, while the hydrophobic grades are commonly applied in more polar systems with low solvent content (low Volatile Organic Compounds) or in resin types bearing

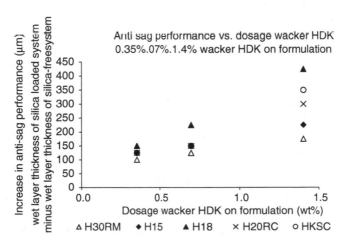

FIGURE 15.4 Evolution of sag resistance with increasing silica weight fraction of different silica grades.

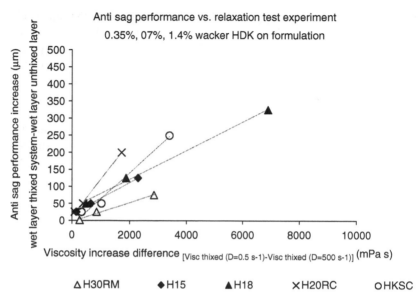

FIGURE 15.5 Effect of viscosity difference at low and high shear rates on sag resistance.

very polar groups such OH functions. Since for a given silica mass used as rheological additive any prior surface treatment should not significantly influence the absolute phase volume of the SiO_2 silica particles, the difference between the rheological properties of the untreated and surface-treated grades must stem from particle–particle and/or particle–resin interactions.[8]

In order to understand the influence of these different kind of interaction forces on the rheology of pyrogenic silica dispersions, and to develop a comprehensive model which at least allows a rough prediction of rheological properties, a rheological study has been performed.

Two grades of pyrogenic silica with a different degree of surface treatment and different polarity have been studied: a nontreated, hydrophilic HDK® N20 (BET surface area 200 m^2/g, 1.8 SiOH/nm^2 equivalent to 100% residual SiOH) and a fully silylated hydrophobic HDK® H18 (carbon content 4.5% C, 15% residual SiOH). The latter is covered by a chemically grafted polydimethylsiloxanes (PDMS) layer. Both silicas are products from Wacker–Chemie GmbH, Germany. The resins used in this study are an unsaturated polyester resin (UP resin), a co-condensate of a diol, maleic acid and *ortho*-phthalic acid; and a vinylester resin (VE resin), a co-condensate of glycidine, methacrylic acid, and bisphenol A bearing pending OH groups. Both resins have a styrene content of 35% and were provided by DSM, NL. Additionally, the rheology of both silica types in pure styrene was studied.

For the evaluation of the rheology of the silica dispersions, different test methods were applied: (a) steady shear test for the apparent viscosity at shear rates D of $1\ \mathrm{s}^{-1}$, $10\ \mathrm{s}^{-1}$, and $100\ \mathrm{s}^{-1}$, each 120 s, respectively; (b) a shear rate controlled relaxation experiment at $D = 0.5\ \mathrm{s}^{-1}$ (conditioning), $500\ \mathrm{s}^{-1}$ (shear thinning), and $0.5\ \mathrm{s}^{-1}$ (relaxation) to evaluate the apparent viscosity, the relaxation behavior, and thixotropy; (c) shear yield-stress measurements using a vane technique introduced by Nguyen and Boger[9]; and (d) low deformation dynamic tests at a constant frequency of $1.6\ \mathrm{s}^{-1}$. All samples contained 3 wt% of pyrogenic silica.

Figure 15.6 illustrates the differences of the rheological properties of hydrophilic silica N20 and hydrophobic silica H18 in unsaturated polyesters and vinylesters.

The step profile reveals that N20 and H18 dispersed in UP resin exhibit an almost identical rheological steady shear behavior, whereas the apparent relative viscosities of N20 and H18 dispersed in VE resin at low and moderate shear rates are distinctly different. To explain this behavior, it is necessary to consider the polarity, functional groups, and chain length of the resins, but also the surface properties of the pyrogenic silica. All parameters together will influence the nature and strength of the colloidal forces, which govern the rheology of the mixtures.

Commonly, the thickening of liquids by hydrophilic silica is explained by the formation of H-bonds between the silanol groups of silica particles.[10]

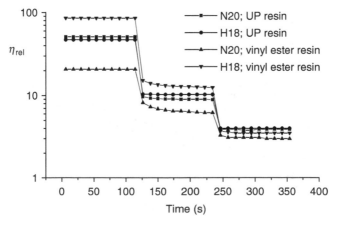

FIGURE 15.6 Steady shear step profile of HDK N20 and H18 in UP and VE resin.

According to this model, the stability of a silica gel in styrene and toluene, two fluids with comparable dielectric properties, should be more or less identical. Figure 15.7 depicts the shear yield-stress experiments using the vane geometry of HDK® N20 in styrene and toluene. The critical shear stress, where the gel structure of the dispersion collapses, for the system N20/styrene is significantly increased compared to N20/toluene dispersion. The critical shear stress is correlated to the strength of the particle–particle interactions. Hence, the attractive forces between silica particles in styrene are supposed to be stronger compared to toluene. However, this result cannot be explained by particle–particle interactions solely based on H-bonds. Their strength in styrene and toluene should be comparable due to the similar dielectric properties of both liquids and therefore result in comparable network stability.

A straightforward explanation for the differences in gel stability is that the attractive particle–particle interactions are not mainly driven by interparticulate H-bonds, but by van der Waals forces. This idea is supported by pair potential calculations using an approximate expression given by Israelachvili based on the Lifshitz theory of van der Waals forces.[11] The results of these calculations are shown in Figure 15.8.

The calculations for silica particles with a diameter of 12 nm reveal that the attractive interaction potential E_{att} for silica dispersed in styrene is stronger by a factor of ca. 4 compared to silica dispersions in toluene. This result demonstrates that the common concept of explaining the rheological properties of hydrophilic silica dispersions by means of interparticulate H-bonds is questionable. In silica–resin mixtures, adsorption of resin oligomers due to the

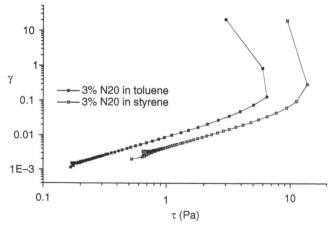

FIGURE 15.7 Yield stress test (vane geometry) of HDK N20 in styrene and toluene.

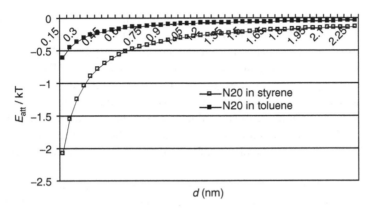

FIGURE 15.8 Van der Waals pair potential of N20 in styrene and toluene.

high surface energy of the silica particles has to be taken into account. The adsorption will just hamper the formation of interparticulate H-bonds by sterical constraints.

In Figure 15.9, the influence of the styrene content on the relative viscosity of UP and VE dispersions containing hydrophilic and hydrophobic silica using a shear rate controlled relaxation experiment is shown.

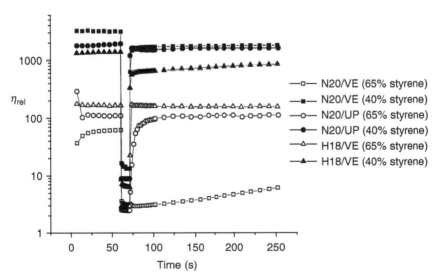

FIGURE 15.9 Influence of styrene content on relaxation behavior of different HDK N20 and H18 resins mixtures.

In general, the increase of styrene content from 35 to 60 wt% results in an increase of the relative viscosity of the N20 and H18 dispersions. This effect is more pronounced for dispersions containing the hydrophilic silica grade N20 than for H18. Even more interestingly, the relaxation rate of N20 dispersions after applying high shear forces strongly depends on the styrene content. Dispersions containing low volumes of styrene exhibit a retarded relaxation, which can be in the range of several minutes up to hours for N20 dispersed in VE resin. However, addition of styrene increases the relaxation rate in such a way that the low shear end-viscosity is almost instantaneously approached after finishing the high shear phase.

In order to evaluate the rest structure of the silica–resin dispersions, small amplitude dynamic measurements have been performed. Figure 15.10 depicts the frequency sweeps of HDK® N20 and H18 dispersed in UP and VE resins, respectively.

The storage modulus of H18 dispersions in UP resin with low styrene content exhibit high values and are almost independent of the frequency. This is indicative of a viscoelastic solid with percolating gel structure.[12] However, addition of styrene results in a pronounced decrease of the storage modulus and a strong dependence on the applied frequency. In fact, the recorded G' is smaller than the corresponding G'' (omitted for clearance), indicating the loss of the gel structure and evolution to a viscoelastic liquid.

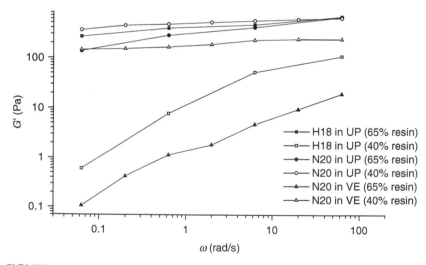

FIGURE 15.10 Influence of styrene content on the frequency spectra of different HDK N20 and H18 resins mixtures.

For N20 dispersions in UP and VE resins, a different behavior is observed. N20-VE resin dispersions containing 65% of resins show liquid behavior with low storage modulus, increasing with frequency. Addition of styrene increases the storage modulus, which is now independent of the frequency. Hence, a viscoelastic solid is obtained. N20-UP resin dispersions with resin content of 65% reveal a relatively strong frequency dependence of the storage modulus G'. At low frequencies is $G' > G''$, which inverts to $G'' > G'$ at higher frequencies. This is indicative of a viscoelastic liquid containing isolated flocs of pyrogenic silica. Addition of styrene increases the storage modulus, which becomes less sensitive to frequency. Hence, a viscoelastic solid is obtained.

15.3 SURFACE CHEMISTRY

The surface of a colloidal system influences most strongly on the viscosity of a fluid where the colloidal solid is dispersed. Empirically, it is well known that when reducing the particle size d of a powder of density ρ, hence increasing its specific surface SA area by $SA = 6/d\rho_{silica}$, will lead to markedly higher viscosities of the dispersion.[13] This makes it understandable why the surface and surface chemistry has to get most of the attention when trying to develop comprehensive models of the thickening action of pyrogenic silica in adhesives, composites, and coatings.

The chemistry of a finely divided solid with high surface area is at the borderline between molecular chemistry and physics of solid surfaces. The surface of a finely divided amorphous silica may be understood as a two-dimensional projection of the three-dimensionally linked silicon dioxide tetrahedra. By this, the chemistry of the pyrogenic silica surface is dominated by Si–O–Si units, and in particular, dangling surface Si–O bonds, which will create silanol groups under the influence of humidity at ambient temperature.

The density of silica surface silanol groups strongly depends on the specific production process of the synthetic silicon dioxides. Wet processed silica surfaces like that of silica gels show high and up to full coverage by silanols, with densities of 4–5 $SiOH/nm^2$ and higher,[1] precipitated silica exhibit 2–4 $SiOH/nm^2$. In contrast to silica products from wet processes, pyrogenic silica shows only a low surface density of surface silanols. Approximately every second silicon atom on its surface bears a silanol group.[14] A part of these silanols are not hydrogen-bonded but isolated and give rise to a sharp infrared band at 3750 cm^{-1}.[1] There is no strong evidence that pyrogenic silica contains larger fractions than 10% of internal silanols. There is some good evidence that the surface silanols on pyrogenic silica are not statistically distributed over the surface but clustered on the surface due to

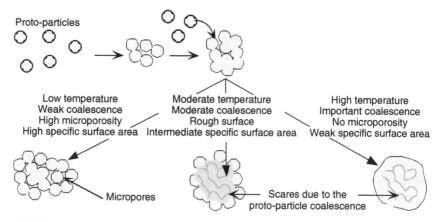

FIGURE 15.11 Sintering an coalescence of silica particles during flame synthesis.

the genesis of primary particles from smaller proto-particles. Remaining surfaces from protoparticles exhibit higher SiOH densities, scarves, or regions of fusion lower SiOH contents, providing pyrogenic silica a raspberry-like structure (see Figure 15.11).[15] Typically, surface silanols on pyrogenic silica are highly reactive and undergo various chemical reactions. Due to its silanol-covered oxide nature, pyrogenic silica exhibits a high surface energy γ and is wetted by water—nontreated silica therefore usually shows a hydrophilic character.

One of the most important chemical reactions on pyrogenic silica surfaces is the silylation of the surface silanol groups. The deactivation of the surface silanols and coverage by nonpolar alkyl groups lowers the surface energy γ of the oxide. The most common silylating agents are dichlorodimethyl silane and hexamethyldisilazane.

$$SiO_{4/2}\text{--}SiOH + 1/2[(CH_3)_3Si]_2NH + H_2O$$

$$\rightarrow SiO_{4/2}\text{--}Si\text{--}O\text{--}Si(CH_3)_3 + 1/2NH_3 \tag{15.1}$$

$$SiO_{4/2}\text{--}SiOH + (CH_3)_2SiCl_2 + H_2O$$
$$\rightarrow SiO_{4/2}\text{--}Si\text{--}O\text{--}[Si(CH_3)_2] + 2HCl \tag{15.2}$$

hydrophilic: $\gamma_{silica} > 72$ mJ/m^2, fully hydrophobic: $\gamma_{silica} \sim 33$ mJ/m^2

Hydrophilic silica is wetted by water and therefore has a surface energy of: $\gamma_{silica} > 72$ mJ/m^2. Interestingly, the surface energy γ_{silica} even of highly silylated silica rarely reaches 30 mJ/m^2, and never comes close to 21 mJ/m^2, which would be the surface tension of PDMS. This must be explained as a strong remaining influence of the inner oxide bulk of the silica particle, and reflects the strength of the dipolar forces related to the silicon dioxide core.

The chemical reaction of the silica surface with hexamethyldisilazane leads to a coverage with trimethylsiloxy groups. The molecular area of a trimethylsiloxy group has been reported to be $a_{Me3SiO} = 0.38$ nm^2.[16] The limiting carbon content (% C) at full surface coverage will then be % C = 1.6 per 100 m^2/g of surface area.

Although Equation 15.3 looks simple and straightforward and seems to result in a well-defined monolayer of trimethylsiloxy units on the silica, more complex reactions develop. Experimentally, a monolayer coverage stops at about 70%–65% of the surface (molar coverage times the molecular area of the trimethylsiloxy groups compared to the total area) or above a content of residual SiOH of about 30%–35% (of the initial SiOH of the bare silica), respectively. Higher coverage led to a breaking up of the silica core, formation of polysilicic acid chains (which can be silylated further on), and finally low molecular weight siloxanes of type $M_aQ_b(OH)_c$ (with M = $(CH_3)_3Si-$, Q = $SiO_{4/2}$). Semi-empirical quantum-chemical modeling reveals stress effects on the silica core and distortion of the SiO_2 carcass, followed by bond cleavage of Si–O–Si in the presence of water for degrees of silylation close to 100%.[17] It has been shown by further modeling studies, using a path-of-reaction approach, that a 66% coverage of the surface by trimethylsiloxy groups, forming hexagonally group six-member rings on the silica surface, are most stable against water attack and re-hydrolysis; whereas lower (50%) and higher (100%) degrees of coverage are less stable and sensitive to bond opening of $SiO_{4/2}$–Si–O–Si$(CH_3)_3$.[18]

Recently, an extended series of studies using different kinds of adsorption techniques including inverse gas-chromatography and various adsorbates, comprising n-alkanes, branched alkanes, alkenes, 2-, 3-, 4-, 5-, and 6 member oligosiloxanes (of dimethylsiloxy and trimethylsiloxy units), and also polar probes like isopropanol and water could demonstrate that the silylation process by far is not linked to a linearly advancing hydrophobicity. In contrast, hydrophobicity, e.g., loss of water wettability and change from a polar oxide surface to a nonpolar methylsiloxy surface occur in two markedly different stepwise processes.[15] Silylating a hydrophilic silica step by step, using a controlled silylation procedure and varying from the bare oxide to more and more surface treated products, a first step order phase change is observed at around 1/3 of coverage of the surface: the silica surface looses its oxide surface connectivity, and water as a bulk liquid is no longer able to wet it. Balard could show that this step depends on the grade or surface area of the pyrogenic silica itself, as the microscopic structure of the silica changes from smaller to larger particle size due to [1]longer resistance time in the flame and [2]higher reaction temperatures.

A second step is observed when the degree of silylation exceeds 66% of surface coverage. Up to 66% of coverage of trimethylsiloxy (TMS) groups, or 33% of remaining oxide surface, polymers like siloxanes have access to free oxide surface, hence to patches of high adsorption energy. Above 66% of TMS coverage, sterical constraints of the TMS grafts shield adsorbates to interact with the remaining oxide surface, and the silica behaves fully nonpolar and inactive with respect to dipolar interactions.[15]

Due to its high surface area, surface chemistry and physics dominate the properties of pyrogenic silica. The distance O–Si–O being 0.3 to 0.4 nm, estimate that there are only about 20 silicon dioxide units spanning the diameter of a primary particle of amorphous silica. Pyrogenic silica therefore has an extremely high surface to bulk ratio—up to about 10%. This is why even bulk methods of chemical analysis are suitable to follow chemical reactions on its surface: elemental analysis (like carbon content), infrared and near infrared methods, etc.

The completion of the reaction is easily followed by standard carbon analysis. Suitable methods for estimating the surface energy γ of hydrophilic or silylated silica are wetting tests or gas adsorption techniques. Acid–based reactions are of particular analytical interest.[19] Titration of pyrogenic silica with sodium hydroxide bears the advantage of an acid–base reaction, and distinguishes easily between the acidic silanols of silica and any other kind of monomeric, polymeric, or resinous carbinols or silanols. Titration is a powerful tool to follow the degree of reaction monitoring the residual amount of silica silanol groups after silylating reaction.

$$SiO_{4/2}\text{–Si–OH} + NaOH \leftrightarrow SiO_{4/2}\text{–Si–O}^-Na^+ + H_2O \qquad (15.3)$$

Under suitable reaction conditions, the acid–base reaction provides a hydroxyl capacity or silanol group density of hydrophilic pyrogenic silica of about 1.7–1.9 SiOH per nm^2. This value falls well between the total amount of silanol groups on pyrogenic silica of about 2.5 SiOH per nm^2 and 1.7 SiOH per nm^2, as reported for the content of reactive isolated silanol groups.[10,20] The silanol density of pyrogenic silica does not vary from grade to grade within the span of regular nonmicroporous particles. Additionally, it has been shown that this value does change during >10 years of storage in ambient and dry conditions.[21]

15.3.1 SURFACE AREA

The surface area is one of the most important and most commonly used parameters to characterize a highly dispersed oxide like pyrogenic silica.[22] Particle sizes seem not be an intrinsic value that really characterize the properties of the colloidal product, as the recorded data rather depend on the concentration, the shear rate, and the technique used itself than on the pyrogenic silica under study.[23]

A most common technique to determine surface areas above 100 m^2/g is using the adsorption of gases, e.g., nitrogen at 78 K, to calculate the BET surface area. This calculation is based on the monolayer capacity as evaluated from the adsorption isotherm and the molecular area of the adsorbate. When working with silylated silica samples, this may lead to questionable results. Silylation and various other surface treatments of pyrogenic silica clearly will change its surface properties, reduce its silanol content and the area of accessible oxide surface, which is replaced by methyl or other alkyl groups, and diminishes its surface energy. Koberstein[24] reported that even nitrogen may exhibit specific polar interactions due to its quadrupole moment. Kiselev[25] pointed out that the apparent molecular area of the adsorbate may depend on its surface interactions with the adsorbent.

A more detailed study on silylated pyrogenic silica that used nitrogen and argon at 78 K and also neo-pentane at 273 K as adsorbates revealed a distinct decrease of the BET surface area with growing carbon loading and decreasing content of surface silanol groups.[8] The BET surface area of a silylated silica, determined by adsorption of nitrogen, shows a pronounced decrease from about 200 m^2/g down to <80 m^2/g with increasing carbon content (Figure 15.12). This decrease cannot be explained simply by a diminution of the geometrical surface area. Pyrogenic silica is typically a nonporous solid. A loss of measured surface area therefore does not result from micro-pore filling by silylation.

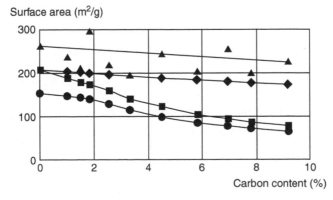

FIGURE 15.12 Influence of carbon content on the BET surface area.

Increasing carbon content (% C) of silylated silicas (i.e., the increasing thickness of the silylation layer) may lead to an increased size of the silica primary particles. A simple calculation of the related reduction of the specific surface area A may be written as

$$A_{silylated} = A_{silica}/(1 + \%C/(100 - \%C) * 74/24 * \rho_{silica}/\rho_{PDMS})^{1/3} \quad (15.4)$$

where the silylation layer is assumed to be poly(dimethylsiloxane) (PDMS), the density of silica $\rho_{silica} = 2.2$ g/ml and that of the silylation layer $\rho_{PDMS} = 1.0$ g/ml. The result is that the geometrical surface area is affected only about 20% at a carbon content of 10% (Figure 15.12). Since the geometrical surface area of the primary silica particles will not change significantly with silylation, the decrease of BET surface area must be due to an increase in the apparent molecular area of the adsorbate.

Obviously, the apparent molecular area of the adsorbate strongly depends on the surface energy of the solid as determined by wetting tests or the equilibrium spreading pressure from gas adsorption. On a high energy surface, like hydrophilic pyrogenic silica, the molecular area of nitrogen $a_{N_2} = 0.162$ nm^2, (adopted from the liquid density), and therefore the related BET surface area from nitrogen adsorption will provide reliable values. Whereas, it seems that on a low energy surface, as for example silylated silica, no ordered monolayer is formed and therefore the monolayer capacity and consequently the BET surface area are evaluated markedly too low. In general, it turns out that a good compromise is to use the nonsilylated hydrophilic surface area as a reference. The main contribution to a decrease of the BET surface area is due to a shielding of the dispersion forces of the oxide surface by the silylation layer. Thus, a simple BET surface area determination may provide a relative measure of the surface energy changes caused by the silylation of the silica surface.

15.4 PYROGENIC SILICA SURFACE STRUCTURES

Pyrogenic silica aggregates are obviously linear and branched particle structures with a mean size of about 100 to 500 nm. By TEM, we derive the size of the partially fused primary particles of about 5 to 30 nm. This very small particle size correlates well with the high surface area of pyrogenic silica, which usually is larger than 100 m^2/g, as determined by nitrogen adsorption at 78 K according to BET. Adsorption techniques and electron microscopy provide very close values of surface areas. This indicates that pyrogenic silica typically exhibits a smooth particle surface in the range of nanometers, and apparently its surface is free of micropores.

This lack of micro-pores is further confirmed by small angle x-ray scattering (SAXS) and nitrogen adsorption data. A very helpful concept is the surface fractal dimension D_s. A smooth, nonporous surface is described by a surface fractal dimension of $D_s = 2.0$, but a totally porous body will reach a D_s of up to nearly 3. SAXS and adsorption measurements on pyrogenic silica result in a surface fractal dimension D_s very close to 2.0.[26] The fact of a nonporous smooth surface of the pyrogenic silica particles is a most important characteristic, as it will simplify the interpretation of chemical surface reactions.[27]

For studying the surface roughness of the silica, common approaches are the fractal dimension as determined by relating the CTAB to the N_2 surface area, evaluation of the nitrogen adsorption isotherms, and SAXS.

15.4.1 CTAB Surface Area

The surface area from the adsorption of the bulky cationic surfactant cetyltrimethyl ammonium bromide (CTAB) is, according to standard methods,[28] considered as a measure of the "outer" surface of the filler particle thought to be accessible to polymer adsorption. Avnir et al.[27] proposed to evaluate the surface fractal dimension D_s by relating surface areas of differently bulky probes, i.e., in our case the adsorbates nitrogen (BET) and CTAB. Assuming $a_{N_2} = 0.162$ nm^2, we find a reasonable range of the surface fractal dimension $2 < D_s < 3$ and $D_s = 2$ for HDK® S13 only if $a_{CTAB} = 0.345$ nm^2.[29]

15.4.2 Small Angle X-Ray Scattering[30]

Typically, the linear portion of the SAXS log–log plot intensity I versus wave vector k for pyrogenic silica is in the range 0.5 nm $< 1/k < 2$ nm and provides a slope of $-m = 6 - D_s$. Silica grades with a low surface area such as HDK® S13, V15, and N20 show a smooth surface with $D_s = 2.0$. HDK® T30 gives $D_s = 2.2$ and may be classified as a mainly smooth surface, too. But HDK® T40 provide $D_s \gg 2$, and thus exhibit a fractal rough surface of the primary particles.[29]

15.4.3 Nitrogen Adsorption—Fractal Frenkel–Halsey–Hill Isotherm

Pfeifer et al.[27] derived the surface fractal dimension from a single adsorption isotherm based on the Frenkel–Halsey–Hill (FHH) isotherm—the classical FHH isotherm $N_{ads} = C (\ln p_0/p)^{-1/s}$ is expanded into a fractal FHH isotherm $N_{ads} = C' (\ln p_0/p)^{-(3-D)/s}$. However, the FHH exponent s may, according to Halsey et al.,[8] be taken as a rough guide to the strength of the adsorbate–solid interaction. For nonporous pyrogenic silica (HDK® S13 to T30), we propose to use $s = 2.5$ (at $D_s = 2$). Hence we interpret larger values of $s/(3 - D_s)$ as a fractally rough surface with $D_s > 2$.[29]

15.4.4 NITROGEN ADSORPTION—FRACTAL BRUNAUER–EMMETT–TELLER– ISOTHERM

Fripjat and VanDamme[31] derived a surface fractal dimension D_s using the Brunauer–Emmett–Teller (BET) approach: The classical BET isotherm is expanded into a fractal BET one, taking into to account that only on a flat surface an infinite number of layers of N_2 molecules can be stapled, whereas on a curved surface this is no longer possible for sterical reasons. The fractal BET of Fripjat and VanDamme uses the classical BET constant and monolayer capacity and introduces a number of finite layers of N_2, together with a surface fractal dimension D_s. We could show that this approach works reasonably well if we consider the surface fractals of primary particles and the mass fractals of the aggregates separately. In addition we assume that micro-pores on primary particle surfaces contain a maximum number of 5 layers of N_2-molecules and that meso-pores between the primary particles of the aggregates have a diameter of 250 times the thickness of a N_2 molecule (see Figure 15.13).

15.4.5 SURFACE STRUCTURE AND THICKENING PERFORMANCE

As the BET surface area and also the CTAB surface area increase from HDK® S13 up to T40, one should expect an increasing thickening efficiency. This expectation is valid for the grades HDK® S13 to N20 but does not hold for

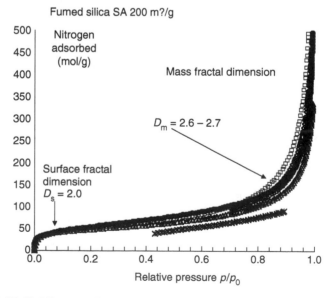

FIGURE 15.13 Nitrogen adsorption isotherm of HDK N20.

TABLE 15.1
Fumed Silica Thickening Efficiency: Unsaturated Polyester Resin

Sample	Surface Area SA (m^2/g)	Thickening Efficiency 3% HDK in UP Resin η/η_0 at $\varDelta = 10\,s^{-1}$
HDK D05	50 ± 15	1.3 ± 0.3
HDK C10	$100 + 15$	2.5 ± 0.5
HDK S13	125 ± 15	4.0 ± 0.5
HDK V15	150 ± 20	$5.5 \perp 0.5$
HDK N20	200 ± 30	6.8 ± 0.5
HDK T30	300 ± 30	6.2 ± 0.5
HDK T40	380 ± 40	5.2 ± 0.5

65% Solid (maleinic + phthalic acid anhydride, 65:35, propanediol) in 35% solvent styrene; $\eta_0 = 1000$ mPas at 25°C; measurement: rotation, spindel-beaker.

grades HDK® T30 and HDK® T40. HDK® N20 achieves the highest thickening of the liquid. But the liquid loaded with HDK® T40 achieves viscosities only comparable to those filled with HDK® S13, i.e., a grade of a third of the surface area of T40 (see Table 15.1).

Hurd et al.[26] could demonstrate by using small angle x-ray and neutron scattering that grades of higher surface area may exhibit surface roughness. Transmission electron microscopic images show very similar particle size distributions of HDK® N20, T30, and T40. We therefore assume the surface roughness of HDK® T40 to be responsible for the observed deterioration of particle–particle and particle-polymer interactions.

For an understanding of pyrogenic silica as an effective thickener in fluid systems, the knowledge of its particle structure and surface texture is essential. Finely divided silica fillers consist of primary particles which are attached to larger aggregates. The interaction of the polymer and resin oligomer chains with these primary particles depends on the availability of the particle surface. Surface roughness or surface microporosity weakens the silica–silica and silica-polymer interaction, as demonstrated in Table 15.2 and Figure 15.14.

15.4.6 POLYMER-LIKE SILYLATION LAYER OF DIMETHYLSILOXY GROUPS

In the foregoing section, the silylation layer has been regarded as behaving as a dense and rigid monolayer of methyl groups on the solid silica surface. As a consequence, silica particles modified by monofunctional trimethylsiloxy groups and silica particles modified by difunctional dimethylsiloxy

TABLE 15.2
Surface Structures of Fumed Silica: Surface Area, Surface Microporosity, and Surface Fractal Dimension D_s of Fumed Silica; Smooth →2,0≤ D_s <3← Rough, Porous

Sample	Acid–Base Titration SiOH Groups per nm²	Nitrogen Adsorption Isotherm BET Surface Area[a] (m²/g)	Pore Volume[b] (% SA)	D_s (FHH)[c]	D_s (Fractal BET)[d]	CTAB Adsorption CTAB Surface Area[e] (nm²)	D_s (CTAB/N₂)[f]	SAXS D_s (SAXS)[g]
HDK S13	1.7	131	<1	2.00	2.1	131	2.00	2.00
HDK V15	1.8	152	<1	2.01	2.1	150	2.04	2.00–2.05
HDK N20	1.8	199	<1	2.00	2.1	200	2.01	2.00–2.10
HDK T30	1.8	300	<1	2.04	2.1	248	2.51	2.15–2.25
HDK T40	1.9	380	<1	2.11	2.1	261	2.99	2.40–2.50
Porous FS1	1.8	185	12	2.05	2.1	175	2.14	2.16
Porous FS2	1.9	204	11	2.01	2.1	184	2.28	2.21
Porous FS3	2.4	249	31	2.20	2.2	173	2.96	2.20
Porous FS4	2.0	278	24	2.17	2.3	176	3.21	2.58
Porous FS5	3.2	336	60	2.54	2.9	83	5.69	2.83

[a] Monolayer BET range: $0.06<p/p_0<0.3$; molecular area $a_{N_2} = 0.162$ nm².
[b] T-plot vs. standard isotherm of colloidal silica (Singh et al.).
[c] FHH multilayer range (Pfeifer et al.), with $s = 2.5$: $0.1<p/p_0<0.8 \Rightarrow 0.8$ nm $<l_s<8.5$ nm.
[d] Fractal BET surface (Fripjat and VanDamme et al.): $0.05<p/p_0<0.6 \Rightarrow 0.6$ nm $<l_s<3.7$ nm; multilayer $n_{N_2} = 5 \rightarrow$ thickness $t_{layer} = 1.77$ nm.
[e] Adsorption of hexadecyl(= cetyl)trimethylammonium bromide CTAB; molecular area $a_{CTAB} = 0.345$ nm².
[f] Surface area SA vs. molecular area a: SA $\sim a^{D_s}$.
[g] D_s-SAXS: $-0.5<\log k_s<0 \Rightarrow 0.316<k_s$ [nm⁻¹]$<1 \Rightarrow \pi/0.316<l_s$ [nm]$<1/1$ nm $<l_s<10$ nm.

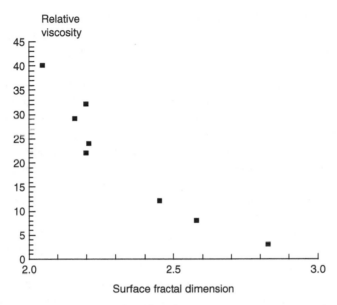

FIGURE 15.14 Effect of the surface fractal dimension on relative viscosity of silica dispersions.

groups should behave in the same manner. At higher carbon contents, pyrogenic silica modified by dimethylsiloxy groups shows a greater thickening ability than the trimethylsiloxy groups-modified silica. This greater thickening is found in particular when a silicone fluid is used as a medium and also in an unsaturated polyester resin; it occurs pronouncedly only for a silica carbon content higher than 3% C and vanishes in a highly polar liquids, as water-based solvents.[32]

At a high coverage of the silica by dimethylsiloxy groups (carbon content of 6% C, i.e., a siloxane content of about 18 wt%), the mean thickness of the silylation layer is about 0.6 nm. Even at this high siloxane content, the mean chain length of grafted siloxane oligomers is not larger than four or five dimethylsiloxy units.[33–35] Therefore, particle bridging by polymer chains or polymer entanglements should be of minor importance.

It has been proposed that the underlying additional particle–particle interactions arise from interpenetration of particle silylation layers.[32] A silylation layer of dimethylsiloxy groups at low coverage will behave like a rigid surface layer, as does a layer of trimethylsiloxy groups. However, with increasing loading, the silylation layer will consist more and more of polymer-like chains of dimethysiloxy groups, and will thus behave like a bonded phase of quasi-liquid silicone.

Solid-state ^{2}H NMR techniques have shown that octadecyl chains grafted onto silica are highly mobile in a wetting nonpolar solvent, but they are forced to lie flat on the solid silica surface when immersed in water.[36] SAXS of HDK® N20, BET surface area 200 m^2/g provides a fractal surface dimension of $D_s = 2.0$, as the surface of pyrogenic silica is smooth and free of micropores (Figure 15.15). Silica surfaces modified by long alkyl chains have been reported to give values of D_s markedly below 2 (Figure 15.16). It has been demonstrated that these unexpected data do not result from surface fractal character, but they may be explained by not a sharp but rather a fuzzy particle boundary due to silylation.[37] Data from highly silylated pyrogenic silica, like HDK® H18 (dimethylsiloxy groups-modified HDK® N20), recorded by SAXS measurements also yield a value of $D_s < 2$ (at carbon content $> 3.5\%$ C, $D_s = 1.8$) (Table 15.3).

It has been shown elsewhere that the monolayer capacities of nitrogen (at 78 K), argon (at 78 K), neo-pentane (at 273 K), and hexamethyldisiloxane (at 303 K) on pyrogenic silica decrease distinctly with increasing silylation.[8] Surprisingly, the monolayer capacities of neo-pentane and hexamethyldisiloxane on silica increases again at high loadings of dimethylsiloxy groups (Table 15.3).

Balard reported absorption rather than adsorption of oligomers of cyclic and linear dimethylsiloxane on highly silylated dimethylsiloxy groups-

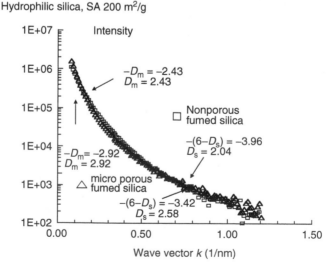

FIGURE 15.15 SAXS scattering plot of HDK N20.

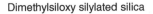

Dimethylsiloxy silylated silica

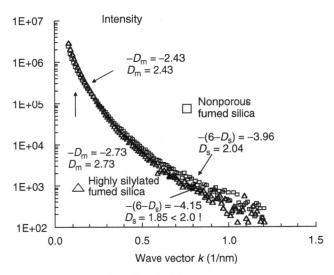

FIGURE 15.16 SAXS scattering plot of silylated HDK N20.

modified silicas, recorded by Inverse Gas-Chromatography at temperatures from 60 to150°C.[15,38]

This increase is observed with neither nitrogen nor argon, nor with trimethylsiloxy group-modified silica. It is assumed that absorption effects additional to adsorption are responsible for the increase in sorption capacity. A silylation layer of dimethylsiloxy groups can be compared to a grafted phase of short PDMS chains. PDMS shows a melting point at 210 K and a glass transition temperature of about 150 K. At a temperature of 78 K, nitrogen or argon is, therefore, adsorbed onto a solid surface of silylated silica, regardless of the thickness of the grafted PDMS layer. However, at the adsorption temperatures of neo-pentane and hexamethyldisiloxane, 273 and 303 K, a thick PDMS layer will be quasi-liquid and thus able to dissolve adsorbate molecules. Both absorption and adsorption will occur and enhance the apparent monolayer capacity.

15.5 PYROGENIC SILICA STRUCTURE OF AGGREGATES AND AGGLOMERATES

An overview of pyrogenic silica structure is given in Figure 15.17. Obviously, pyrogenic silica particle sizes cover a multiple of orders of magnitudes,

TABLE 15.3
Surface Structures of Surface Silylation Layers

Sample	Carbon Content (%)	BET Surface Area → Adsorption Isotherm				N₂ Adsorption	SAXS
		N_2 at 78 K (nm²)	Ar at 78 K (nm²)	$C(CH_3)_4$ at 273 K (nm²)	$((CH_3)_3Si)_2O$ at 303 K (nm²)	D_s (FHH)	D_s (SAXS)
HDK N20	0.0	199	153	118	142	2.00	2.04
N20DMS1.3	1.3	193	147	70	98	2.00	2.27
N20DMS1.5	1.5	189	140	65	89	2.00	2.21
N20DMS3.3	3.3	151	115	47	53	1.98	2.17
N20DMS7.0	7.0	105	83	71	90	2.02	1.95
N20DMS9.2	9.2	72	68	84	105	2.01	1.85

$D_s > 2$, Hard and rigid surface: hydrophilic silica monomer-like silylated silica (e.g. trimethylsiloxy groups-modified); $D_s < 2$, Soft and fuzzy surface: polymer-like silylated silica.

FumedSilica

-Structures-

Unit	Size	Characteristics	Structure	Density
$SiO_{4/2}$-tetrahedron	0.5 nm	Broad distribution of Si-O-Si und O-Si-O bond angles	Amorphous	
Primary particle surface	1-10 nm	Smooth surface microporous surface highly silylated surface	$D_s = 2.0$ $2 < D_s < 3$ "$D_s < 2$"	
Primary particle	10 nm	Not isolated part of aggregates	$D_m = 3$	2200 g/l
Aggregate, inner core	100 nm	Dense core	$D_m = 2.5$	700 g/l
Aggregate, outer shell	500 nm	Open, space-filling	$D_m \leq 2.1$	??? (<100 g/l)
Agglomerates	1-250 μm	Mainly spherical	Homogenous	75 g/l
Flocks, bulk	> 0.25 mm	White, powdery solid	Fluffy	20–30 g/l

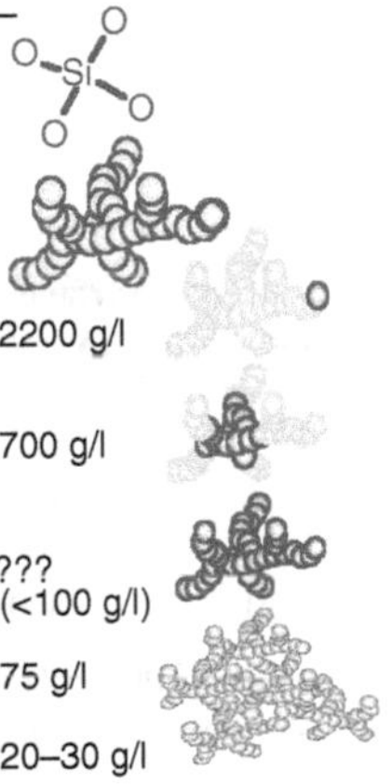

FIGURE 15.17 General structure characteristics of pyrogenic silica.

ranging from 1 nm of the primary particle pores of high surface area products via primary particles of 5–50 nm, aggregates of 100–500 nm, agglomerates of 1–100 μm, and flocks of 0.1–1 mm.

Particle sizing techniques provide additional information. Classical techniques are laser light diffraction and light scattering. For the latter measurement, both static light scattering and dynamic light scattering (photon correlation spectroscopy) can be applied to pyrogenic silica. Both techniques, if used in the classical equipment, require highly diluted suspensions (<0.01 wt%) in order to suppress particle interactions and multiple scattering. The range of particle sizes for Fraunhofer light diffraction (FHLD) typically goes from some 100 μm down to approximately 1 μm, and at smaller sizes Mie scattering should also be taken into account. As a result, FHLD records agglomerate sizes. However, as agglomerates sizes are shear and concentration is sensitive, FHLD provide particle size values which are strongly depending on the measurement protocol, hence its strength is a relative monitor for quality control, but not absolute values suitable for technical and R&D purposes. Interestingly, by FHLD, using exactly the same protocol, we measure higher particle sizes for HDK® N20 than for HDK® T30 or HDK® V15, similarly to a highest thickening efficiency of HDK® N20.

More basic information may be received from light scattering (LS). The range of LS typically goes from 1 to 1000 nm. LS therefore gives access to

TABLE 15.4
BET Surface Area—Calculated Sizes of "Primary Particles"

| Sample | BET Surface Area Nitrogen Adsorption at 78 K: Calculation of "Primary Particle" Sizes | | | TEM Thickness of Aggregate Branches ≈ Size of "Primary Particles" | |
	Surface Area SA (m²/g)	Diameter $d = 6/\rho*SA$ (nm)	Molecular Weight (10^6 g/Mol)	Mean Diameter (nm)	Minimum–Maximum Diameter (nm)
HDK S13	131	22	4.1	32 ± 15	11–105
HDK V15	152	18	2.4	29 ± 14	8–97
HDK N20	199	14	1.0	22 ± 7	7–50
HDK T30	300	9	0.3	20 ± 5	9–39

Estimation of "primary particle" size distributions in fumed silica aggregates, comparing "primary particle sizes from BET surface area and TEM "primary particle" size distribution.

pyrogenic silica aggregate sizes—Table 15.4 and Table 15.5 summarize some data. Pyrogenic silica of lower specific surface areas shows larger aggregates, an observation which is confirmed by TEM of pyrogenic silicas in a silicone rubber (Table 15.6). Comparison of smaller dynamic and larger static LS sizes reveals a high porosity of the silica aggregates. Stintz reported a factor of 4

TABLE 15.5
Fumed Silica Aggregates Light Scattering: Hydrodynamic Size and Radius of Gyration

Sample	Surface Area SA (m²/g)	Hydro-dynamic Size (nm)	2* Radius of Gyra-tion[a] (nm)	Molecular Weight (10^6 g/Mol)	Number of Primary Particles	Aggregate Size Unit: (Primary Particle)
HDK S13	131	290 ± 10	510 ± 40	892 ± 60	≈ 220	16
HDK N20	199	216 ± 8	440 ± 20	1160 ± 45	≈ 1170	28
HDK T30	300	204 ± 6	440 ± 40	476 ± 15	≈ 1620	31

[a]2* Radius of gyration from static light scattering; solvent methanol, p.a., dispersion by magnetic bar stirring (24 h) and ultrasonic bath (1 h), stabilization by KOH, pH = 9 (stable > 1 week), concentration: 0.01%, 0.005%, 0.0025%, 0.001% → extrapolation to concentration 0%, multi-angle measurement → extrapolation to angle 0: ⟹ Isolated "primary particles" do not exist in fumed silica.

TABLE 15.6
Fumed Silica Aggregates by Transmission Electron Microscopy:
Geometrical Sizes in Silicone Rubber

Sample	Surface Area SA (m^2/g)	Mean Geometrical Size (nm)	Maximum Geo-metrical Size (nm)
HDK V15	152	380	560
HDK N20	199	345	510
HDK T30	300	275	380

Mixing of 30 wt% silica in a HTV silicone rubber polymer; dilution of the mixture to 1 wt% of silica by silicone polymer; adding a peroxide catalyst and curing by temperature; semi-quantitative TEM image analysis of HDK aggregate sizes.

between geometrical and aerodynamic sizes, which also should hold for liquid systems.[23]

Ulrich gave a description of the pyrogenic silica flame process based on the immediate formation of proto-particles directly related to the chemical reaction rather than surface deposition.[2] At high flame temperatures, collision and coalescence of proto-particles leads to the formation of primary particles. The rate of coalescence depends on the viscosity of the molten oxide, which is exceedingly high for silicon dioxide at a flame temperature of about 1500 K. Therefore, the size d of the pyrogenic silica primary particles, i.e., the surface area $SA = 6/d\rho_{silica}$ is strongly related to the flame temperature. At lower temperatures, collision and sticking of primary particles only results in partial fusion and stable particle aggregates are formed. They leave the flame and cool, and agglomerates of aggregates are formed by collision, which are held together by physico–chemical surface interactions (Figure 15.1).

One approach to quantitatively describe the structure of a finely divided solid is the concept of mass fractal dimension D_m relating the size d of a particle cluster to its mass M.[6]

$$M \sim d^{D_m} \text{ with } 1 < D_m < 3 \tag{15.5}$$

The formation of aggregates may be understood as a diffusion-limited aggregation (DLA) of primary particles, since the sticking coefficient is virtually unity. Reaction-limited cluster aggregation (RLCA) controls the agglomeration of aggregates, as aggregates will rearrange after collision in order to optimize surface interactions. DLA and RLCA provide mass fractal dimensions of $D_m^{DLA} = 1.8$ for aggregates and $D_m^{RCLA} = 2.0$ for agglomerates.[6]

SAXS or nitrogen adsorption at 78 K exhibit a mass fractal dimension of $D_m = 2.5$–2.7 for the inner core of aggregates <50 nm.[39] But for the whole aggregate size up to 500 nm, small angle neutron scattering (SANS) provide $D_m = 2.2$[40a], and also rheology investigations show a mass fractal $D_m = 2.1$,[40b] confirming aggregate mass fractal dimensions as revealed by DLA.

Pyrogenic silica appears as a fluffy white powder characterized by an extremely low bulk density down to the range of about 20–50 g/L. In contrast, the submicron pyrogenic silica particle consists of amorphous silicon dioxide, and hence its true density is about 2200 g/L.

Any discussion of pyrogenic silica particle structure has to take into account this enormous difference. The approach of mass fractal dimension may provide a rough estimation. A real mass fractal is limited by the size of the cluster as an upper limit and the size of the particles as a lower limit. Then, the density of the cluster $\rho_{cluster}$ may be calculated from the true density of the particle $\rho_{particle}$, the ratio of the cluster size $d_{cluster}$ to the particle size $d_{particle}$, and the mass fractal dimension D_m of the cluster (Equation 15.6):

$$\rho_{cluster} = \rho_{particle}(d_{cluster}/d_{particle})^{D_m-3} \tag{15.6}$$

The size of the primary particles in the aggregates is estimated to be about 10 nm, the size of aggregates are in the range of 100 nm, and the sizes of agglomerates are about 5 μm.

By Equation 15.6 and $D_m = 2.5$ from DLA, we estimate the apparent density of the aggregate to be 700 gL[1]. This is the density found experimentally, when powdered pyrogenic silica is pressed to a solid disc. $D_m = 2.1$ from RLCA then gives an agglomerate density of about 20 g l[1]— close to the bulk density of the freshly produced fluffy product.

15.6 SILICA RESIN INTERACTION

15.6.1 NMR SPECTROSCOPY

A deeper understanding of the nature of the interactions between resin molecules and pyrogenic silica particles is provided by ^{13}C CPMAS studies at different cross polarization times τ_{cp} from 0.5 to 8 ms.[41] At short τ_{cp}, spectra intensities are enhanced by ^{13}C resonances of the least mobile chain fragments. This study includes two different pyrogenic silicas of surface area of 200 m²/g a hydrophilic silica HDK® N20 and fully silylated HDK® H18, with an unsaturated polyester resin Palatal P4 and a vinyl ester resin Atlas 590, respectively. Figure 15.18 depicts the ^{13}C CPMAS spectra of N20 and H18, respectively, in Palatal P4, 35 wt% in styrene at different cross polarization time τ_{cp}.

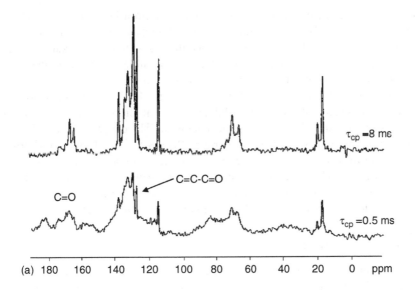

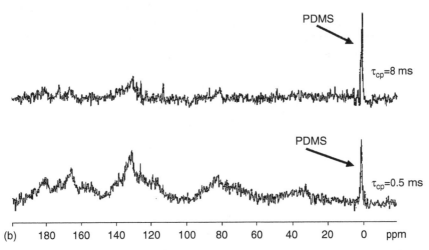

FIGURE 15.18 CP-MAS NMR spectra of HDK N20 and HDK H18 in UP resin at different CP contact times.

The N20/Palatal P4 ^{13}C CPMAS spectra exhibits an enhanced intensity of the C=O and C=C signals at short τ_{cp}, whereas Palatal P4 without added N20 shows no signals in the same experiment. This indicates that specific interactions of carbonyl and C=C–C=O groups of the resin oligomers with the pyrogenic silica surface result in an immobilization of the resin molecules.[42,43] This finding supports our interpretation of the rheological relaxation experiment, where we suggested that the increased relaxation times in the presence of Palatal P4 are related to the reversible adsorption of the resin oligomers at the silica surface after shear deformation.

The ^{13}C CPMAS spectra of H18 in Palatal P4/35% styrene show an enhanced intensity of the grafted PDMS chains at short τ_{cp}. In a ^{13}C MAS experiment, no signals for the PDMS chains of H18 could be detected. The fact that it was possible to detect the grafted PDMS by ^{13}C CPMAS indicates a strong immobilization of the chains, which is in agreement with our suggestion of H18 silica–silica network formation by a combination of hydrophobic interactions and chain entanglement. Both mechanisms are supposed to immobilize the PDMS chains in the grafted layer.

15.6.2 IR SPECTROSCOPY

Further support for a specific interaction between the silanol groups of hydrophilic pyrogenic silica and the C=O groups of resin molecules comes from IR spectroscopy, where in addition to the carbonyl band of free Atlac 590 oligomers at $1724\ cm^{-1}$, a second band at $1704\ cm^{-1}$ appears under adsorption at the N20 surface (Figure 15.19). This indicates an interaction of the C=O function with the silica surface and particularly with the silanol groups of the silica.[44] At very low resin concentrations of less than 1.0 wt%,

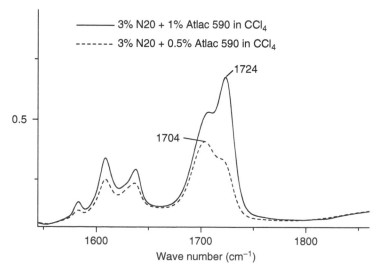

FIGURE 15.19 IR spectrum of HDK N20 in VE resins Atlac 590 (carbonyl region).

the fraction of adsorbed oligomer is approximately 60% of the total resin, as seen from the IR intensities. Upon increasing the amount of resin, the fraction of adsorbed resin oligomer remains small and does not exceed 5–8 wt% relative to the silica. The fact that only a small portion of the resin molecules is immobilized is in accordance with our model of thickening by hydrophilic pyrogenic silica based on (a) direct particle–particle interactions, (b) polymer bridging, and (c) steric stabilization.

15.7 QUANTUM CHEMICAL CALCULATIONS

Quantum chemical modeling is a suitable tool to elucidate the microscopic mechanisms of the adsorption processes of polymers on silica surfaces.[45] For modeling of hydrophilic pyrogenic silica particles, a hydroxylated silica cluster $[SiO_248–OH9]$[45] containing 48 silicon dioxide units and 9 surface silanol groups has been used. Grafting two five member dimethylsiloxy (DMS) chains lop-wise (bonded at both ends) on it provided the model for fully PDMS-silylated pyrogenic silica, silica cluster $[SiO_248–OH5–DMS_52]$. Two models of resin molecules have been simulated representing all typical functional groups of unsaturated polyester resins (UP) and vinyl ester resins (VE).

A special study has been dedicated to the nature of bonds in the system, in particular, in the silica–resin system.[18] The interaction energies of different kinds of H-bonds in the systems decrease in the order C–O–H···O(H)–Si > C=O···H–O–Si > C–(H)O···H–O–Si (see Table 15.7); –C–OH···O(H)–Si forms the strongest H-bond, but which is observed only in the VE system. Surprisingly, the carbonyl functions of the resin oligomers also show a rather

---

TABLE 15.7
Energies of Specific Interactions: Silica Model [SiO₂18–OH5] and VE Model: Methylether of Glycidine — Methacrylic Acid

System	Energy of Interaction (kJ/mol)	Bond Distance (nm)
–C–H···O(H)–Si	−12.1	0.283
–C–(H)O···HO–Si	−19.3	0.185
–C=O···HO–Si	−24.2	0.183
–C–OH···O(H)–Si	−41.1	0.177

Source: Balard, H., Papirer, E., Khalfi, A., Barthel, H., and Weis, J., *Organosilicon Chem. IV*, Weis, J., Ed., Wiley-VCH, Weinheim, p. 773, 2000.

high H-bond energy. Carbonyl groups seem to be active sites of interactions in the system resin–silica, leading to immobilization of the carbonyl groups of resins on silica surfaces, as seen by ^{13}C CPMAS NMR,[41] and red shift of carbonyl bands, as observed by IR. In summary, this indicates that a VE-like and also a UP-like resin structure shows strong adsorption affinity towards hydrophilic pyrogenic silica.

Figure 15.20 shows fully optimized structures of adsorption complexes of VE and UP resin models with the [SiO$_2$48–OH9] and [SiO$_2$48–OH5–DMS$_5$2] silica clusters. Calculated interactions energies of silica–silica, silica–resin, and resin–resin adsorption complexes are given in Figure 15.21.

15.7.1 Vinylester System

Hydrophilic silica particles interact strongly with the VE molecule (complex [SiO$_2$48–OH9]/VE) due to H-bonds of type –C–OH⋯O(H)–Si–, H–bonds of type C=O⋯H–O–Si–, and also dispersion interactions. The silica–VE–resin

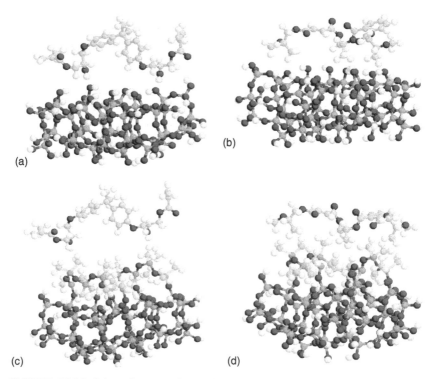

FIGURE 15.20 Adsorption complexes of VE and UP resin models on silica model fragments.

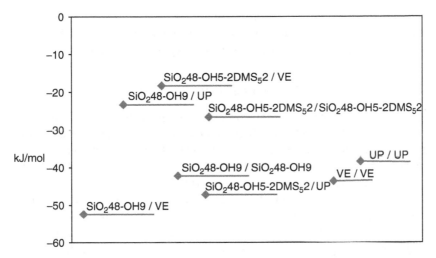

FIGURE 15.21 Calculated interaction energies of different adsorption complexes.

interaction is stronger than the silica–silica interaction (complex [$SiO_2$48–OH9]/[$SiO_2$48–OH9]), leading to adsorption and sterical stabilization of the colloidal system, hampering direct particle–particle contacts. In terms of rheology, we would interpret this as low thickening efficiency of hydrophilic silica in VE-type resins—in fact, experiments show that HDK® N20 is not a stable thickener for VE resin systems.

The weakest interaction in Figure 15.21, in terms of adsorption enthalpy, is that of silylated silica with a VE molecule (complex [$SiO_2$48–$DMS_5$2]/VE). It is even weaker than the interaction of silylated silica with itself (complex [$SiO_2$48–$DMS_5$2]/[$SiO_2$48–$DMS_5$2]). Modeling makes evident a particle-to-particle entanglement of the 5-membered-DMS chains on the silica clusters, additional to nonspecific dispersion interaction. Additional immobilization of PDMS chains on HDK® H18 was seen also by ^{13}C CPMAS NMR.[41] The strongest complex in the system VE resin and silylated silica is that between VE resin molecules. In consequence, the interface of highly silylated silica towards a VE resin phase is energetically less favorable. In order to optimize VE–VE contacts, the silylated silica particle surfaces should separate from the VE resin phase by minimizing VE resin to silylated silica contacts but forming direct silylated silica-to-silylated silica particle contacts—a phenomenon which is know as "hydrophobic interaction."[46] However, real systems contain monostyrene, which is not at all or only very weakly adsorbed on a silica surface according to calculations and spectroscopic experiments. Hence, styrene is enhancing the effect of silylated silica separation from the VE–styrene phase, and strength of hydrophobic interactions. In fact, HDK® H18 is an excellent and stable thickener and rheology control additive for all VE and epoxy resin-like systems.

15.7.2 UNSATURATED POLYESTER SYSTEM

The calculated enthalpies of the adsorption complex UP resin on hydrophilic silica (Figure 15.21, (complex [SiO$_2$48–OH9]/UP)) show rather weak adsorption energy and the interaction of two hydrophilic silica cluster results in higher interaction energies (complex [SiO$_2$48–OH9]/[SiO$_2$48–OH9]), the resin–resin interaction giving an intermediate strength of adsorption (complex UP/UP). Additionally (see Table 15.8), an extensive quantum-chemical modeling study shows that styrene is not at all or only weakly adsorbed on hydrophilic silica. In consequence, hydrophilic pyrogenic silica may show strong particle–particle interactions in a UP resin dissolved in monostyrene: HDK® N20 is a good thickener and thixotropic agent for UP resins.

TABLE 15.8
Energies of Interactions: Styrene–Styrene, Styrene–Resin, Styrene–Silica

System	Energy of Interaction (kJ/mol) per 1 Molecule of Styrene
2*styrene s1	−10.26
2*styrene s2	−5.89
6*styrene s1	−8.29
VE/styrene-s1	−20.34
VE/styrene-s2	−19.91
VE/styrene-s3	−23.71
VE/styrene-s4	−22.84
UP/styrene-s1	−16.93
UP/styrene-s2	−29.79
UP/styrene-s3	−20.17
UP/styrene-s4	−14.98
[SiO$_2$48–OH9]/styrene-s1	−2.12
[SiO$_2$48–OH9]/styrene-s2	−11.23
[SiO$_2$48–OH9]/styrene-s3	−11.01
[SiO$_2$48–DMS$_2$DMS$_3$]/styrene-s1	−7.73
[SiO$_2$48–DMS$_2$DMS$_3$]/styrene-s2	−8.85

However, the adsorption complex of a UP molecule on the silylated silica cluster (complex [SiO$_2$48–DMS$_5$2]/UP) shows a distinctly high interaction energy. Following the discussion given above, this should result in wetting, adsorption, and sterical stabilization, a weak colloidal network and unstable thickening. However, as shown by the rheological experiments (see above), highly PDMS silylated silica HDK$^{®}$ H18 is an excellent thickener for UP resins also.

This demonstrates that not only resin–silica, resin–resin, and silica–silica interactions control the properties of the system. It seems that the solvent styrene plays a key role. As seen from Table 15.8 (styrene), quantum-chemical modeling shows that styrene only badly wets a hydrophilic silica surface (complex [SiO$_2$48–OH9]/styrene) and is only weakly adsorbed on a silylated silica surface (complex [SiO$_2$48–DMS$_2$DMS$_3$]/styrene). Being a bad or worse-than-θ solvent for (the) PDMS (layer on silica), styrene is the driving force of hydrophobic interactions. In the case of silylated silica in UP resins, the adsorption of UP molecules on the silylated silica surface seems not to hinder hydrophobic interaction—this is a reasonable assumption, as the latter is not particle–particle contact but phase-separation driven.

15.8 SILICA–SILICA INTERACTIONS

The origin of the linkage of silica aggregates to agglomerates and to particle networks are particle–particle interactions. Interparticle hydrogen bonds have been discussed as a main force to bind pyrogenic silica particles together.[10] Most recently, an infrared spectroscopy and inelastic neutron scattering investigation, together with semi-empirical quantum-chemical modeling, on the adsorption interactions of PDMS oligomers/polymers with nontreated and trimethylsiloxy-treated pyrogenic silica have demonstrated that silica interactions are mainly controlled by dipolar interactions of the polar $\equiv$Si$^{\delta+}$–O$^{\delta-}$– bonds of the silicon dioxide backbone and the PDMS polymer.[47] This reveals that interactions of pyrogenic silica particles are dominated by interactions of type van der Waals fluctuating dipoles (London's dispersion energy), induced dipoles (Keesom-type), and permanent dipoles (Debye-type).

Study of the thickening effect of hydrophilic and silylated hydrophobic pyrogenic silica in liquids of different polarities revealed a generalized picture of particle interactions and thickening ability. It turned out that a similar system of silica surface and liquid medium will not show pronounced thickening, whereas if the surface of the silica is markedly different from the liquid, then in consequence a high thickening effect will occur (Table 15.9).[32,33]

TABLE 15.9
Principles of Particle–Particle Interactions

	Silica Polar/ Hydrophilic	Silica Nonpolar/ Hydrophobic
Polar liquid	No silica interaction	Hydrophobic silica interaction
	No thickening	High thickening
Nonpolar liquid	Dipolar, H–bonding silica interaction high thickening	No silica interaction
		No thickening

Pseudoplastic flow of noncolloidal hydrophobic silica dispersions in liquids of high polarity has been related to hydrophobic particle coagulation as a result of nonwettability by the liquid,[48] high hydrophilic/lyophilic balance values of the liquid,[49] and protruding polymer chains from the silylation layer.[50] The thickening behavior of colloidal hydrophobic pyrogenic silica in water has been explained by dispersion interactions between alkyl chains grafted onto the silica surface. [51,52]

Pyrogenic silica bearing a monolayer of trimethylsiloxy groups on the surface shows the same efficiency of thickening in the glycerin mixture as silica modified by a polymer-like layer of dimethylsiloxy groups. Thickening of the polar liquid by the hydrophobic particles is related to the content of residual silanol groups, i.e., hydrophobicity, rather than to different structures of the silylation layer. The origin of the underlying particle–particle interactions are so-called hydrophobic interactions.[11] By means of this solvent mediated interaction, the hydrophobic silica particles build up stable agglomerates and particle networks. Highly hydrophobic pyrogenic silica thicken a polar liquid medium as effectively or stronger than a hydrophilic silica thickens a nonpolar one.

It has to be mentioned that the understanding of particle interactions in liquid media is not restricted to hydrogen bonding. Qualitatively, London dispersion forces also lead to particle–particle interactions and thickening[53] if the medium and the particles differ in their Hamaker constants, but the London dispersion interaction energy will tend to zero if these Hamaker constants become equal.[54]

Given a primary particle radius of $R = 5$ nm and Hamaker constants of $A_{11} = 12 \times 10^{-20}$ J (silica) and $A_{22} = 6.3 \times 10^{-20}$ J (PDMS), the interaction of two silica particles separated at a distance $d = 0.1$ nm (surface to surface distance) by a silicone fluid yields a nonretarded dispersion energy of attraction $\Delta G_{121}^d = -A_{121}R/12d,$[54] of about 5–15 kT, where kT is the thermal energy at a temperature of 298 K. The interaction energy due to hydrogen bonding of hydrophilic silica particles in a silicone fluid may be roughly estimated as follows: A primary particle of $R = 5$ nm with a silanol density of 2 SiOH nm^{-2} bears about 600 silanol groups.[14] However, owing to the particle structure and steric constraints at the solid surface, only a minor part of these silanol groups participate in strong hydrogen bonds with a neighboring particle. Assuming that 1% of the total amount of silanol groups interacting provides an energy of interaction of two particles, ΔG_{SiOH}^p, of about 50–100 kT. In water, the hydrophobic interaction energy of two particles, hydrophobized by hexadecyltrimethyl ammonium bromide (CTAB), was reported to be $\Delta_{CTAB}^h = 40Re^{-d/d_0}$ where $d_0 = 1$ nm.[55] A particle distance $d = 0$, i.e., particles in contact, yields a hydrophobic interaction energy of about 80 kT. Taking into account the higher solubility of CTAB in water, compared to that of a silicone fluid, the silylation of the surface (PDMS layer) should result in more pronounced hydrophobicity than CTAB coverage. Thus, hydrophobic thickening by silylated silica should be greater than hydrogen bonds-based thickening by hydrophilic silica.

The rheological properties of highly hydrophobic polydimethylsiloxane-modified pyrogenic silica HDK® H18 can be readily explained by treating the particles as a composite material of a SiO_2 core and a thin grafted PDMS shell.[56] The thickness of the PDMS layer strongly depends on the solvent quality of the liquid medium. Liquids with a good solvency for PDMS (better than THETA-conditions) will cause a swelling of the PDMS layer. This results in weaker attractive particle–particle interactions due to the barrier function of the PDMS layer. Additionally, the swelling of the PDMS layer by the diffusion of solvent molecules into the layer will cause a better matching of the dielectric and optical properties of the liquid medium and the adlayer. This results in similar Hamaker constants of the homogenous phase and the layer. Hence, according to Equation 15.7, any contribution of the PDMS layer to the attractive net interaction force based on van der Waals interactions will be reduced. [11]

$$A_{131} \approx ((A_{11})^{0.5} - (A_{33})^{0.5})^2 \qquad (15.7)$$

This means for the system H18/UP resin, the addition of styrene, which shows reasonable solvency for PDMS, will cause swelling of the grafted PDMS surface layer and therefore reduce the net interaction force between the dispersed particles. This is demonstrated by the dynamic rheological tests,

which reveal a decrease of the storage modulus upon addition of styrene. This effect is visualized in Figure 15.21.

However, for hydrophilic silica N20, an explanation of the rheological effects by means of particle interactions is less straightforward. The high surface energy of hydrophilic silica will result in complex adsorption phenomena. Hence, the constitution, thickness, and segment density gradient in the surface adsorption layer is difficult to predict.

In the case of low molecular weight polar resins such as VE resins, relatively thin and dense adsorption layers can be assumed. This should result in low viscosities due to low effective phase volumes of the dispersed phase and weak interparticulate interactions forces according to steric stabilization. However, addition of a solvent-like styrene will influence the Hamaker of the liquid medium, the structure of the adlayer in terms of swelling and/or multilayer formation, and the Hamaker constant of the adlayer. In particular, any multilayer formation could result in surface layer entanglement depending on the solvency of the liquid medium expressed in terms of the Flory-Huggins parameter χ.[57] These effects should dramatically influence the viscosity and rest structure of the dispersion, as seen in the experiments.

For long-chained resin oligomers, the situation is even more complex. In addition to the adsorption phenomena resulting in steric stabilization, polymer bridging also has to be taken into account. In general, polymer bridging should be more relevant in diluted systems because of a less dense adlayer, which might facilitate the simultaneous adsorption of polymer coil at two different particles.[58] This model is at least in general supported by our results of the rheological tests.

In order to prove the suggested interaction models, more detailed calculations of the pair potentials between silica particles are required. For hydrophobic silica HDK® H18, an approach suggested by Vincent based on the original work of Vold can be used.[59]

In the case of hydrophilic silica HDK® N20, a more elaborate approach will be needed in order to take into account the chain entanglement of the adsorption layer containing long-chained polymer molecules depending on the solvent quality of the homogeneous medium. These effects can be treated by applying approximations given by Napper and Vincent, respectively.[57,60]

15.9 PARTICLE STRUCTURE AND RHEOLOGY

The viscosity, η, of a dispersion of solid particles can be related to the viscosity of the pure liquid, η_0, and the volume fraction of the dispersed solid,

Φ, by a modified relation of Einstein,[13] taking into account the polydispersity of the particles:

$$\eta/\eta_0 = (1 - \Phi)^{-2.5} \tag{15.8}$$

The true density of the pyrogenic silica primary particles $P_{silica} = 2200$ g/L (amorphous silicon dioxide) yield a volume fraction of $\Phi = 0.009$ in a silicone fluid of $\rho = 970$ g/L at a loading of 2 wt% pyrogenic silica. However, dispersed in air, pyrogenic silica agglomerates show a markedly lower density of $\rho_{agglomerate} = 20$ g/L (Equation 15.6). Therefore, assuming that the silica agglomerates show a similar fractal space-filling mass structure in a fluid to that in air and assuming that the agglomerates behave like rigid spheres at the applied shear rate owing to dipolar forces between the hydrophilic pyrogenic silica aggregates, a much higher effective agglomerate volume fraction of $\Phi > 0.9$ has to be considered at a loading of 2 wt% silica.

Indeed, higher loadings of hydrophilic silica than 2 wt% in a nonpolar silicone fluid of 1000 mPas lead to a gel-like consistency accompanied by a yield value, indicating a percolating particle structure and an effective volume fraction of virtually unity. At a silica loading of 2 wt%, resulting in a relative viscosity $\eta/\eta_0 = 35$ in the silicone fluid, at a shear rate of 1 s^{-1}, Equation 15.6 and Equation 15.8 provide a mean size of spherical agglomerates of about 4 μm, which is close to values obtained from light diffraction. An increase of the shear rate to $D = 10$ s^{-1} decreases η/η_0 to about 7, revealing a reduction in the mean agglomerate size to 2–3 μm. The reduction in the relative content of residual silanol groups of the silica to less than 30%, by silylation with trimethylsiloxy groups decreases η/η_0 to < 1.2, related to an aggregate size of less than 250 nm (Figure 15.22).

This simple calculation shows that there is a strong relationship between silica particle structure and thickening efficiency. However, a deeper insight requires further detailed and sophisticated rheological and particle sizing approaches.

15.10 CONCLUSION

Pyrogenic silica is a well-known rheological additive to impart high viscosities, shear thinning, thixotropy, or even a yield point to liquid media such as coatings, adhesives, and composite resins.

The rheological behavior of colloidal dispersions is mainly driven by the phase volume of the dispersed phase and particle–particle interaction forces.

The phase volume of particulate mater dispersed in a liquid is usually calculated from the applied mass and density of the solid material. However,

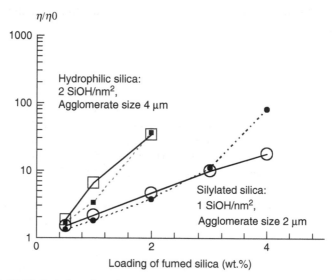

FIGURE 15.22 Relative viscosity of hydrophilic and silylated silica of increasing weight fractions in a silicone fluid.

this approach does not hold for pyrogenic silica. The high temperature production process of pyrogenic silica results in a material which already exhibits a space filling, fluffy structure without being dispersed in a liquid. This fluffy structure can be assigned to the existence of stable sinter aggregates, which resemble the ultimate particles of pyrogenic silica. The space filling properties of the sintered aggregates and subsequently also of agglomerates formed by association of aggregates can be described by means of the mass fractality concept. In order to compute the effective phase volume of dispersed pyrogenic silica, the sinter aggregates have to be used as the actual rheological relevant particles. By using this concept, phase volumes and therefore volume fractions can be calculated, which in principle can be used as a parameter in phenomenological equations such as the Einstein equation or the Krieger-Dougherty equation.

Due to the high surface to mass ratio of pyrogenic silica, particle–particle interactions are a decisive factor for the rheology of silica dispersions. Silica–silica interactions are modulated by the specific surface area, the surface roughness, which can be expressed in terms of a surface fractality, and last but not least, the kind and degree of chemical surface modification.

Most of these properties are directly relevant for the adsorption of organic molecules. These adsorption processes strongly govern the rheological properties of hydrophilic silica. In fact, the theory of steric stabilization can be applied to understand the rheological behavior of hydrophilic silica dispersions.

However, the solvency of the liquid medium does not only influence the particle–particle interactions via the Flory-Huggins parameter in the osmotic term of the net interactions equations. Even more important seems to be the influence of the mean solvents quality on adsorption processes, which can change the thickness of adsorption layers and segment density functions. An additional contribution to the net interaction potential presumably stems from polymer bridging between two adjacent silica particles. However, polymer bridging requires a minimum oligomers/polymer size which should be in the order of the primary particle diameter. Hence, this mechanism can be expected to be relevant only for relatively long-chained (nonpolar) prepolymers.

In order to understand silica–silica interaction in the case of hydrophobic silica particles, a simplified approach of steric stabilization is sufficient. Here, only van der Waals forces are taken into account. Any surface layer is only seen as a kind of barrier. The thickness of this barrier depends on swelling effects and is directly responsible for the minimum distance that two particles are able to approach.

However, in order to at least semi-quantitatively predict technically relevant properties such as sag resistance or a yield stress, a more detailed qualitative and quantitative understanding of pair potentials between hydrophilic and hydrophobic silica particles is required. For this purpose, more data concerning the structure and composition of surface layers on silica particles are necessary. Suitable tools to gather this information can be advanced spectroscopic methods or quantum chemical calculations.

REFERENCES

1. Iler, R. K., *The Chemistry of Silica*, Wiley, New York, 1979.
2. Ulrich, G. D., *Chem. Eng. News*, 62, 22–29, 1984.
3. Quemada, D., *Prog. Colloid Polym. Sci.*, 79, 112–119, 1989.
4. Russel, W. B., *J. Rheol.*, 24, 287–317, 1980.
5. Tadros, T. F., *Chem. Ind.*, 7, 210–218, 1985.
6. Meakin, P., In *The Fractal Approach to Heterogeneous Chemistry, Surfaces, Colloids, Polymers*, Avnir, D., Ed., Wiley, New York, pp. 131–160, 1989.
7. Larson, R. G., *The Structure and Rheology of Complex Fluids*, Oxford University Press, New York, 1999.
8. Barthel, H., In *4th Symposium on Chemically Modified Surfaces*, Mottola, H. A. and Steinmetz, J. R., Eds., Elsevier, Amsterdam, pp. 243–256, 1992.
9. Nguyen, Q. D. and Boger, D., *J. Rheol.*, 29, 335–347, 1985.
10. Michael, G. and Ferch, H., *Schriftenreihe Pigmente*, Vol. 11, Degussa, AG.
11. Israelachvili, J., *Intermolecular and Surface Forces*, 2nd ed., Academic Press, London, 1991.
12. Tadros, T. F., *Adv. Colloid Interface Sci.*, 68, 97–100, 1996.

13. Hiemenz, P. C., *Principles of Colloid and Surface Chemistry*, Dekker, New York, 1977.
14. Tertykh, V. A., Mashchenko, V. M., and Chuiko, A. A., *Dokl. Akad. Nauk SSSR*, 200, 865–868, 1971.
15. Balard, H., Papirer, E., Khalfi, A., Barthel, H., and Weis, J., In *Organosilicon Chemistry IV*, Weis, J., Ed., Wiley-VCH, Weinheim, pp. 773–792, 2000.
16. Larsen, P. and Schou, O., *Chromatogr.*, 16, 204–207, 1982.
17. Khavryutchenko, V. D., Niktina, E. A., Sheka, E. F., Barthel, H., and Weis, J., *Phys. Low-Dim. Struct.*, 5/6, 1–8, 1998.
18. Nikitina, E. and Barthel, H., In *2nd International Conference on Silica*, Mulhouse, p. 121, 2001.
19. Sears, G. W., *Anal. Chem.*, 28, 1981–1983, 1956.
20. Evans, B. and White, T. E., *J. Catalysis*, 2, 336–341, 1968.
21. Barthel, H., *Unpublished Results*, 2006.
22. Brunauer, S., Emmett, P. H., and Teller, E., *J. Am. Chem. Soc.*, 60, 309–311, 1938.
23. Barthel, H., Heinemann, M., Stintz, M., and Wessely, B., *Chem. Eng. Technol.*, 21, 745–752, 1998.
24. Koberstein, E. and Voll, M., *Phys. Chem.*, 71, 275, 1970.
25. Kiselev, A. V., Korolev, A. Y., Petrova, R. S., and Shcherbakova, K. D., *Kolloid Zh.*, 22, 671, 1960.
26. Hurd, A. J., Schaefer, D. W., and Martin, J. E., *Phys. Rev. A*, 35, 2362–2364, 1987.
27. Avnir, D., Farin, D., and Pfeifer, P., *J. Chem. Phys.*, 79, 3566–3571, 1983.
28. ASTM, *Designation D*, 3765.
29. Barthel, H., Achenbach, F., and Maginot, H., *Proc. Interface Symp. on Mineral and Organic Functional Fillers in Polymers (MOFFIS 93)*, 301, 1993.
30. Bale, H. D. and Schmidt, P. W., *Phys. Rev. Lett.*, 53, 597–599, 1984.
31. Fripjat, H. and VanDamme, J., In *The Fractal Approach to Heterogeneous Chemistry, Surfaces, Colloids, Polymers*, Avnir, D., Ed., Wiley, New York, p. 242, 1989.
32. Barthel, H., *Colloids Surf., A: Physicochem. Eng. Aspects*, 101, 217–226, 1995.
33. Barthel, H., Roesch, L., and Weis, J., In *Organosilicon Chemistry II*, Weis, J., Ed., Wiley-VCH, Weinheim, pp. 763–778, 1995.
34. Litvinov, V. M., *Polm. Sci. USSR*, 30, 2250, 1988.
35. Litvinov, V. M., In *Organosilicon Chemistry II*, Auner, N. and Weis, J., Eds., VCH, Weinheim, pp. 779–814, 1995.
36. Zeigler, R. C. and Maciel, G. E., *J. Am. Chem. Soc.*, 113, 6349–6358, 1991.
37. Schmidt, P. W., Avnir, D., Levy, D., Hohr, A., Steiner, M., and Roll, A., *J. Chem. Phys.*, 94, 1474–1479, 1991.
38. Khalfi, A., Doctoral thesis, Univ. Mulhouse (Mulhouse), 1997.
39. Barthel, H., *Unpublished Results*, 1994.
40. Barthel, H., *Unpublished Results*, 2005; Gottschalk-Gaudig, T., *Unpublished Results*, 2004.

41. Barthel, H., Dreyer, M., Gottschalk-Gaudig, T., Litvinov, V., and Nikitina, E., In *Organosilicon Chemistry V*, Auner, N. and Weis, J., Eds., Wiley-VCH, Weinheim, pp. 752–776, 2003.
42. Cory, D. G. and Ritchey, W. M., *Macromolecules*, 22, 1611–1615, 1989.
43. Litvinov, V. M., Braam, A. W. M., and van der Ploeg, A. F. M. J., *Macromolecules*, 34, 489–502, 2001.
44. Joppien, G. R. and Hamann, K., *J. Oil Colour Chem. Assoc.*, 60, 412–423, 1977.
45. Nikitina, E., Khavryutchenko, V., Sheka, E., Barthel, H., and Weis, J., In *Organosilicon Chemistry IV*, Auner, N. and Weis, J., Eds., Wiley-VCH Verlag GmbH, Weinheim, pp. 745–762, 2000.
46. Israelachvili, J. N. and Pashley, R. M., *J. Colloid Interface Sci.*, 98, 500–514, 1984.
47. Nikitina, E. and Barthel, H., *Silicon Chemistry*, submitted.
48. Kao, S. V., Nielsen, L. E., and Hill, C. T., *J. Colloid Interface Sci.*, 53, 358, 1975.
49. Diemen, A. J. G. v., Schreuder, F. W. A. M., and Stein, H. N., *J. Colloid Interface Sci.*, 104, 87–94, 1985.
50. Schreuder, F. W. A. M. and Stein, H. N., *Rheol. Acta*, 26, 45–54, 1987.
51. Lee, G., Murray, S., and Rupprecht, H., *Colloid Poly. Sci.*, 265, 535–541, 1987.
52. Lee, G. and Rupprecht, H., *J. Colloid Interface Sci.*, 105, 257–266, 1985.
53. Tsai, S. C. and Zammouri, K., *J. Rheol.*, 32, 737–750, 1988.
54. Ross, S. and Morrison, L. D., *Colloid Systems and Interfaces*, Wiley, New York, 1988.
55. Israelachvili, J. N., In *Proceedings of Symposium on Complex and Supermolecular Fluids*, Clark, N. A. and Safran, S., Eds., Wiley, New York, p. 101.
56. Bevan, M. A., Petris, S. N., and Chan, D. Y. C., *Langmuir*, 18, 7845–7852, 2002.
57. Vincent, B., Edwards, J., Emmett, S., and Jones, A., *Colloids Surf.*, 18, 261–281, 1986.
58. Otsubo, Y., *Adv. Colloid Interface Sci.*, 53, 1–32, 1994.
59. Osmond, D. W. J., Vincent, B., and Waite, F. A., *J. Colloid Interface Sci.*, 42, 262–269, 1973.
60. Napper, D. H., *J. Colloid Interface Sci.*, 58, 390–407, 1977.

Section VI

Computer Simulation of
Solid Surfaces and Interfaces

16 Silicate Glass Surfaces, Their Heterogeneity, and Computer Simulations

Victor Bakaev, Carlo Pantano, and William Steele

CONTENTS

16.1 INTRODUCTION

The surface atomic structures of silicate glasses are the basis for understanding their surface properties on an atomic level. However, the experimental methods for exploring the surface properties of silicate glasses and their atomic structures are limited. A major reason for this limitation is the fact that silicate glasses cannot be obtained (at present) in high area forms. Only silica can be obtained in a form (such as, for example, silica gel) that has a specific surface of $100 \ m^2/g$ or more. A high surface area form of a multi-component silicate glass—for example, E-glass, which is used for the production of glass fibers—is a glass fiber typically of about $10 \ \mu m$ diameter that has a specific surface of about $0.16 \ m^2/g$. (Of course, one can crush any silicate glass into the colloidal state with very large specific surface. However, as explained below, such a fracture surface has an atomic structure very different from that of glass fibers.) Since specific surfaces of silicate glasses are conventionally much smaller than those of silica, the methods that have been used for the study of the silica surface are much less effective when applied to such surfaces. For example, the main source of information on the nature of the silica surface is physical adsorption studies. The number of studies where these techniques were applied to glass fibers is of a smaller magnitude than those applied to the surfaces of high surface area silica.

We have tried to deal with this problem with the help of inverse gas chromatography (IGC). In contrast to conventional gas chromatography, which is used for separation and analysis of gas mixtures, IGC is used (at least in this case) for the analysis of surfaces. In this application, IGC has only one component (solute) in the carrier gas and yields information about the surface of the chromatographic column packing from the shape of the elution profile of this solute. We can apply IGC to the study of the surface heterogeneity of low surface area solids such as glass fibers. This heterogeneity is closely related to surface atomic structure. The information on the latter is very scarce for silicate glasses and even for amorphous silica. The reason for that is again the very small specific surface of silicate glasses. This is aggravated by the fact that atomic structures of glasses are amorphous and the conventional methods of atomic structures analysis, such as x-rays or neutron scattering, are not yet sufficiently developed for the study of such surfaces. Even if these methods do give reliable information on the bulk structure, which is the case for silica and some silicate glasses, they still cannot resolve their surface atomic structures. To approach this problem, we resort to computer simulation of glass surfaces and the adsorption on them. Thus, we are concerned in this review with two closely connected subjects: atomic structure and surface heterogeneity of silicate glass surfaces. However, in an attempt to solve these specific problems, we have developed new methods which are more general and deal with problems of computer simulation of all glass surfaces and surface heterogeneity in general.

To understand the meaning of the surface heterogeneity dealt with here, consider first an idealized model of a surface as a plane border separating a uniform continuous medium from a vacuum. This is a perfectly homogeneous surface. All other solid surfaces are more or less heterogeneous. This means that with respect to their interaction with external molecules at sufficiently low temperatures, they can be considered as composed of adsorption (active) sites or simply sites. Molecules adsorb only on these sites (which is, of course, no more than a useful model). If the sites form a regular pattern on the surface, as for faces of ideal crystals, the surface can be called regularly heterogeneous. Regularly heterogeneous surfaces are also often called homogeneous, otherwise the surface is described as heterogeneous. The literature on adsorption on heterogeneous surfaces is extensive. It suffices here to refer to a comprehensive review of earlier (still useful) papers [1], two monographs [2,3], and a recent review considering surface heterogeneity and IGC [4].

A typical example of a surface that can be strongly heterogeneous is that of an amorphous solid. In this case, the sites do not display a regular pattern as on homogeneous (regularly heterogeneous) surfaces, but are irregularly scattered over the surface and have various adsorption energies due to the irregular atomic structure of an amorphous surface. The isotherms, heats, and other characteristics of adsorption on heterogeneous surfaces deviate considerably from what is expected on theoretical grounds (and observed in specially designed experiments) for homogeneous surfaces. Such adsorption characteristics are frequently observed not only for amorphous solids, but also for high surface area crystalline powders. For example, isotherm, heats, and heat capacities of argon on industrially prepared high surface area crystalline rutile (TiO_2) and some other tiny crystals are very close to those on amorphous silica (which suggests that they might be covered by amorphous layers [4–6]). Another possible explanation for the heterogeneity of surfaces of almost all industrially prepared high area oxides is a high concentration of surface defects that make their surface atomic structure closer to the irregular atomic arrangement of amorphous solids than to the regular atomic structure of an ideal crystal [5].

A glass atomic structure is amorphous, that is, irregular. Molecules adsorbed at different points of such a surface see different environments. Thus one should expect heterogeneity of silicate glass surfaces, which may be considered from an adsorption and chromatographic point of view as composed of adsorption sites. The basic (and fundamental for what follows) characteristic of an adsorption site is the time of adsorption (residence time) which can be described by the Frenkel equation [7]:

$$\tau = \tau_0 \exp\left(-\frac{E}{RT}\right) \qquad (16.1)$$

Here, τ is the time of adsorption, which is the mean time a molecule spends in an adsorbed state on a site; the time itself being, of course, random. In Equation 16.1, E is the energy of adsorption—that is a difference between energies of the adsorbed molecule in the adsorbed and the gas state (see Section 16.2); R is the gas constant; T is temperature; and τ_0 is a constant connected to the entropy of adsorption (see at the end of Section 16.2). In the Langmuir model, which is the most widely used in the theory of adsorption on heterogeneous surfaces [2–4,8], the time of adsorption determines both the equilibrium isotherm and the kinetics of adsorption. This model assumes that each site can adsorb only one molecule and adsorption on different sites occurs independent of one another. If sites are identical, the surface is homogeneous; otherwise it is heterogeneous. Also, in the molecular theory of chromatography on homogeneous surfaces, the time of adsorption is the most fundamental characteristic of a site. It determines both the net retention time and the shape of an elution profile of a chromatographic column [7,9]. When a molecule is carried along the column by a carrier gas it will be adsorbed on a series of sites. The net retention time is the total time the molecule spends in an adsorbed state during its travel along the column. Clearly, the net retention time depends not only on the time of adsorption on a site, but also on the number of sites in the column. In conventional gas chromatography, a column is usually packed with high surface area solids to increase the number of adsorption sites and correspondingly the net retention time that increases the effectiveness of mixture separation.

In contrast to this, the specific surfaces of glass fibers and correspondingly the numbers of adsorption sites in columns packed with such fibers are relatively small. This decreases the net retention times of such columns and puts a limit on the ability of IGC to study these surfaces. When the net retention time gets close to the retention time (the full time of travel from injector to detector), the shape of the elution profile is determined by various smearing factors [10] and not by the adsorption characteristics of the surface. This is certainly true for homogeneous surfaces but not necessarily so for strongly heterogeneous surfaces which are composed of various adsorption sites. Some of these sites are characterized by very long times of adsorption with respect to some probe molecules at sufficiently low temperatures, which produce very long net retention times for those molecules that happen to adsorb on them. Thus the elution profiles of some probe molecules (solutes) on glass surfaces have long tails characteristic of strong adsorption sites. In the version of IGC described below, we pay special attention to those tails. Therefore, the peculiarity of our analysis of low surface area solids by IGC is that we try to characterize not all the adsorption sites, but only the strongest.

One reason for our special interest in the strongest sites for physical adsorption is the concept of precursor state, introduced by Langmuir and Lennard-Jones [11]. It is known that in many cases, chemisorption, and in particular dissociative chemisorption, occurs in two stages: first a molecule adsorbs in a precursor state, and then it dissociates. In the majority of cases, the precursor state is a physically adsorbed molecule. It is, for example, the first stage of chemical reaction of siloxane with silica [12]. For this reason, we believe that the strong adsorption sites for physical adsorption of water might be favorable places for the dissociative chemisorption of water on glass. It is known that surface interactions with water, in particular water chemisorption and dissociation on glass surfaces, are the reason for glass corrosion [13]. Thus the stronger sites might be the places where glass corrosion or crack growth might follow the physical adsorption of water molecules. In general, the stronger the site, the longer it holds a physically adsorbed molecule. Since the chemisorption and dissociation of molecules on the surfaces are relatively slow activated processes a molecule adsorbed on a weak site for a very short time simply does not stay on the surface long enough to experience chemisorption. Moreover, a strong site can distort a physically adsorbed molecule like water, and in this sense is fundamentally related to chemisorption. Strong adsorption sites for physical adsorption may also play an important role in other interactions of glass surfaces with their environment, such as biomolecule immobilization, polymer adhesion, and stress-corrosion cracking. Another reason for emphasizing the role of the strong adsorption sites is that they conform better to the Langmuir model of adsorption than the weak ones, thus validating the choice of the Langmuir model of adsorption as the basis of the treatment of surface heterogeneity discussed below (see the end of Section 16.2).

A surface is heterogeneous from the chromatographic point of view if the times of adsorption vary from site to site. Since energy of adsorption E in Equation 16.1 depends both on the surface and the probe molecule, the surface heterogeneity, as determined by variation of τ, is characteristic not of a surface but of a system: surface/probe molecule. The variation in adsorption time also depends on the temperature, as seen from Equation 16.1. At sufficiently high temperatures, the exponential function in Equation 16.1 is close to unity and the whole concept of adsorption sites breaks down. In fact, it can be shown that at sufficiently high temperatures, a heterogeneous surface effectively becomes homogeneous with respect to its adsorption properties [14]. In what follows, we always refer the strength of a site to an appropriate molecule at a particular low temperature.

In the next section, the method for analysis of surface heterogeneity developed [15–17] is described. In fact, this is an extension of the method widely used in IGC on homogeneous surfaces to heterogeneous surfaces. The application of IGC to the analysis of homogeneous surfaces usually consists in a determination of the Henry constant K that is also called the Langmuir constant or a distribution coefficient. This constant is proportional to the time of adsorption τ in Equation 16.1 and totally determines adsorption on a site in the Langmuir model. However, as seen in Equation 16.1, τ and therefore K depend exponentially on E/T, which allows one to determine E from IGC experiments performed at different temperatures. Simultaneously with E, one determines by this method the entropy of adsorption on a site S. This is a standard procedure for IGC analysis of homogeneous surfaces composed of only one type of site. (The concept of adsorption site is not that important for homogeneous surfaces, nevertheless we use this terminology to preserve uniformity with the treatment of heterogeneous surfaces.) For a heterogeneous surface composed of various sites, one would expect a distribution of sites both in E and S. However, the modern theory of adsorption on heterogeneous surfaces is couched only in terms of the energy distribution of sites [4]. This is achieved by approximate theoretical estimate of the pre-exponential factor in the Langmuir constant K, which is equivalent to the estimate of the entropy of adsorption on a site. This estimate may be reasonable for small molecules, such as nitrogen, that were used in earlier studies of surface heterogeneity but hardly can be applied to complex organic molecules usually used as probe molecules in IGC. Instead, we describe in the next section a method which allows one to determine the energy and entropy distributions of sites without any estimates. The method makes no assumptions except for the validity of the concept of adsorption sites (patches of homogeneous surfaces). It also provides a consistency check that allows one to determine if the standard procedure of IGC, which is widely used for obtaining isotherms of adsorption, and the adsorption energy distributions is consistent. It turns out that this check fails for strong adsorption sites, which raises the problem of modifying the standard IGC procedure. This is discussed in Section 16.3, where we also present and discuss some results of the application of the method described in Section 16.2.

Finally, in Section 16.4 we consider computer simulation of the surface atomic structure and the surface heterogeneity of silicate glasses. The basics of computer simulations of silicate melts and solid amorphous silicates were reviewed in Ref. [18]. Such standard items of computer simulations of silicates as inter-atomic potentials are thoroughly discussed in this review and will not be taken up here. Instead, we focus on the problem of time in computer simulations of glasses. In our view, this has not been paid due attention in the literature, especially in the papers concerned with simulations of silicate glass surfaces. The importance of this problem is due to the fact that the glass is an essentially nonequilibrium atomic structure, which means that its structural or thermodynamic characteristics, for example the atomic structure of its surface, depend on the history of the glass formation. (In contrast with this, a silicate melt is an equilibrium structure which does not depend on the history of its formation but only on such thermodynamic parameters as pressure and temperature.) For example, consider two silicate glass surfaces: a fracture surface obtained by crushing a solid piece of glass in a mortar and a solid surface obtained from a silicate melt at its glass transition temperature. We will call the latter a quasi-equilibrium glass surface. (One cannot call it an equilibrium surface because the glass is an essentially nonequilibrium solid state.) As shown in Section 16.4, these two glass surfaces have qualitatively different atomic structures and heterogeneities in contrast with the equilibrium liquid surface that does not depend on the way it was obtained. Because of the great difference between the time scales of computer simulation and the real glass transition experiments, computer simulations of a fracture glass surface are relatively easy to perform compared to simulation of the quasi-equilibrium glass surface.

16.2 HETEROGENEITY OF SURFACES

The conventional method for the determination of adsorption thermodynamic functions in IGC is based on the following equations (see, e.g., Ref. [19], p. 231):

$$\Delta G^{\ominus} = \Delta H^{\ominus} - T \Delta S^{\ominus} = -RT \ln K = -RT \ln V_S \qquad (16.2)$$

where $\Delta G^{\ominus}$, $\Delta H^{\ominus}$, and $\Delta S^{\ominus}$ are the standard molar Gibbs free energy, enthalpy, and entropy of adsorption, respectively; K is the Henry constant mentioned above (the slope of an isotherm at zero coverage), and V_S is the net retention volume divided by the total surface area of the column packing. The values of $\Delta G^{\ominus}$, $\Delta H^{\ominus}$, and $\Delta S^{\ominus}$ (to simplify the notations, we will designate them below as G, H, and S) are the differences between the values of the corresponding molar thermodynamic functions in the standard adsorption state and those in the standard gas state (see the definition of these standard states below). In fact, G is the reduced chemical potential of the adsorbate—which in adsorption equilibrium is equal to that of gas—so that we will call G chemical potential below. In Equation 16.2, thermodynamic functions have nonphysical dimensionality since the arguments of logarithmic functions are not dimensionless. This, however, does not prevent successful use of Equation 16.2 in IGC. The inconsistency can be removed by multiplying V_S in Equation 16.2 by a factor making the argument of logarithm dimensionless [17]. It is usually assumed that H and S do not depend on temperature (at least in a relatively narrow temperature interval of a conventional chromatographic experiment) and can be obtained from the temperature dependence of G or $RT \ln V_S$:

$$dG/dT = -S \qquad (16.3)$$

Equation 16.2 and Equation 16.3 are the basic equations that are widely used in IGC for characterization of homogeneous surfaces. On a homogeneous surface, the elution profile is an approximately symmetric peak, such as that of octane in Figure 16.1. (The slight asymmetry of the methane and octane peaks is mainly due to a very low efficiency of the short column packed with low surface area glass fiber.) The retention time of octane in Figure 16.1 is obtained from the position of the maximum of its peak. At sufficiently small injected amounts, it does not depend on the injected amount of octane, and after subtracting the retention time of a nonadsorbing gas (methane), it can be used to determine [19,20] the retention volume V_S in Equation 16.2 as well as G, and finally, using

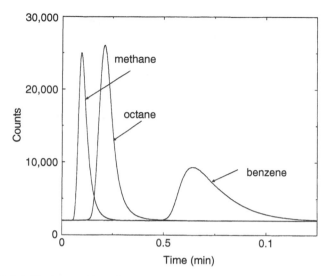

FIGURE 16.1 Elution profiles of non-polar solutes on a coloumn packed with 14μm E-glass fibers at 60°C counts are internal units of detector.

Equation 16.3, the standard enthalpy H and entropy S of adsorption of octane on a glass surface.

The fact that an elution profile should be approximately symmetric on a homogeneous surface follows from the molecular theory of gas chromatography [7,9]. The position of the maximum of such a peak that can be measured at any temperature (in a certain temperature interval) determines the pair of constants $\{H, S\}$, which thus determines the adsorption properties of a homogeneous surface. However, one frequently observes other types of elution profiles in IGC. Two examples are the elution profiles of benzene in Figure 16.1 and of butanol in Figure 16.2. The latter is a strongly asymmetric peak with a sharp jump called "shock" followed by a long tail. This tail can last for hours and still there will be some butanol molecules on the surface, which are not eluted by the carrier gas (helium) at 100°C. To remove them, we increase the temperature after 30 min of elution and obtain, in fact, a thermo-desorption peak such as that shown in Figure 16.2 for 250°C. Its area gives the number of molecules which were not eluted from the surface at 100°C after 30 min. In contrast to the peak for octane in Figure 16.1, the maximum of the left-hand peak in Figure 16.2 (the position of shock) does not determine the adsorption properties of butanol on an E-glass surface because it depends mainly on the injected amount of butanol (0.02 μL in Figure 16.2). If one injects slightly less butanol, shock will shift to the right but its tail will coincide with that corresponding to the larger injected amount and the

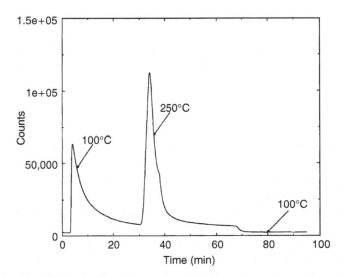

FIGURE 16.2 Elution profile of butanol on the some coloumn as in Figure 16.1 at indicated temperatures (see text).

thermo-desorption peak will not change. Since the position of the maximum of the elution profile in Figure 16.2 does not determine the adsorption propertics of the surface, the analysis based on Equation 16.2 and Equation 16.3 is not applicable and another approach is necessary. This is based on the assumption that the E-glass surface is strongly heterogeneous with respect to the adsorption of butanol at 100°C, weakly heterogeneous with respect to adsorption of benzene at 60°C [8], and homogeneous with respect to adsorption of octane at 60°C.

As noted previously, the adsorption properties of a homogeneous surface can be described by a pair of constants $\{H, S\}$. As a first approximation, a heterogeneous surface can be considered as a collection of patches of homogeneous surfaces, each patch being described by a unique pair of constants $\{H, S\}$. Thus a heterogeneous surface is described by a collection of such pairs, which gives rise to the problem of finding a bi-dimensional distribution of the number of patches with respect to these pairs. Let $n(H, S)$ be the bi-dimensional density of patches, $n(H, S)\mathrm{d}H\mathrm{d}S$ being the number of patches in the rectangle of area $\mathrm{d}H\mathrm{d}S$ around the point (H, S) in the plane of H and S. Consider the function $N(G)$ defined by [17]

$$N(G) = \int\int \chi(H - TS - G)n(H, S)\mathrm{d}H\mathrm{d}S \qquad (16.4)$$

where T is a parameter and $\chi(x)$ is a unit step function.

$$\chi(x) = 1 \quad \text{if } x < 0 \quad \text{and} \quad \chi(x) = 0 \quad \text{if } x > 0 \tag{16.5}$$

The integration in Equation 16.4 is over the domain of definition of H and S, but the step function in the integrand selects only those pairs of H and S which satisfy the inequality

$$H - TS < G \tag{16.6}$$

which means that $N(G)$ is the total number of patches that have chemical potential smaller than G. In other words, $N(G)$ is the distribution function for the number of patches with respect to G. (The derivative of $N(G)$, i.e., distribution density [21], is usually called a distribution function in adsorption literature [2–4]; in this case, one should call $N(G)$ in Equation 16.4 the cumulative distribution function.)

The easiest way to obtain an equation similar to Equation 16.3 from Equation 16.4 is by using generalized functions [22]. It is seen from Equation 16.4 that N is, in fact, a function of G and T that can be differentiated to give

$$\left(\frac{\partial G}{\partial T}\right)_N = -\frac{\left(\frac{\partial N}{\partial T}\right)_G}{\left(\frac{\partial N}{\partial G}\right)_T} \tag{16.7}$$

Now one can formally differentiate Equation 16.4 by using the generalized functions equality [22].

$$\frac{d\chi(x)}{dx} = -\delta(x) \tag{16.8}$$

where $\delta(x)$ is the Dirac delta function. Then

$$\left(\frac{\partial N}{\partial T}\right)_G = \iint \delta(H - TS - G)Sn(H, S)dHdS \tag{16.9a}$$

$$\left(\frac{\partial N}{\partial G}\right)_T = \iint \delta(H - TS - G)n(H, S)dHdS \tag{16.9b}$$

In Equation 16.9a and Equation 16.9b, the integration is again over the domain of definition of H and S, but the delta function in the integrand selects only those values of H and S, which satisfy the equality

$$H - TS = G \tag{16.10}$$

This means that the right-hand side of Equation 16.9b gives the number of sites, which have different values of H and S but the same value of G and the right-hand side of Equation 16.9a gives the total value of S for these sites. Thus the right-hand side of Equation 16.7 gives the mean entropy of adsorption $\langle S \rangle_G$ averaged over all of the sites with various values of H and S, but the same value of G:

$$\left(\frac{\partial G}{\partial T}\right)_N = -\langle S \rangle_G \tag{16.11}$$

(Equation 16.11 is discussed from another point of view in [41].)
For another interpretation of $\langle S \rangle_G$, consider a particular case when $n(H, S)$ in Equations 16.9 is a two-dimensional normal distribution density (normal density) [17,21]:

$$n(H, S) = \frac{N_0}{2\pi\sigma_H\sigma_S\sqrt{1-\rho^2}}\exp\left[-\frac{1}{2(1-\rho^2)}\left(\frac{\Delta H^2}{\sigma_H^2}-2\rho\frac{\Delta H\Delta S}{\sigma_H\sigma_S}+\frac{\Delta S^2}{\sigma_S^2}\right)\right] \tag{16.12}$$

Here, $\Delta H = H - \langle H \rangle$ and $\Delta S = S - \langle S \rangle$ where $\langle H \rangle$ and $\langle S \rangle$ are mean values of H and S, respectively; N_0 is the total number of sites on the surface; σ_H and σ_S are standard deviations of H and S, respectively; and ρ is the correlation coefficient between H and S. Thus the normal density determined by Equation 16.12 depends on six constant parameters: $\langle H \rangle$, $\langle S \rangle$, σ_H, σ_S, ρ, and N_0. Substitute Equation 16.12 into Equations 16.9, and integrate with respect to H or S. The result will again include the normal densities obtained by substitution $(G+TS)$ for H or $(H-G)/T$ for S in Equation 16.12. The argument of the exponential function in this normal density can be written as

$$y(x) = A(x-\langle x \rangle_G)^2 + B \tag{16.13}$$

where A, B, and $\langle x \rangle_G$ are constants depending on the parameters of Equation 16.12 as well as on G and T, x being S or H. It follows from Equation 16.13 that

$$\langle x \rangle_G = -\frac{\left(\frac{dy}{dx}\right)_{x=0}}{\left(\frac{d^2y}{dx^2}\right)} \tag{16.14}$$

Thus, to obtain expressions for $\langle S \rangle_G$ or $\langle H \rangle_G$, one substitutes $G+TS$ for H or $(H-G)/T$ for S in Equation 16.12 and uses Equation 16.14. The result is [17]

$$\langle S \rangle_G = \langle S \rangle + \frac{\sigma_S}{\sigma_H} \frac{(G - \langle G \rangle)(\rho - a)}{1 - 2\rho a + a^2} \qquad (16.15a)$$

$$\langle H \rangle_G = \langle H \rangle + \frac{(G - \langle G \rangle)(1 - \rho a)}{1 - 2\rho a + a^2} \qquad (16.15b)$$

where

$$\langle G \rangle = \langle H \rangle - T\langle S \rangle \quad \text{and} \quad a = T\sigma_S / \sigma_H \qquad (16.15c)$$

Two relationships follow from Equations 16.15. The first one is general,

$$\langle H \rangle_G - T\langle S \rangle_G = G \qquad (16.16)$$

and can also be obtained by averaging Equation 16.10 at constant T and G. The second relationship is

$$\frac{\langle S \rangle_G - \langle S \rangle}{\langle H \rangle_G - \langle H \rangle} = \frac{\sigma_S(\rho - a)}{\sigma_H(1 - \rho a)} = \frac{1}{T_i} \qquad (16.17)$$

Generally, $\langle S \rangle_G$ is a function of G, as seen from Equation 16.11, Equation 16.7, and Equations 16.9. However, the left-hand side of Equation 16.17 does not depend on G, which means that there is a linear relationship between $\langle S \rangle_G$ and $\langle H \rangle_G$ that does not depend on G, but only depends on T

$$\langle S \rangle_G = \frac{\langle H \rangle_G}{T_i} + S_0 \qquad (16.18)$$

If one knows the values of T_i and S_0 at a given temperature, one can determine the energy distribution directly from the chemical potential distribution at this temperature (see discussion of Figure 16.4). Linear dependences of entropy upon energy (enthalpy) similar to Equation 16.18 have been observed in various fields of physical chemistry, with the constant T_i being sometimes called isokinetic temperature (see [5] and references therein). In our particular case, it can be interpreted in the following way: consider an adsorption site (or group of identical sites) chosen at random on a heterogeneous surface. It will have random values of G, S, and H. The mean values $\langle H \rangle_G$ and $\langle S \rangle_G$ are also random because G is random. They can be called regressions on G of enthalpy and entropy of adsorption, respectively [17]. However, one can expect a correlation between the entropy and energy of a molecule adsorbed on this randomly chosen site—the larger the magnitude of adsorption energy, the more restricted is the freedom of the molecule adsorbed on the site and the less is the adsorption entropy. This does not mean that the relationship between the entropy and the energy of adsorption is necessarily linear, as indicated in Equation 16.17 and Equation 16.18. Their linearity is the direct consequence of the fact that we assumed the normal density for the site distribution, and the regressions of the normal distribution are linear [21].

Although Equation 16.11 is similar to Equation 16.3, there are two differences between them. Firstly, the full derivative on the left-hand side of Equation 16.3 is substituted by the partial derivative at constant value of the cumulative distribution of sites in Equation 16.11. Secondly, the meanings of the right-hand sides in Equation 16.3 and Equation 16.11 are different: while S in Equation 16.3 is the standard entropy of adsorption for the homogeneous surface (equal for all the identical sites in the Langmuir model), $\langle S \rangle_G$ in Equation 16.11 is entropy averaged over the sites with almost equal values of G. For a system adsorbate/homogeneous surface, S is constant; for a system adsorbate/heterogeneous surface, $\langle S \rangle_G$ depends on G or on N since N is a function of G, Equation 16.4.

Therefore, an analysis of an elution profile such as that shown in Figure 16.2 requires first the distribution of patches (sites) in G. The determination of this chemical potential distribution function of adsorption sites from experimental isotherms of adsorption or from an elution profile in IGC is a long-standing problem in the theory of adsorption on heterogeneous surfaces. Many different methods have been developed over the last 70 years which allow one to determine $n(G)$ or $N(G)$ from experimental data. These methods are reviewed in Refs. [1–4] (see also references in [16]). When $N(G,T)$ is obtained from experiments at different temperatures, one uses Equation 16.11 and Equation 16.16 to obtain N vs. $\langle H \rangle_G$ and N vs. $\langle S \rangle_G$. These two functions correspond to the (cumulative) energy and entropy distributions of homogeneous patches on a heterogeneous surface. Strictly speaking, the energy distribution of patches is N vs. E and not N vs. $\langle H \rangle_G$. However, the distinction between $\langle H \rangle_G$ and E is very small. Firstly, $H = E - RT$ ($E < 0$), because H and E are, as mentioned above, differences between the corresponding molar thermodynamic functions in the standard adsorbed and gas states. In an adsorbed state, the energy and the enthalpy coincide, and in the ideal gas state they differ by RT, but usually $RT \ll |E|$. Secondly, $\langle H \rangle_G$ is not the enthalpy of adsorption on a patch but the average enthalpy of adsorption on a group of patches having (almost) the same values of G. This is because on a heterogeneous surface, the patches having the same values of the chemical potential $G = H - TS$ can have different values of H and S since small deviation of H from its mean value can be compensated by the corresponding deviation of TS.

In the adsorption literature, the problem of the separation of the chemical potential on a heterogeneous surface into energy and entropy has been solved mainly on the basis of assumptions that values of S on all the patches are either the same or are given by some other approximations (see Introduction in [15] and Chapter 3, Section 2 in [16]), whereas the separation of the chemical potential into the enthalpy and entropy of adsorption on a homogeneous surface is an accurate procedure based on Equation 16.2. The method described above, in fact, extends this procedure to heterogeneous surfaces. It is still relatively accurate in the frame of the model of adsorption sites (patches). One limitation is that, strictly speaking, we obtain as a main result of this method not the energy distribution of homogeneous patches on a heterogeneous surface, but distribution of patches with respect to $\langle H \rangle_G$.

Another limitation of this approach follows from the fact that it is an extension of the method epitomized by Equation 16.2. The latter considers adsorption on homogeneous surfaces in the Henry region, i.e., in the limit of infinitely small coverage where the linear chromatography approximation (see discussion of Equation 16.21) is valid. In the Henry region, the interactions between adsorbed molecules can be neglected since they are far apart at very low coverage. Accordingly, in our extension of Equation 16.2, to heterogeneous surfaces we neglect the interaction between adsorbed molecules and characterize any homogeneous patch of a heterogeneous surface by two constants $\{H, S\}$ as if they were in the Henry region. But in this case, the total (global) adsorption isotherm on a heterogeneous surface would be linear and the elution profile would be a symmetrical band. This obviously is not the case for butanol in Figure 16.2. We resolve this contradiction by accepting the Langmuir model of adsorption for heterogeneous surfaces. In this model, even a nonlinear isotherm of adsorption on a homogeneous surface is described by the pair $\{H, S\}$ that account for the interaction of a single molecule with a surface and the interaction between adsorbed molecules being dealt with by assuming that each site can adsorb only one molecule. Thus, the Langmuir model takes account of the strong repulsion between molecules that does not allow more than one to be adsorbed on a single site and assumes that the much weaker interactions between molecules adsorbed on different sites can be neglected. In our case, this approximation of the Langmuir model is at least partly justified by the fact that, as mentioned in the Introduction and as shown in the next section, we consider not all the adsorption sites on a heterogeneous surface but only less than 20% of the strongest ones. These sites obey the Langmuir model better then the weaker ones, either because they are separated by relatively large distances so that interaction of molecules adsorbed on them is small, or because this interaction, even for neighboring sites, is much smaller than that of the molecules interacting with their respective surface sites. In the both cases, the interaction of molecules adsorbed on different sites can be neglected for the strongest sites. Another reason for considering only strong sites is that we study low surface area adsorbents and can obtain reliable experimental information only from the strongest sites (see the Introduction for still another reason).

In the conclusion of this section, we consider the choice of adsorption and gas standard states. These are, of course, arbitrary, but should be carefully fixed mainly to fix the entropy of adsorption. We chose these standard states on the basis of the Langmuir model of adsorption. The standard entropy of adsorption in the Langmuir model is [8]

$$S = R \ln(\nu \tau_0 C_0) \tag{16.19}$$

Here, C_0 is the concentration of the ideal gas chosen as a standard gas state; νC_0 is the number of impacts of gas molecules from this standard state on an adsorption site of area σ in unit time and

$$\nu = \sqrt{RT/2\pi M} \sigma \alpha N_A \tag{16.20}$$

Here, M is the molecular mass of an adsorbed molecule; α is the accommodation coefficient; and N_A is the Avogadro number. Finally, τ_0 in Equation 16.19 is the pre-exponential factor in Equation 16.1 and thus is the mean time that an adsorption site holds a molecule at a temperature extrapolated to an infinitely high value. To fix these standard states, we chose the following values in Equation 16.19 and Equation 16.20 [17]: $\tau_0 = 10^{-12}$ s, $\alpha = 0.1$, and σ equal to the molecular area of the adsorbed molecule that is used in the BET method for the determination of surface area (e.g., 0.28 nm^2 for butanol [17]). Finally, we choose a value of C_0 in Equation 16.19 that nullifies the standard entropy of adsorption S: $C_0 = 1/\nu\tau_0$. With such a choice of parameters, the value of C_0 at 300 K equals about 800 mol/dm^3 corresponding to a pressure of the ideal gas of about 2 GPa. Since the gas in the standard state is ideal, the standard enthalpy of adsorption H does not depend on the gas pressure. As mentioned above, the standard entropy of adsorption is the difference between the absolute value of the entropy of adsorbate and that of gas in their respective standard states. The entropy of adsorbate in its standard state is equal to the thermal entropy in the Langmuir model [8]. The high pressure of the gas standard state has been chosen to make its molar entropy equal to that of adsorption standard state at the values of parameters τ_0, α, and σ mentioned above. (Since values of α and τ_0 for real adsorption sites are different from those chosen above, the entropy of adsorption deviates from zero.)

16.3 GC ON GLASS SURFACES

Now we consider the application of the method described in the previous section to the characterization of glass surfaces. Inverse gas chromatography has already been applied to the study of glass fiber surfaces [23–25]. However, probe molecules and temperatures were chosen such that the majority of elution profiles were symmetric, like those in Figure 16.1. Thus the surfaces were considered to be homogeneous and analyzed on the basis of Equation 16.2 and Equation 16.3. Even in the case of asymmetric elution profiles, the surfaces were treated (approximately) as homogeneous. In contrast to this, we focused on glass surface heterogeneity with respect to a polar probe molecule as butanol. The reason for this choice is explained at the end of this section.

Chromatographic columns for our studies were mainly prepared by pulling about 3 g of glass fibers (usually about 10 μm diameter) through a glass tube (ca. 23 cm long, ca. 4 mm internal diameter) [17]. Since the glass surface in such a column was about 0.4 m^2, the column is very ineffective from the conventional gas chromatography point of view. One can see in Figure 16.1 how small its retention times for octane and even benzene are. (Of course a longer column would be more effective, but it would be more difficult to pack with continuous glass fiber. The ease of a column preparation is not that important in conventional chromatography, where one column can be used for analysis of many samples, but is critical for IGC where usually each analysis includes preparing a new column.) However, the situation is very different for butanol, as shown in Figure 16.2. In this case, elution profiles have very long tails and one can use the methods developed in the theory of adsorption on heterogeneous surfaces to determine from these tails the distribution of (strongest) sites (patches) with respect to chemical potential. We emphasize that one cannot directly determine the energy distribution of sites but only their chemical potential distribution. This standard procedure, as well as details of experimental techniques, are described elsewhere [16,17]. The chemical potential distributions depend on temperature, and one can use Equation 16.11 and Equation 16.16 to split it into two separate distribution for energy and entropy: the distribution of patches (sites) with respect to $\langle H \rangle_G$ and $\langle S \rangle_G$ will be called the energy and entropy distributions of sites. The details of the analysis are described elsewhere [17] and the energy distribution obtained by this method is shown in Figure 16.3.

The distribution there has physical meaning only in the interval of fraction of sites larger than 0.04 and less than 0.12. Beyond these borders, two values of fraction of sites can correspond to one energy, which is not what one would expect from a (cumulative) distribution function. To understand the reason for these inconsistencies, we describe the standard procedure for obtaining the chemical potential distribution of sites in IGC. First of all, it is based on the

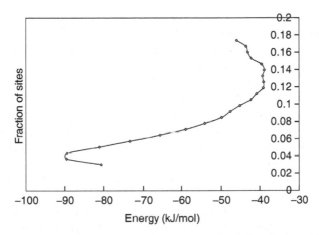

FIGURE 16.3 The energy distribution of sites as obtained directly from IGC.

mathematical model of a chromatographic column that can be described in terms of the following balance equation [10]:

$$\frac{\partial C}{\partial t} + \frac{\partial q}{\partial t} + \frac{\partial uC}{\partial z} = D\frac{\partial^2 C}{\partial z^2} \qquad (16.21)$$

Here, t is the time, z is the coordinate along the column, C is the concentration of the solute in the carrier gas, q is its concentration in the stationary phase in the column, and u is the average (over a cross-section of the column) linear velocity of the carrier gas. Finally, D (which is sometimes called axial or longitudinal diffusion coefficient) takes into account the fluctuation of linear velocity of the carrier gas in a packed column (eddy diffusion) and the molecular diffusion of the solute in the carrier gas. In gas chromatography, u depends on z due to the compressibility of the carrier gas, which does not allow one to take u out of the partial derivative in the third term on the left-hand side of Equation 16.21 as is usually done in liquid chromatography [10]. To simulate a process in a column requires solving Equation 16.21 numerically under appropriate initial and boundary conditions to obtain the elution profile. Since Equation 16.21 is also the mathematical model of industrial adsorbers for purification and separation of various compounds, its numerical solution has been considered almost since the advent of computers. Some of these simulations, specifically related to chromatography, are described in [10].

There are, however, two main cases when Equation 16.21 can be solved analytically. One case is called linear chromatography [10]. It is applicable when q in Equation 16.21 is proportional to C. For homogeneous surfaces, this regime can be achieved at very small values of C. It is widely used in IGC and leads to Equation 16.2. Theoretically, the regime of linear chromatography should be also achievable for heterogeneous surfaces. For this to take place, adsorption on all the sites, even the most active ones, should be in linear regime (in the Henry region). However, it was shown many years ago that even in an ultrahigh vacuum (down to 10^{-11} mmHg) isotherms of adsorption of nitrogen and rare gases at their respective boiling temperatures on glass surfaces are still strongly nonlinear, which has been explained by glass surface heterogeneity (see review in Ref. [3], pp. 42–51). The same situation can easily occur for IGC with organic molecules. Of course, one can always reach the Henry region that is to make an isotherm linear even on a heterogeneous surface by raising temperature. The slope of the linear isotherm obtained in this way on a heterogeneous surface will be the Henry constant averaged over all the patches [26]. In principle, this averaged Henry constant contains information on the distribution of patches with respect to their Henry constants, which according to Equation 16.2 is equivalent to the distribution with respect to the chemical potential and can be extracted either by the method outlined in [26] or by the method of moments developed in [8]. Both of these methods, however, are considerably less informative than that described above. Besides, as explained in Ref. [16], the temperature at which the linear regime is achieved for high boiling point molecules, such as butanol in Figure 16.2, can be prohibitively high for gas chromatography.

In a nonlinear case, Equation 16.21 can be solved analytically under the following approximations:

$$\text{(i) } D = 0; \quad \text{(ii) } u \text{ does not depend on } z; \quad \text{(iii) } q = q(C). \tag{16.22}$$

Here, $q(C)$ is an equilibrium isotherm of adsorption, so that relationship (iii) in Equation 16.22 means that adsorption equilibrium is reached instantly at any point of a column, i.e., the kinetics of adsorption is neglected. Under these assumptions, Equation 16.21 simplifies to the following form:

$$\left(1 + \frac{\partial q}{\partial C}\right) \frac{\partial C}{\partial t} + u \frac{\partial C}{\partial z} = 0 \tag{16.23}$$

This can be immediately solved, since

$$\left(\frac{\partial z}{\partial t}\right)_C = -\frac{\frac{\partial C}{\partial t}}{\frac{\partial C}{\partial z}}$$

The solution of Equation 16.23 is:

$$\left(\frac{\partial z}{\partial t}\right)_C = \frac{u}{K(C)} \quad \text{where } K(C) = \left(1 + \frac{\partial q}{\partial C}\right) \tag{16.24}$$

This solution is the basis for the method called elution by characteristic point (ECP), which is widely used in IGC [19,20]. A meaning of Equation 16.24 is that each concentration (characteristic point) moves along the column independently with a velocity determined by u and $K(C)$. This gives the following equation [19,20]:

$$\frac{dN}{dC} = V_N(C) = jt(C)F \tag{16.25}$$

Here, $C(t)$ is a function corresponding to the tail of the left-hand peak in Figure 16.2 (the function on the right side of Equation 16.25 is inverse to $C(t)$), C being proportional to the detector signal at time t. In Equation 16.25, F is the volume flow rate of the carrier gas; j is the factor [20] that makes an approximate correction for the compressibility of the carrier gas; and V_N is called the net retention volume [20]. The left-hand side of Equation 16.25 is the derivative of the isotherm of adsorption, the latter being the amount adsorbed per the total amount of adsorbent in the column vs. concentration. If one converts concentration of the ideal gas into its chemical potential and uses the so-called condensation approximation of the theory of adsorption on heterogeneous surfaces [2–4], one readily obtains the chemical potential distribution density of adsorption sites in the column (normalized to the total number of sites in the column) from the left-hand side of Equation 16.25. (The condensation approximation becomes almost exact for sufficiently strong adsorption sites and needs only very small correction [16]). One must then integrate starting from the number of sites given by the thermo-desorption peak in Figure 16.2 to obtain the chemical potential distribution of sites.

It should be emphasized here that the chemical potential distribution obtained in this way is a normal single-valued plot. The energy distribution obtained from it by the approximate methods mentioned above also looks normal. The problem of the multi-valued distribution function shown in Figure 16.3 arose only when we applied the method described in the previous section for obtaining the energy and the entropy distributions of sites from the temperature dependence of the chemical potential distribution of sites. However, we do not believe that the problem arises due to the inconsistency of the method described in the previous section, but rather that the application of this relatively accurate method uncovered the problem, which existed but was overlooked by less accurate methods used earlier. (This is an example of the consistency check mentioned in Section 16.1.) It seems likely that the inconsistency of the analysis presented in Figure 16.3 is due to approximations in the ECP method. In particular, the unrealistic behavior of the energy distribution in Figure 16.3 in the region of the strongest sites (fraction of sites less than 4%) is due to neglect of kinetics of adsorption in the ECP-method (see (iii) in Equation 16.22). The stronger a site is, the slower the kinetics of adsorption will be. This is both because the time of adsorption on a stronger site is longer and the equilibrium gas phase concentration is lower so that the transport from the carrier gas to the surface is slower. ECP does not take the relatively slow kinetics of adsorption on the strongest sites into account but assumes instant adsorption on each site. The reason for unrealistic behavior of the energy distribution in Figure 16.3 in the region of the weakest sites (fraction of sites above 12%) is probably due to neglect of axial diffusion ((i) in Equation 16.22) by ECP. This diffusion is known to be one of the main smearing factors that determines the shape of a narrow lines in chromatography [10] and correspondingly the weak site part of the left-hand peak in Figure 16.3. To solve these problems, one has to resort to an exact (without the ECP approximations of Equation 16.22), numerical solution of Equation 16.21.

In the meantime, a palliative solution of this problem can be found by plotting the entropy vs. energy of adsorption, as suggested by Equation 16.18 and as shown in Figure 16.4. In the central region of this plot, which corresponds to the physically meaningful part of the distribution in Figure 16.3, the relationship between entropy and energy is linear as predicted by Equation 16.18. We simply found coefficients of Equation 16.18 for the middle part of the straight line in Figure 16.4; ($S_0 = 132 \text{ J mol}^{-1}\text{K}^{-1}$, $T_i = 380$ K). The line has been extrapolated to the domain where ECP give

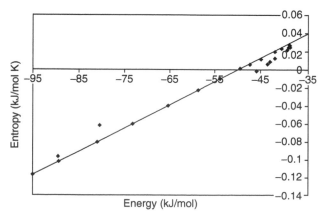

FIGURE 16.4 The entropy/energy relation.

unrealistic results. Equation 16.16 and Equation 16.18 allow one to find $\langle H \rangle_G$ from any value of G; i.e., to formally convert the chemical potential distribution of sites obtained by ECP into the energy distribution. (This is valid only for butanol in the limited temperature interval around 50°C; for 100°C the values of S_0 and T_i are slightly different [16].) The results are qualitatively correct, but the energy scale is distorted due to the inadequacy of the ECP method for strong adsorption sites. One can even argue that such an extrapolation overestimates the energies of strong adsorption sites. Indeed, it is known that tailing of peaks in chromatography can be caused not only by surface heterogeneity but also by slow kinetics (see, e.g., Ref. [20], pp.37, 53). Since ECP neglects the kinetics of adsorption, it ascribes tailing caused by slow kinetics on strong adsorption sites to their energy.

The distribution of sites with respect to their energy of interaction with butanol also dramatically depends on the modification of the surface. The most widely used modification of the glass fiber surface is silylation, which is treatment of the surfaces by organosilanes. Usually organosilanes are used as coupling agents in glass reinforced polymer composites—the inorganic parts of organosilanes strongly interact (chemically) with glass while their organic parts strongly interact with organic polymers. Thus, silylation covers a glass surface by the organic groups of organosilanes. This considerably reduces the interaction of the surface with the OH-group of butanol and dramatically changes the energy distribution with respect to this probe molecule. The resulting site energy distributions also depend on the temperature of silylation (curing temperature), because this temperature determines both the kinetics of chemisorption of organosilanes on a glass surface and the temperature stability of the organic film formed on a glass surface [17].

Finally, we will discuss the choice of butanol as a probe molecule. The primary reason for this is that on one hand, the most popular in IGC flame ionization detector (FID) is quite sensitive to butanol. However, the alkyl group of butanol that is responsible for this sensitivity has practically no influence on the retention time of butanol on glass fiber surfaces, which is almost totally determined by the OH-group of this molecule. One can see from comparison of Figure 16.1 and Figure 16.2 that octane has a much smaller retention than butanol (on the strongest sites), and butane should have retention time even smaller (somewhere between that of methane and octane). Thus, one may argue that it is not the alkyl but the OH-group of butanol that primarily determines the interaction of this molecule with a glass surface. (We have also shown experimentally that alcohols with somewhat larger or smaller alkyl groups give about the same energy distributions as butanol on E-glasses.) Thus, we use butanol as a probe molecule because it, in fact, probes the interaction of an OH-group with a polar surface. As mentioned above, the observed surface heterogeneities depend not only on the surface but also on the probe molecule or even on the functional group that determines the surface/molecule interaction (as well as on temperature). The choice of OH-group for probing the silicate glass heterogeneity is justified because this group determines the interaction of glass surfaces with many important molecules, particularly water, which was mentioned in the Introduction as the prime reason for silicate glass corrosion.

16.4 ATOMIC SIMULATIONS OF GLASS SURFACE HETEROGENEITY

In computer simulations of the atomic structures associated with surface heterogeneity of silicate glass surfaces, the goal is to understand the interactions of these heterogeneous surfaces with adsorbed substances from the atomic point of view. Since the glass is an essentially nonequilibrium state of matter, we consider first a parameter called the Deborah number (De), which quantifies the deviation of a system from equilibrium. Deborah is a prophetess (after Judges V:5) who (in the original version) says, "The mountains flowed before the Lord" [27]. This is interpreted as: what appears to be stationary to mortals is not necessarily so to an eternal deity. Thus a Deborah number is defined as a ratio of the timescales of the observed and the observer [28]. When $De \ll 1$ a condensed system is relaxed, i.e., in an equilibrium state, whereas the situation when $De \gg 1$ pertains to nonequilibrium states, for instance, a polarized interface in electrochemistry [27]. In the case of rheology (where the Deborah number first appeared [27]) or in glass formation, $De \ll 1$ means that the system is fluid whereas $De \gg 1$ means that it is solid. At the border between these extremes, $De \approx 1$ defines the glass transition temperature [27,28].

The timescale of formation of a glass fiber surface may be estimated as the time necessary for a liquid thread flowing from an orifice at the bottom of a crucible (called bushing) to solidify in air. The estimate for this time is 0.1–1 ms [29] for an E-glass thread about 10 μm diameter, and this is probably the smallest timescale for the glass surface formation from the melt. The surface atomic structure of the basic silicate glass—amorphous silica—has been simulated many times by molecular dynamics [30–35]. The timescale of these simulations was at most 1 ns. The time of formation of the glass fiber surface mentioned above can be taken as a timescale of observed and 1 ns as that of observer (simulator). Thus, for the computer simulations of glass surfaces described in the literature, the Deborah number is higher than 10^5 (an exception is in Ref. [35], where, in fact, the surface of liquid silica was simulated). This means that these studies did not, in fact, simulate the glass surface solidified from the melt at the real glass transition temperature. As mentioned in the

Introduction, we call such surfaces quasi-equilibrium. They are formed at $De \approx 1$ [28] and cannot be simulated at the present time by conventional means. What was probably simulated in the papers cited above was another type of glass surface called the fracture surface (see Section 16.1).

We concluded that this might be the case while trying to simulate a hydrophobic silica surface [36]. The work was suggested by the well-known experimental fact that surfaces of dehydroxylated silica gels are hydrophobic (see [36] and references therein). (Usually, silica glass surfaces are covered by silanol groups as a result of water chemisorption and thus are hydrophilic.) The surfaces of amorphous silica gels are usually obtained at conditions close to equilibrium, at least in comparison to fracture silica surfaces. For this reason, we consider them to be quasi-equilibrium surfaces together with silicate glass surfaces solidified from melts. This suggests that any quasi-equilibrium silica surface not covered by silanols, in particular a simulated one, should be hydrophobic. On the other hand, it is also known that the surface of silica fractured in ultrahigh vacuum avidly adsorbs water and certainly is not hydrophobic [37]. (A surface or some of its patches may be hydrophobic if the heat of water adsorption on the patch is less than the heat of water liquefaction, i.e., less than 44 kJ/mol at room temperature.)

These experimental facts are in line with the computer simulation results presented in Figure 16.5 [36]. The shading in the two parts of this figure represents values of the minimum energy (in kJ/mol) of a water molecule interacting with the simulated silica surface, as shown in the key to the right of the figure. The energy minimization was carried out both over all orientations of the molecule and its distance from the surface. (Distances along the x- and y-axes are in Å.) The upper part of the figure represents a fracture surface that contains many coordination defects (dangling bonds). This simulated surface was obtained in a molecular dynamics simulation by a method similar to others described in the literature [30–34]. In this method, one proceeds by first simulating the bulk silica atomic structure. Details, which are not very different from those considered in the review [18], are described in [36]. Then, one cuts this atomic structure by a plane and all atoms in the half-space above the plane are removed (see details in [36]). This creates an initial surface that contains a large concentration of under-coordinated atoms ("dangling" bonds) because some of the atoms that had been bonded to those remaining in the lower half-space were removed with the upper half-space. Usually in computer simulations of atomic structures, the initial state is arbitrary and does not necessarily model a real atomic structure. This is also the case for this initial surface since its artificial construction corresponds neither to a quasi-equilibrium surface nor, strictly speaking, to a fractured one.

To convert this initial surface into a realistic one (quasi-equilibrium or fractured), it is necessary to make it possible for the simulated atomic structure

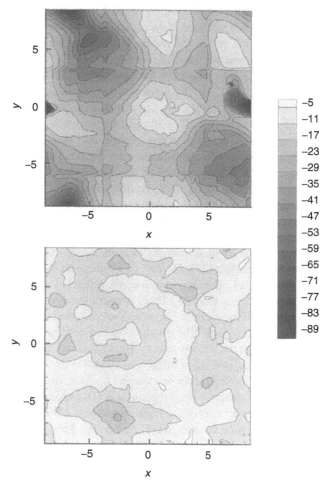

FIGURE 16.5 Adsorption energies of a water molecule over a fracture and a quasi-equilibrium silica surfaces (see text).

to rearrange itself according to the prescribed inter-atomic interactions, assuming, of course, that the latter are realistic. We will call this process annealing because both real and simulated annealing includes an initial elevated temperature followed by a slow cooling. However, the simulated annealing is very different from the real annealing of silicate glasses. The latter uses a very slow (on the order of an hour) cooling of a macroscopic piece of glass to remove residual stresses while the former uses about 12 orders of magnitude faster (on the order of a nanosecond) simulated cooling intended to rearrange the model atomic structure of several hundred atoms. The annealing of the upper structure in Figure 16.5 was performed by raising the temperature of the initial surface up to 4000 K and then cooling it at an exponentially decreasing rate down to 270 K for 20 ps [36].

It is seen from Figure 16.5 that this surface is hydrophilic. Indeed, notice the dark areas in the upper part of Figure 16.5 surrounded by closed equipotential lines. The former are adsorption sites—the most favorable sites for water adsorption and the basic elements of the discussion in Section 16.2. (Due to periodic boundary conditions, some areas on the opposite sides of Figure 16.5 are parts of the same site). The magnitudes of the adsorption energy of water molecules on these sites are larger than the heat of liquefaction of water, which means hydrophilicity. Therefore it does not correspond to a quasi-equilibrium, bare (not covered by silanols) silica surface, which should be hydrophobic, as mentioned above, but rather it corresponds to a fracture silica surface that is hydrophilic. Detailed analysis of the atomic configurations around the hydrophilic sites in Figure 16.5 shows that these sites are situated near structural defects such as nonbridging oxygen atoms [36]. These defects produce an especially strong electrostatic field in the neighborhood and this field attracts polar water molecules, which means that elimination of these structural defects might make a silica surface hydrophobic. This was accomplished by a special method of computer simulation— a combination of molecular dynamics and Monte Carlo. This method simply does not allow the number of structural defects to increase but only to decrease [36]. As a result, one obtains a silica surface without dangling bonds, which corresponds to the lower part of Figure 16.5. One can readily see that this surface is hydrophobic as a real dehydroxylated silica surface is observed to be, which confirms the assumption that the reason for the hydrophilicity of a fracture silica surface is the presence of dangling bonds.

Since the surface shown in the lower part of Figure 16.5 was obtained by the artificial procedure mentioned above, it does not necessarily have the same atomic structure as a real hydrophobic silica surface. To simulate the latter, one has to model the process of glass surface formation as realistically as possible. As explained above, the main problem here lies with the time scale of the computer simulation, i.e. situations where De is too large. We partially resolved this problem by using nonlinear optimized annealing [38]. When one decreases the temperature, the time of structural relaxation of a glass atomic structure increases, which inhibits the ability of the atomic structure to adapt itself to the changing temperature. Thus one has to steadily decrease the rate of cooling, i.e., to use a nonlinear temperature-time profile. This is what was done in [36], however, the temperature profile used there was not optimal. It was shown later that there exists an optimal nonlinear temperature profile that minimizes the concentration of structural defects in an annealed atomic structure for a given total time of simulated annealing [38]. The reason for that is the following: the equilibrium (not real) concentration of structural defects in the simulated atomic structure corresponds to the final temperature reached at the end of annealing. Starting from the same initial temperature, the final temperature and the equilibrium concentration corresponding to it will be minimal for a constant time of annealing when the average rate of cooling is the fastest. However, the faster the rate of cooling is, the larger the deviation of real concentration of structural defects from its equilibrium value. Therefore one chooses an optimal temperature profile to produce the minimal concentration of *real* structural defects in the simulated system for the given time of simulation. To determine such an optimal temperature profile, one has to know parameters that determine the equilibrium concentration of structural defects and time of structural relaxation for the simulated system.

At present, these parameters are known or can be estimated only for the most studied silicate system—amorphous silica [38]. Thus, the simulation of the nonlinear optimized annealing was performed for an initial silica surface mentioned above starting at about 4500 K. The final temperature reached after 100 ps of this simulated annealing was about 2800 K, half of the time being spent in the temperature interval below 3000 K [38]. The simulated silica surface obtained after this annealing had no structural defects. This means that the continuous random network of corner-sharing SiO_4 tetrahedral units, which is the basis of all silicate glass atomic structures, extends up to the surface of amorphous silica without any structural defects (dangling bonds) [38]. Indeed, what in fact was shown in Ref. [38] is that these dangling bonds disappear already at 2800 K when real silica is still liquid (its glass transition temperature is about 1500 K). Moreover, the deviation of the simulated silica surface from equilibrium (as, for example, quantified by the Deborah number) was certainly larger than that for the real silica surface and correspondingly one should expect more structural defects at the simulated surface than at a real one. Since there were none of these defects at the simulated silica surface, the same is *a fortiori* true for the real one.

Hydrophobicity is a qualitative characteristic relevant to physical adsorption of water at surfaces. A more complete characterization of physical adsorption is given by the isotherms of adsorption of various gases. A set of isotherms measured at different temperatures contains in itself all the thermodynamics of adsorption, including the property of being hydrophobic or hydrophilic. Isotherms of adsorption can be simulated and compared with experimental data. For that, one has to first simulate the surface atomic structure and then simulate isotherms for adsorption on that model surface. Comparisons of simulated and experimental isotherms can validate simulations of atomic structure because isotherms of physical adsorption are very sensitive to the details of atomic structure. At present time, when direct methods of probing surface atomic structures of amorphous surfaces are very scarce, this method can be useful for validation of computer simulations of these surfaces. We have simulated (by grand canonical Monte Carlo) isotherms of adsorption of CO_2 on simulated (by molecular dynamics) silica and silicate glass surfaces [29,39]. The reason for the choice of CO_2 as a probe molecule is that experimental isotherms of adsorption of this molecule on glass fibers (glass wool) have been measured [15] and its isotherms on high surface area silica can be obtained from the literature. Isotherms of the physical adsorption of water on silicate glass surfaces are more difficult to measure and to simulate than that of carbon dioxide due to much stronger interaction of water molecules with these surfaces. On the other hand, a molecule of carbon dioxide has a considerable quadrupole moment and therefore interacts with electrostatic field at silicate glass surfaces. Although this interaction is not as strong as in the case of water, one can use it to probe the surface heterogeneity and the associated electrostatic field that determines also water adsorption at glass surfaces.

It was found that simulated isotherms of adsorption of CO_2 on simulated silica surfaces depend strongly on the degree of simulated annealing, absolutely not annealed (initial) or slightly annealed (fractured) surfaces being much more active with respect to adsorption of CO_2 than a deeply annealed (by the methods mentioned above) simulated silica surface. The latter produces simulated isotherms of CO_2, which agree with its experimental counterparts, albeit after some empirical scaling of the inter-atomic potentials. It was also found that the simulated isotherms of adsorption of Ar on all simulated surfaces lie close to one another and to the experimental isotherm of Ar on high surface area silica [39]. Thus, Ar does not feel the differences in annealing of simulated surfaces, while CO_2 is very sensitive to them. This is because simulated annealing removes structural defects from the surface while leaving its basic atomic structure essentially intact. The Ar atom has neither dipole nor higher multipole moments and therefore its interaction with the surface electrostatic field is only through induced moments and thus is relatively weak. The adsorption energy of Ar on silica surface depends on the basic surface atomic structure, mainly through the van der Waals interaction, and relatively small concentration of structural defects does not change this energy considerably. This is why simulated annealing makes no big changes to adsorption isotherms of Ar on simulated silica surfaces [39].

In contrast with this, the considerable quadrupole moment of a CO_2 molecule interacts strongly with an inhomogeneous electrostatic field produced by structural defects, which is the reason for the fracture silica surface heterogeneity with respect to carbon dioxide [39]. Thus fracture silica surfaces are strongly heterogeneous with respect to water, less but still considerably heterogeneous with respect to carbon dioxide, and almost homogeneous with respect to argon. This once more emphasizes the fact that surface heterogeneity depends both on the atomic structure of a surface and on a probe molecule.

Annealing of simulated silicate glass (xNa$_2$O$(1-x)$SiO$_2$; $x = 0.245$) surfaces influences simulated isotherms of CO_2 far less than annealing of silica glass surfaces [29]. Both slightly and deeply annealed surfaces generate simulated isotherms of CO_2, which are relatively close to each other and are much higher than the isotherm on a deeply annealed silica surface, but lower than that on an unannealed silica surface. Thus, the quasi-equilibrium silicate surface is more active with respect to physical adsorption of carbon dioxide than the quasi-equilibrium silica surface. This is ascribed to the fact that sodium is a network modifying cation. That is, when Na$_2$O is added to SiO$_2$, the oxygen anions surrounding Na enter SiO$_4$ tetrahedral units because the affinity of oxygen to Si is larger than its affinity to Na. This disrupts (modifies) a continuous random network of corner sharing SiO$_4$ tetrahedral units (the atomic structure of pure silica) by substituting one bridging (shared by two tetrahedral units) oxygen by two nonbridging oxygens (NBOs) whose dangling bonds are saturated by two sodium atoms. Thus, the bulk atomic structure of a sodium silicate glass contains NBOs in a concentration equal to that of the sodium ions. This holds basically true also for the quasi-equilibrium surface of sodium silicate, albeit the relation between sodium and NBO concentrations may be slightly different than in the bulk. Thus, in contrast with the silica glass surface, where NBOs are structural defects that are present in considerable concentration only on strongly nonequilibrium (fractured) surfaces, NBOs at the sodium silicate glass surfaces should not be considered as a structural defect but as normal components of the atomic structure. Nevertheless, both at the silica and silicate glass surfaces, NBOs are the sources of strong electrostatic fields.

In all silicate glass surfaces, the formal charge of oxygen, bridging or NBO, is -2 and the formal charge of silicon is $+4$. In bulk silica, each Si has coordination 4 with respect to O and each O has coordination 2 with respect to Si. Thus, the SiO-bond in silica can be considered as a formal dipole consisting of two elementary charges separated by the relatively short distance of the SiO-bond (about 0.16 nm). (Since the distance between charges is finite, this structure has not only a point dipole but also higher multipole moments.) In a regular SiO$_4$ tetrahedron, four such dipoles form a symmetric structure so that the electrostatic field of each of them is partially compensated by the others. Thus, regular SiO$_4$ units composed only of bridging oxygen atoms (Q$_4$-type) generate only relatively weak electrostatic fields in the surrounding space and their interaction with water or carbon dioxide is also weak. This means that regular SiO$_4$ tetrahedral units of the Q$_4$-type as well as surface atomic structures composed of such units are hydrophobic [36]. However, if one of the oxygen atoms of such a SiO_4 unit is an NBO (the unit is of Q_3-type), one formal charge of NBO is not compensated by Si so that formally the NBO bears a negative charge. On a fracture silica surface where NBOs may appear as structural defects, the negative charge of an NBO is compensated by the positive charge of a triply-coordinated oxygen that also has to be present in the structure to maintain electro-neutrality [36]. However, the dipole moment of a paired NBO/triply-coordinated oxygen is larger than that for a SiO-bond due to the larger distance between the charges in the first pair. Besides, a NBO disturbs the tetrahedral symmetry of the SiO_4. Thus, one may consider NBOs (or their neighboring triply-coordinated oxygen) as a source of the electrostatic field at the fracture silica surface [36]. The same reasoning can be applied to the sodium silicate surface, except that one should consider sodium cations instead of triply-coordinated oxygen atoms as the bearers of positive formal charges that compensate negative formal charge of the NBO; and also, the sodium silicate surface need not necessarily be fractured to display a strong electrostatic field (i.e., to be hydrophilic) because NBOs are essential elements of its atomic structure.

Thus the presence of NBOs at surface increases the activity of a silicate glass with respect to physical adsorption of polar or quadrupolar molecules. An even deeper correlation is the correlation between the relative durabilities of silicate glass surfaces [13] and their activities with respect to the physical adsorption of CO_2 that was observed for silica, sodium silicate, and E-glass surfaces [29]. The relative durability of silicate glass surfaces is intimately connected to the corrosion of glass surfaces in the presence of water [13]. The latter is usually explained by the chemisorption and dissociation of water molecules on glass surfaces [13]. As mentioned in the Introduction, there is reason to believe that the physical adsorption of water molecules on glass surfaces precedes its chemisorption and dissociation. This suggests that the sites on silicate glass surfaces that are most active in physical adsorption of water also determine the relative durability of those surfaces. As mentioned above, the activity of those sites can be explained by the strength of electrostatic field, which interacts with polar water molecules. The same field should interact with quadrupolar molecules such as carbon dioxide, so that such gases may be used as probes for the sites most susceptible to water corrosion. This is a qualitative explanation of the correlation mentioned above.

This correlation suggests further analysis of surface heterogeneity of silicate glass surfaces by computer simulation. This analysis clarifies atomic structure of adsorption sites on sodium silicate glass surfaces [40]. In the future, we intend to compare the energy distributions obtained for simulations of silicate glasses with that obtained from IGC.

ACKNOWLEDGMENT

V.A.B. and C.G.P. acknowledge financial support by Johns Manville Inc.

REFERENCES

1. Honig, J. M., *Ann. NY Acad. Sci.*, 58 (Art 6), 741–797, 1954.
2. Jaroniec, M. and Madey, R., *Physical Adsorption on Heterogeneous Solids*, Elsevier, Amsterdam, 1988.
3. Rudzinski, W. and Everett, D. H., *Adsorption of Gases on Heterogeneous Surfaces*, Academic Press, London, 1992.
4. Charmas, B. and Leboda, R., *J. Chromatogr. A*, 886, 133–152, 2000.
5. Bakaev, V. A., *Surf. Sci.*, 198, 571–592, 1988.
6. Bakaev, V. and Steele, W., In *Adsorption on New and Modified Inorganic Sorbents, Studies in Surface Sciences and Catalysis*, Dabrowski, A. and Tertykh, V. A., Eds., Vol. 99, Elsevier, Amsterdam, pp. 335–355, 1996.
7. Dondi, F., Cavazzini, A., and Remelli, M., *Adv. Chromatogr.*, 38, 51–74, 1998.
8. Bakaev, V. A., Bakaeva, T. I., and Pantano, C. G., *J. Chromatogr. A*, 969, 153–165, 2002.
9. Giddings, J. C. and Eyring, H., *J. Phys. Chem.*, 59, 416–421, 1955.
10. Guiochon, G., Shirazi, S. G., and Katti, A. M., *Fundamentals of Preparative and Nonlinear Chromatography*, Academic Press, Boston, 1994.
11. Kang, H. C. and Weinberg, W. H., *Surf. Sci.*, 299/300, 755–768, 1994.
12. Hertl, W., *J. Phys. Chem.*, 72, 1248–1253, 1968.
13. Varshneya, A. K., *Fundamentals of Inorganic Glasses*, Academic Press, Boston, 1994. Chap. 17.
14. Bakaev, V. A. and Chelnokova, O. V., *Surf. Sci.*, 215, 521–534, 1989.
15. Bakaeva, T. I., Bakaev, V. A., and Pantano, C. G., *Langmuir*, 16, 5712–5718, 2000.
16. Bakaeva, T. I., Pantano, C. G., Loope, C. E., and Bakaev, V. A., *J. Phys. Chem. B*, 104, 8518–8527, 2000.
17. Bakaev, V. A., Bakaeva, T. I., and Pantano, C. G., *J. Phys. Chem. B*, 106, 12231–12238, 2002.
18. Poole, P. H., McMillan, P. F., and Wolf, G. H., In *Structure, Dynamics and Properties of Silicate Melts, Reviews in Mineralogy*, Stebbins, J. F., McMillan, P. F., and Dingwell, D. B., Eds., 32, pp. 563–616, 1995.
19. Paryjczak, T., *Gas Chromatography in Adsorption and Catalysis, Ellis Horwood Series in Physical Chemistry*, Ellis Horwood/Halsted Press, Chichester West Essex/New York, 1986.
20. Conder, J. R. and Young, C. L., *Physicochemical Measurement by Gas Chromatography*, Wiley, Chichester, 1979.
21. Feller, W., *An Introduction to Probability Theory and its Applications*, Vol. 2, 2nd Ed., Wiley, New York, 1971.
22. Hoskins, R. F., *Generalized Functions, Ellis Horwood Series in Mathematics and its Application*, Halsted Press, New York, 1979.

23. Saint Flour, C. and Papirer, E., *J. Colloid Interface Sci.*, 91, 69–75, 1983.
24. Papirer, E. and Balard, H., *J. Adhesion Sci. Technol.*, 4, 357–371, 1990.
25. Osmont, E. and Schreiber, H. P., In *Inverse Gas Chromatography: Characterization of Polymers and Other Materials*, Lloyd, D. R., Ward, T. C., Schreiber, H. P., and Pizana, C. C., Eds., American Chemical Society, Washington, DC, pp. 230–247, 1989.
26. Jaroniec, M., In *Adsorption on New and Modified Inorganic Sorbents, Studies Surface Science and Catalysis*, Dabrowski, A., Tertykh, V. A., Eds., Vol. 99, Elsevier, Amsterdam, pp. 411–433, 1996.
27. Lyklema, J., *Fundamentals of Interface and Colloid Science, Volume I: Fundamentals*, Academic Press, London, p. 2.6, 5.76, 1991.
28. Hodge, I. M., *J. Non-Cryst. Solids*, 169, 211–266, 1994.
29. Bakaev, V. A., Steele, W. A., and Pantano, C. G., *J. Chem. Phys.*, 114, 9599–9607, 2001.
30. Garofalini, S. H., *J. Chem. Phys.*, 78, 2069–2072, 1983.
31. Feuston, B. P. and Garofalini, S. H., *J. Chem. Phys.*, 91, 564–570, 1989.
32. Trioni, M. I., Bongiorno, A., and Colombo, L., *J. Non-Cryst. Solids*, 220, 164–168, 1997.
33. Timpel, D., Schaible, M., and Scheerschmidt, K., *J. Appl. Phys.*, 85, 2727–2735, 1999.
34. Wilson, M. and Walsh, T. R., *J. Chem. Phys.*, 113, 9180–9190, 2000.
35. Roder, A., Kob, W., and Binder, K., *J. Chem. Phys.*, 114, 7602–7614, 2001.
36. Bakaev, V. A. and Steele, W. A., *J. Chem. Phys.*, 111, 9803–9812, 1999.
37. D'Souza, A. S. and Pantano, C. G., *J. Am. Ceram. Soc.*, 82, 1289–1293, 1999.
38. Bakaev, V. A., *Phys. Rev. B*, 60, 10723–10726, 1999.
39. Bakaev, V. A., Steele, W. A., Bakaeva, T. I., and Pantano, C. G., *J. Chem. Phys.*, 111, 9813–9821, 1999.
40. Leed, E. A. and Pantano, C. G., *J. Non-Cryst. Solids*, 325, 48–60, 2003.
41. Bakaev, V. A., *Surf. Sci.*, 564, 108–120, 2004.

Index

A

Abraham approach, 77
Abrasion resistance, 150–152
  carbon black, 152–157
    filler surface, chemical adsorption of, 155
    heat treatment, 154–155
    mixing procedure, 156–157
  silica, 158–163
Acid–base
  approaches, 357–359
  index Ω
    fiber interaction capacity and, 72–73
    powder interaction capacity and, 72–73
  scales
    fiber interaction capacity and, 73–74
    powder interaction capacity and, 73–74
Addition reactions, 226–228
Adhesion
  acid–base approaches, 357–359
  formation and destruction, 32–40
  two liquid phase method, 355–357
  wettability and, 351–359
Adsorbed polyelectrolyte interfacial reconformation, 546–549
Adsorption
  energy distribution functions, 89–94
  kinetics of, 535–536
  isotherm, 401, 537
    determination, 87–89
  permanent, 399–402
  process
    hydrolyzed polyacrylamide/aluminosilicate/water system and, 544–546

polyacrylamide/aluminosilicate/water system and, 538–539
protein, biocompatibility and, 468–469
temporal, 399–402
AFMs, *see* atomic force microscopies
Agglomerate structure, pyrogenic silica and, 589, 591–594
Aggregate structure, pyrogenic silica and, 589, 591–594
Airplane industry, inverse gas chromatography and, 103
Aluminosilicate water system, 537–543
  hydrolyzed polyacrylamide, 544–550
Amorphous fume silicas, hydroxylation, 253–254
Atomic force microscopies (AFMs), 420–427
  force spectroscopy, 425–427
  polymer fibers, 432
  principles of contact-AFM, 420–422
  techniques, other, 422–423
  tips and cantilevers, 423–425
Atomic
  layer deposition, infrared spectroscopy and, 192–200
  simulations, silicate glass surfaces and, 634–642

B

Bad wetting, 317–318
Basic Stern model, 509
Biochromatography, 264–269
Biocompatibility
  human immune response, 454–464
  powdered materials, 453–473
    composition of, 467–468
    crystal surface, 471
    magnetism, 473

nonconventional, 80–86
powder interaction capacity,
68–80
surface energy, 54–59
surface nano-roughness, 59–68
polymer-filler, 131–141
powder values, 57–58
rubber reinforcement
application of, 123–164
filler–filler interactions, 131–140
fillers surface energies, 141–146
polymer-filler, 131–140
silicate glass surfaces and,
614–619
solids, surface energy, 125–127
thermodesorption, combining of,
94–97
IR spectroscopy, silica resin interaction
and, 596–597
Isoteric heats of adsorption, 22
Isotherms
adsorption, 401
polymer adsorption and, 401

J

JKR theory, 37

K

Kamlet and Taft method, 75–76
Kawabata Evaluation System for Fabrics
(KES-F), 334
Kelvin effect, 22–24
interfacial curvature, 22–24
Ostwald ripening, 24
KES-F, *see* Kawabata Evaluation System
for Fabrics
Kinetic aspects, vapor–liquid transition
and, 310–312
Kinetics of
adsorption, 535–536
interfacial transfers, 536–537
Kovats retention indices, powder
interaction capacity and, 71–72

L

Laminates case, 406–407
flow micro-calorimetry (FMC) and,
397–399
Liquid–glass transition
bad wetting, 317–318
free volume picture, 314–316
good wetting, 318–319
interface cases, 317–318
polymer
films, experiments, 312–313
reinforcement, 312–320
confined polymer layers,
319–320
nanocomposites, 319–320
thin polymer films, 316–317
Liquids, solid–solid confinements and,
287–321
Low temperature IGC-ID, 83–85

M

Magnetic resonance spectroscopies,
245–281
nuclear magnetic resonance (NMR),
246–247
powders or fibers, 249–281
biological interfaces, 271–279
ceramic powders, 270–271
microporous carbons,
279–281
silica, 249–270
spin labels, electron paramagnetic
resonance (EPR), 247–249
Magnetism, powder composition and,
biocompatibility of, 473
Mechanical aspects, fabrics and,
334–340
Metal complex catalysts, 232
Metallic
nanowires, 441–442
tubing, 85–86
specific interaction energy, 86
Metastability, vapor–liquid transition
and, 307–310

Milton Keynes UK
Ingram Content Group UK Ltd.
UKHW020003071024
449327UK00031B/2633